Dragonflies of North America

Dragonflies of North America

Male Blue Dasher in obelisk position: Some dragonflies will perch with their abdomens pointed upward as a means of thermoregulation. The position is typically used for cooling, lowering the amount of body surface exposed to the overhead sun. The position can also be deployed for warming, shifting the body toward the sun when it is close to the horizon.

Written and illustrated by **Ed Lam**

PRINCETON UNIVERSITY PRESS
Princeton and Oxford

Published by Princeton University Press
41 William Street, Princeton, New Jersey 08540
99 Banbury Road, Oxford OX2 6JX
press.princeton.edu

All Rights Reserved

ISBN (pbk.) 9780691232874
ISBN (e-book) 9780691245140
British Library Cataloging-in-Publication Data is available

Editorial: Robert Kirk and Megan Mendonça
Production Editorial: Natalie Baan
Cover Design: Wanda España
Production: Steven Sears
Publicity: Matthew Taylor
Copyeditor: Charles J. Hagner
Cover Credit: Ed Lam
This book has been composed in Goudy Oldstyle and Myriad Pro

The publisher would like to acknowledge the author of this volume for acting as the compositor for this book.

Printed in Italy

10 9 8 7 6 5 4 3 2 1

Contents

Introduction

In North America, dragonfly-watching is seasonal. While a few species may be found year-round in the southernmost United States, only the warmth of spring brings dragonflies to the rest of the continent. Which species is seen first often depends on location. In parts of the West, the first dragonfly may be an early-emerging California Darner or a Variegated Meadowhawk that survived through the winter. In the South, the Blue Corporal or one of the species of baskettails is typically first on the wing. However, for much of North America, it is the Common Green Darner that marks the beginning of the dragonfly year.

It appears suddenly like a living comet, a welcome sight after a long still winter. A migratory species, the Common Green Darner flies hundreds of miles northward each spring to recolonize the slowly awakening wetlands. At three inches in length, it is an impressive insect, with males resplendently marked in sky-blue and green. As its name implies, it is a common species and, being the only dragonfly recorded in all 50 United States, among the most widespread. After mating and laying eggs, these spring migrants die off, but then separate resident populations of the species emerge to fly throughout the summer. Another emergence of Common Green Darners occurs at the end of summer into the fall. These are offspring of migrants from previous springs, having reached adulthood after two to three years of aquatic life. This population migrates southward in great numbers, so in addition to being among the first dragonflies recorded each year, the Common Green Darner is also among the last.

Whichever dragonfly is first of the year, other species quickly follow. There are over 300 species of dragonflies in North America, and they are richly diverse. Many are large on average for insects, making them eye-catching and conspicuous. Our longest dragonfly measures over four inches in length, but the smallest is less than an inch long. Some are abundant, others rare and range-restricted. They are found in a wide variety of habitats, from wild rivers to human-made ponds, from cold, mossy pools at the edge of the Arctic tundra to the tropical swamps of the Florida Keys. Dragonflies are marked by every color of the rainbow. Some are showy and boldly patterned, others more cryptic and camouflaged. They vary in form, the shapes and proportions of their wings and bodies adapted to their particular habits.

This wealth of diversity is both inviting and challenging. Identifying dragonflies requires careful observation and a keen attention to detail. In return, there is hardly a better subject to study in the natural world. Dragonflies are beautiful insects with fascinating behaviors and life stories, many aspects of which are observable. Identification is an invitation to look closer at these charismatic creatures.

Although dragonfly study has a long history in North America, dragonfly-watching is a relatively new popular activity. It has been only in the last two decades that field guides have become widely available, and our identification tools continue to evolve and improve. As the number of dragonfly observers has increased, so has our knowledge. More interest has led to vastly more observations, furthering our understanding of species occurrence throughout the continent. With much of North America undersurveyed, there remains ample opportunity for observers of all levels of experience to contribute.

What Are Dragonflies?

Dragonflies are ancient creatures. The fossil record reveals their ancestors took flight more than 300 million years ago. Some are among the largest insects known, with wingspans of more than two feet. "Modern" dragonflies, recognizable to families that exist today, arose during the Mesozoic Era, some 200 million years ago. Dragonflies flew above the heads of dinosaurs and survived when the dinosaurs did not. As the continents divided and drifted apart, dragonflies witnessed the rise of mammalian life and, more recently, all of human history, evolving into apex insect predators with extraordinary visual acuity and flight capabilities.

adult dragonfly

Dragonflies are classified in the insect order Odonata. The name *Odonata* is derived from the Greek word for "tooth," in reference to the serrated jaws of the adult insects. While the Odonata use their jaws to feed on insect prey, having toothed mandibles is not their defining characteristic, as many other insects have similar mouthparts. The Odonata are best distinguished from other insects by a number of anatomical features: small, inconspicuous antennae, large compound eyes, four large veined wings of similar length, a long ten-segmented abdomen, and the unique copulatory organs of the male.

The Odonata are divided into three suborders. Dragonflies make up the suborder Anisoptera, meaning "unequal wings." Dragonfly wings differ in shape and size with the hindwing broader than the forewing, particularly at its base. Damselflies make up the suborder Zygoptera, which means "similar wings." Damselflies have wings that are similar in shape, and if there is a difference in length, the hindwing is smaller. A tiny third suborder, the Anisozygoptera, consists of three Asian species that exhibit characteristics of both dragonflies and damselflies and is thought to be an evolutionary link between the suborders. The term *dragonfly* is often used to refer to all Odonata, particularly outside of North America. To avoid confusion, in this book, *dragonfly* refers only to the Anisoptera, *damselfly* to the Zygoptera. The term *odonate* will be used for any reference to the entire order.

Although closely related, dragonflies and damselflies differ in ways other than wing shape. Damselfly eyes are separated by the width an eye, while dragonfly eyes touch or are separated by less than an eye's width. Dragonflies at rest hold their wings out flat like an airplane or somewhat angled up or downward but nearly always open. Some damselflies also perch with their wings open, but most hold their wings closed over their abdomen while at rest. Dragonflies are typically larger and stouter than damselflies, and their flight is more powerful. Damselflies are often less conspicuous, tending to fly low in the shelter of vegetation or skimming along the surface of the water.

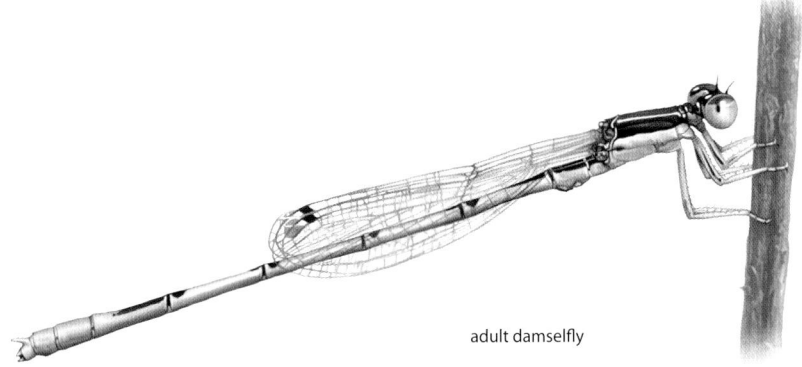

adult damselfly

All odonates are carnivorous, feeding on a wide variety of insects, including other odonates. They provide the benefit of consuming large numbers of "pest" insects, such as flies, midges, mosquitoes and their larvae. Despite given folk names like "devil's darning needle" and "horse stinger," dragonflies have no stinger and are in no way harmful or aggressive to humans and non-prey animals.

Life Cycle

Odonates spend most of their lives as aquatic larvae, hatching from eggs laid in or near water. The larvae, also called nymphs or naiads, are armed with an articulated lower jaw (*labium*) that can be rapidly projected forward to capture prey. Their prey largely consists of other insects, although larger dragonfly larvae may take small fish and tadpoles. Most larvae are ambush predators, lying in wait amid aquatic vegetation or hidden in the bottom substrate. Most detect their prey by touch, but some are able to track and pursue prey visually.

The larva is encased in a rigid exoskeleton, the confines of which the larva eventually outgrows. In order to develop, the larva undergoes a series of molts, each time breaking

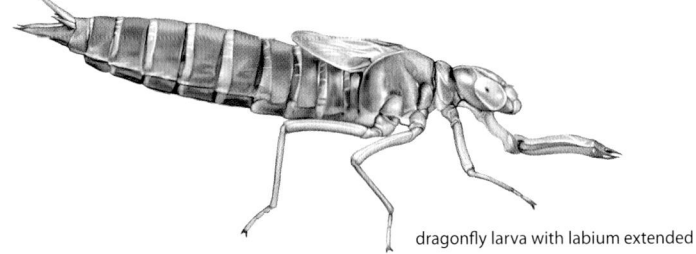

dragonfly larva with labium extended

through and casting off its previous exoskeleton. Changes after each molt include further development of the eyes, labium, and wing buds, along with an increase in the number of abdominal segments. Damaged and lost limbs may also be regenerated. Each step between molts is known as an *instar*, the number of which varies according to species. Temperature, food supply, and other conditions affect the time spent as a larva. Most of our species have a single generation per year, with several months of larval development. In colder, higher latitudes and elevations, development can span years. Species with the quickest development are those that breed in temporary pools, able to progress from egg to adult in as little as 40 days.

In preparation for the final molt (called *emergence*), the larva seeks a suitable site to exit the water. Most species require a vertical surface or support such as a plant stem, while some, like members of the clubtail family, will emerge on a flat surface. Internal pressure splits the exoskeleton and the adult dragonfly pushes outward, slowly freeing its head, thorax, and part of the abdomen. After a brief pause to allow some drying and hardening of the body, the dragonfly swings forward to grasp the empty larval exoskeleton with its legs. After withdrawing the remainder of the abdomen, the dragonfly is fully free of the larval shell (called *exuvia*). Blood pressure then expands the abdomen and wings to their full length. Because the body and wings are soft and fragile during emergence, difficulty in breaking free of the exuvia or encountering an obstruction can cause deformities. It is not unusual to see adult odonates with a permanently bent abdomen or deformed wings. Being knocked down at this critical stage by a wave or bad weather is often fatal.

Newly emerged adults (called *tenerals*) can be recognized by their overall pale coloration and by an opacity or glistening sheen (similar to crinkled plastic wrap) on their wings. Tenerals often initially hold their wings together over their abdomen. Their first flight is away from the water to the refuge of surrounding vegetation. To avoid predation, many but not all species attempt to complete emergence before sunrise, but scores of weak-flying tenerals fall prey to birds, frogs, spiders, ants, and other insects, including other odonates.

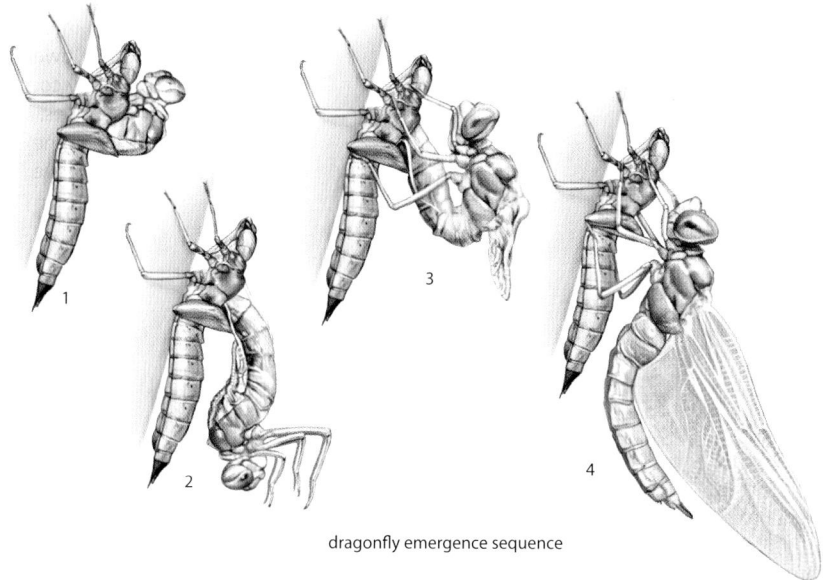

dragonfly emergence sequence

During a period of feeding and maturation, immature odonates keep away from the water. Many stay near their breeding habitat, but some may disperse widely. Males usually reach maturity before females, some developing coloration and patterning that is dramatically different from the females of their species. Eventually both sexes must return to suitable habitat in order to breed, but observers will almost always find more males than females at the water. Males are often territorial, jockeying, clashing, and chasing each other as they try to claim the best breeding sites. Meanwhile, females remain in the surrounding countryside, feeding, maturing, and developing eggs.

Females arriving at breeding habitat are quickly detected by the males. With a few exceptions, there is little recognizable courtship among dragonflies, and the male often appears to attack and tackle the female. Mating begins with the male grabbing the female's thorax with his legs. He then uses a set of appendages at the tip of his abdomen to grasp the back of the female's head. These appendages (or claspers) include a pair of *cerci* above and the *epiproct* below. After taking hold, the male releases his legs and the joined pair is described as being in tandem. Often other males will fly at tandem pairs, attempting to separate the female for themselves.

dragonfly pair flying in tandem

As in other insects, sperm is produced toward the tip of the male's abdomen. Other insects typically mate with their abdomens end to end with their genitalia in contact. However, the tandem position does not allow sperm to be taken by the female directly. Usually prior to mating, the male curls his abdomen forward to transfer sperm from the end of his abdomen to a secondary copulatory organ located at the base of the abdomen near the thorax.

This secondary genitalia is unique to odonates. It includes one or two pairs of hook-like structures, the anterior and posterior *hamules*, which the male uses to clasp onto the female's genitalia during mating (see page 15). Nestled inside a groove called the *genital fossa* is a four-segmented penis designed to both deposit and remove sperm. Its tip is equipped with short spines that the male uses to scrape out any sperm remaining in the female from previous matings. Shielding the penis is the genital *ligula*, a rigid spoon-like structure.

The female completes the mating by curling her abdomen until her genital aperture near her abdomen tip contacts the male's secondary genitalia. The resulting coupling is called the "wheel," a position unique to odonates. Mating may occur in flight or at rest, its duration varying from a few seconds to over an hour depending on the species.

To further ensure that only the male's sperm fertilizes the female's eggs, he may continue to hold the female in tandem throughout egg-laying (called *oviposition*). In other cases, the male guards the female by hovering above her and chasing off any intruders. Females of some species oviposit alone.

Females of some dragonfly families have an egg-laying organ called the ovipositor located at the underside of the abdomen toward the tip. The ovipositors of petaltails and darners have sharp blades used to cut plant tissues into which the eggs are laid. Females of other families have a structure called the *subgenital plate* (or *vulvar lamina*), which varies widely in form. In some families, the subgenital plate is spout- or spike-like and used to drive eggs into aquatic vegetation or into a bottom substrate. In other families, the subgenital plate is reduced to small lobes or flap-like structures. These females release eggs directly into the water while in flight, the tip of their abdomens dabbing the water's surface.

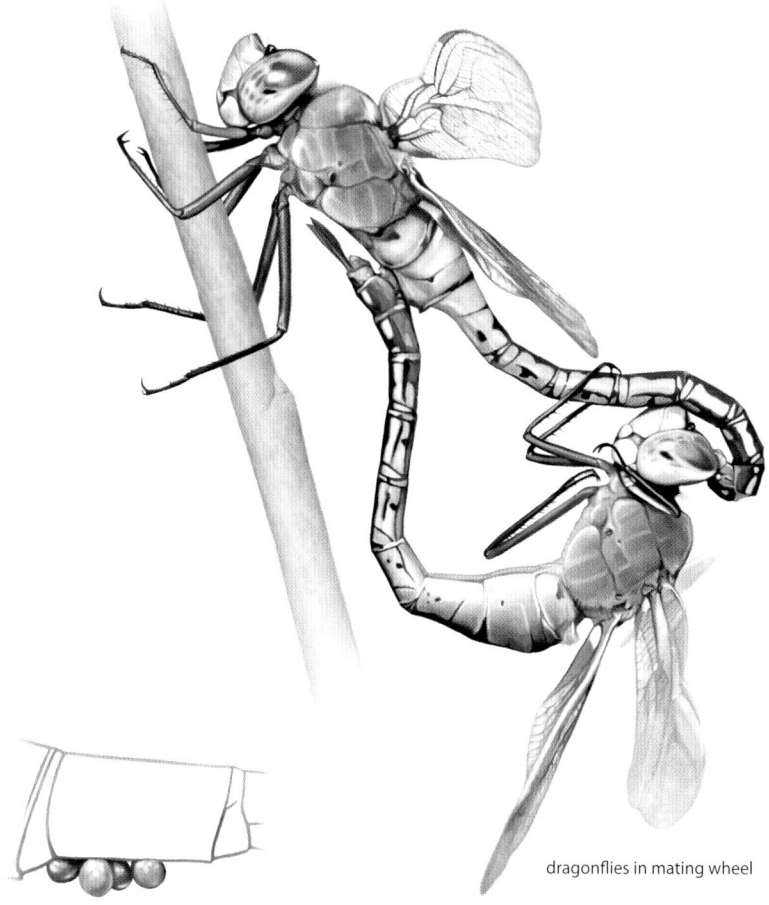

dragonflies in mating wheel

Not dragonfly eggs: Bead-like, reddish, or blackish parasitic mites commonly infest dragonflies. The mites eventually detach, falling back into the water to complete their life cycle.

Dragonfly Anatomy

Dragonflies share with all insects three major body sections: head, thorax, and abdomen.

Head
The head is dominated by two large compound eyes that are keenly developed to detect and track prey. The eyes' size and prominence enables sight in all directions except directly behind. Dragonflies can see ultraviolet and polarized light in addition to the entire visible light spectrum. Each compound eye has thousands of facets, so many that some dragonfly species have the largest eyes in the insect world.

Much less conspicuous are three simple eyes (called *ocelli*) arranged in a triangle in front of the compound eyes. These simple eyes may function to measure light intensity.

Flanking the ocelli are two antennae. In comparison to other insects, dragonfly antennae are very small, limiting their use as tactile or olfactory receptors. Instead, they function to monitor air currents and flight speed.

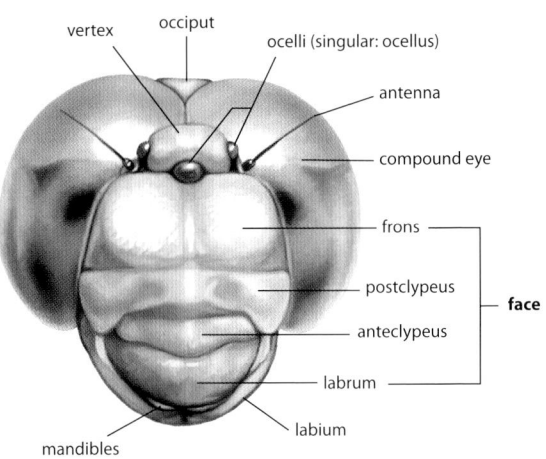

In front of the eyes is the face. The color and pattern of the face can be important for species identification. The top of the face is called the *frons* with the *clypeus* below. Enclosing the chewing mandibles are the *labrum* (suggestive of an upper lip) and the chin-like *labium*.

Thorax
The thorax contains the dragonfly's powerful flight muscles, and the size of the thorax corresponds to the overall size of the dragonfly. The thorax consists of three segments, each bearing a pair of legs. The first segment (the *prothorax*) is small and neck-like, while the second and third are much larger and fused together, forming the *synthorax*. Because of its prominence, the synthorax is referred to simply as the thorax in this book. Markings on the thorax are critical for the identification of many species.

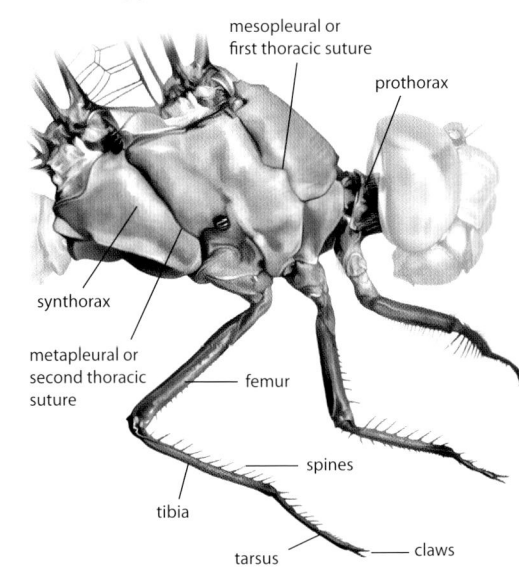

The thorax is shaped like an obliquely angled box, and its slant thrusts the legs forward unlike any other insect. The forward-facing legs are armed with spines and can be deployed to form a basket to catch prey. When the dragonfly is not actively feeding, the legs are held folded and tucked against the body while in flight. At rest, the legs are used for perching, as odonates typically do not walk. Some dragonflies often hold their perch using only the two hind pairs of legs, their forelegs tucked against the front of the thorax.

Wings

All dragonflies have four large veined wings. Unlike most other insects, the fore- and hindwings are not connected, allowing each wing to beat independently for increased maneuverability. Dragonflies are remarkably agile, able to hover, fly forward and backward, and change direction freely. Compared to other flying insects, they beat their wings slowly, too slowly to produce a buzz or audible sound. However, the dragonfly is one of the fastest-flying insects; some species are capable of exceeding 35 miles an hour.

To withstand the stresses of such speed and acrobatics, dragonfly wings are very strong. They consist of a double membrane supported by a dense network of veins. The thickest and strongest veins arise from the base of the wing and run lengthwise. A multitude of cross-veins serve to strengthen the structure. While all dragonflies share numerous venation features, each family, and often each genus, has its own characteristic venation pattern. Details in venation can also aid in separating similar species.

Parts of the wings (skimmer family)

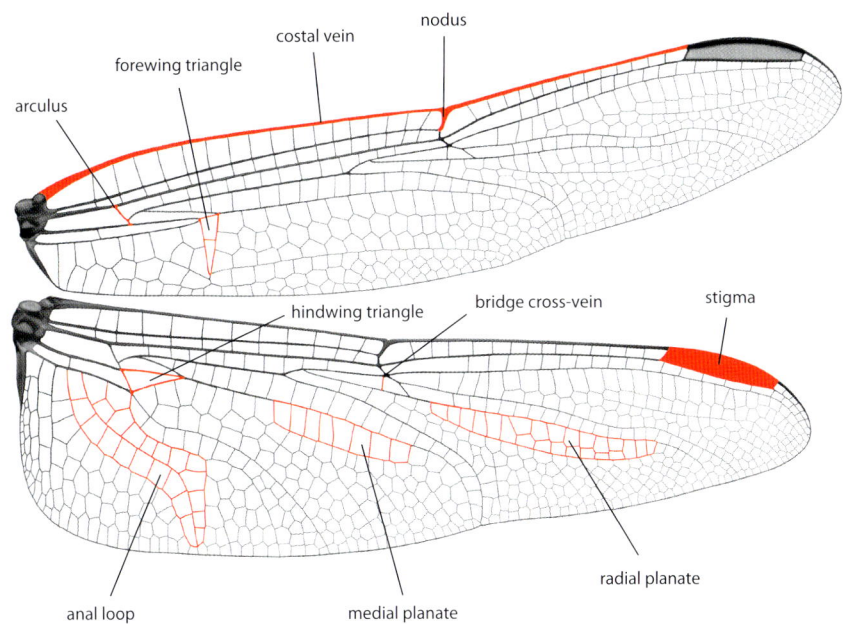

Stigma or **pterostigma** is a specialized single cell near the wing tip that differs in color and texture from the rest of the wing

Abdomen

Dragonfly abdomens are typically long and slender, functioning to counterbalance the heavy thorax and head. As a consequence, shorter abdomens tend to be wider. The abdomen always has 10 segments, numbered from segment 1 (S1) at the base of the abdomen to segment 10 (S10) at the tip.

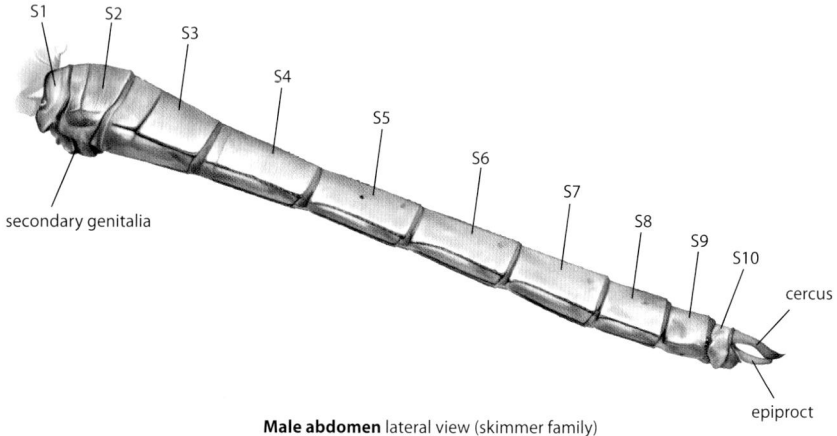

Male abdomen lateral view (skimmer family)

Sexing Dragonflies

Learning to recognize the sex of a dragonfly is very important. Males and females, even of the same species, may differ greatly in appearance, and mistaking the sex can lead to misidentifications. The abdomen is usually the best indicator, as it bears the dragonfly's reproductive organs. In general, the female abdomen tends to be stouter and more cylindrical than the male abdomen, which is usually slimmer and can be noticeably constricted at S3.

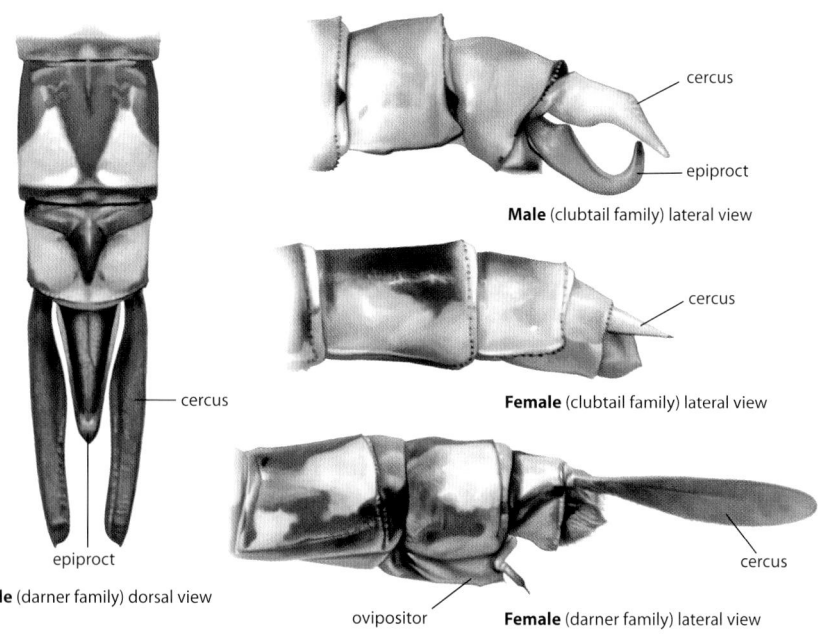

Male (clubtail family) lateral view

Female (clubtail family) lateral view

Male (darner family) dorsal view

Female (darner family) lateral view

Males typically have three terminal appendages at the tip of the abdomen: a pair of cerci above and the epiproct below. These structures vary widely in form, and their shapes can be unique to each species, making them very useful for identification. Female cerci are usually simpler in form. In most families, the female's cerci are short and pointed, but in the darner family, the cerci can be long and leaf-like. The epiproct is present but insubstantial in females.

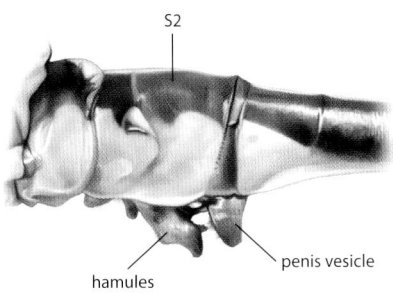

Male secondary genitalia (clubtail family)

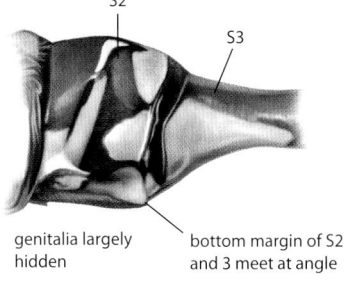

Male secondary genitalia (darner family)

The male's secondary genitalia, located beneath S2, are often visible in profile view. However, these structures may be largely hidden in some species, but the bottom margins of S2 and 3 often meet at an angle in males. The margin formed by these segments is usually straighter or smoothly curved in females. Many females can also be recognized in profile view by the presence of an ovipositor or subgenital plate on the underside of the abdomen near the tip.

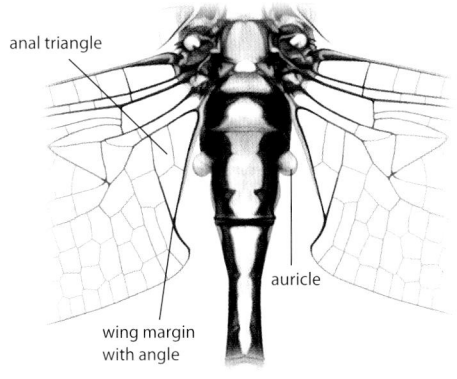

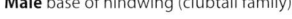

Male base of hindwing (clubtail family)

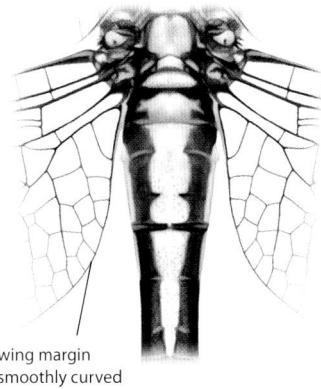

Female base of hindwing (clubtail family)

Wing shape can also help determine sex. Male dragonflies, with the exception of skimmers and green darners, have a small flap-like structure (*auricle*) on each side of S2. To accommodate these structures, the margin at the base of the male's hindwing is concave, resulting in venation differences, such as the presence of an anal triangle. The female hindwing is smoothly curved at the base.

Finding Dragonflies

Dragonflies are usually best observed on warm, sunny days. The majority of species are most active in direct sunlight, and even passing clouds may decrease activity. Low temperatures, persistent cloudiness, wind, and rain are poor conditions to find odonates. Activity begins when dragonflies are sufficiently warm to fly. Mornings can be rewarding, as species are less active and may be found perched warming in the sun. Midday often marks the peak of activity, but some dragonflies, like darners, typically feed in the late afternoon into evening. A few, such as shadowdragons, compress all their hunting and reproductive activities into a brief period after dusk. On very hot days, activity can begin and end early, as high temperatures force dragonflies to take shelter.

With the ability to fly and disperse freely, dragonflies can be found almost anywhere, but the best places to look for them are the wetlands in which they breed. These waters are their home throughout their larval stages, with different species having different habitat requirements. Some prefer temporary or fishless waters. The larvae are often adapted to dwelling in a specific substrate, such as gravel or sand. Some species are associated with certain types of water plants, but the presence of any vegetation, including submerged, emergent, or floating, is often attractive to dragonflies. Plants provide perch and oviposition sites for adults and shelter for the larvae. Breeding habitats are primarily freshwater, although a few species are tolerant of brackish conditions. Only one species of dragonfly, the Seaside Dragonlet, can breed in salt water. Some species prefer habitat with specific chemistries—acidic or alkaline waters, for example. Water quality is one of the most important variables, as most species require clean water. Dragonfly population and diversity are often good indicators of water cleanliness, as polluted habitats support very few dragonflies.

Which species may be found also depends on the water's motion. Almost all dragonfly species can be divided into those that breed in still water and those that require some flow. Most of our North American species are found at still or slowly flowing waters, a reflection of the wide variety of water bodies that make up this habitat. Still waters include ponds, lakes, marshes, swamps, bogs, fens, and temporary pools filled by rainwater, as well as human-made bodies, such as ditches, canals, and borrow pits. Running waters such as streams, creeks, and rivers support a different set of dragonflies. The larvae of these species often require higher amounts of dissolved oxygen and are adapted to the substrates found in running waters, such as sand and stones. However, some species typical of running water may be found at large lakes, where wind and wave action provide suitable oxygenation for their larvae. Conversely, along slow portions of streams and rivers, still-water species may be found. Visiting a variety of habitat will often yield a greater diversity of species. Dragonfly species have different flight periods, so revisiting a location at different times of the year can be productive.

Dragonfly habitat also includes areas beyond the water's edge. These surrounding areas provide necessary food and shelter for maturing adults and should not be overlooked. With careful observation, individuals may be found roosting in trees and vegetation. Foraging dragonflies are often attracted to open areas, such as clearings, meadows, and forest edges. Roads and trails crossing through these habitats provide us access but also function as sheltered corridors along which dragonflies patrol and feed.

Beginners should consider visiting a local pond on a warm, sunny day. Ponds often feature an array of common species, making them a good place to start.

Identifying Dragonflies

To identify dragonflies, it is essential to get a good look, but seeing them well enough can be challenging. Dragonflies are swift and agile, and our views of them can be fleeting. When they do linger, they may not cooperate, flying high or resting on an obscured or distant perch. Although many are large for insects, they are still relatively small creatures to observe, and the features that may identify them are correspondingly small.

Until recently, it was widely doubted that free-flying dragonflies could be reliably identified at all. Many species are similar in appearance, and the known differences between them seemed too difficult to discern in the field. To identify a dragonfly, one usually had to catch it. With a specimen in hand, identification was made by examining anatomical features such as wing venation and reproductive structures, minute characters requiring close examination. These identifying features were often the primary characters used when each species was described to science.

Prior to 2000, the principal references on dragonfly identification were technical manuals aimed at identifying collected specimens. These emphasized anatomy over other attributes, such as color and pattern. This emphasis may be traced to the practice of working with collected specimens, standard methodology in the study of natural history but limiting in the case of dragonflies. When dragonflies die, they quickly lose their color, and decomposition darkens and obscures their patterns. Eye color, often distinctive and striking in life, is completely lost. With specimens lacking the features of living dragonflies, those pioneers working to differentiate and describe species could do so only through anatomy.

Fortunately, modern optics have enabled us to observe free-flying dragonflies better and lessened the need for capture. While binoculars have long been utilized for viewing distant subjects, recent designs are better suited for insect-watching. The best binoculars for observing dragonflies are those that close-focus and can be used to view subjects six feet away or closer. These provide views of small subjects at a range close enough to reveal minute characters and structures. Advancements in digital photography have also proven to be very useful. Cameras can now capture highly detailed images, even of dragonflies in flight, that can be enlarged, studied, and shared afterward. With these tools, we are better able to study living dragonflies, leading to the discovery of identifying field marks based on living color and pattern. These field marks do not replace anatomical characters but supplement them, providing us more pieces to complete the identification puzzle. This approach, fueled and refined by the publication of better and better field guides, has led to great advances in field identification.

When encountering an unfamiliar dragonfly, try to get as close as possible. Use optics and study it from as many viewpoints as you can. Take photographs from multiple angles. Note its general appearance, but also look for details on specific parts of the dragonfly: head, thorax, abdomen and wings. Not all dragonfly identification puzzles are the same, as the identifying marks and where they are located vary greatly among species.

Dragonflies should always be identified by a combination of characters, working from general to more specific. First impressions, such as size and color, can be valuable but are rarely conclusive. However, in combination with overall body shape, general pattern, and also behavior, we may be able to assign a dragonfly to its family. Recognizing family immediately narrows down the choice of possibilities. Further identification to genus and species usually requires closer looks at specific details of pattern and structure.

Dragonfly Families

The first step in identifying a dragonfly is to determine to which family it belongs. There are seven families in North America, each with its own set of attributes. Some of these are anatomical, including wing venation, the arrangement of the eyes, and reproductive structures. Other clues, such as flight and perching behavior, overall shape, pattern, and color, are often observable by naked eye or through binoculars, so most dragonfly families can be recognized in the field.

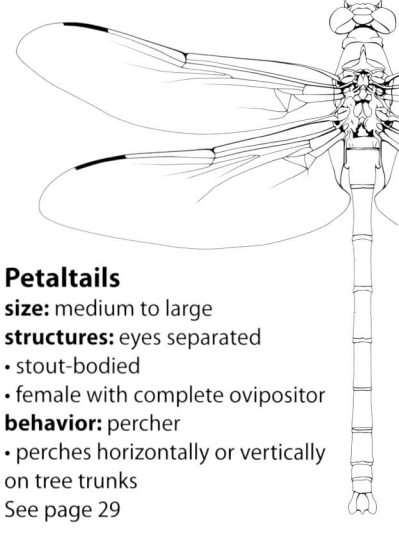

Petaltails
size: medium to large
structures: eyes separated
• stout-bodied
• female with complete ovipositor
behavior: percher
• perches horizontally or vertically on tree trunks
See page 29

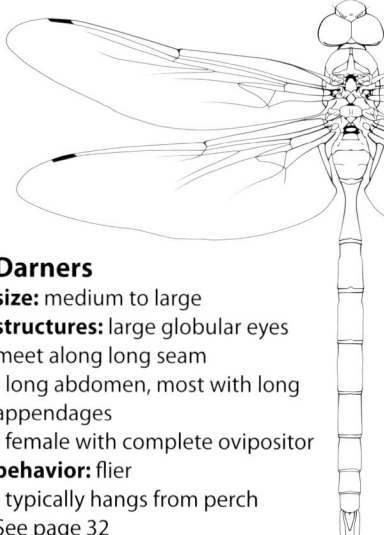

Darners
size: medium to large
structures: large globular eyes meet along long seam
• long abdomen, most with long appendages
• female with complete ovipositor
behavior: flier
• typically hangs from perch
See page 32

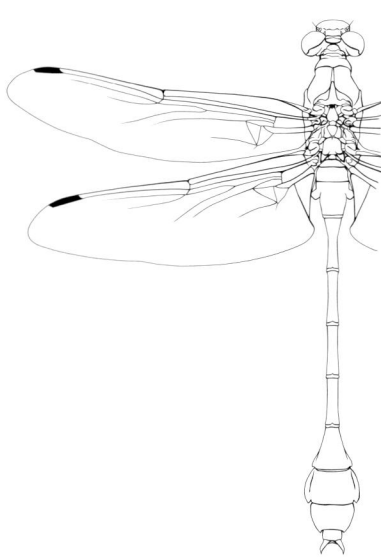

Clubtails
size: small to large
structures: eyes separated
• abdomen often clubbed
behavior: mostly perchers
• often perches horizontally, flat on rocks and ground
See page 80

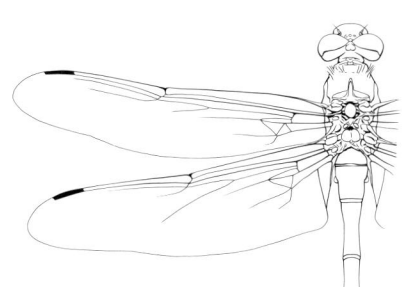

Spiketails
size: medium to large
structures: eyes meet at single point
• female with horizontal spike-like
subgenital plate
pattern: two pale lateral stripes
on thorax
behavior: flier (flight low)
• perches vertically or obliquely
See page 191

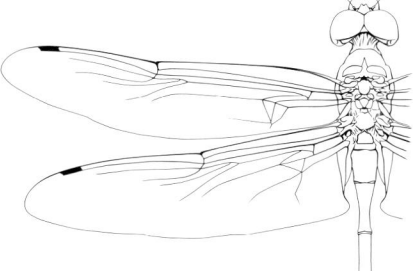

Cruisers
size: medium to large
structures: eyes meet along
seam
• legs long
pattern: single pale lateral
stripe on thorax
behavior: flier
• hangs vertically or perches
obliquely
See page 202

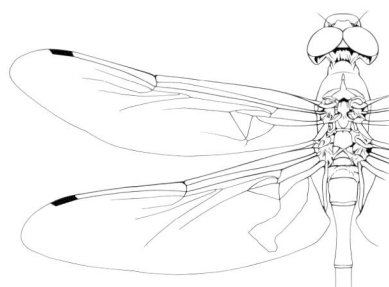

Emeralds
size: small to large
structures: eyes meet along seam
pattern: most species with emerald-
green eyes
• some species with distinct wing
markings
behavior: mainly fliers, few perchers
• most hang vertically or perch
obliquely
See page 214

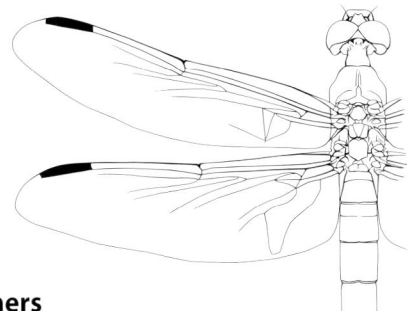

Skimmers
size: small to large
structures: eyes meet along seam
pattern: sexes often different in
appearance when mature
• many species with distinct wing markings
behavior: mostly perchers, some fliers
• most perch horizontally or obliquely, fliers
often hang vertically
See page 268

Identification Basics

Behavior. One of the most important clues when we first spot a dragonfly is its behavior. Even when seen at a distance, how a dragonfly flies can suggest its family. Constant flight is characteristic of certain families like darners, cruisers, and emeralds. Members of these families rarely land while foraging and patrolling. These families may form feeding swarms often consisting of mixed species. Cruiser flight can be noticeably rapid while patrolling up and down a river. Spiketails are also fliers, but their flight is typically slow and low, following the course of their breeding ground streams. The other families—the petaltails, clubtails, and skimmers—are mostly perchers (although many male clubtails patrol over water and some genera of skimmers are constant fliers). When flushed, some clubtails fly a rapid series of U-shaped maneuvers, up and down like a roller coaster.

A dragonfly's posture at rest can also be helpful. Petaltails perch horizontally on logs and other flat surfaces but will land vertically on trees. Some clubtails hang on vegetation, but most tend to sit horizontally, including flat on the ground. With few exceptions, any dragonfly sitting on a rock in the middle of a stream is likely to be a clubtail. Some skimmers also are ground perchers, but a vast majority hunt from perches, usually the stems of vegetation, sallying out after prey, then returning to their perches. At rest, darners typically hang vertically, but some, particularly northern or early-season species, will land flat on the ground. Cruisers and emeralds hang vertically or perch obliquely from stems or branches. In general, constant fliers hang vertically, while perchers tend to rest more horizontally.

Structure. A dragonfly's shape and flight silhouette can also help determine identity. Long abdomens give darners a distinctive profile, with most species having long terminal appendages. Cruisers are large and slender and have long narrow wings, and almost all males have slightly clubbed abdomens. Flying overhead, most emeralds, particularly the males, can be recognized by their narrow abdomens constricted at the base. See flight silhouettes of some constant fliers on page 22.

Size. Overall size can be helpful, particularly when considering families that are fairly consistent in size. Most, but not all, petaltails, darners, cruisers, and spiketails are large dragonflies. Size is more variable within other families, but some species are conspicuously large while others are notably small. Size, however, should be assessed with caution, as it is often difficult to judge in the field. Novice observers in particular tend to overestimate insect size.

Color. Color is another initial clue. Some families and species groups are more likely to feature certain colors. Blue markings are common among darners and skimmers but rare in other families. Bright red is common only among skimmers. Most spiketails and cruisers are marked with bright yellow. Noting the dominant colors of a dragonfly is always useful, but be aware that our perception of color can vary depending on light conditions and the background against which it is viewed. All dragonflies also undergo color changes as they age. Tenerals are initially very pale, and immatures can be less bright (or brighter) in color than when mature. Immature males often resemble females, but many males (and some females) undergo changes in both color and pattern, some dramatically. See illustrated examples, pages 23–24.

Pattern. More important than color alone is pattern. While colors may be variable even within a single species, the species' pattern usually remains consistent. When studying a dragonfly, concentrate on the pattern of light and dark areas. Pattern typically consists of stripes on the thorax, face, and legs, and also spots, stripes, dashes, and rings on the abdomen. Terminal appendages can be noticeably pale or contrasting. Wing markings should always be noted and include patches, bands, and spots. At a distance, only general pattern may be discernible, but these can still aid in identification. For example, having a single pale stripe on the side of the thorax can separate a cruiser from a similarly colored spiketail having two pale stripes.

While family identification can often be made quickly, further identification to species is more demanding. Still, provided a good view, many species can be identified in the field, their diagnostic features visible to the naked eye or through binoculars. With experience, some species may become familiar enough to be identified even while in flight, but most will require greater scrutiny. In the field, the opportunity to view any individual dragonfly is usually brief, so it is best to be prepared. Try to develop familiarity with the various groups of similar species within each family. The field marks that best separate the species of one group of similar species may not be the same for another. For example, mosaic darners are often best identified by the pattern of stripes on the side of the thorax, but green darners are more separable by abdomen color and pattern. Knowing where to look and what features to look for is challenging but becomes easier with practice.

There will always be limitations to field identification, and even the most experienced observer will not be able to identify every dragonfly. Sometimes a dragonfly just will not land or otherwise allow us a good look. From field observation, we may be able to place it into its family, genus, and even a small set of species but can go no further. The most demanding instances often require examination of anatomical features that are too small or hard to see to be accurately observed in the field. In these cases, the identity of the dragonfly can be confirmed only after catching it.

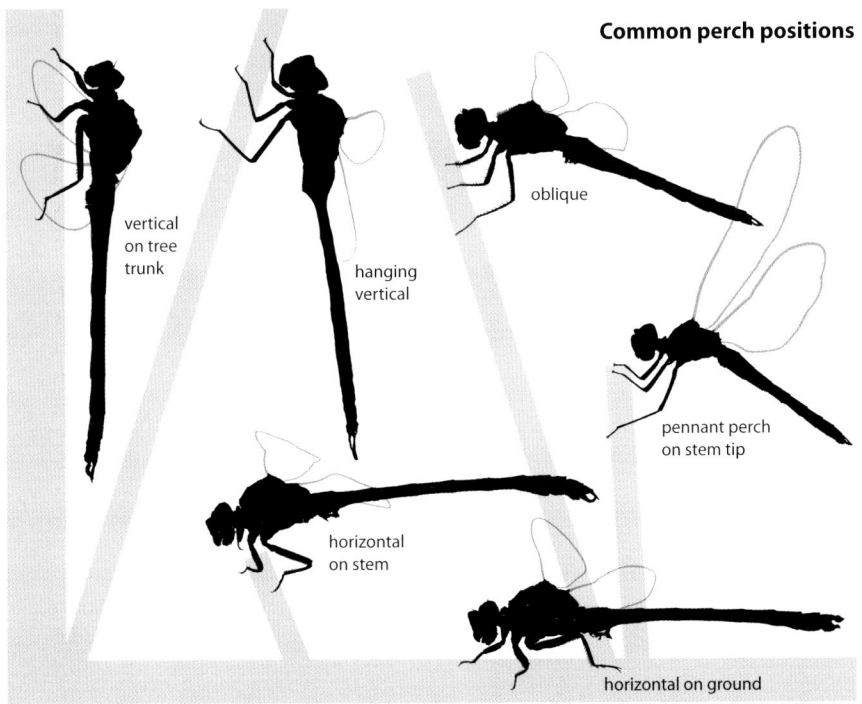

Common perch positions

vertical on tree trunk

hanging vertical

oblique

pennant perch on stem tip

horizontal on stem

horizontal on ground

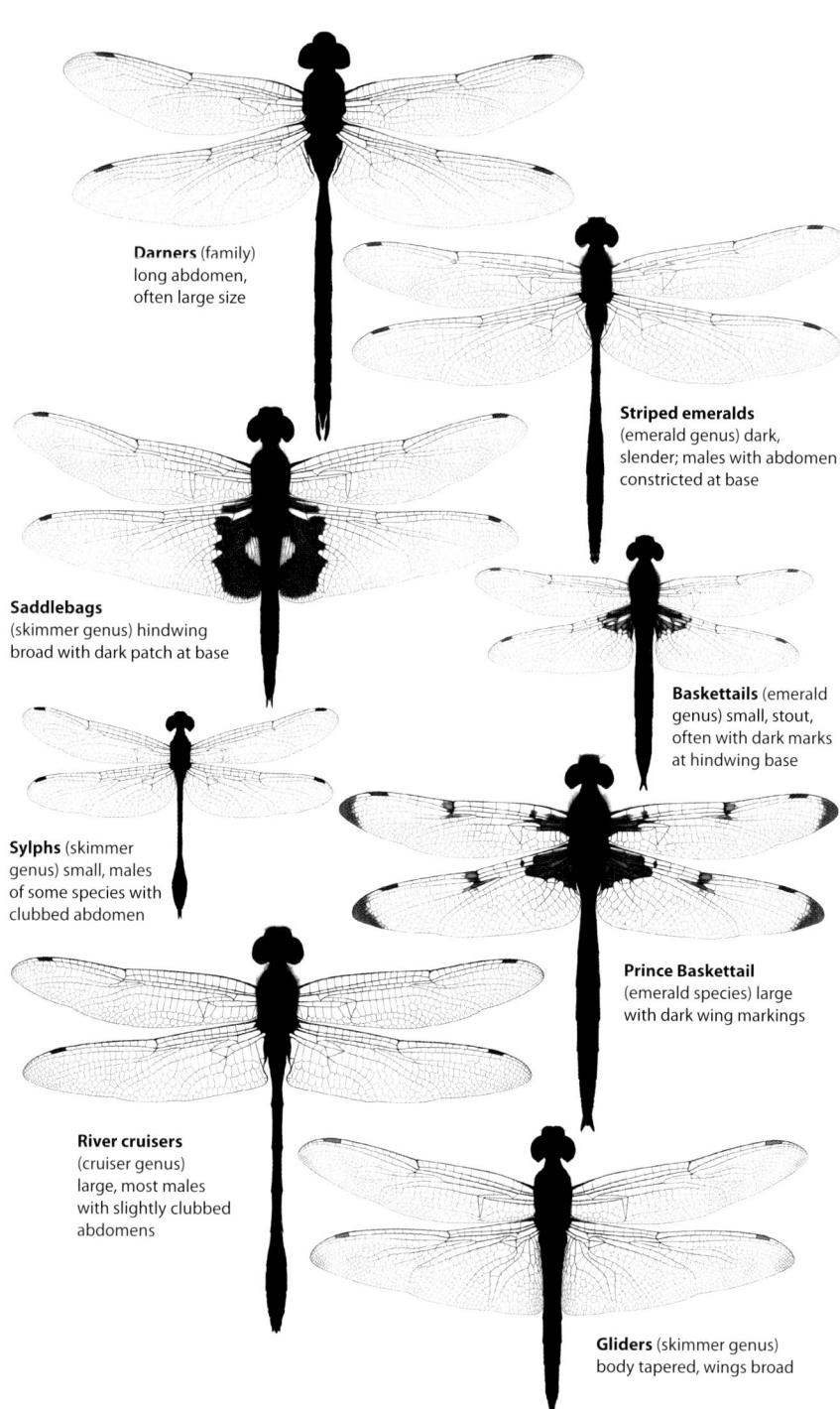

Darners (family)
long abdomen,
often large size

Striped emeralds
(emerald genus) dark,
slender; males with abdomen
constricted at base

Saddlebags
(skimmer genus) hindwing
broad with dark patch at base

Baskettails (emerald
genus) small, stout,
often with dark marks
at hindwing base

Sylphs (skimmer
genus) small, males
of some species with
clubbed abdomen

Prince Baskettail
(emerald species) large
with dark wing markings

River cruisers
(cruiser genus)
large, most males
with slightly clubbed
abdomens

Gliders (skimmer genus)
body tapered, wings broad

Overhead flight silhouettes of some constant fliers

teneral pattern pale and low contrast; eyes gray lacking
pseudopupil; wings initially clouded and held over
abdomen, then become shiny like cellophane

immature eyes gray; pattern contrast
strong with pale areas yellow

Pattern > color
pattern is often more consistent than color,
making it a better clue for identification

mature eye color blue; color of face and thorax greenish

older thorax color grayish; pattern less
contrasting as pale areas darken; wings
may become tinted and tattered

Example of color changes over lifetime (male Lancet Clubtail)

immature male resembles female with yellow-striped abdomen, reddish eyes and pale face

maturing male begins to develop pruinosity on wings, abdomen and front of thorax; face, eyes, and pale markings on thorax and abdomen darken

pruinosity is a waxy bloom that many odonates develop and can appear white, pale blue, or gray

mature male with dark eyes and face; abdomen and parts of wings and thorax pruinose white

pruinosity on abdomen often abraded by female's legs during mating, resulting in dark areas

Example of color and pattern change (male Widow Skimmer)

Catching Dragonflies

Despite the progress in field identification, catching a dragonfly remains the most reliable way to confirm its identity. However, catching dragonflies is not easy, and many observers choose not to do so. Even when practicing catch and release, the act of swinging a net at a dragonfly can result in harm to the insect. Be aware that many parks and preserves prohibit the capture of any type of wildlife. Some species are rare in certain areas and can be state or locally protected (the Hine's Emerald is federally protected), so it is advisable to observe all regulations and consult proper authorities prior to collecting.

The best way to catch dragonflies is with an insect net. A dragonfly net should be strong but lightweight with a wide net ring and a long handle. At minimum, the handle should be 3–4 ft long, but additional reach is often desired when pursuing wary or high-flying insects. Some handles are designed to telescope, while others can be extended by adding additional segments. Too much length can be unwieldy and difficult to swing. The net ring should be no smaller than 12 in in diameter; 18–24 in is recommended. Mesh openings should measure about 2 mm, fine enough to hold the smallest dragonfly yet wide enough to reduce air resistance. When approaching a perched dragonfly, it is best to move slowly and avoid casting any shadow on the insect. If the subject is perched very low, quickly slap the net over it, bringing the net's rim flat against the ground or water's surface. Continue to hold the net down with one hand and lift the mesh upward with the other to allow the dragonfly to fly upward into the net's pocket. Dragonflies perched in vegetation can be caught with a similar downward swing, but an upward sweep can be productive. It is not unusual for the net to get snagged. Horizontal swings tend to swat too much vegetation, often resulting in the escape of the dragonfly.

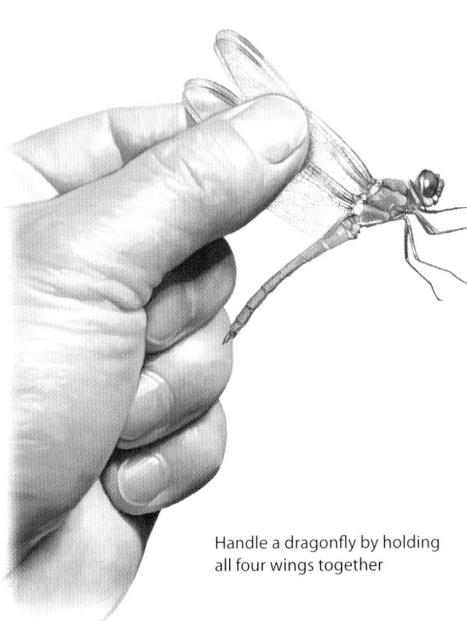

Handle a dragonfly by holding all four wings together

Catching flying dragonflies is particularly challenging. Dragonflies are wary; just getting in range to swing can be trying. Patrolling dragonflies often follow similar routes, so by observing their behavior it is sometimes possible to anticipate their movements. Quick reflexes and a fast swing of the net are required. It is best to swing from behind and below the dragonfly, where its vision is the poorest. Follow through on the swing to sweep the insect into the net's pocket and quickly twist the handle to close the net. Low-flying dragonflies can sometimes be caught with a downward slap or with the net initially held low, and then a quick upward scoop as the dragonfly flies by.

After catching a dragonfly, carefully reach a hand in from below to remove it from the net. Grasp the dragonfly by holding all four wings together over the abdomen with the thumb and forefinger. Before removing the dragonfly from the net, take care that it is not entangled or grasping the mesh with its legs or jaws.

Holding the wings should not harm the insect. Keep your hands as clean and dry as possible. Moisture and sweat can cause the wings to stick together and become damaged. Insect repellants and sunscreens may also cause harm. Limit the amount of time handling the dragonfly, and keep it out of direct sun to avoid overheating. Avoid catching tenerals. Any type of handling can permanently damage their soft bodies and wings.

A magnifying lens is often necessary to view many of the anatomical features needed for species identification. A hand lens with a magnification of at least 8x is recommended. Even with common and field-identifiable species, it can be rewarding to examine a dragonfly in the hand. It will enable confirmation of the dragonfly's identity and provide an opportunity to appreciate its structure, pattern, and beauty up close. With careful handling, the dragonfly can be released unharmed.

hand lens magnifier

Collecting Dragonflies

A well-maintained collection can be useful for in-depth study, recordkeeping, and future reference for oneself and others. The collecting of a few specimens is unlikely to harm any given population. For most observers, however, collecting is unnecessary. The catch-and-release method is practiced by many dragonfly-watchers and is sufficient for identification in a great majority of cases.

The dragonfly to be collected should be placed in a small container or in an envelope with its wings folded together. It should be placed in a cooler or other crush-proof container and kept from heat. Care should be taken to keep the dragonfly alive. Upon death, dragonflies quickly lose their color and begin to darken from decomposition.

The dragonfly is killed by dropping it into a jar of acetone. Acetone works to preserve the specimen by killing bacteria, displacing water, and dissolving fats that can discolor the body. Care must be taken, as acetone is highly flammable and harmful to breathe, so it must be used in a well-ventilated area or outdoors away from flame and heat sources. Acetone is a strong solvent, so the jar must be glass or some nonreactive substance. After the dragonfly dies, remove the specimen to arrange the wings and straighten the abdomen. The dragonfly can then be placed in a paper envelope to hold its position, then dropped back into the acetone. Tightly cap the jar and allow the specimen to soak for a day. After removing the specimen from the acetone, remove it from the envelope and allow it to dry.

Acetone largely preserves the dragonfly's pattern, but most colors eventually fade. Eye color is not preserved, and pruinosity is often lost. Preserved dragonflies are very fragile and not suited to being spread and pinned like other insects. Instead, they should be stored in paper, glassine, or plastic envelopes accompanied with a card or label recording the date and exact location of collection. The specimen envelope should then be filed in an air-tight container along with a few mothballs to ward off pests that would feed on the specimens. Packets of silica are useful to protect the collection from moisture.

Species Accounts

This guide follows taxonomic order, beginning from the most primitive family, the petal-tails (Petaluridae), and concluding with the family considered the most recently evolved, the skimmers (Libellulidae). It is important to review the introduction to each family, as the defining features of each are listed and illustrated. These are typically a combination of anatomical and pattern characters along with family members' general behavior.

Each family is divided into smaller groups of related species called genera (singular: genus). Similar genera are grouped together, and a description of each genus is provided with guidance on how to separate genera. Within each genus, similar-appearing species have been placed together, with the species most alike presented on opposite pages to facilitate comparison.

Common and scientific names: North American dragonflies have only recently been given common names. The names used in this book have been established by the Dragonfly Society of the Americas. Many beginners find common names more descriptive and easier to remember than scientific names, but veteran workers may not recognize them. As the popularity of odonate-watching has grown, common names have become better established, but it is useful to be familiar with both. Scientific names are italicized, with the first word denoting genus, followed by species and subspecies. For example, *Erpetogomphus lampropeltis natrix*: *Erpetogomphus* is a genus of clubtails called the ringtails, *lampropeltis* is the species name of the Serpent Ringtail, and *natrix* is the inland desert subspecies of Serpent Ringtail.

Size: Range of body length throughout North America.

Species text: Habitat, behavior, and descriptive and comparative features are highlighted for each species.

♂ **male**
♀ **female**

Illustrations: With few exceptions, both sexes of each species are depicted from above (dorsally) and from the side (laterally). Dorsal views of the male include his wings. The wings of the female may also be shown in full or in part if they are significantly patterned. Views of subspecies and variations are also presented. Enlargements and additional details are included when necessary. Key markings and features are captioned for each species, often with lines pointing directly at those characters.

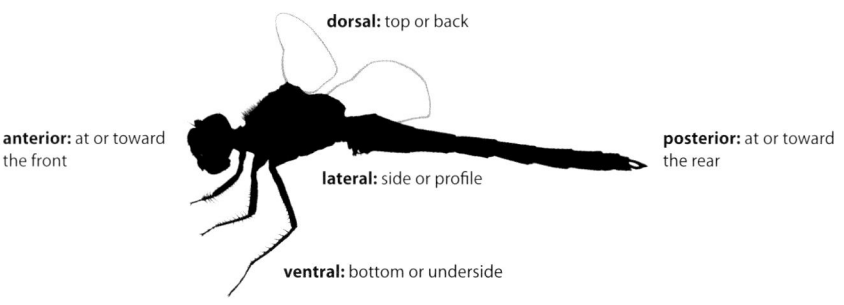

dorsal: top or back

anterior: at or toward the front

posterior: at or toward the rear

lateral: side or profile

ventral: bottom or underside

Similar species: The species that are most likely to be confused are listed along with notes on how the similar species may be separated.

Range maps: Each map shows a species range as currently known in the continental US and Canada. Distribution information comes from a variety of sources, primarily records compiled by Nick Donnelly for his North American Dot Map Project, along with more recent records reported to Odonata Central, iNaturalist, and other groups and websites. The maps should be used with caution. Many species are local and require specific habitats, and it should not be interpreted that they can be found everywhere within their marked range. Some records are historical; others are of vagrants. The ranges of species are constantly in flux, often due to habitat changes both natural and human-made. In many cases, areas between records have been included in a species range. While this "filling in the blanks" helps visualize where records are clustered, applying this approach to areas where records are sparer required a fair measure of speculation.

Dragonfly distribution: There is still much to be learned about dragonfly distribution and species' current and evolving ranges. In recent years, organized surveys by a few states have added many species to their state and county lists. Our knowledge of other states, provinces, and regions comes from the work of dedicated individuals and small groups, but vast areas remain undersurveyed. There is ample opportunity for beginners to contribute. Undoubtedly, new records will quickly show these maps to be less than complete. When using these maps, keep in mind that much territory is largely unexplored. It would be unwise to automatically exclude a species from consideration because it has not been recorded in a certain area.

Dates: A range of recorded dates is listed under the range maps. With wide-ranging species, northern populations usually emerge later and have briefer flight periods than southern populations. The dates of a month are described as early, mid, and late when falling within days 1–10, 10–20, and 21–31, respectively.

In-Hand Characters

Following the species accounts is an appendix of illustrated anatomical details. These are provided to aid the identification of the most similar species and groups. Identification of these dragonflies may require close examination of specific features that are difficult to see well in the field. Capture is usually required, and magnification is recommended to view these features, which include terminal appendages, hamules, subgenital plates, and other characters.

Petaltails • Family Petaluridae

Petaltails are large dragonflies with stout abdomens. Males of most species have wide petal-like cerci from which the family gets its name. Their wings feature long narrow stigmas. Their eyes are widely separated similar to clubtails, and females have large bladed ovipositors similar to the darners. Their breeding habitats are primarily seeps and spring-fed bogs, where their larvae are semi-terrestrial, dwelling in mud, under leaves, or in burrows dug in wet areas.

The petaltails are considered our most primitive dragonflies. Their fossil remains indicate they arose in the Jurassic Era and were once the most dominant dragonfly family on Earth. Only 11 species remain worldwide, two in North America.

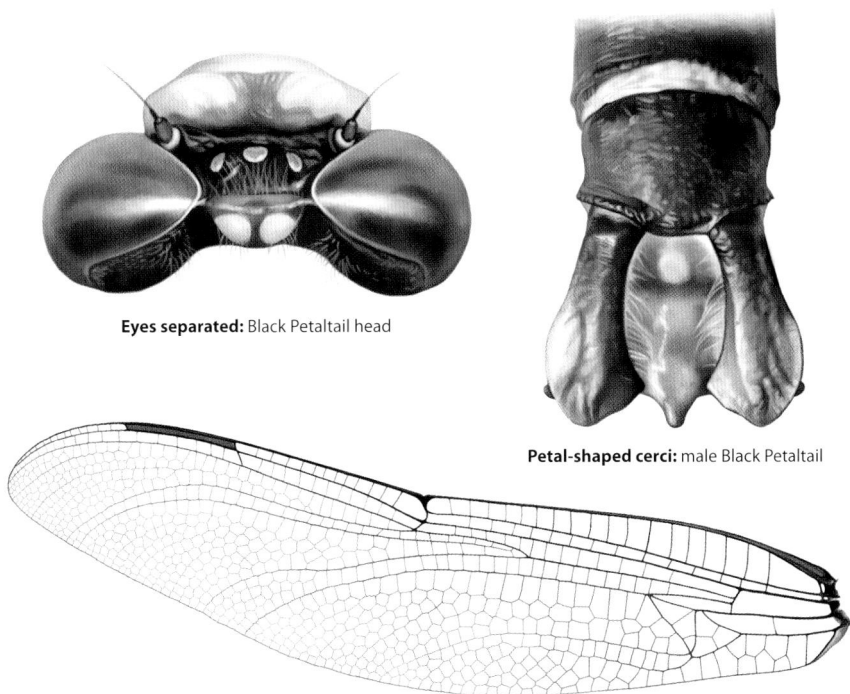

Eyes separated: Black Petaltail head

Petal-shaped cerci: male Black Petaltail

Long narrow stigma: Gray Petaltail wing

Gray Petaltail, genus *Tachopteryx,* page 30
This genus consists of a single eastern species, the Gray Petaltail. Its large size and overall gray coloration are distinctive, but it could be mistaken for a darner in flight. Unlike darners, which typically hang vertically, Gray Petaltail often perches horizontally on flat surfaces or vertically on tree trunks.

Dark Petaltails, genus *Tanypteryx,* page 31
Genus consisting of two mostly black species, the Black Petaltail of the Pacific Northwest and another found in Japan. Black Petaltail's yellow and black coloration, separated eyes, and habitual perching on flat surfaces suggest a clubtail, but its spotted pattern is unique.

Gray Petaltail · *Tachopteryx thoreyi*

71–80 mm, 2.8–3.1 in. Large eastern petaltail, local at hillside seeps in deciduous forests. Large size and mostly gray coloration distinctive. Eyes separated, deep brown in color, grayer when mature. Stigmas long and linear. Often perches vertically on tree trunks where well camouflaged. Tame, will land on light-colored clothing.

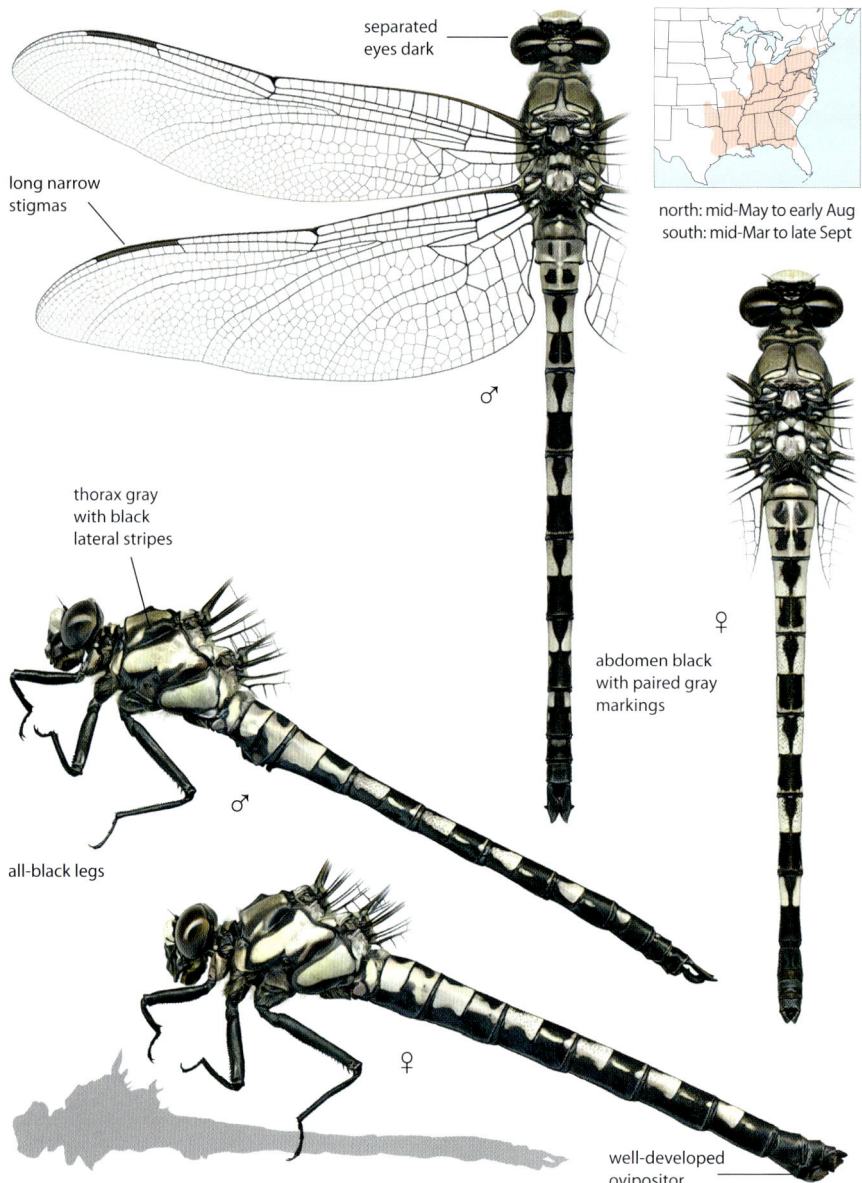

separated
eyes dark

long narrow
stigmas

north: mid-May to early Aug
south: mid-Mar to late Sept

♂

thorax gray
with black
lateral stripes

♀

abdomen black
with paired gray
markings

♂

all-black legs

♀

well-developed
ovipositor

Similar species: No other large dragonfly is mostly gray. Immature **Mosaic darner species** may be dully colored but have larger eyes that meet along seam. **Clubtail species** have separated eyes like petaltails but can be distinguished by pattern and behavior.

Black Petaltail • *Tanypteryx hageni*

54–57 mm, 2.1–2.25 in. Local dark northwestern petaltail inhabiting permanent mossy fens and bogs. Thorax and abdomen black with pale yellow spots. Eyes separated, dark brown in color. Stigmas long and linear. Adults tame and approachable, often perching low on bog vegetation or flat against logs and stones.

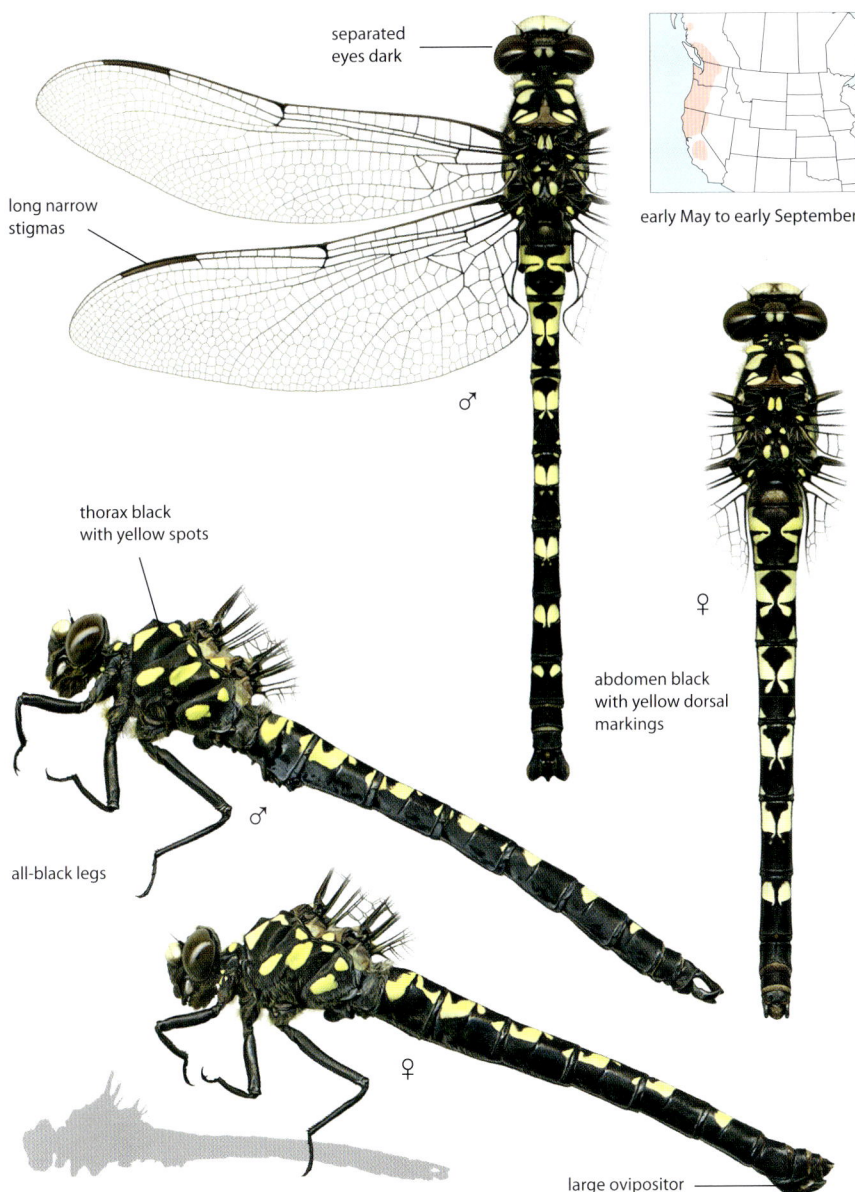

separated eyes dark

long narrow stigmas

early May to early September

♂

♀

thorax black with yellow spots

abdomen black with yellow dorsal markings

♂

all-black legs

♀

large ovipositor

Similar species: Pacific Spiketail has lateral thoracic stripes and blue eyes. **Mosaic darner species** have larger eyes that meet along a seam. **Clubtail species** have striped thoraxes, most species with row of pale markings along top of abdomen.

Darners • Family Aeschnidae

The darners are a sizable family of dragonflies with a worldwide distribution. Most species are large; some are among our biggest and lengthiest dragonflies. The head is dominated by very large eyes that meet broadly along a seam. The abdomen is characteristically long and slender, males and many females often having long terminal appendages contributing to their overall length. Females have a well-developed, bladed ovipositor used to cut and insert their eggs into plant tissues. The name "darner" likely originates from "devil's darning needle," a name for dragonflies from Old World folklore.

Darners are fliers, most often seen while patrolling and feeding in flight. At rest, they hang vertically. Some species will sit flat on the ground or vertically on tree trunks to absorb warmth from the sun, particularly early-season species or dragonflies at cooler latitudes. Many of our darners are predominately brown or black in color, often with blue or green markings. Others are predominately green. Only a few species have red, yellow, or white markings. The wings are generally clear, although they may be tinted in portion or entirely. Only a couple of our species have dark wing markings, but these are small and restricted to the wing base. 494 species worldwide, 42 North America.

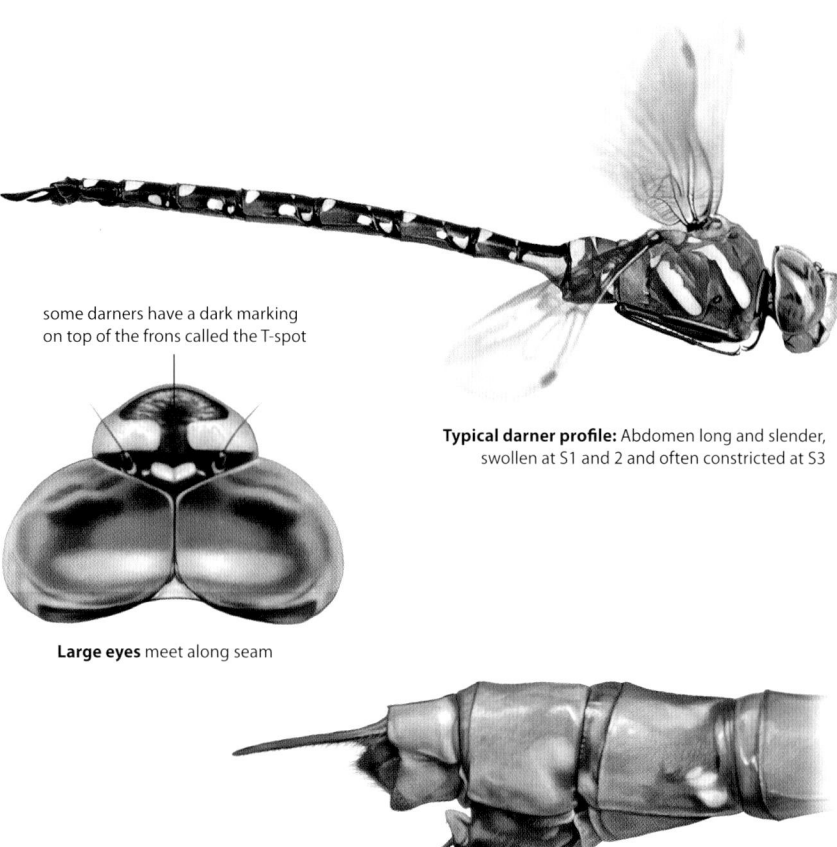

some darners have a dark marking on top of the frons called the T-spot

Typical darner profile: Abdomen long and slender, swollen at S1 and 2 and often constricted at S3

Large eyes meet along seam

Female with large bladed ovipositor

Green Darners, genus *Anax,* page 37
A genus of large, robust darners. North American species are characterized by having un-marked green thoraxes. Unlike other darners, males lack projecting auricles on the side of S2. Correspondingly, the base of the hindwing is smoothly curved along its margin similar to the shape of the female's wing. 32 species worldwide, five North America.

The solid green thorax separates green darners from most other darners. Some pilot darners have mostly green tho-raxes, but their abdomens are dark with narrow green markings. The species of green darner are most separable by the pattern, color, and proportion of the abdomen. Mature males can often be identified by color pattern even while in flight. Three of our species have a bull's-eye-shaped marking on top of the frons, while the other two do not. Because males lack an inner angle at the base of the hindwing, males and females cannot be separated by wing shape.

Bull's-eye marking
on top of frons
(Common Green Darner)

Pilot Darners, genus *Coryphaeschna,* page 42
Large Neotropical dragonflies with long abdomens and short legs. Eight species world-wide, four North America.

Our species of pilot darners have bright green thoraxes, some with dark thoracic stripes. Their abdomens are mostly black or brown with narrow green rings. Males and females are similar in pattern. Both sexes have very long appendages, but those of the female break off with age. Top of frons with black T-spot.

Malachite Darners, genus *Remartinia,* page 45
A small Neotropical genus, closely related to pilot darners, but females have very short cerci. Four species worldwide, two North America.

The thorax is bright green with wide dark stripes. The abdomen is dark with small green dorsal markings. The frons lacks a T-spot. The sexes are similar in pattern. Our two North American species are very similar, best separated by details in thoracic pattern and the structure of the male appendages.

Three-spined Darners, genus *Triacanthagyna,* page 47
Crepuscular Neotropical forest species characterized by very large eyes and broad wings. They rest by day, hanging in dense woodland, and forage primarily at dusk. Females have three ventral spines on S10, unique among dragonflies. Male and female have very long cerci; those of the female break off with age. Nine species world-wide, three North America.

Our species have mostly green thoraxes and dark slender ab-domens. The sexes are similar in pattern. A large head combined with a relatively short thorax help separate three-spined darners from the larger pilot darners.

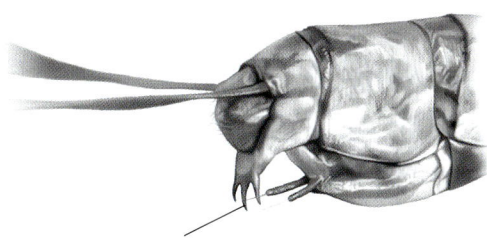

Three-spined darner female
has three ventral spines on S10

Two-spined Darners, genus ***Gynacantha,*** page 50
Large genus of tropical darners found in both the New and Old World. Closely related to three-spined darners, but females have only two ventral spines on S10. Crepuscular, hangs in dense woodlands during the day, forages at dusk or under cloudy conditions when the temperature is sufficiently warm. 101 species worldwide, two North America.

Our two species have small dark lateral thoracic spots that other similar darners lack. Both species are rather plain in appearance, subtly patterned in dull brown and green. Their ranges are not known to overlap, but they can be separated by thoracic pattern.

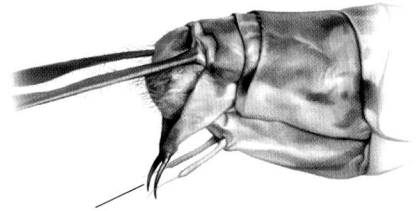

Two-spined darner female has two ventral spines on S10

Swamp Darner, genus ***Epiaeschna,*** page 52
The single species in this genus, the Swamp Darner, is one of our largest dragonflies. It is a brawny, strong-flying dragonfly with a long thick abdomen and long appendages. The male abdomen has a triangular-shaped spine on top of S10. Its overall appearance is

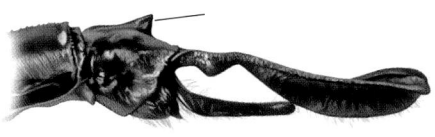

Swamp Darner male has triangular spine on top of S10

dark, having only narrow green abdominal rings and green thoracic stripes. Pilot darners are the most similar in pattern and size but have slimmer profiles and predominately green thoraxes.

Cyrano Darner, genus ***Nasiaeschna,*** page 53
This genus consists of a single species, the Cyrano Darner, a large, stocky, thick-waisted darner. The species is named for its projecting frons, suggestive of the prominent nose of the literary character Cyrano de Bergerac. Most similar to Swamp Darner but has blue-green dorsal abdominal stripes and a blue face.

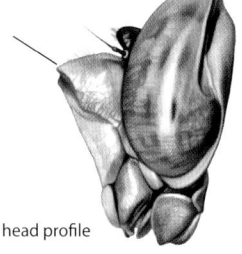

Cyrano Darner head profile

Mosaic Darners, genus ***Aeshna,*** page 54
Large genus of colorfully patterned darners, mostly northern in distribution but widespread throughout the Northern Hemisphere. They utilize a wide variety of breeding habitat, including lakes, ponds, bogs, marshes, and slow streams, but can be found well away from water. Mosaic darners commonly gather in mixed-species feeding swarms. 31 species worldwide, 15 North America.

The thorax and abdomen are dark brown and black with contrasting blue and green markings. Thoracic markings typically include at least two pale lateral stripes and two pale frontal stripes. The abdomen features spots arranged in rows suggestive of a mosaic pattern. Seen mostly in flight, mosaic darners are recognizable by their size, shape, and bright coloration, but most species are similar in appearance, making field identification difficult. Species are best identified by their thoracic patterns, particularly the shape and color of the lateral thoracic stripes. These are best observed in the hand or through binoculars while the darner is perched. It is sometimes possible to discern the thoracic pattern on a flying insect through binoculars if the dragonfly flies a predictable beat or hovers. Due to the similarity of pattern, dorsal views are less helpful, although abdominal pattern details can provide some identification clues. The male cerci have two basic forms, described as simple and wedge-shaped. The face is pale with a dark T-spot on top of the frons. Several species have a narrow dark cross-line on the face. Female mosaic darners often have different color forms, but the light/dark pattern of each species remains consistent.

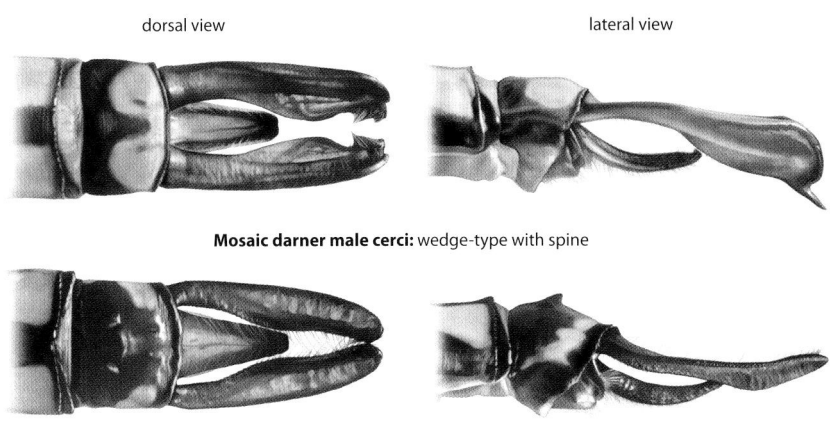

dorsal view

lateral view

Mosaic darner male cerci: wedge-type with spine

Mosaic darner male cerci: simple type

Neotropical Darners, genus *Rhionaeschna,* page 70

Large group of mostly Neotropical species recently separated from the genus *Aeshna*, which they closely resemble. Both sexes have a small projection (tubercle) on the underside of the first abdominal segment, a feature all other darners lack. 42 species worldwide, five North America.

Neotropical darners have obvious pale areas adjacent to the T-spot on the frons. Three of the five North American species (Blue-eyed, Spatterdock, and Arroyo Darners) are similar in appearance, with males having distinctive sky-blue eyes and blue body markings. California Darner is similar in pattern, but its eyes are darker blue and the face has a narrow dark stripe. Turquoise-tipped Darner is quite different from the others, showing green thoracic stripes and inconspicuous abdominal markings.

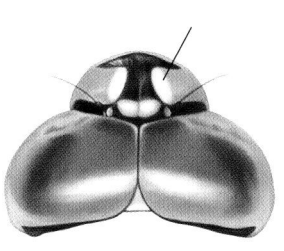

Neotropical darner head
with pale areas adjacent to T-spot

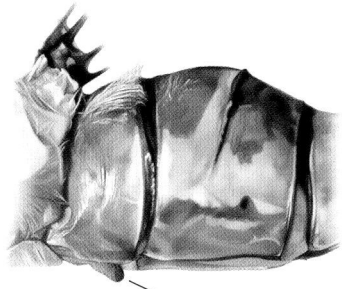

Neotropical darner: tubercle under S1

Riffle Darners, genus *Oplonaeschna,* page 75

Genus consists of two large species, only one ranging into North America in the Southwest. They are similar in appearance to mosaic and Neotropical darners but considered to be more closely related to two- and three-spined darners. The male has a small finger-like projection on top of S10. Riffle darners are best identified by the shape and coloration of the lateral thoracic stripes, along with abdomen pattern.

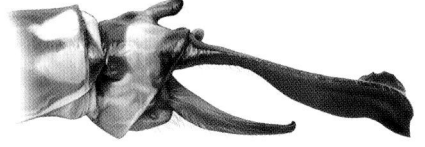

Riffle Darner: S10 with long dorsal projection

Springtime Darner, genus *Basiaeschna,* page 69

This genus consists of a single species, the Springtime Darner, an eastern early-season dragonfly. Eyes are smaller than most other darners, so the seam where the eyes meet is comparatively short. Wing venation is sparser and more primitive than other members of the family. With a blue-spotted abdomen, its pattern is suggestive of a mosaic darner, but spring flight season is earlier than most other darners. Springtime Darner is one of the few darners with wing markings, having small dark dashes at the base of each wing.

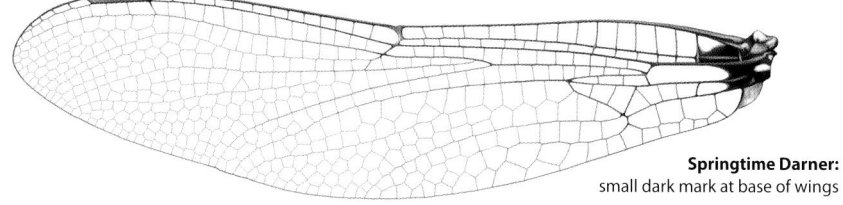

Springtime Darner:
small dark mark at base of wings

Spotted Darners, genus *Boyeria,* page 76

Genus of large brown darners with broad wings and large eyes. Their wings have cross-veins in the basal space, unlike other North American darners. Seven species worldwide, two North America.

Spotted darners are often recognizable in flight, as they fly slowly and low up and down shaded woodland streams. Both of our species are eastern and have two contrasting pale yellow spots on each side of the thorax.

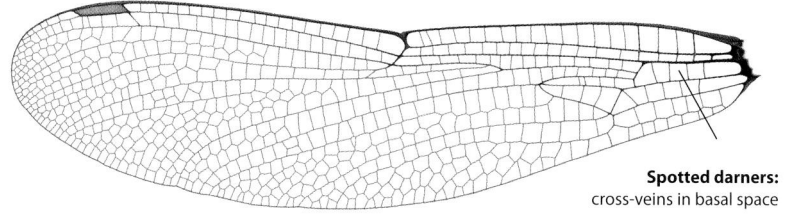

Spotted darners:
cross-veins in basal space

Pygmy Darners, genus *Gomphaeshna,* page 78

This genus consists of two small eastern North American species. They are slender, dark-appearing dragonflies, their intricate patterns revealed at close range. Unlike on other darners, the male epiproct is deeply forked. Wing venation is relatively simple and reduced compared to other darners.

Their complex mottled patterns and small size make pygmy darners unlikely to be confused with other darners. Both species regularly perch vertically on tree trunks.

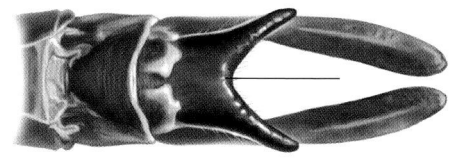

Pygmy darner male has forked epiproct

Amazon Darner • *Anax amazili*

70–75 mm, 2.75–2.95 in. Darner with solid green thorax and white-spotted abdomen. Rare at vegetated open ponds and marshes in southern Texas and Florida; single Louisiana record. Sexes similarly patterned: abdomen dark brown with large whitish spots, white vertical stripe on the side of S1. Bull's-eye marking on forehead.

bull's-eye marking

eyes gray

early April to late November

♂

blue markings
on top of thorax
and S3

solid green thorax

♀

white-spotted
abdomen

♂

♀

white vertical stripe
on side of S2

Similar species: Common Green Darner has striped abdomen. **Blue-spotted Comet Darner** has pale whitish or greenish abdominal spots when immature but lacks white vertical stripe on side of S2 and bull's-eye marking on forehead.

Common Green Darner • *Anax junius*

68–80 mm, 2.7–3.1 in. The most common and widespread North American darner; found at most still-water habitats. Migratory in both spring and fall. Thorax all green. Bull's-eye mark on forehead. Mature male abdomen blue with black dorsal stripe. Female abdominal markings gray-green; markings red-brown to purple in immatures.

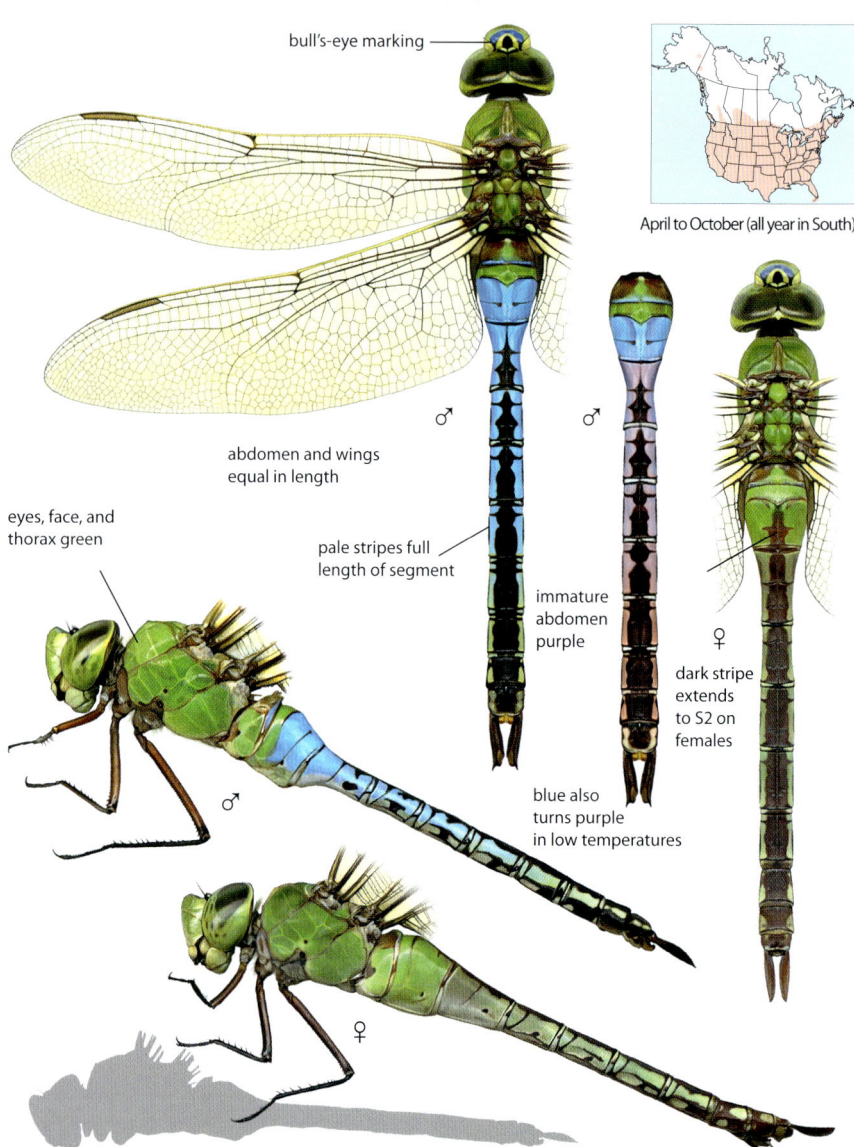

bull's-eye marking

April to October (all year in South)

♂

abdomen and wings
equal in length

eyes, face, and
thorax green

pale stripes full
length of segment

immature
abdomen
purple

♂

♀

dark stripe
extends
to S2 on
females

♂

blue also
turns purple
in low temperatures

♀

Similar species: Giant Darner larger, with abdomen longer than its wings; male flies with abdomen arched. Pale abdominal markings extend two-thirds length of each abdominal segment. **Comet Darner** lacks bull's-eye marking on forehead and has longer abdomen, wings, and legs. **Blue-spotted Comet Darner** lacks bull's-eye marking on forehead, has blue abdominal spots instead of stripes. **Amazon Darner** has white abdominal spots.

Giant Darner • *Anax walsinghami*

100–116 mm, 3.9–4.6 in. Largest North American dragonfly by length. Southwestern, inhabiting slow, spring-fed streams, often in canyons, also ponds and marshes. Similar to Common Green Darner but with longer abdomen. Thorax all green. Both sexes with abdomen longer than wings. Male flies with abdomen arched.

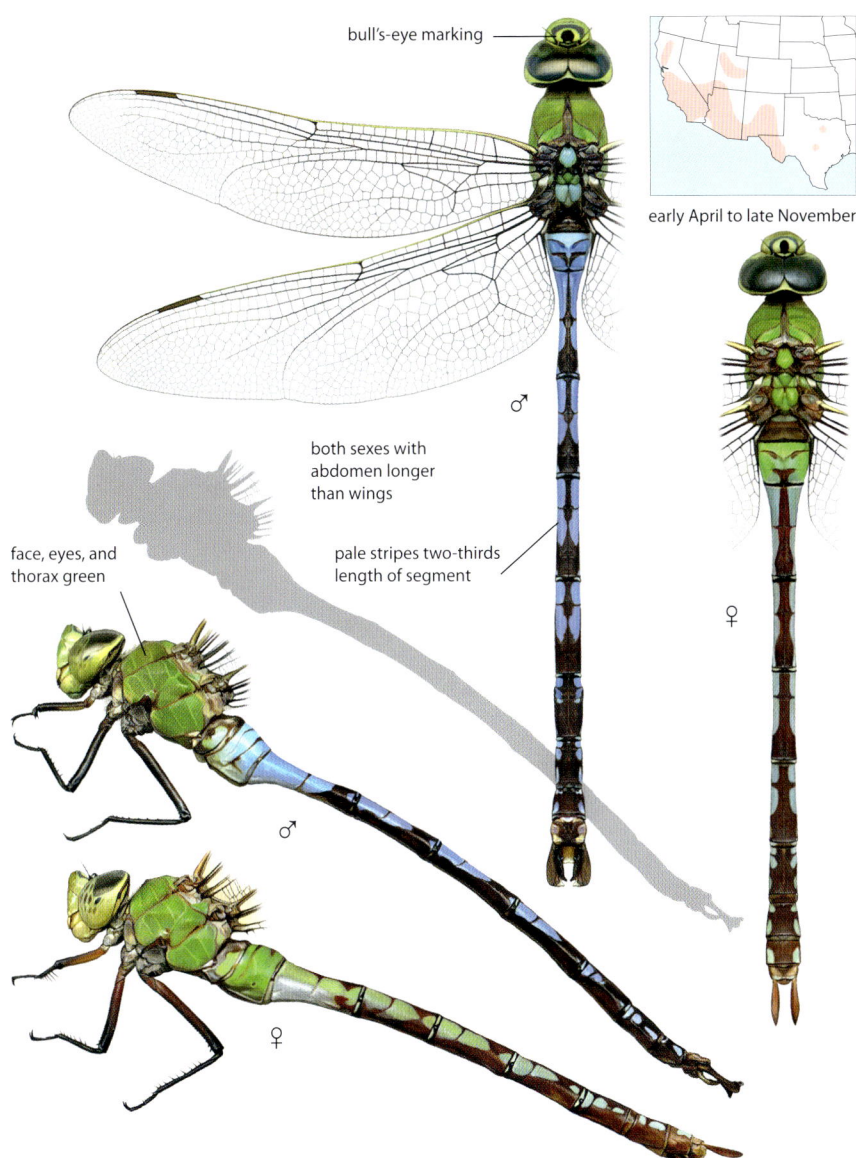

bull's-eye marking

early April to late November

both sexes with abdomen longer than wings

face, eyes, and thorax green

pale stripes two-thirds length of segment

♂

♀

♂

♀

Similar species: Common Green Darner has shorter abdomen; abdomen equal to wings in length. Also pale abdominal stripes run full length of segments. **Comet Darner** has shorter red or reddish abdomen and no bull's-eye marking on forehead. **Blue-spotted Comet Darner** has shorter, blue-spotted abdomen, lacks bull's-eye marking on forehead.

Comet Darner • *Anax longipes*

75–87 mm, 2.95–3.4 in. Large green and red darner inhabiting shallow grassy lakes and ponds. Eastern. Thorax all green. Mature male unmistakable with bright red abdomen. Female abdomen reddish-brown with pale spotting. Face green with no forehead markings. Legs long with reddish-brown femurs.

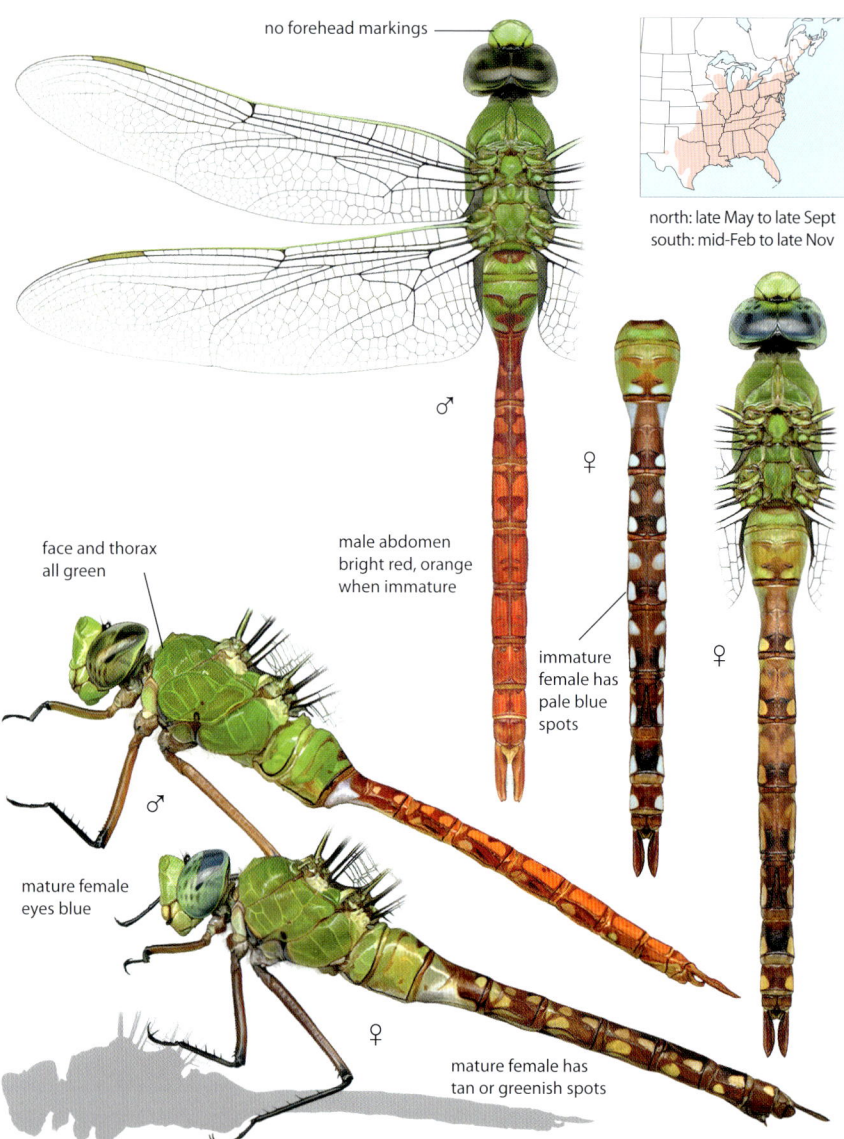

no forehead markings

north: late May to late Sept
south: mid-Feb to late Nov

♂

♀

face and thorax all green

male abdomen bright red, orange when immature

immature female has pale blue spots

♀

♂

mature female eyes blue

♀

mature female has tan or greenish spots

Similar species: Common Green Darner has shorter, striped abdomen and bull's-eye mark on forehead. Mature **Blue-spotted Comet Darner** has blue abdominal spots and blue markings between the wings. Females are more similar, particularly similar to immature female Comet Darner. Compare cerci shape, page 394. **Amazon Darner** has whitish abdominal spots, white vertical stripe on side of S1, and bull's-eye mark on forehead.

Blue-spotted Comet Darner • *Anax concolor*

66–77 mm, 2.6–3 in. Large green darner with blue-spotted abdomen. Rare in South Texas at shallow open ponds. Abdomen of both sexes brown with contrasting blue spots; spots whitish or pale green when immature. Face green with no forehead markings. Blue dorsal thoracic markings at base of wings.

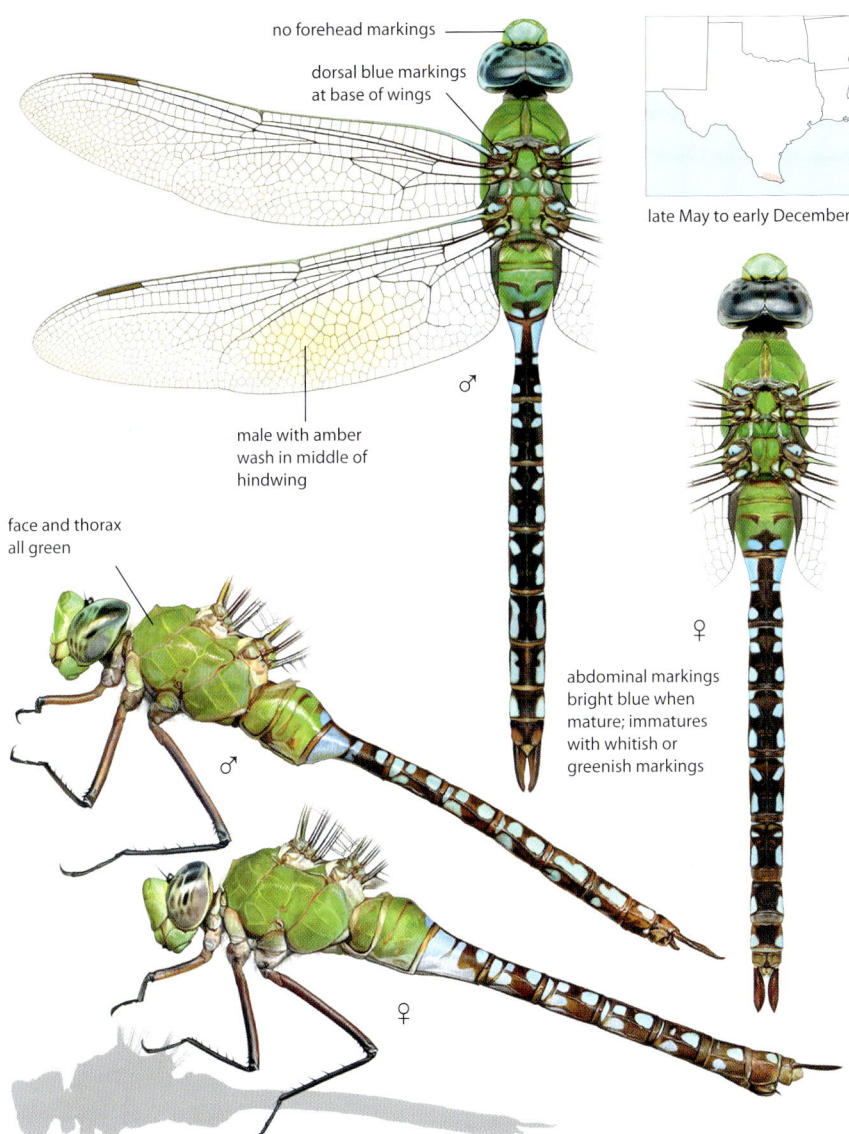

no forehead markings

dorsal blue markings at base of wings

late May to early December

male with amber wash in middle of hindwing

face and thorax all green

♂

♀

abdominal markings bright blue when mature; immatures with whitish or greenish markings

♂

♀

Similar species: Comet Darner abdomen redder with less contrasting spots. Immature females with pale blue spots nearly identical to immature female Blue-spotted Comet Darner. Compare slight differences in cerci shape, page 394. **Amazon Darner** has white vertical stripe on side of S1, whitish abdominal spots, and bull's-eye mark on forehead. **Common Green Darner** has striped abdomen, bull's-eye mark on forehead.

Regal Darner • *Coryphaeschna ingens*

86–90 mm, 3.4–3.5 in. Large, slender, green-striped darner found at vegetated lakes and slow streams in the Southeast. Sexes similar. Thorax green with wide brown stripes, abdomen black with narrow green rings. Face green. Male eyes green, mature female eyes blue. Both sexes with long cerci, often broken on mature female.

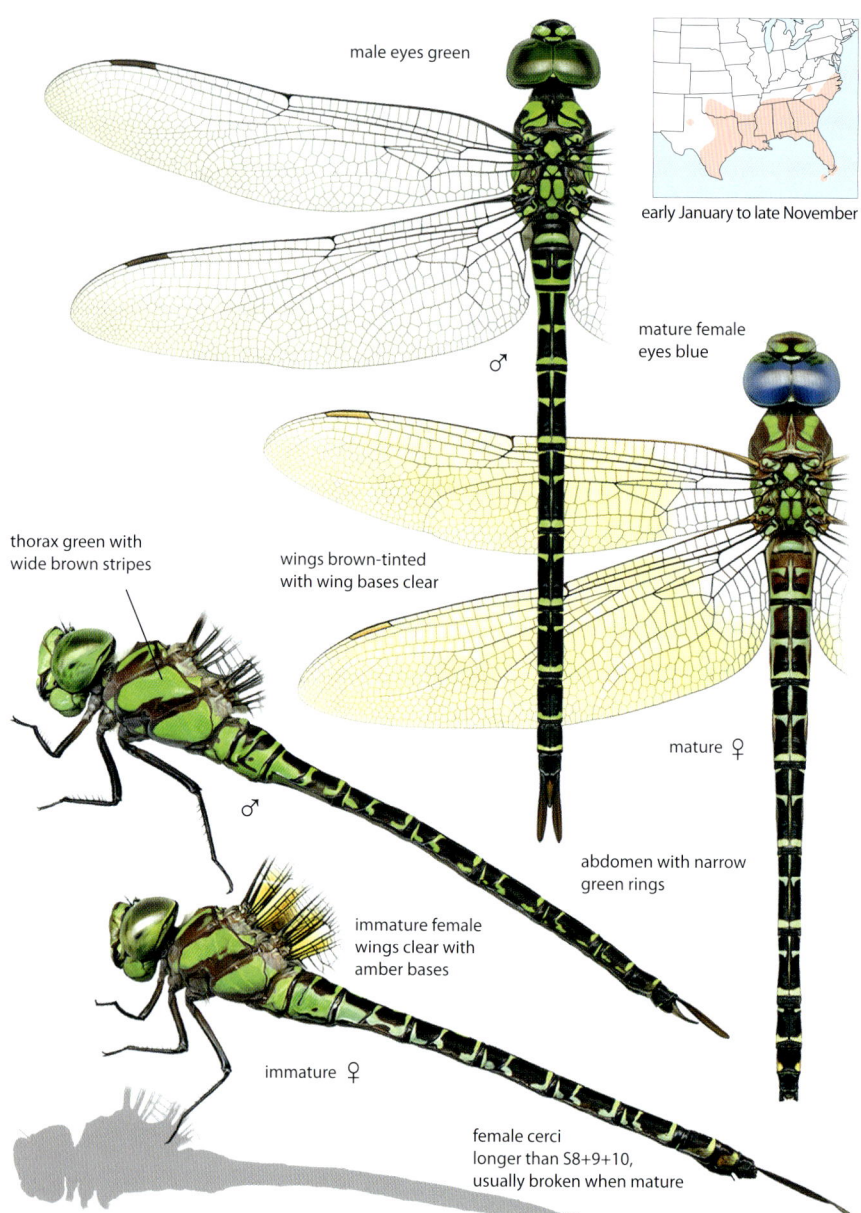

male eyes green

early January to late November

mature female eyes blue

thorax green with wide brown stripes

wings brown-tinted with wing bases clear

♂

mature ♀

abdomen with narrow green rings

immature female wings clear with amber bases

immature ♀

female cerci longer than S8+9+10, usually broken when mature

Similar species: Swamp Darner has darker thorax with narrower green stripes. Top of face dark. **Mangrove** and **Blue-eyed Darners** lack wide dark thoracic stripes.

Mangrove Darner • *Coryphaeschna viriditas*

76–85 mm, 3–3.3 in. Large, slender, predominately green-colored darner inhabiting fresh-water ponds and marshes; associated with mangroves in coastal southern Florida. Thorax green with very narrow brown stripes; abdomen dark with narrow green rings. Face, eyes, and back of the head green.

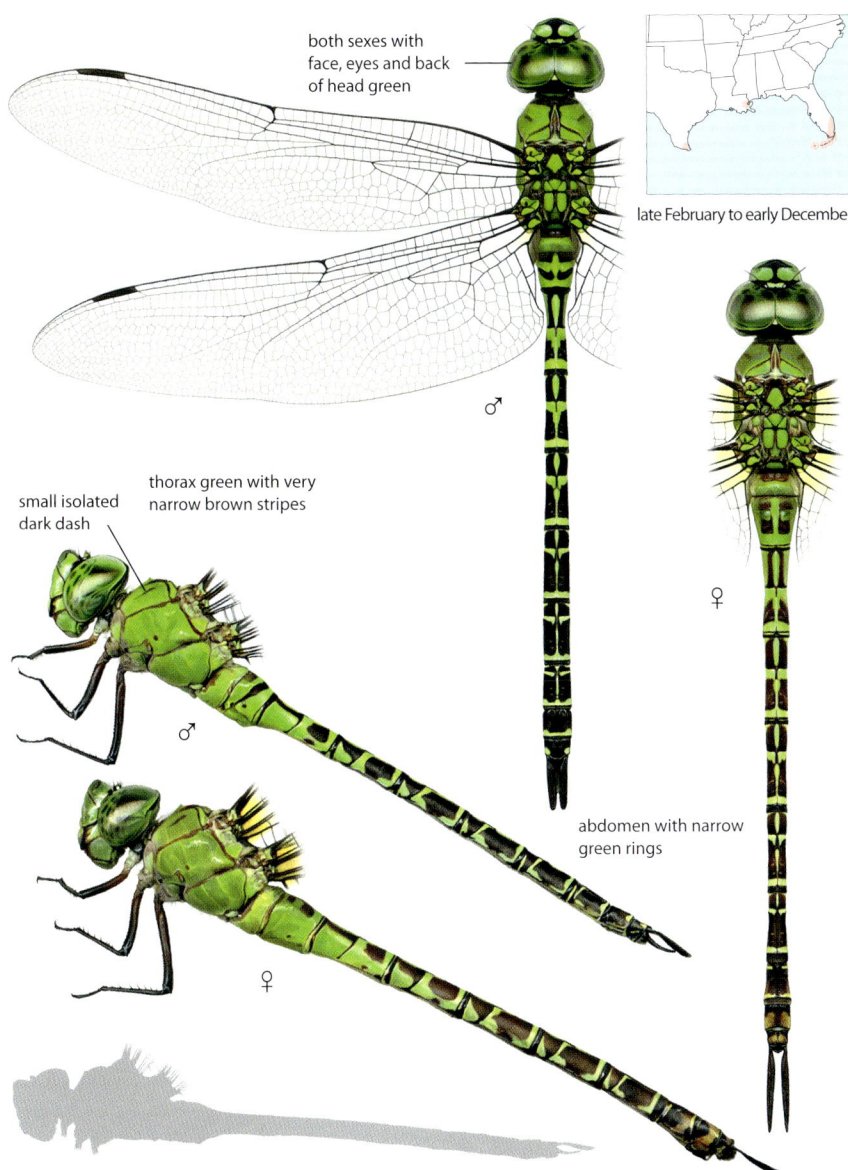

both sexes with face, eyes and back of head green

late February to early December

thorax green with very narrow brown stripes

small isolated dark dash

abdomen with narrow green rings

♂

♀

Similar species: Regal Darner has wide brown thoracic stripes. **Blue-faced Darner** usually lacks small dark dash at front of thorax and has subtly narrower green rings at end of abdominal segments. Face and back of head blue when mature. **Green darners** have different abdomen patterns, spotted or striped but not ringed.

Blue-faced Darner • *Coryphaeschna adnexa*

66–69 mm, 2.6–2.7 in. Small southern darner inhabiting weedy lakes and canals. Thorax green with narrow brown sutures. Abdomen dark with narrow green rings; female abdomen often reddish. Upon maturity, both sexes with rear of head blue, eyes partially blue. Mature male with bright blue face.

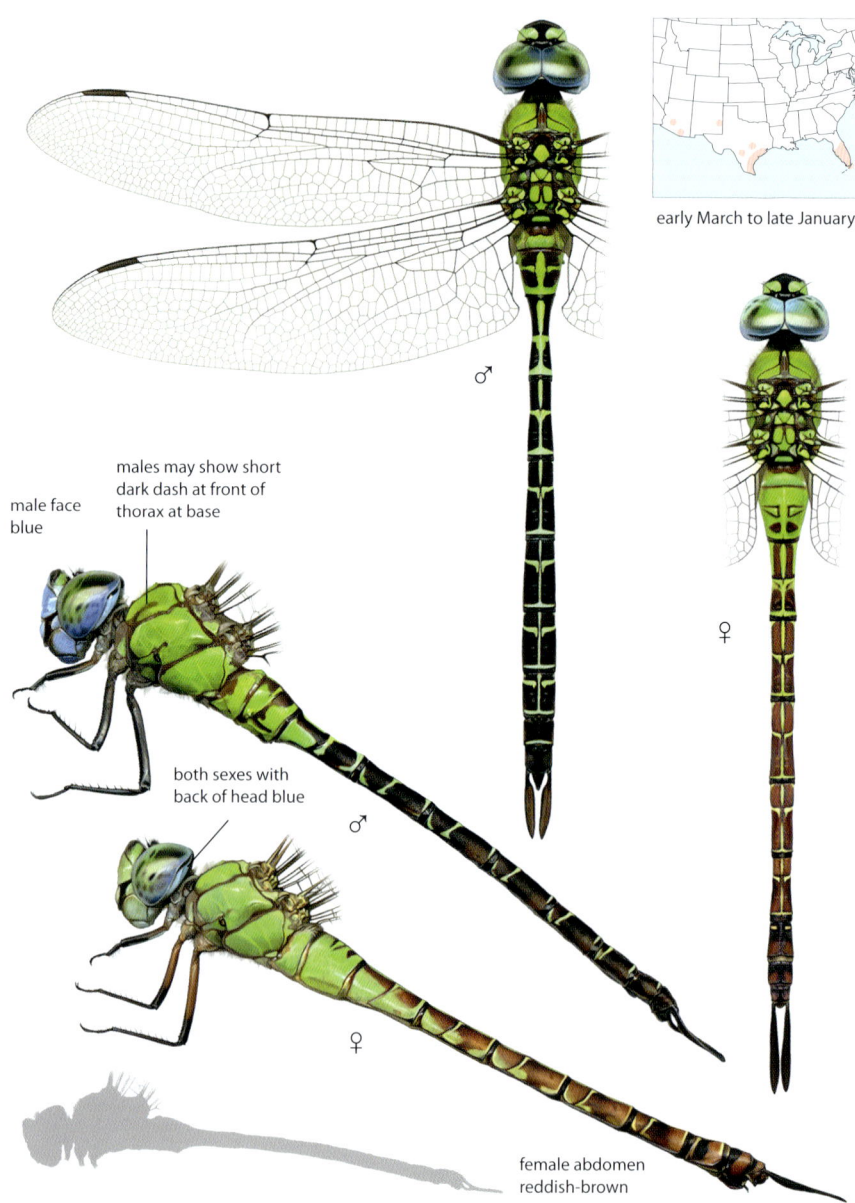

early March to late January

males may show short dark dash at front of thorax at base

male face blue

both sexes with back of head blue

♂

♀

♂

female abdomen reddish-brown

♀

Similar species: Mangrove Darner larger; has small, isolated, dark dash at front of thorax and slightly wider green abdominal rings. Lacks blue on head, face, and eyes. **Regal, Malachite** and **Secretive Darners** are all larger with wide dark thoracic stripes.

Icarus Darner • *Coryphaeschna apeora*

79–85 mm, 3.1–3.3 in. Very large pilot darner recently observed in southern Texas. Breeding habitat unknown. Both sexes have a weakly patterned red-brown abdomen. Thorax mostly green with brown frontal stripes.

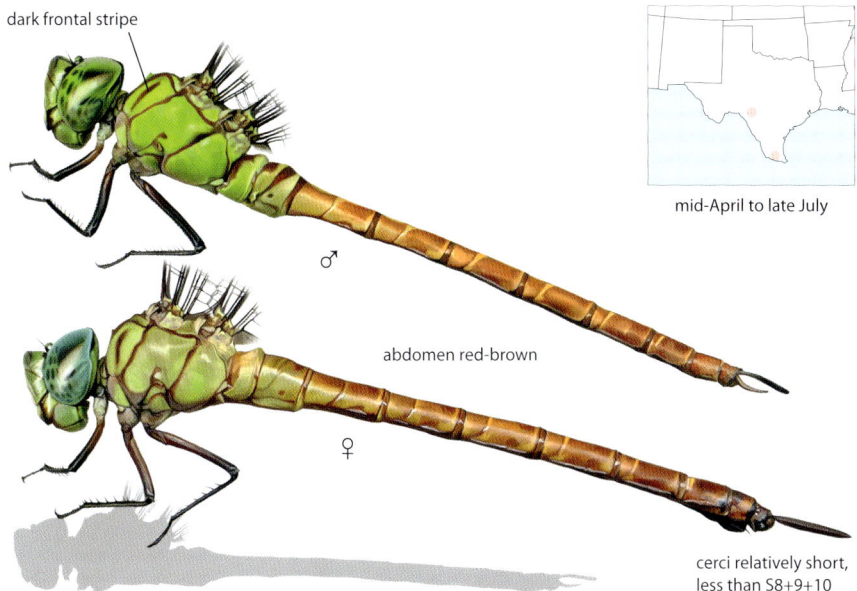

dark frontal stripe

mid-April to late July

♂

abdomen red-brown

♀

cerci relatively short,
less than S8+9+10

Similar species: Blue-faced Darner has narrow green rings and stripes on abdomen; dark frontal thoracic stripes shorter when present. **Comet Darner** lacks dark frontal stripes. Mature male abdomen brighter red; female with pale spots on abdomen.

Secretive Darner • *Remartinia secreta*

78–80 mm, 3.1 in. Large, poorly known darner; single record in South Texas. Thorax bright green with wide dark brown shoulder and lateral stripes. Abdomen with green dorsal markings, side of abdomen mostly dark.

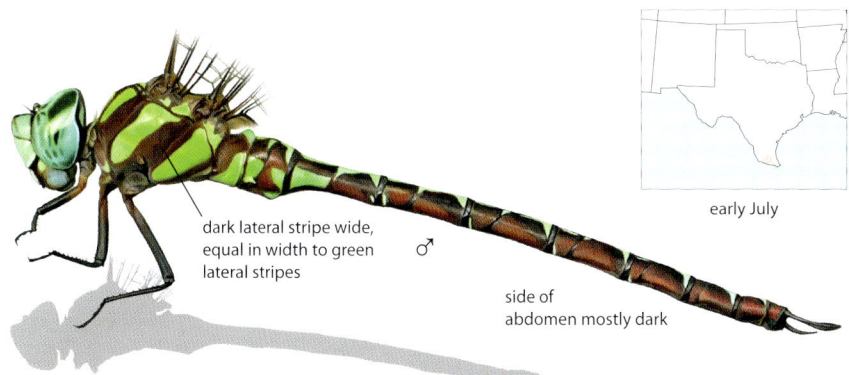

early July

dark lateral stripe wide,
equal in width to green
lateral stripes

♂

side of
abdomen mostly dark

Similar species: Malachite Darner very similar, but dark thoracic stripes narrower. Compare male appendages, page 394. **Pilot darners** have pale markings on side of abdomen.

Malachite Darner • *Remartinia luteipennis*

78–80 mm, 3–3.15 in. Large bright green and brown darner found at southwestern marshy lakes, ponds, and stream pools. Abdomen dark with small green markings largely restricted to top of abdomen. Face pale blue, lacks T-spot. No other darner within its range has a green thorax with wide dark brown stripes.

top of frons
unmarked

late July to late October

thorax bright green
with dark brown stripes

face pale
blue

♂

abdomen black
with green dorsal
markings

♀

side of abdomen
mostly dark

♀

female cerci
very short

Similar species: Secretive Darner very similar, but has wider brown thoracic stripes. Compare male appendages, page 394. **Blue-faced Darner** lacks wide dark thoracic stripes.

Pale-green Darner • *Triacanthagyna septima*

59–66 mm, 2.3–2.6 in. Slender pale crepuscular darner inhabiting seasonal woodland ponds and forested swamps in southern Texas and southern Florida. Thorax pale green with sides largely unmarked. Abdomen long and slender, brown with small green markings. Legs uniformly pale.

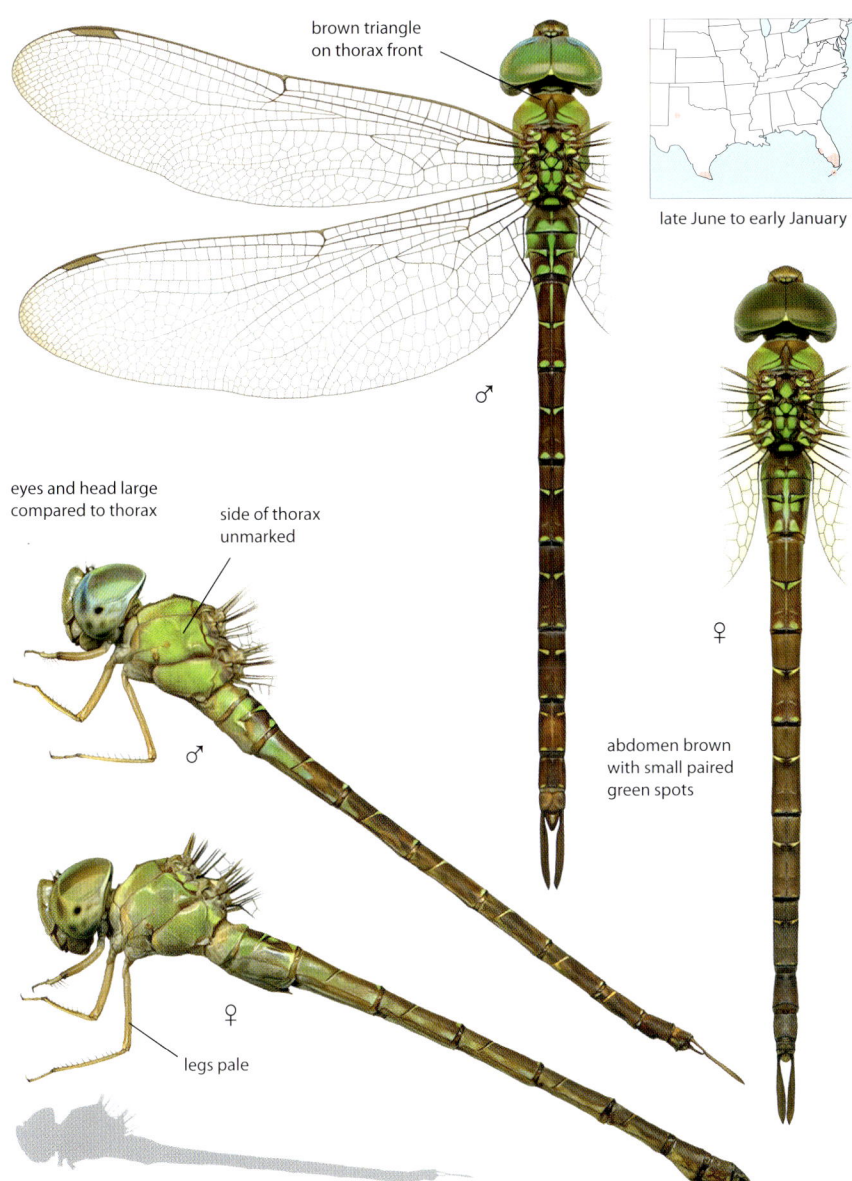

brown triangle on thorax front

late June to early January

♂

eyes and head large compared to thorax

side of thorax unmarked

♂

♀

abdomen brown with small paired green spots

♀

legs pale

Similar species: Phantom and **Caribbean Darners** have brown thoracic stripes. **Blue-faced** and **Mangrove Darners** have black on legs, pale middorsal stripes on abdomen. **Bar-sided Darner** has dark stripe on side of thorax; abdomen constricted at S3.

Phantom Darner • *Triacanthagyna trifida*

62–75 mm, 2.4–3 in. Slender forest-dwelling darner of the southeastern coastal plain; breeds at temporary forest pools. Thorax mostly green with brown stripes; shoulder stripe widest. Abdomen dark with small green markings. Hangs in deep shade during the day, feeding at dusk.

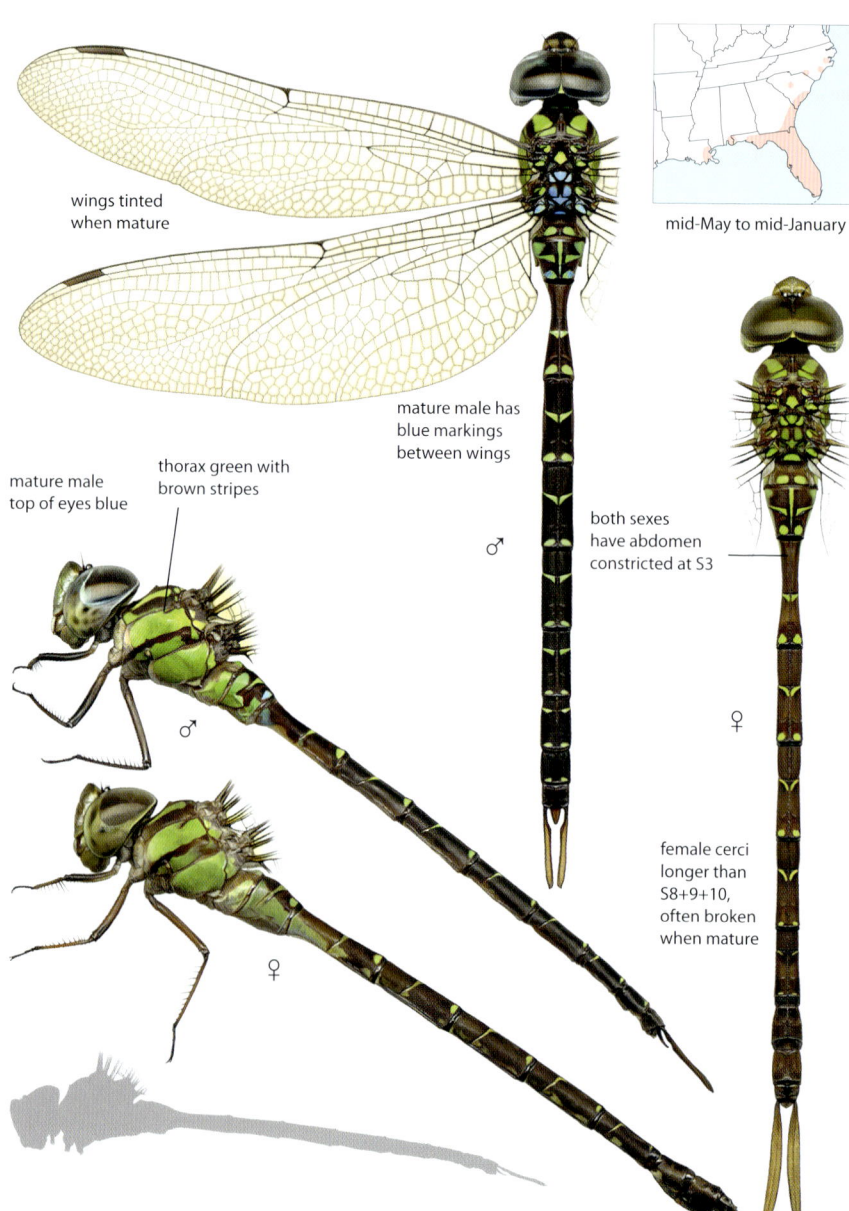

wings tinted when mature

mid-May to mid-January

mature male has blue markings between wings

mature male top of eyes blue

thorax green with brown stripes

both sexes have abdomen constricted at S3

♂

♀

female cerci longer than S8+9+10, often broken when mature

♂

♀

Similar species: Blue-faced and **Mangrove Darners** lack brown shoulder stripes; thoraxes almost entirely green. **Regal Darner** much larger, has wide brown lateral thoracic stripe and abdomen unconstricted. See **Caribbean Darner.**

Caribbean Darner • *Triacanthagyna caribbea*

57–67 mm, 2.25–2.6 in. Slender crepuscular darner inhabiting forest pools and swamps. Rare in South Texas. Similar to Phantom Darner, but tip of male abdomen yellow-green above; female abdomen not constricted. Hangs and perches vertically in shade on branches and tree trunks.

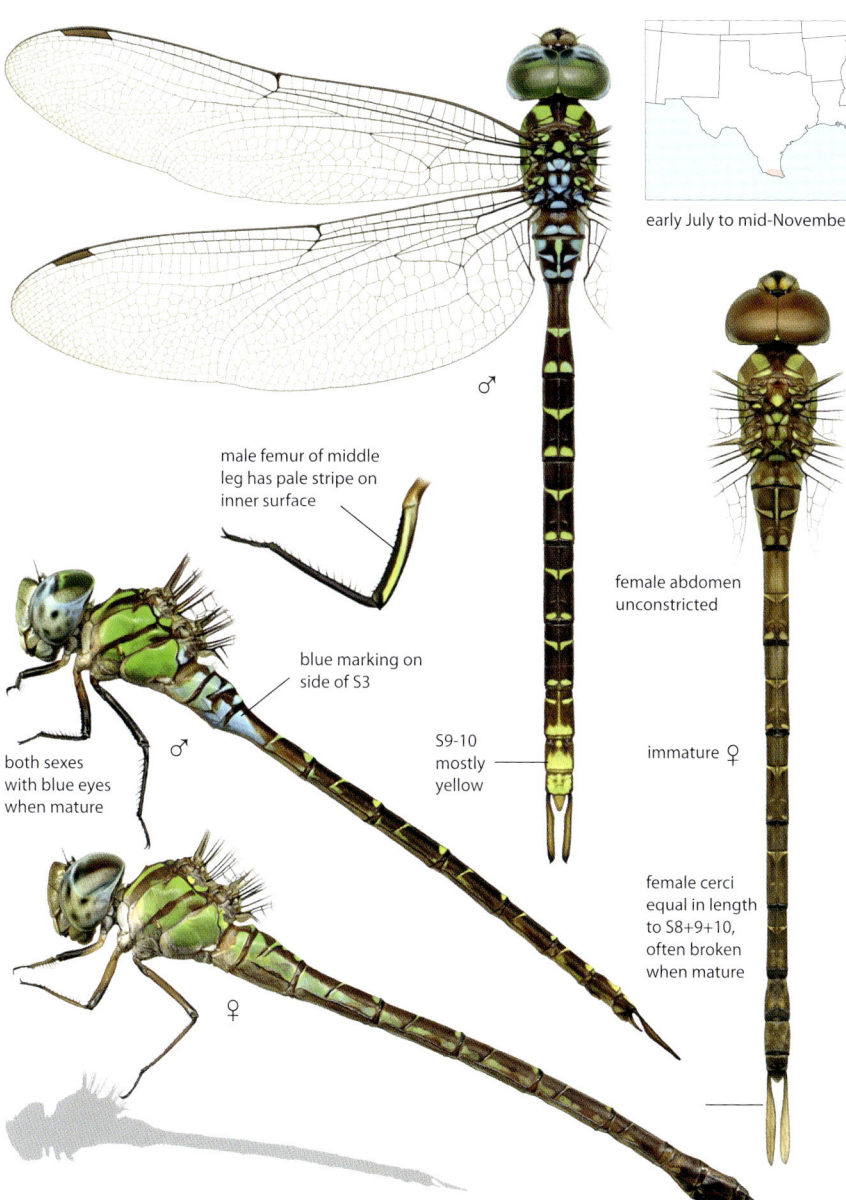

early July to mid-November

male femur of middle leg has pale stripe on inner surface

blue marking on side of S3

both sexes with blue eyes when mature

S9-10 mostly yellow

female abdomen unconstricted

immature ♀

female cerci equal in length to S8+9+10, often broken when mature

Similar species: Phantom Darner very similar, but no known range overlap. Male has tip of the abdomen dark; female with abdomen constricted at S3 and has longer cerci. **Pale-green** and **Blue-faced Darners** lack wide dark thoracic stripes.

Twilight Darner • *Gynacantha nervosa*

75–80 mm, 3–3.15 in. Slender, dull brown darner inhabiting shaded pools and swamps. Southeastern. Thorax plain with only small lateral dots; thorax top and front greenish when mature. Abdomen dull brown with inconspicuous pale dorsal markings. Crepuscular, roosts in forest undergrowth by day but may be active at dawn and on cloudy days.

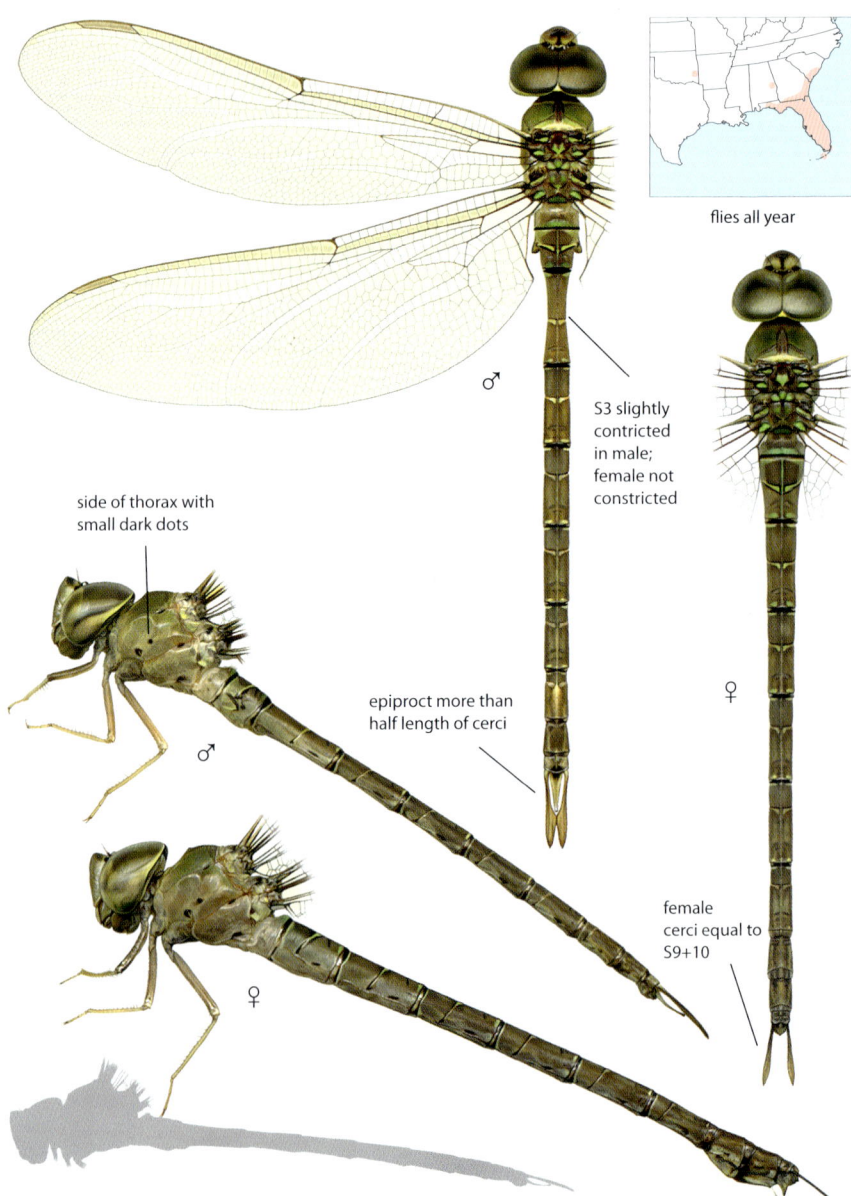

flies all year

S3 slightly contricted in male; female not constricted

side of thorax with small dark dots

epiproct more than half length of cerci

female cerci equal to S9+10

Similar species: Bar-sided Darner has dark lateral thoracic stripe; female with S3 constricted, but range not known to overlap with Twilight Darner in North America. **Palegreen Darner** has a greener thorax with no dark lateral spots.

Bar-sided Darner • *Gynacantha mexicana*

70–76 mm, 2.74–3 in. Slender greenish-brown darner found in southernmost Texas. Crepuscular; roosts in woodlands during the day and breeds in temporary pools. Similar to Twilight Darner but has dark stripe on rear lateral margin of the thorax. Wings often with dark stripe along leading edge.

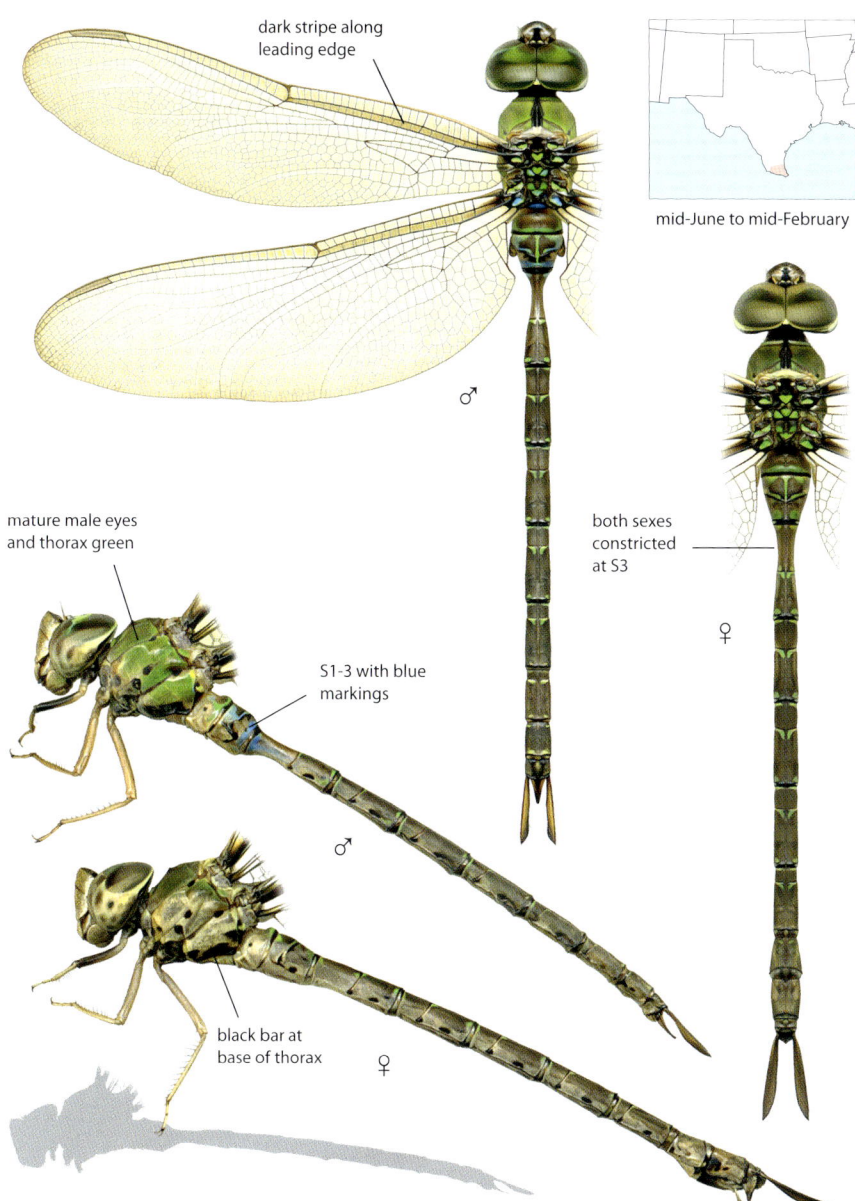

dark stripe along leading edge

♂

mid-June to mid-February

mature male eyes and thorax green

both sexes constricted at S3

S1-3 with blue markings

♀

♂

black bar at base of thorax

♀

Similar species: Twilight Darner lacks dark lateral thoracic stripe, female abdomen not constricted, and range not known to overlap with Bar-sided Darner in North America. **Pale-green Darner** lacks dark lateral thoracic stripe.

Swamp Darner • *Epiaeschna heros*

82–91 mm, 3.2–3.6 in. One of our largest dragonflies; common at swamps, slow streams, and woodland ponds, including temporary ones. Eastern. Thorax dark brown with wide, green, straight stripes; abdomen dark with narrow green rings. Eyes dark blue when mature. Top of face brown.

eyes dark blue

north: mid-Apr to late Oct
south: late Feb to mid-Nov

top of face brown

thorax brown with wide, straight, green stripes

♂

♂

dark abdomen with narrow green rings

♀

female cerci long and leaf-shaped

♀

Similar species: Regal Darner has paler thorax, with wider green stripes and narrower brown ones. Face green. **Cyrano Darner** has striped abdomen and pale blue face.

Cyrano Darner • *Nasiaeschna pentacantha*

62–73 mm, 2.4–2.9 in. Large eastern darner found at swamps, slow streams, also lakes and ponds. Thorax dark brown with jagged green stripes; abdomen with blue-green stripes. Eyes and back of head blue. Face aqua-blue with projecting frons. Males fly short patrols over water, often with abdomen slightly curved, creating a humped appearance.

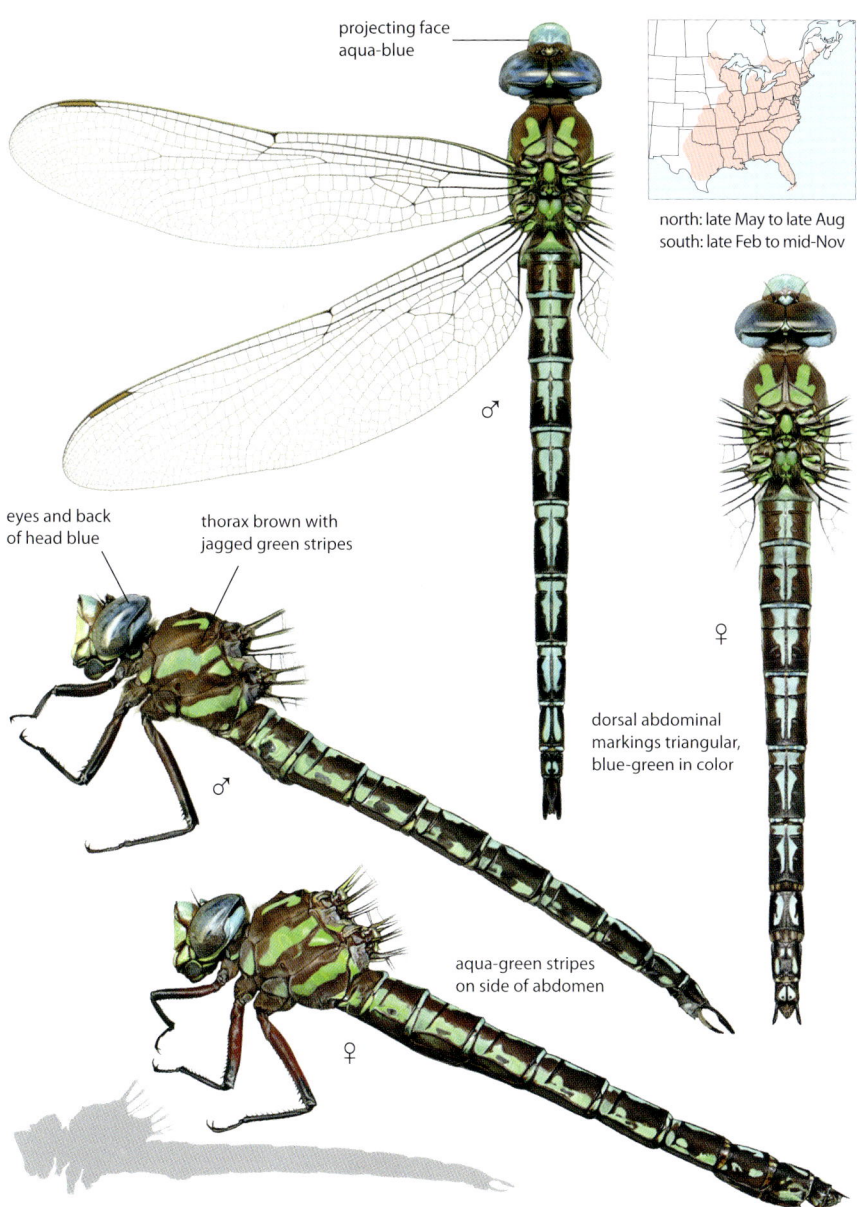

projecting face aqua-blue

north: late May to late Aug
south: late Feb to mid-Nov

♂

eyes and back of head blue

thorax brown with jagged green stripes

♂

dorsal abdominal markings triangular, blue-green in color

♀

aqua-green stripes on side of abdomen

♀

Similar species: Regal and **Swamp Darners** have ringed abdomen. **Mosaic darners** have spotted abdomens.

Canada Darner • *Aeshna canadensis*

68–74 mm, 2.7–2.9 in. Common mosaic darner at vegetated lakes, beaver ponds, and bogs. Shape of lateral thoracic stripes distinctive; anterior stripe with right-angle notch and narrow flag. No dark stripe on face. Male cerci simple-type. Females may have blue or green abdominal markings, often in combination, with green above and blue on sides.

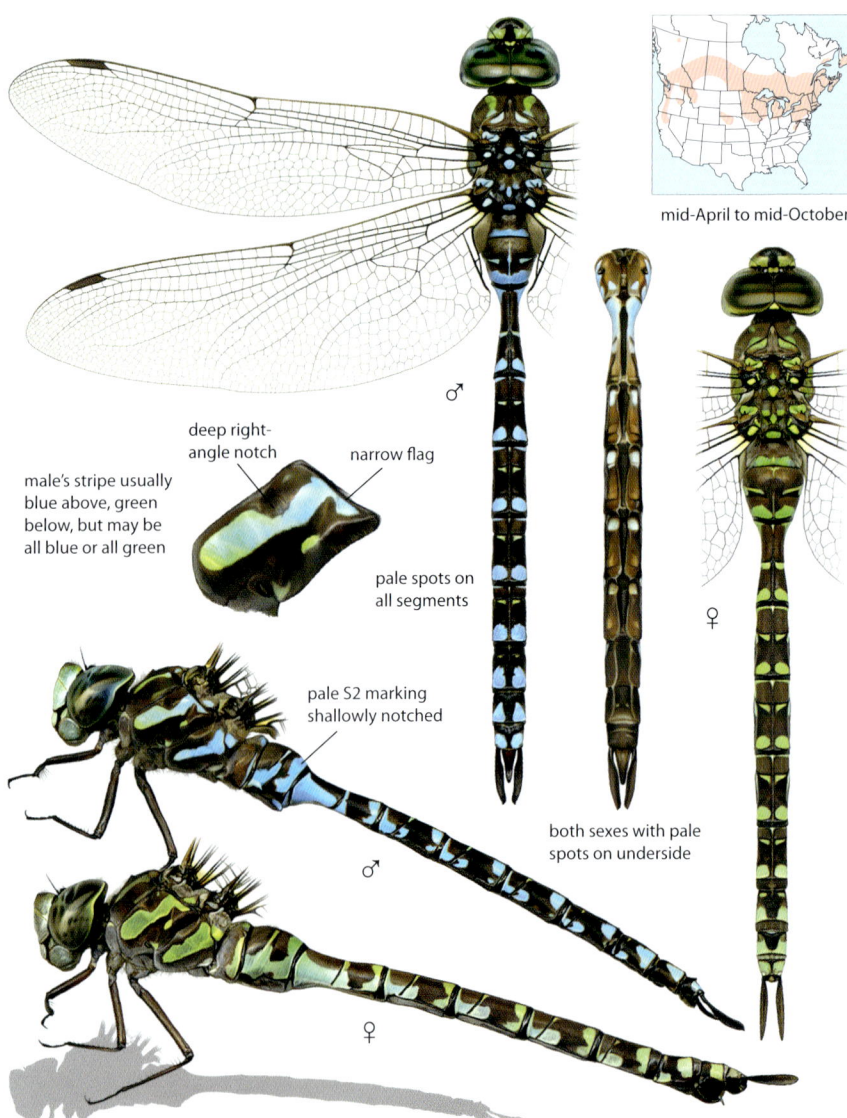

mid-April to mid-October

male's stripe usually blue above, green below, but may be all blue or all green

deep right-angle notch

narrow flag

pale spots on all segments

pale S2 marking shallowly notched

both sexes with pale spots on underside

♂

♀

Similar species: Green-striped Darner nearly identical. Anterior lateral thoracic stripe with shallower notch and wider dorsal flag; male's stripe all green. Pale markings on S2 deeply notched. See Green-striped Darner for further in-hand identification. **Lake Darner** has dark facial stripe. Anterior lateral thoracic stripe constricted in middle; lacks pale spots on underside of abdomen. **Lance-tipped Darner** has anterior lateral thoracic stripe only slightly notched. Male cerci wedge-type; female cerci wide, lance-shaped.

Green-striped Darner • *Aeshna verticalis*

74–79 mm, 2.9–3.1 in. Northeastern mosaic darner found at vegetated ponds, lakes, and slow streams. Nearly identical to Canada Darner, but anterior lateral thoracic stripe entirely green, shallowly notched, and has wide dorsal flag. No dark stripe on face. Male cerci simple-type.

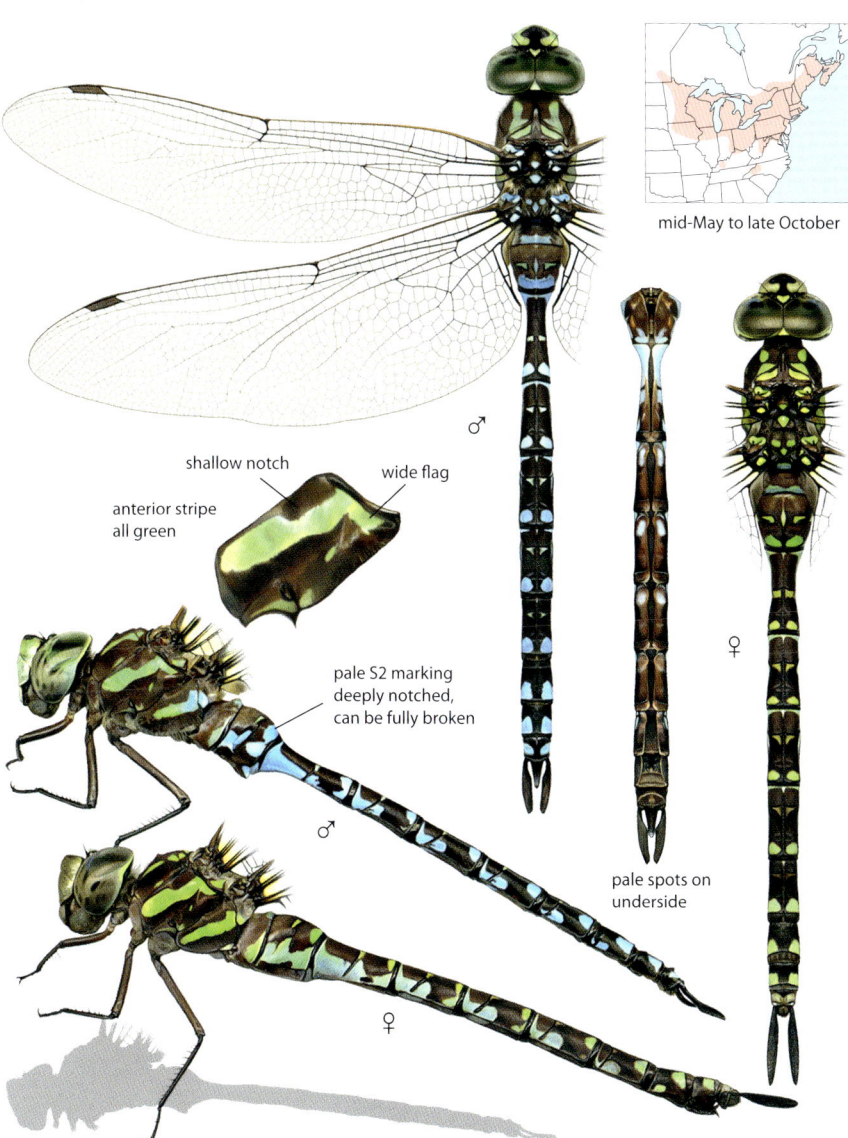

mid-May to late October

shallow notch

wide flag

anterior stripe all green

pale S2 marking deeply notched, can be fully broken

♂

♀

pale spots on underside

♂

♀

Similar species: Canada Darner nearly identical but usually separable by shape of the anterior lateral thoracic stripe. On occasion, pattern and color may be inconclusive, requiring examination of anatomical structures. Male Canada has a row of tiny teeth along the top of the cerci that Green-striped lacks. Also compare shape of hamules and structure of ovipositors, page 393. **Lake Darner** has dark facial stripe. Anterior lateral thoracic stripe constricted in middle; lacks pale spots on underside of abdomen.

Lake Darner • *Aeshna eremita*

74–79 mm, 2.9–3.1 in. Large northern mosaic darner common at wooded and marshy lakes, also large ponds and slow streams. Anterior lateral thoracic stripe deeply notched in the middle. Face with narrow black cross-stripe. Male cerci simple-type.

mid-June to mid-October

black face stripe

anterior stripe deeply notched in middle

male thoracic stripes blue, green, or purplish but lack yellow

pale spots on all segments, but none on underside

♂

♀

♂

♀

Similar species: Canada and **Green-striped Darners** with anterior thoracic stripe more shallowly notched, pale spots on the underside of the abdomen; lack black facial stripe.

Subarctic Darner • *Aeshna subarctica*

66–76 mm, 2.6–3 in. Small pastel-colored mosaic darner found at northern mossy bogs and fens. Anterior lateral thoracic stripe narrowed in the middle, appears bent, and has thin dorsal flag. Face with black cross-stripe. Male cerci simple-type.

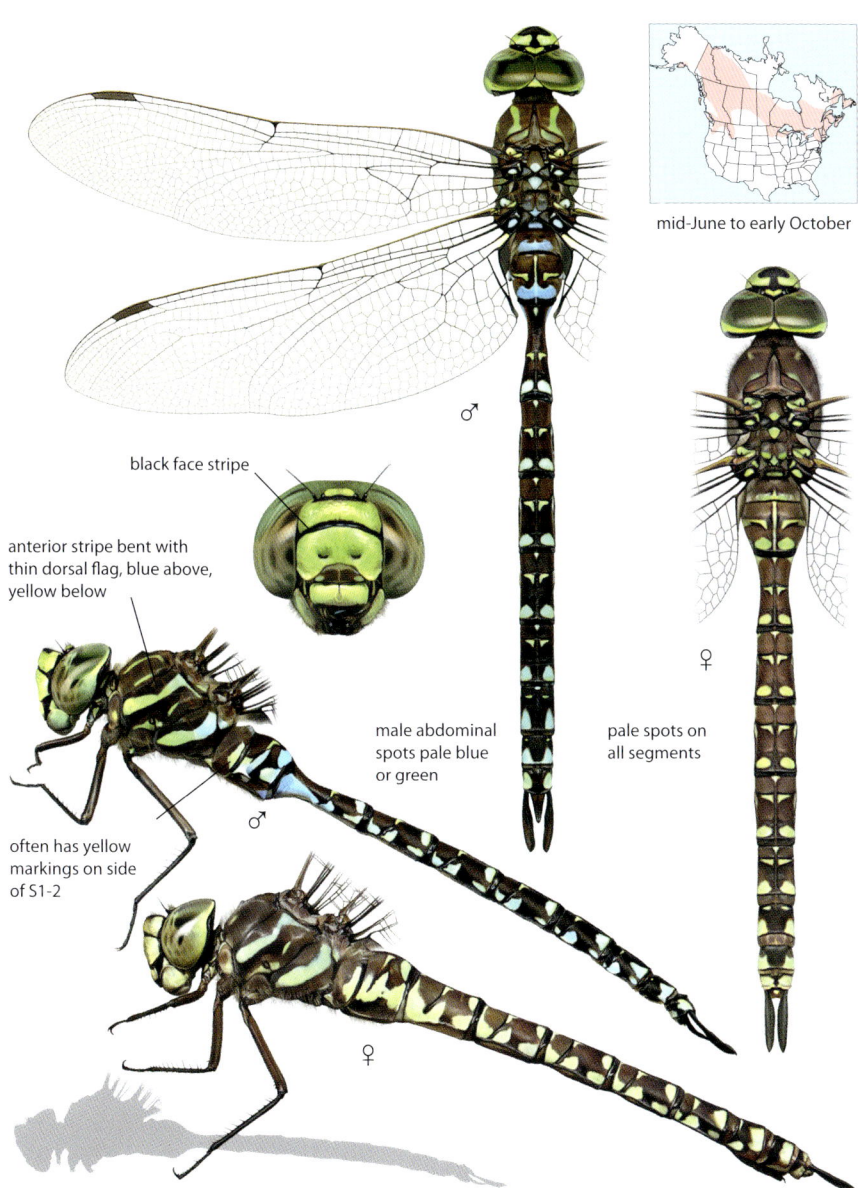

mid-June to early October

black face stripe

anterior stripe bent with thin dorsal flag, blue above, yellow below

♂

often has yellow markings on side of S1-2

male abdominal spots pale blue or green

pale spots on all segments

♂

♀

♀

Similar species: Sedge Darner has wider straight lateral thoracic stripes. **Canada, Green-striped,** and **Lake Darners** have deeply notched anterior lateral thoracic stripes. **Shadow Darner** has straight lateral thoracic stripes; male with wedge-type cerci and no dark facial stripe.

Sedge Darner • *Aeshna juncea*

66–75 mm, 2.6–2.95 in. Common and widespread northern mosaic darner found at sedge marshes, lakes, and peat ponds. Thorax with full-length frontal stripes. Lateral thoracic stripes broad and straight, anterior stripe widest at bottom. Face with black cross-stripe. Male cerci simple-type.

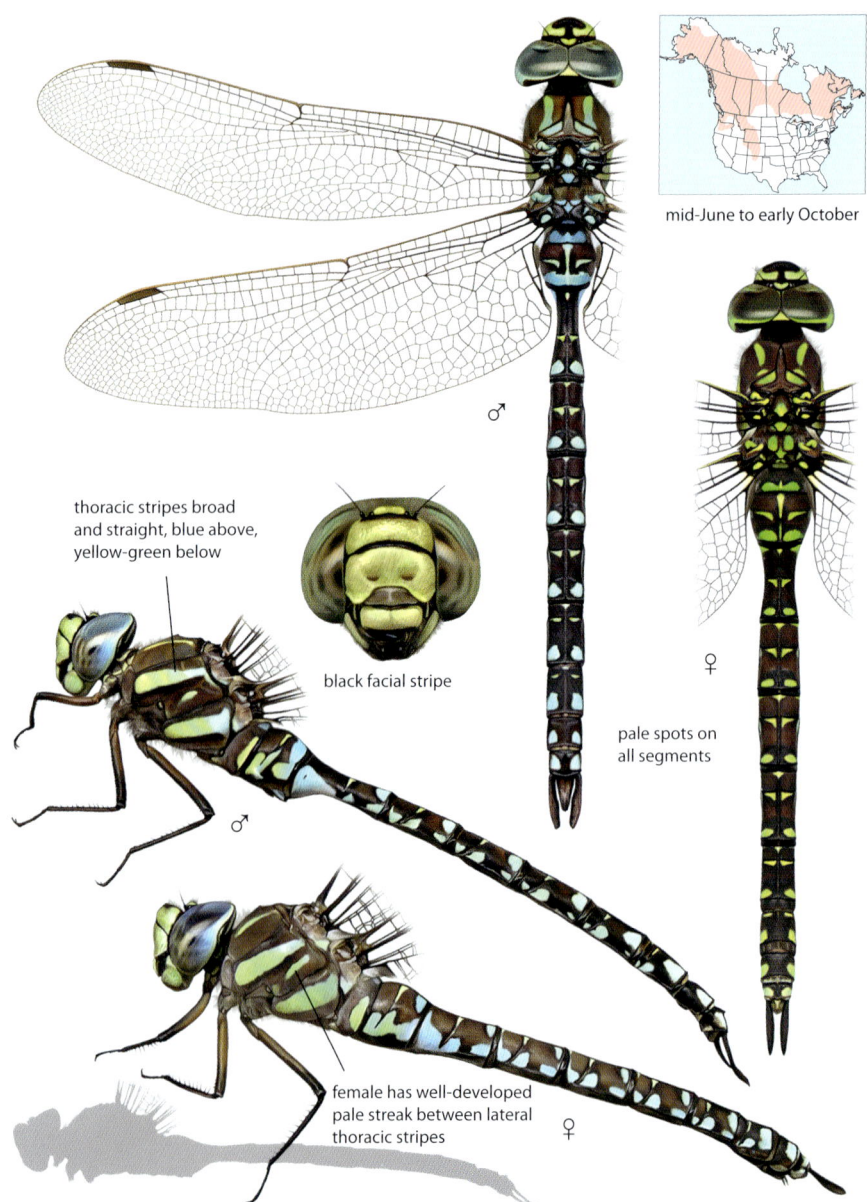

mid-June to early October

thoracic stripes broad and straight, blue above, yellow-green below

black facial stripe

pale spots on all segments

♂

♀

♂

female has well-developed pale streak between lateral thoracic stripes

♀

Similar species: Subarctic Darner has narrower, bent lateral thoracic stripes. **Black-tipped Darner** has S10 black and lacks black facial line. **Shadow** and **Paddle-tailed Darners** have narrower lateral thoracic stripes; males with wedge-type cerci.

Black-tipped Darner • *Aeshna tuberculifera*

71–80 mm, 2.8–3.15 in. Long and slender darner uncommon at bog-edged ponds and lakes, cattail ponds in the East. Lateral thoracic stripes broad and straight, anterior stripe slightly notched. S10 all black. Male cerci simple-type. Female abdomen long and constricted at S3, with large ovipositor and long broad cerci.

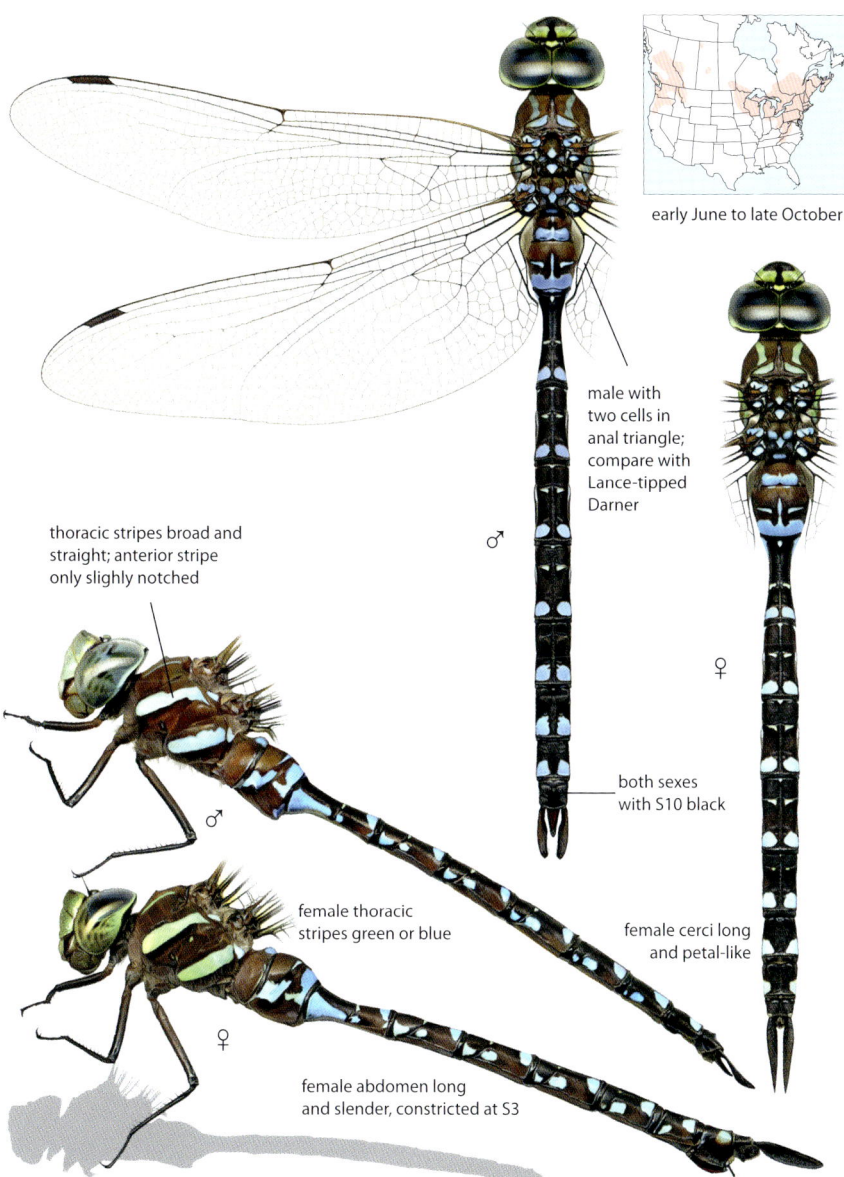

early June to late October

male with two cells in anal triangle; compare with Lance-tipped Darner

thoracic stripes broad and straight; anterior stripe only slightly notched

♂

♀

both sexes with S10 black

female thoracic stripes green or blue

female cerci long and petal-like

♂

♀

female abdomen long and slender, constricted at S3

Similar species: Lance-tipped Darner most similar in pattern, but male has blue spots on S10 and wedge-type cerci; female with large pale spot extending down on side of S9. **Sedge** and **Paddle-tailed Darners** have pale spots on S10 and face with black cross-stripe.

Lance-tipped Darner • *Aeshna constricta*

68–72 mm, 2.7–2.8 in. Common mosaic darner at open marshy ponds and slow streams. Anterior lateral thoracic stripe straight, slightly notched, with small dorsal flag. Male cerci wedge-type with spine. Female abdomen often lacks pale spots on S10, but S9 pale marking large, squarish, and extends down side. Female cerci long and lance-shaped.

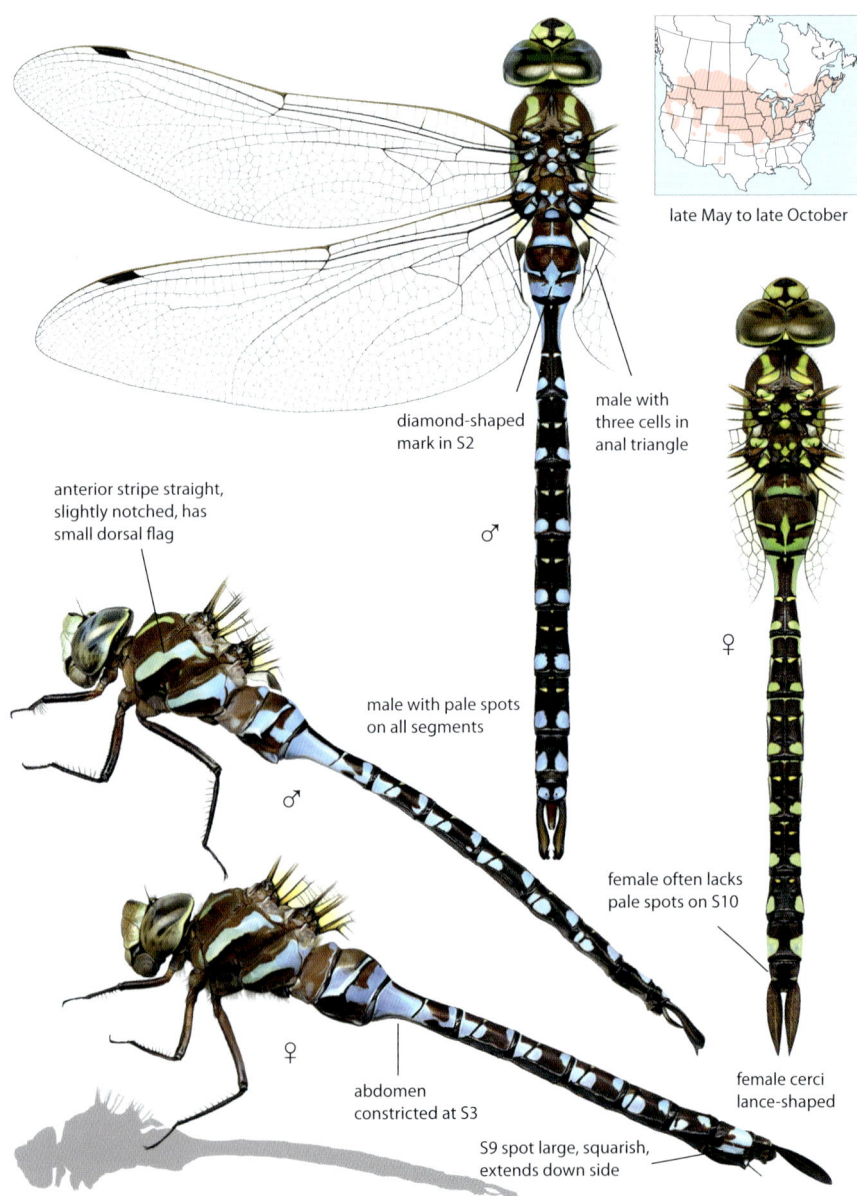

late May to late October

diamond-shaped mark in S2

male with three cells in anal triangle

anterior stripe straight, slightly notched, has small dorsal flag

♂

male with pale spots on all segments

♀

♂

female often lacks pale spots on S10

female cerci lance-shaped

♀

abdomen constricted at S3

S9 spot large, squarish, extends down side

Similar species: Black-tipped Darner male has simple-type cerci and no pale spots on S10; female with only small spot on S9. **Paddle-tailed Darner** has black face stripe, larger abdominal spots with S9 spots fused, pale streak on side of S1.

Paddle-tailed Darner • *Aeshna palmata*

65–75 mm, 2.6–3 in. Common western mosaic darner at ponds, lakes, and slow streams. Anterior lateral thoracic stripe straight, slightly notched. Face with black cross-stripe. Abdomen with large pale markings; S9 spots on male often fused. Pale streak on side of S1. Male cerci wedge-type with spine.

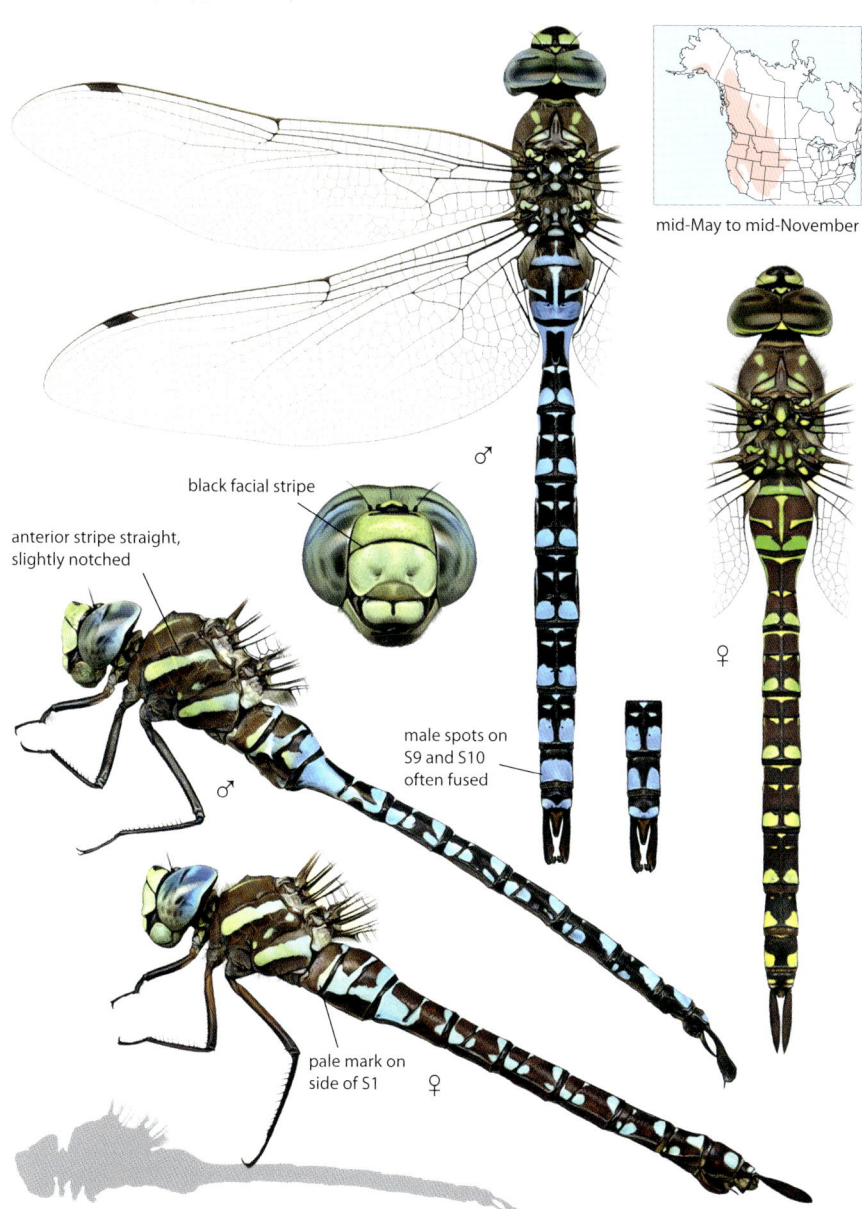

mid-May to mid-November

black facial stripe

anterior stripe straight, slightly notched

♂

male spots on S9 and S10 often fused

♀

pale mark on side of S1 ♀

Similar species: Lance-tipped and **Shadow Darners** have smaller abdominal spots, lack black face stripe and pale streak on side of S1. **Persephone's Darner** has wider thoracic stripes; S10 black.

Persephone's Darner • *Aeshna persephone*

74–79 mm, 2.9–3.1 in. Southwestern species inhabiting mountain streams. Lateral thoracic stripes very wide, straight, and yellow in color. Abdomen dark with ringed appearance; pale abdominal markings largely absent in the middle of segments, reduced laterally, and reduced or absent on S10. Male cerci wedge-type with spine.

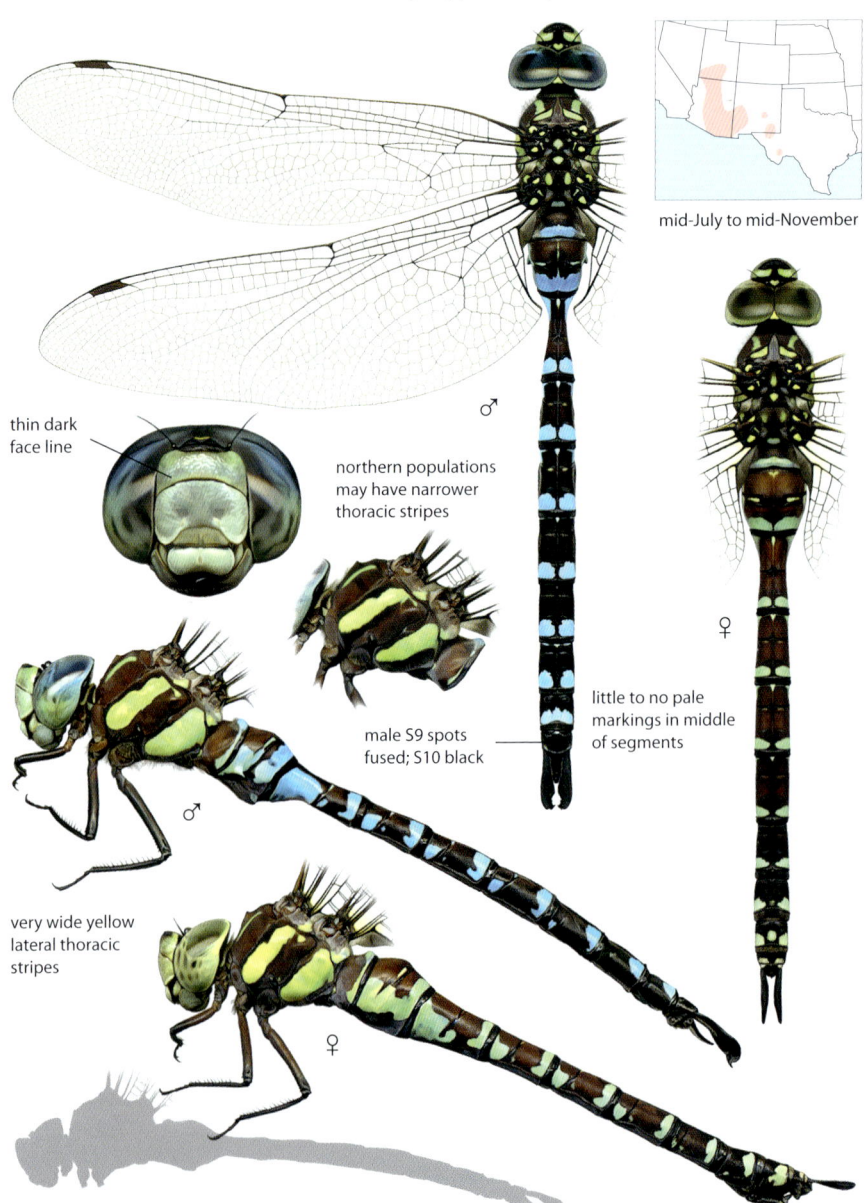

mid-July to mid-November

thin dark face line

northern populations may have narrower thoracic stripes

male S9 spots fused; S10 black

little to no pale markings in middle of segments

very wide yellow lateral thoracic stripes

♂

♀

Similar species: Paddle-tailed Darner has narrower lateral thoracic stripes and pale markings at middle of abdominal segments, on sides of S7-8, and on S10. **Walker's Darner** has narrow, whitish lateral thoracic stripes.

Walker's Darner • *Aeshna walkeri*

64–66 mm, 2.5–2.6 in. Far-western mosaic darner found at foothill streams, mainly in California. Lateral thoracic stripes narrow, straight, and whitish. Male with S9 dorsal pale stripes fused, S10 unmarked. Male cerci wedge-type with spine.

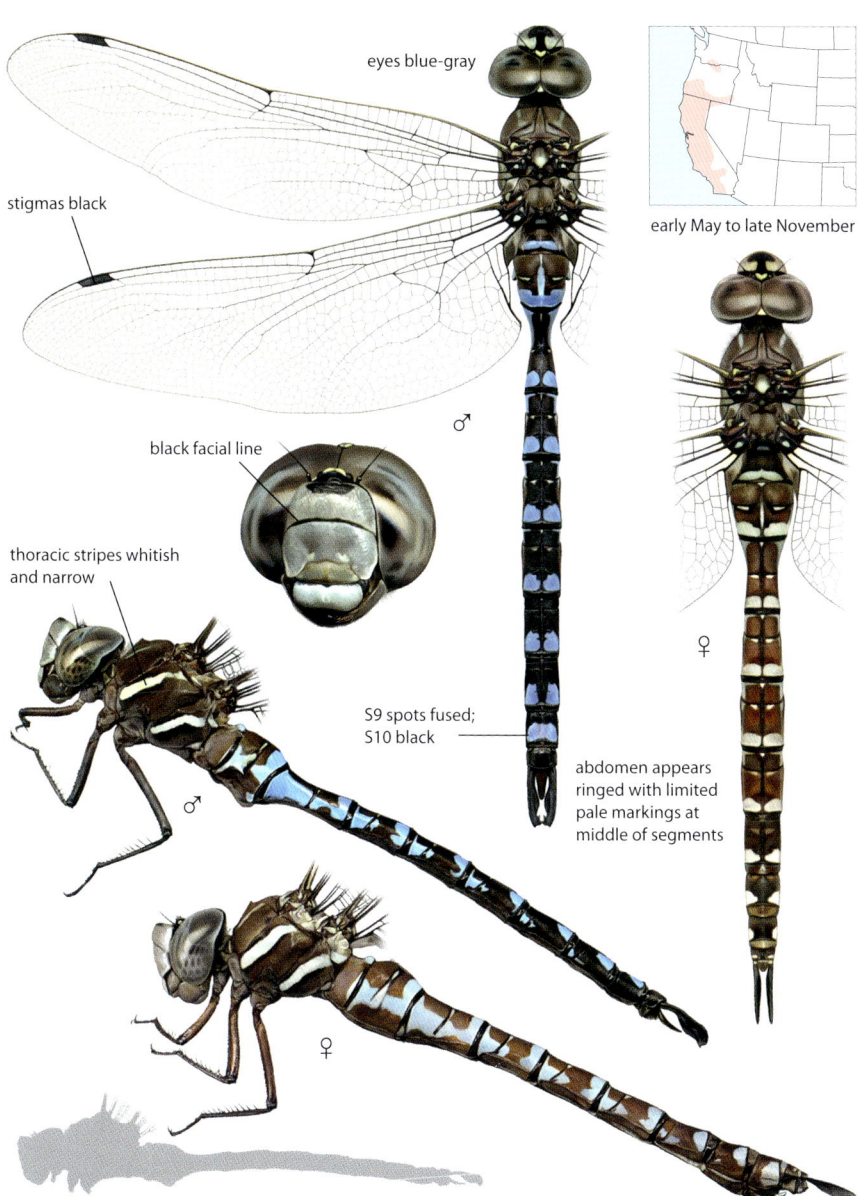

eyes blue-gray

stigmas black

early May to late November

black facial line

thoracic stripes whitish and narrow

♂

♀

S9 spots fused; S10 black

abdomen appears ringed with limited pale markings at middle of segments

♂

♀

Similar species: Blue-eyed Darner has pale markings at middle of abdominal segments. Male has sky-blue eyes and forked cerci; female has tubercle under S2. **Shadow Darner** has yellow-green lateral thoracic stripes, no black facial stripe.

Variable Darner • *Aeshna interrupta*

72–77 mm, 2.8–3 in. Widespread common species at lakes and ponds, often with abundant shoreline vegetation. Reduced lateral thoracic stripes distinctive. Stripes straight and narrow in western Canada, Rocky Mountains, and Great Plains; stripes reduced to spots in the East and along West Coast. Face with black cross-line. Male cerci simple-type.

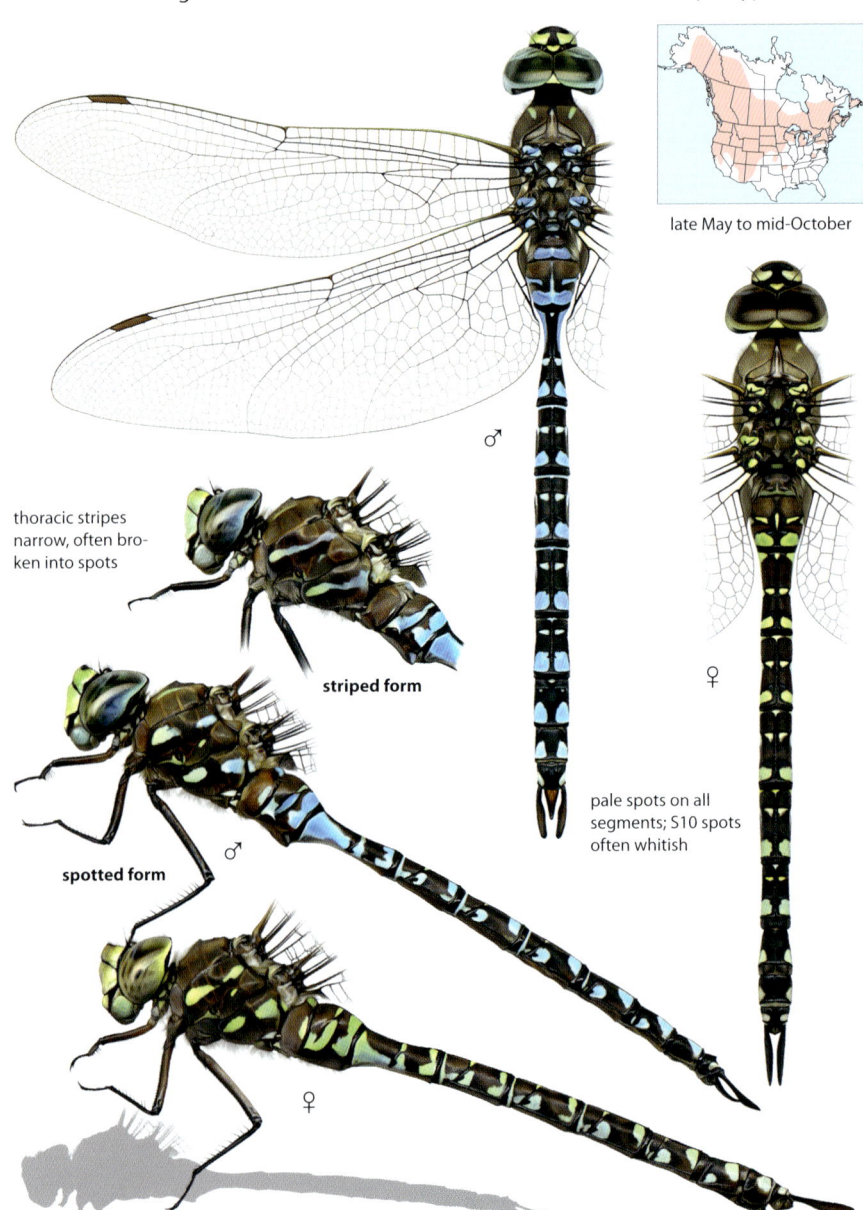

late May to mid-October

thoracic stripes narrow, often broken into spots

striped form

♂

♀

spotted form

♂

pale spots on all segments; S10 spots often whitish

♀

Similar species: Blue-eyed Darner female resembles striped-form Variable but has tubercle under S1. **Shadow Darner** has smaller pale abdominal markings with pale spots on the underside; male cerci wedge-type. Other **mosaic darners** have wider thoracic stripes.

Mottled Darner • *Aeshna clepsydra*

65–70 mm, 2.6–2.75 in. Large common northeastern species at vegetated lakes, beaver ponds, and bogs. Lateral thoracic pattern distinctive with large pale spots between anterior and posterior stripes; anterior stripe deeply notched and extends forward of first thoracic suture. Face with brown cross-line. Male cerci simple-type.

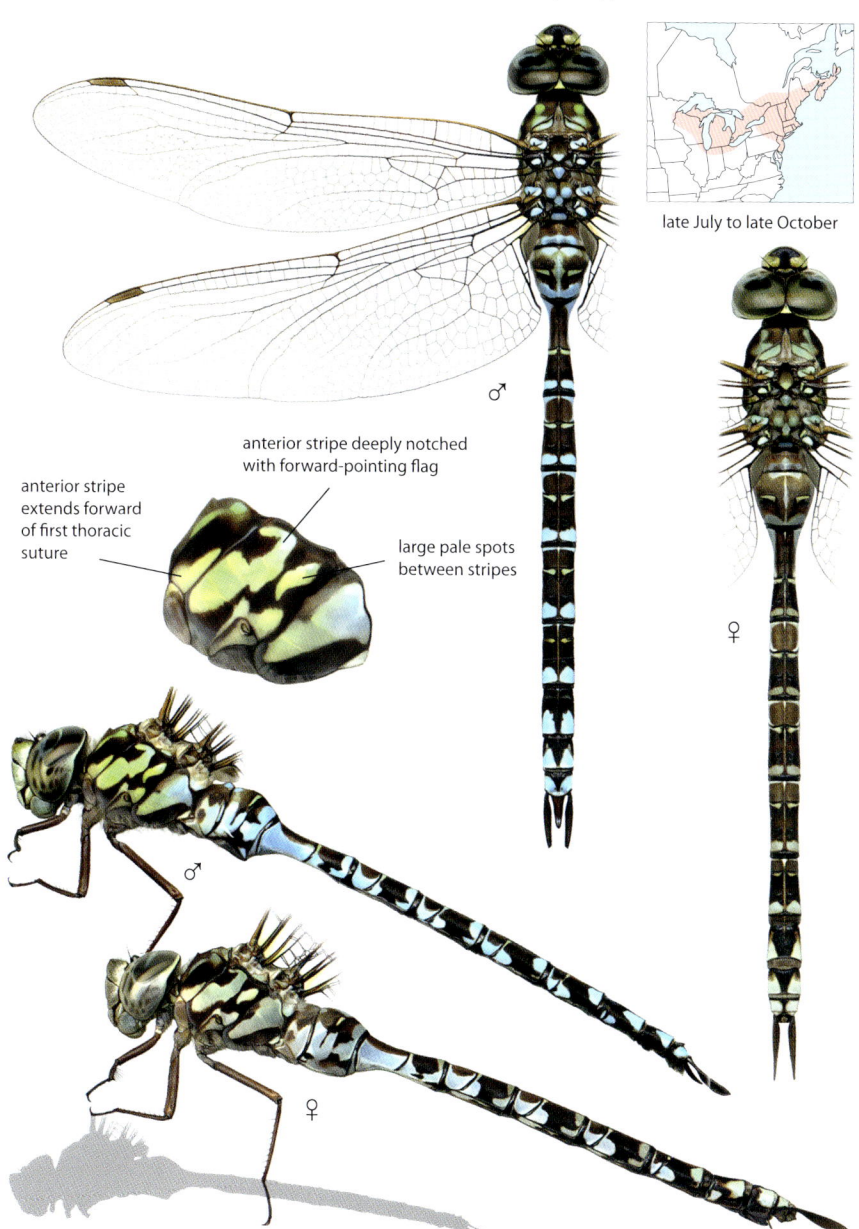

late July to late October

anterior stripe deeply notched with forward-pointing flag

anterior stripe extends forward of first thoracic suture

large pale spots between stripes

♂

♀

♂

♀

Similar species: Among mosaic darners, only **Zigzag** and **Azure Darners** have a pale lateral stripe forward of first thoracic suture, but thoracic stripes much narrower.

Zigzag Darner • *Aeshna sitchensis*

54–64 mm, 2.1–2.5 in. Small northern mosaic darner inhabiting small bog pools and shallow fens vegetated with moss and sedges. Thoracic stripes very narrow; anterior stripe strongly bent, forming a zigzag pattern. Abdominal segments mostly dark above. Face pale yellow with black cross-stripe. Perches low, often on logs or flat on the ground.

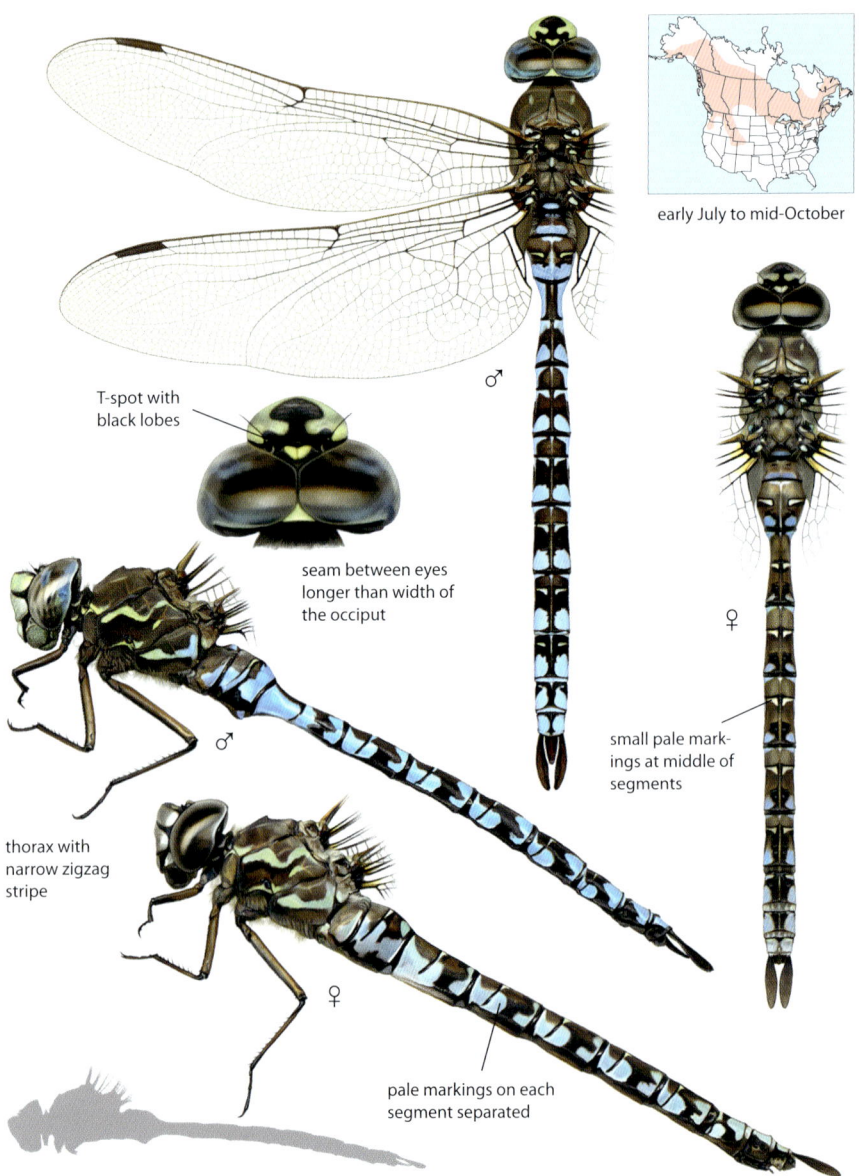

early July to mid-October

T-spot with black lobes

seam between eyes longer than width of the occiput

♂

thorax with narrow zigzag stripe

♂

♀

small pale markings at middle of segments

♀

pale markings on each segment separated

Similar species: Azure Darner has mostly pale abdomen; pale spots often fused, forming stripes. T-spot pattern on frons lacks black lobes.

Azure Darner • *Aeshna caerulea septentrionalis*

53–61 mm, 2.1–2.4 in. Small far-northern mosaic darner inhabiting mossy bogs and fens. Ranges farther north than any other North American dragonfly. Thoracic pattern similar to Zigzag Darner. Abdomen mostly pale above. Face pale blue or green with black cross-stripe. Flies and perches low, often flat on moss mats.

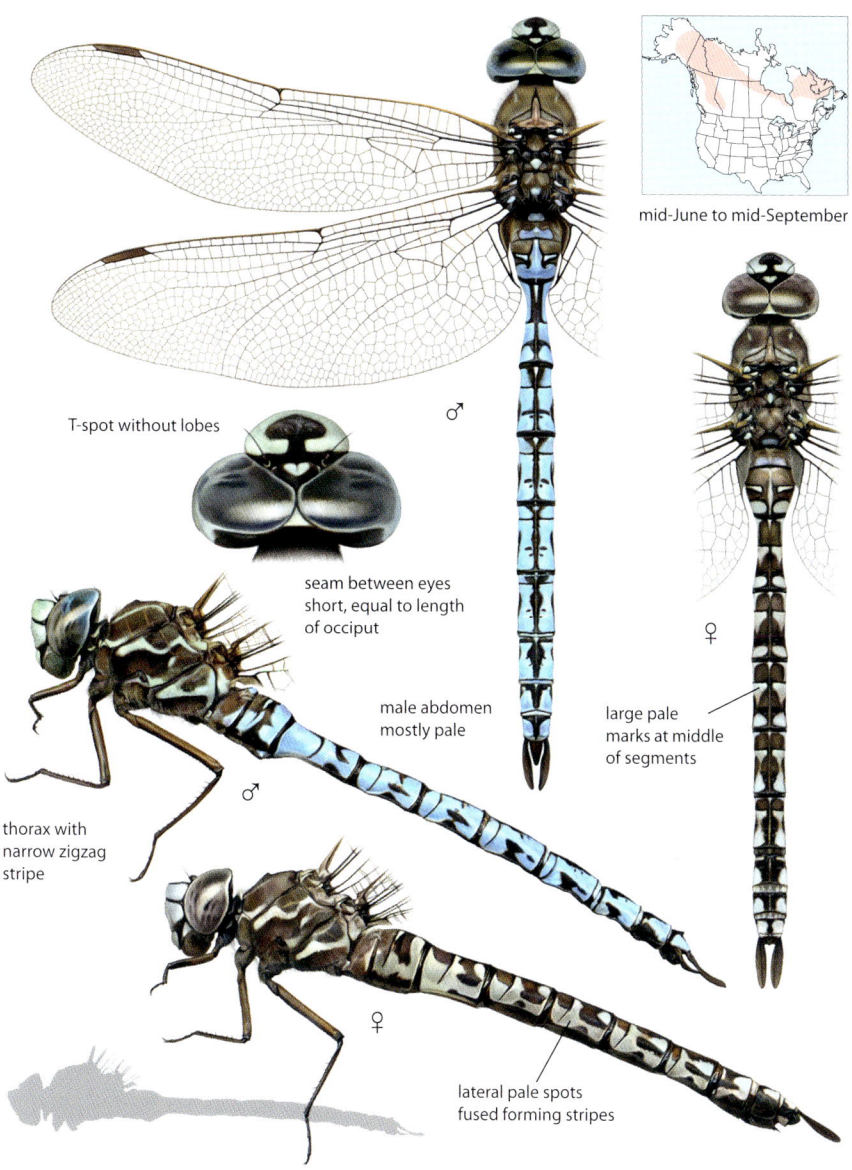

mid-June to mid-September

T-spot without lobes

♂

seam between eyes short, equal to length of occiput

male abdomen mostly pale

♀

large pale marks at middle of segments

thorax with narrow zigzag stripe

♂

♀

lateral pale spots fused forming stripes

Similar species: Zigzag Darner has darker abdomen with separated pale spots. T-spot on frons with black lobes.

Shadow Darner • *Aeshna umbrosa*

68–78 mm, 2.7–3.1 in. Common and widespread mosaic darner inhabiting slow forest streams and ponds. Lateral thoracic stripes narrow, straight, yellow-green in color. Eastern form has small green abdominal spots; western male has larger blue abdominal spots. Male cerci wedge-type.

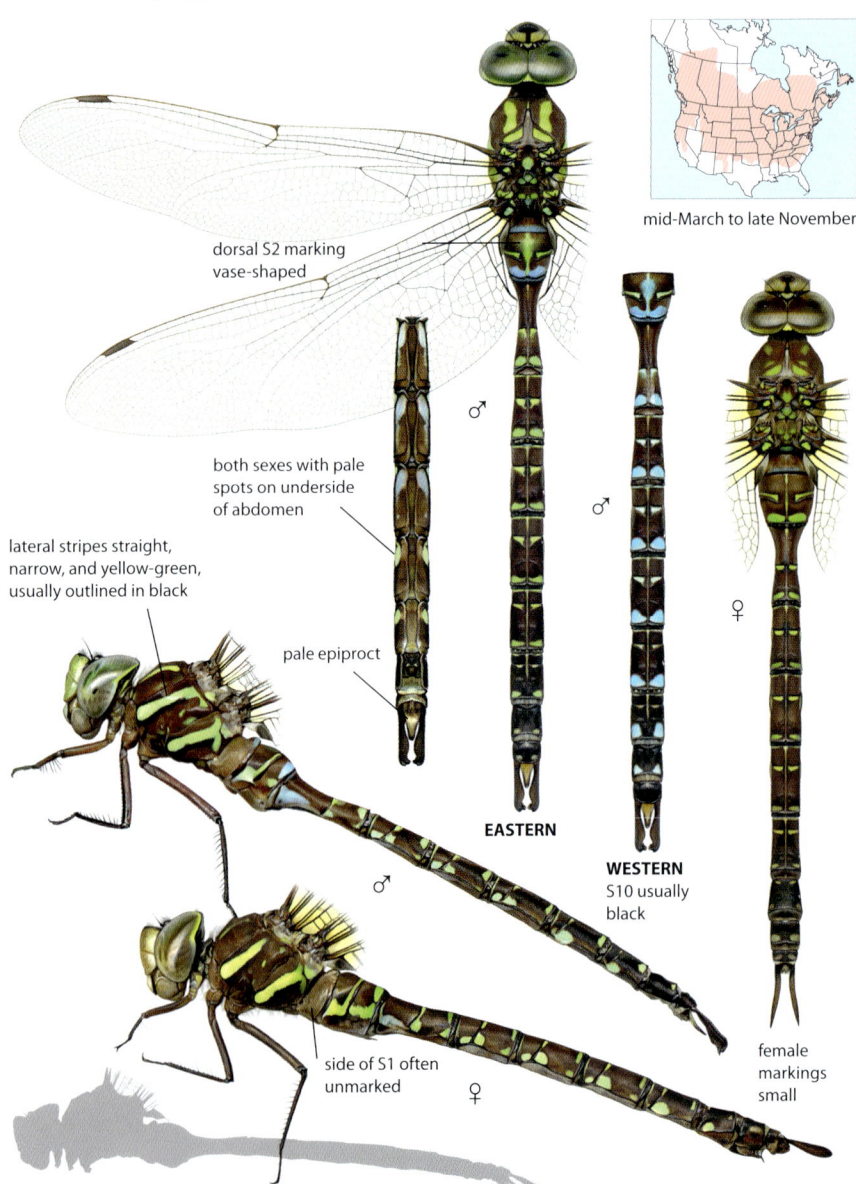

mid-March to late November

dorsal S2 marking vase-shaped

both sexes with pale spots on underside of abdomen

lateral stripes straight, narrow, and yellow-green, usually outlined in black

pale epiproct

♂ EASTERN

♂ WESTERN
S10 usually black

♀

side of S1 often unmarked ♂ ♀

female markings small

Similar species: Combination of straight green thoracic stripes and small abdominal markings distinctive, particularly in the East. **Paddle-tailed Darner** has wider thoracic stripes, larger abdominal pale markings, including markings on S10, and black face stripe; lacks pale markings on underside of abdomen.

Springtime Darner • *Basiaeschna janata*

50–67 mm, 2–2.6 in. Eastern early-season darner found at gentle-flowing rivers and streams. Thorax brown with contrasting pale yellow lateral stripes. Abdomen with pale spots, blue in male, blue or green in female. Small, dark brown marking at the base of each wing.

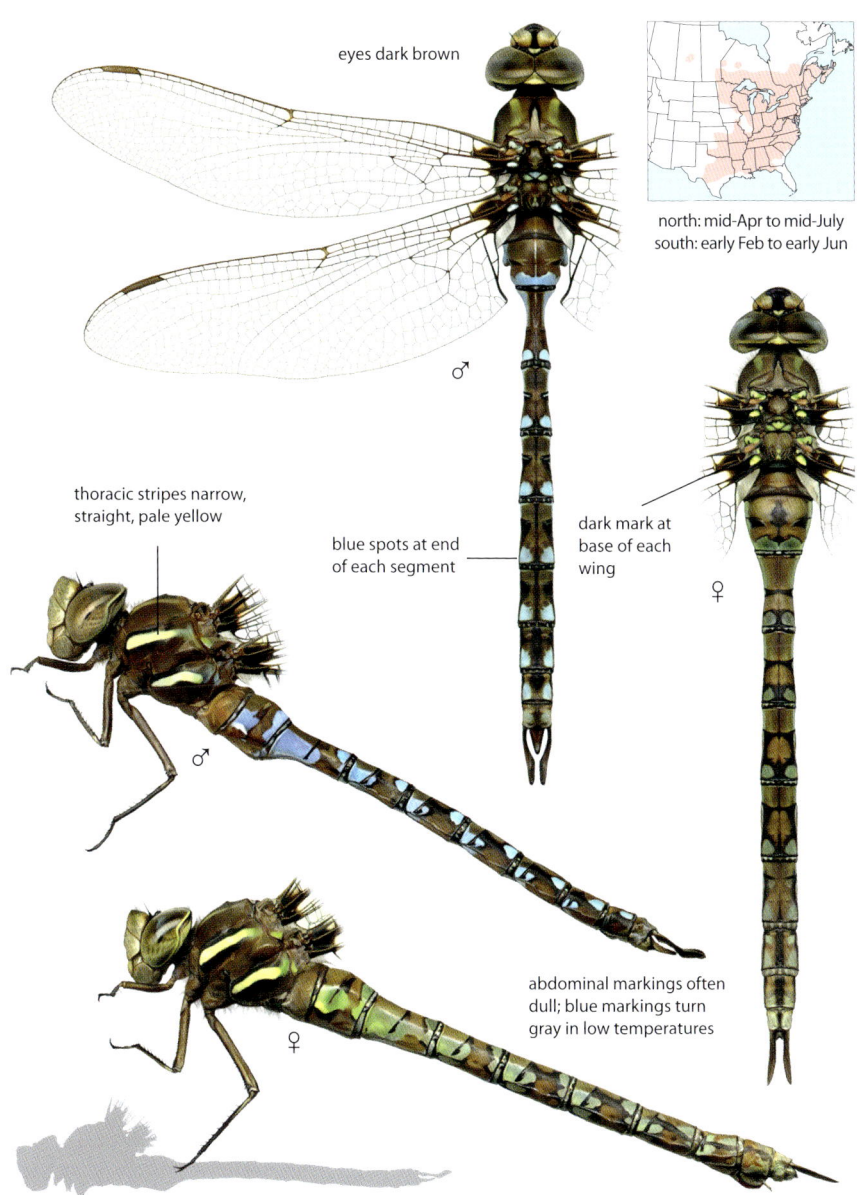

eyes dark brown

north: mid-Apr to mid-July
south: early Feb to early Jun

thoracic stripes narrow, straight, pale yellow

blue spots at end of each segment

dark mark at base of each wing

abdominal markings often dull; blue markings turn gray in low temperatures

♂

♀

Similar species: Mosaic darners have later flight seasons, lack dark markings at base of the wings. **Spatterdock Darner** larger; has blue or green thoracic stripes, pale face and eyes and lacks dark wing markings.

Blue-eyed Darner • *Rhionaeschna multicolor*

68–72 mm, 2.7–2.8 in. Common western darner found at lowland lakes, ponds, and slow streams. Mature male with sky-blue eyes, face, and body markings. Lateral thoracic stripes narrow and straight. Male cerci forked in side view (see page 395). Female may have blue or yellow-green markings.

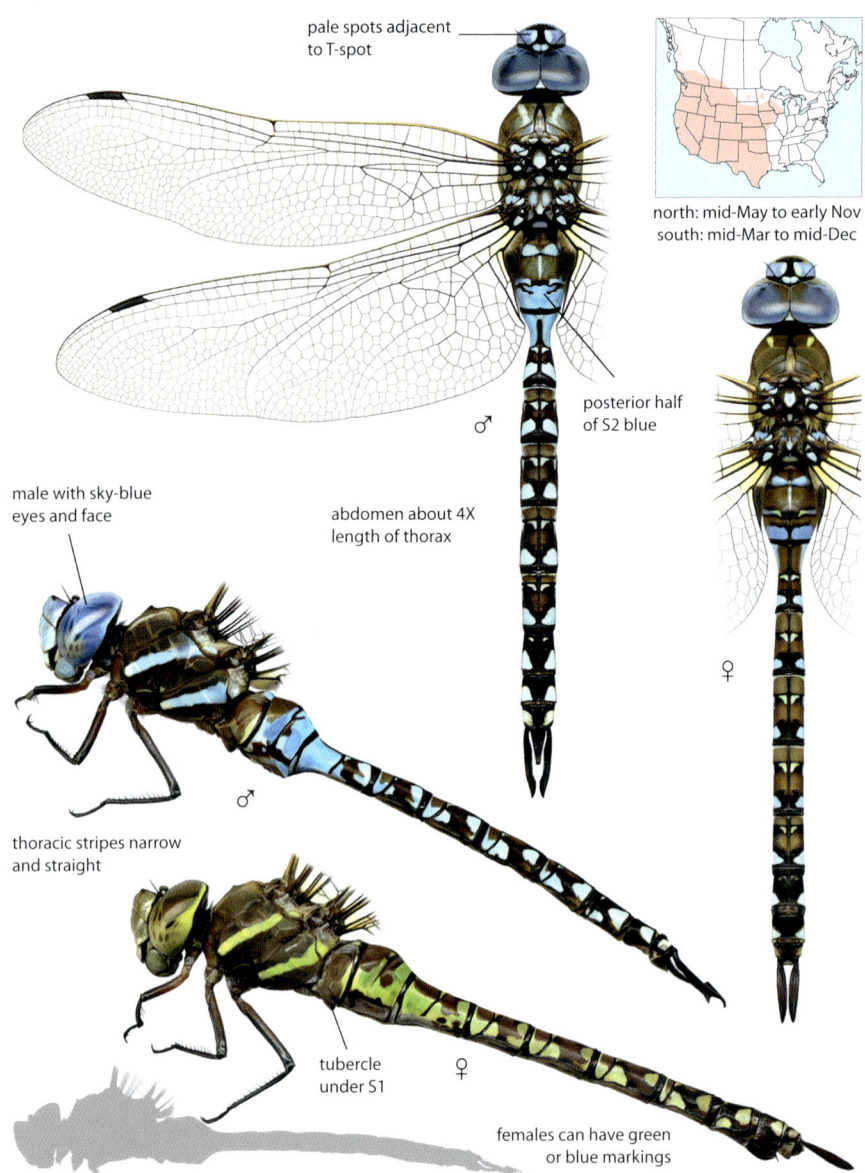

pale spots adjacent to T-spot

north: mid-May to early Nov
south: mid-Mar to mid-Dec

♂

posterior half of S2 blue

male with sky-blue eyes and face

abdomen about 4X length of thorax

♀

♂

thoracic stripes narrow and straight

tubercle under S1

♀

females can have green or blue markings

Similar species: Arroyo Darner has small dorsal extension on anterior lateral thoracic stripe; arrow-shaped mark and additional blue band on S2. Compare male cerci, page 395. Female Blue-eyed similar to **Shadow, Paddle-tailed,** and **Variable Darners** but has tubercle under S1 and pale areas adjacent to T-spot. Also see **California** and **Spatterdock Darners**.

Spatterdock Darner • *Rhionaeschna mutata*

74–76 mm, 2.9–3 in. Nearly identical to Blue-eyed Darner but early-season northeastern species. Inhabits lakes and ponds, usually fishless and with water lilies. Sky-blue eyes and male's all-blue markings distinctive in the East. Mature females develop pale blue eyes; body markings blue or green, often in combination.

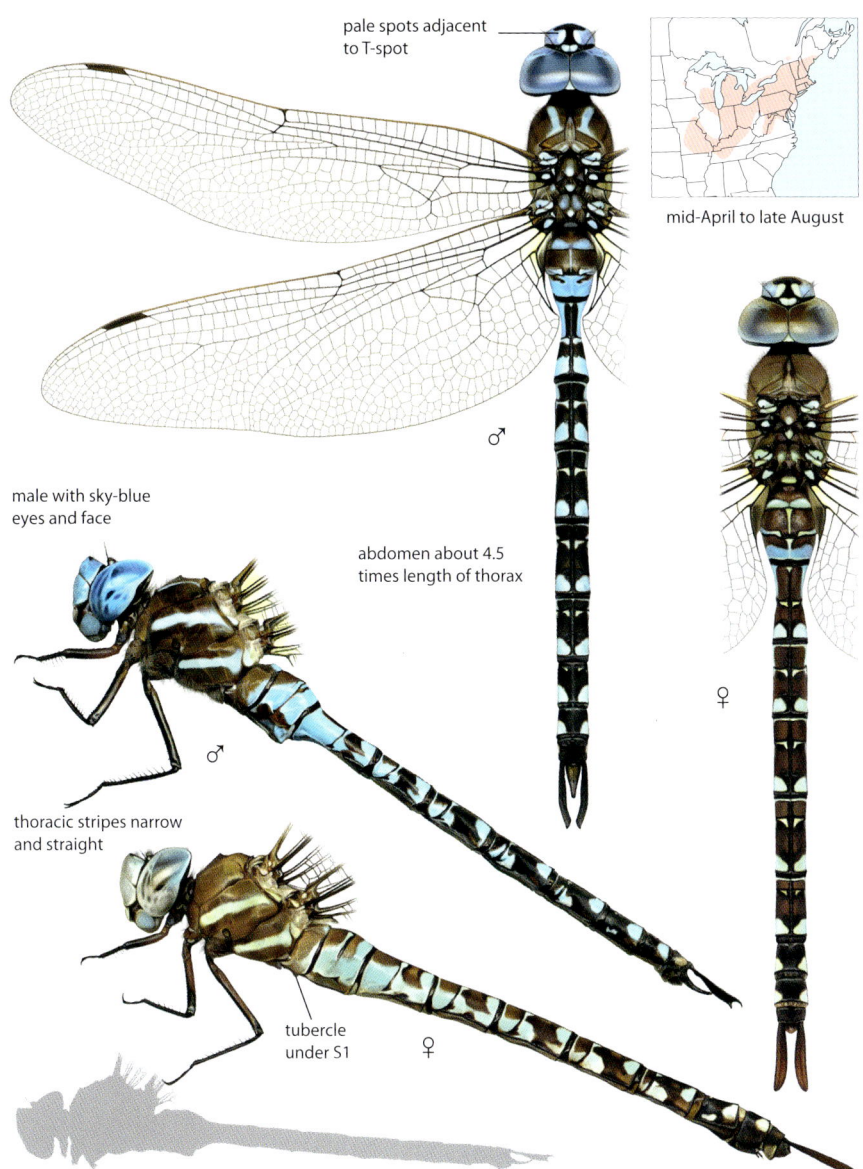

pale spots adjacent to T-spot

mid-April to late August

male with sky-blue eyes and face

abdomen about 4.5 times length of thorax

♂

thoracic stripes narrow and straight

tubercle under S1

♀

♂

♀

Similar species: No other eastern darner has sky-blue eyes and all-blue markings. **Blue-eyed Darner** range not known to overlap; has proportionally shorter abdomen, female with shorter cerci. Also compare male cerci, page 395. All **mosaic darner species** lack pale areas adjacent to T-spot; females lack tubercle under S1.

Arroyo Darner • *Rhionaeschna dugesi*

70–74 mm, 2.75–2.9 in. Uncommon southwestern darner with sky-blue eyes found at slow streams, stream pools, and ponds. Usually found at higher elevations closer to stream headwaters than similar Blue-eyed Darner. Anterior lateral thoracic stripe has small flag-like extension at top.

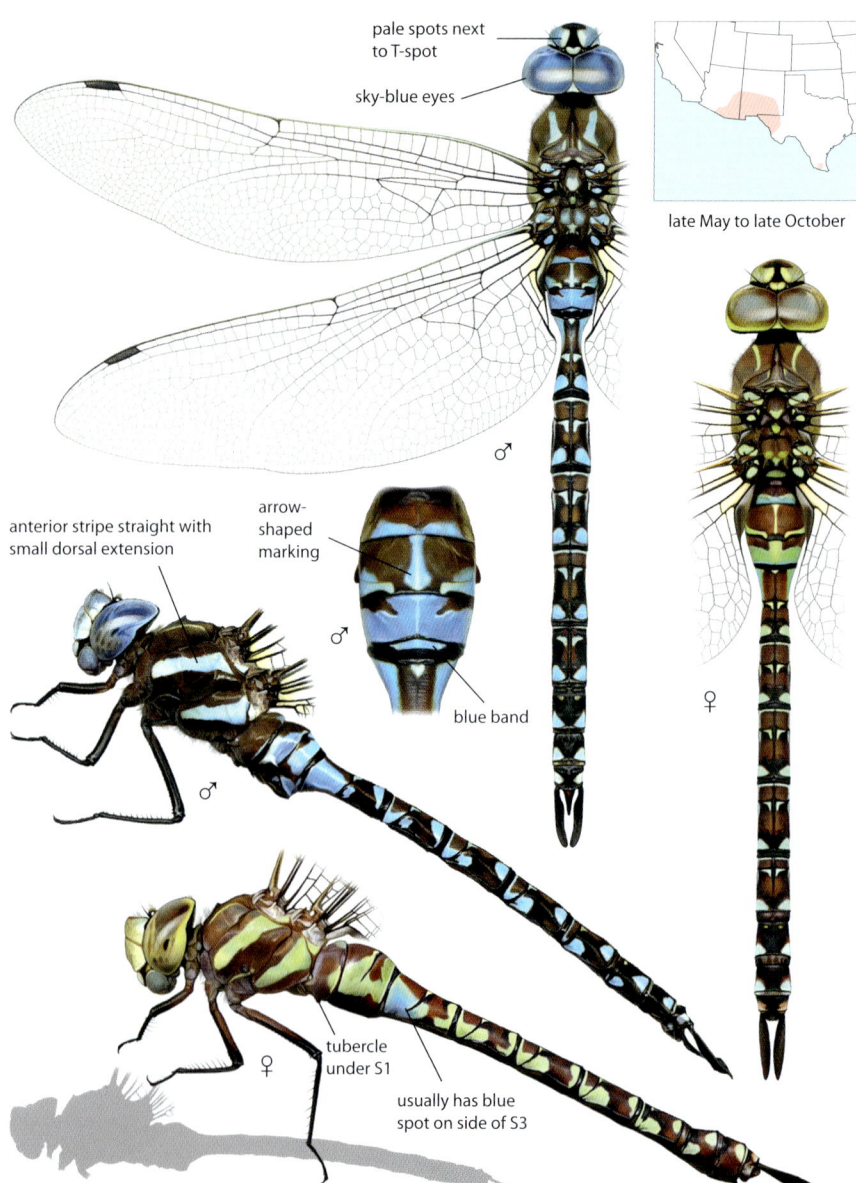

pale spots next to T-spot

sky-blue eyes

late May to late October

anterior stripe straight with small dorsal extension

arrow-shaped marking

♂

blue band

♂

♀

tubercle under S1

♀

usually has blue spot on side of S3

Similar species: Blue-eyed Darner lacks dorsal extension on anterior lateral thoracic stripe, blue band and arrow-shaped mark on S2; male cerci forked (see page 395). Green-form female lacks blue spot on side of S3. **Variable Darner** has narrower lateral thoracic stripes, lacks pale areas adjacent to T-spot; female lacks tubercle under S1. Also see **California Darner**.

California Darner • *Rhionaeschna californica*

57–61 mm, 2.2–2.4 in. Small, common western darner found at lakes, ponds, and slow streams. Flies in early spring, often the first dragonfly seen in its range. Lateral thoracic stripes narrow, anterior stripe tapered at top. Both sexes lack frontal stripes on thorax. Face with black cross-stripe. Male cerci unforked.

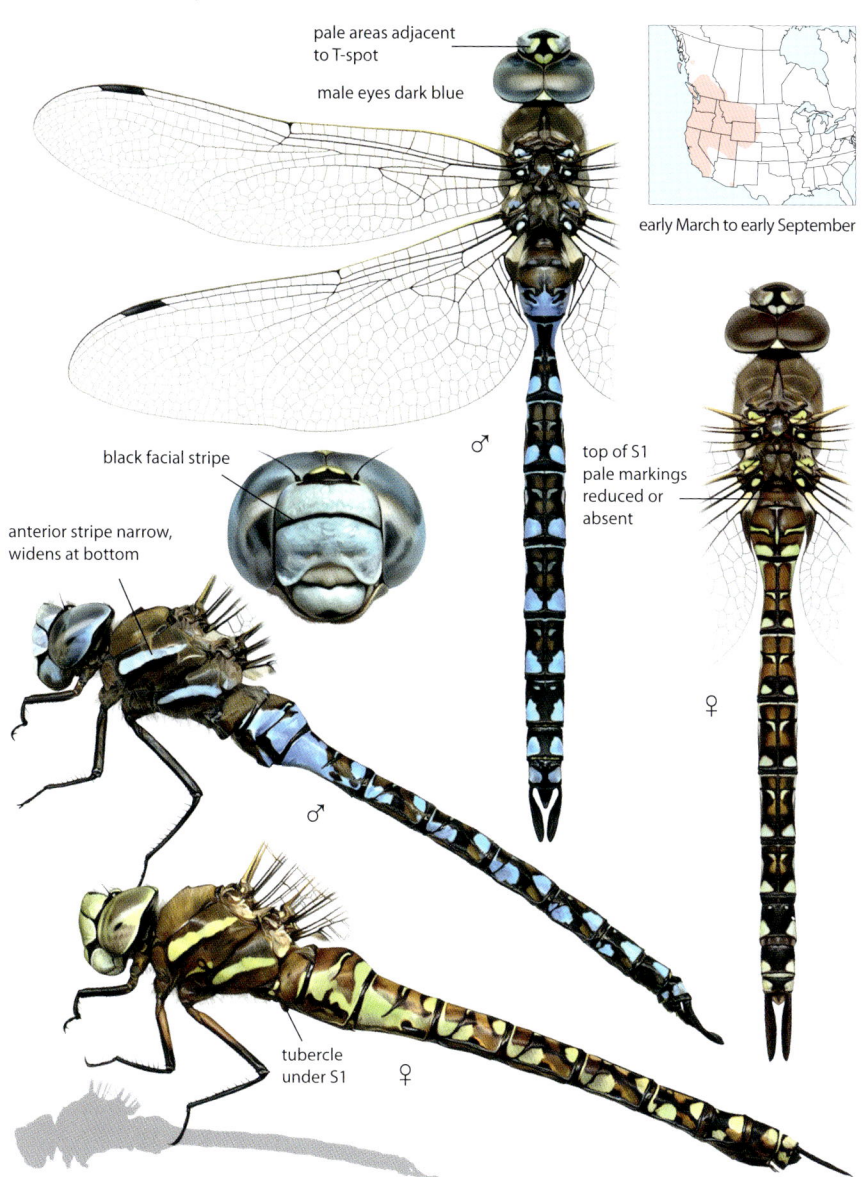

pale areas adjacent
to T-spot

male eyes dark blue

early March to early September

black facial stripe

top of S1
pale markings
reduced or
absent

anterior stripe narrow,
widens at bottom

♂

♂

♀

tubercle
under S1 ♀

Similar species: Blue-eyed and **Arroyo Darners** larger, usually with frontal thoracic stripes and larger pale markings on top of S1; lack black face stripe. Males have brighter blue eyes. Also compare male cerci, page 395. **Variable Darner** larger, lacks pale areas adjacent to T-spot; female lacks tubercle under S1.

Turquoise-tipped Darner • *Rhionaeschna psilus*

58–60 mm, 2.3–2.4 in. Small, dark southwestern darner found at forested ponds and stream pools. Thoracic stripes bold and green, anterior lateral stripe indented. Abdomen dark, appears ringed with narrow pale markings. Male has bright blue marking on underside of S9 and 10—diagnostic when visible. Posterior of S2 with blue markings.

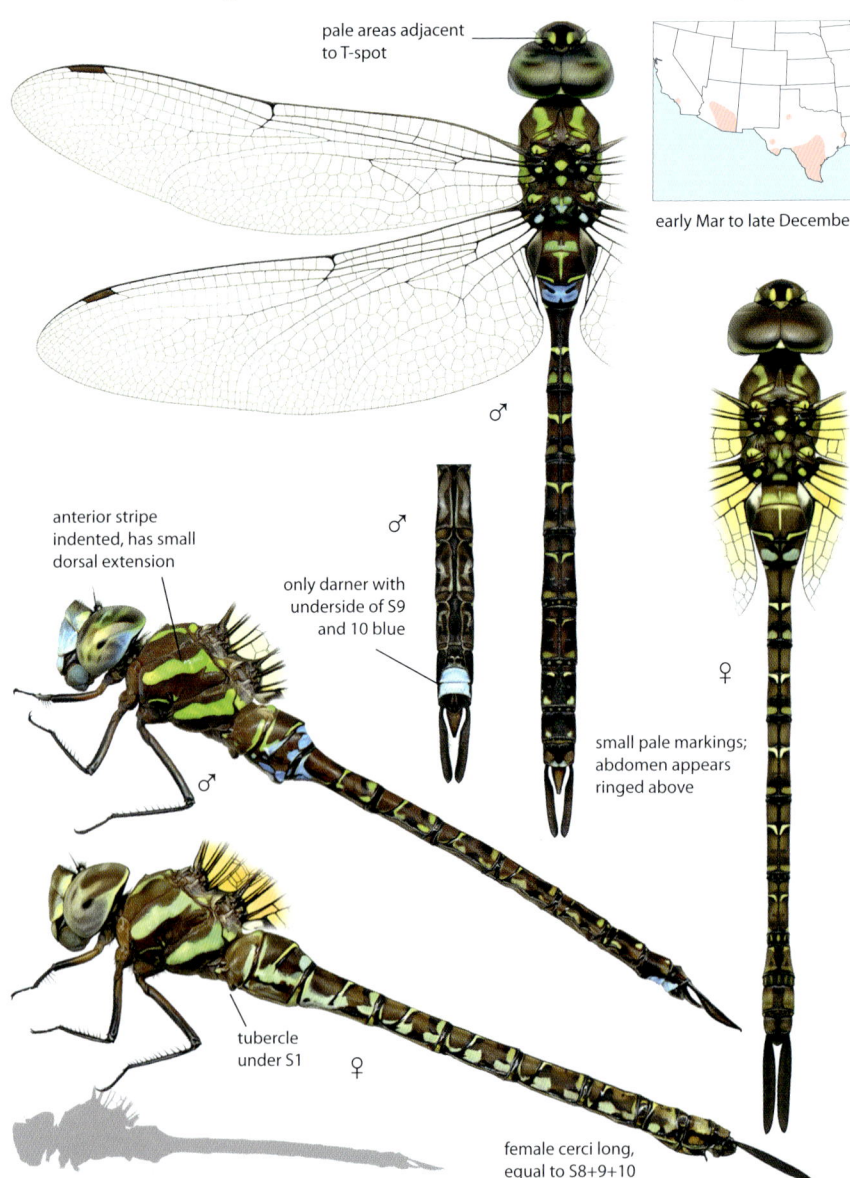

pale areas adjacent to T-spot

early Mar to late December

anterior stripe indented, has small dorsal extension

only darner with underside of S9 and 10 blue

small pale markings; abdomen appears ringed above

tubercle under S1 ♀

female cerci long, equal to S8+9+10

Similar species: Blue-eyed and **Arroyo Darners** have straight lateral thoracic stripes and larger abdominal markings. **Springtime Darner** has narrower yellow lateral thoracic stripes, larger pale abdominal markings, and dark markings at base of each wing. **Pilot, malachite**, and **three-spined darners** have mostly green thoraxes.

Riffle Darner • *Oplonaeschna armata*

66–75 mm, 2.6–2.95 in. Fairly common darner found at rocky southwestern mountain streams. Anterior lateral thoracic stripe deeply notched, sometimes broken. Abdomen with small pale markings. Male cerci similar to the wedge-type cerci of mosaic darner species; has finger-like dorsal projection on S10.

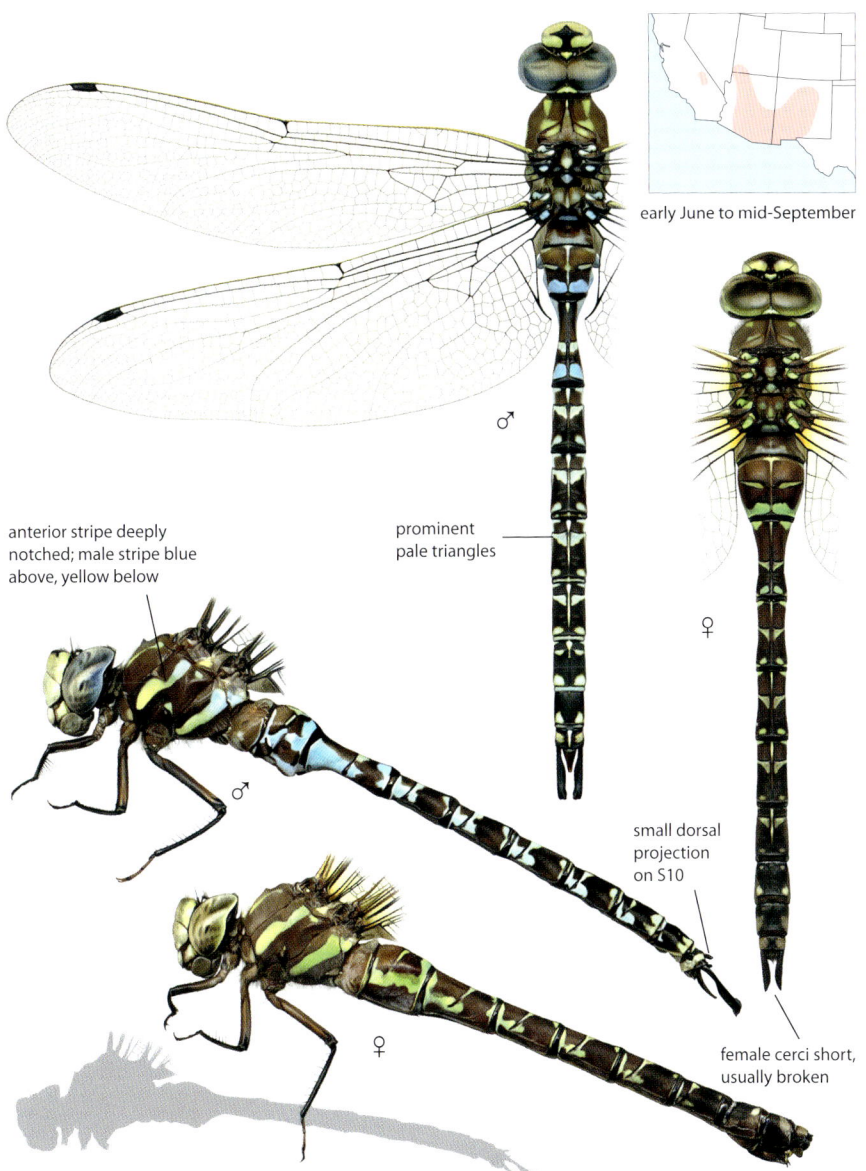

early June to mid-September

anterior stripe deeply notched; male stripe blue above, yellow below

prominent pale triangles

♂

♀

small dorsal projection on S10

♀

female cerci short, usually broken

Similar species: Blue-eyed and **Arroyo Darners** have straight thoracic stripes and larger abdominal markings. **Turquoise-tipped Darner** has pale areas adjacent to T-spot, female with tubercle under S1. **Mosaic darners** within range of Riffle Darner have larger abdominal markings.

Fawn Darner • *Boyeria vinosa*

60–71 mm, 2.4–2.8 in. Dark brown darner of shady forested streams and rivers, occasionally lakes. Eastern. Thorax with two yellow lateral spots; abdomen with small yellow dots. Dark brown streaks at the base of each wing. Males patrol slowly along stream bank, low to the water, typically late in the afternoon and continuing until after dark.

frontal stripes
reduced

north: mid-Jun to late Oct
south: late May to mid-Dec

dark markings
at base of each
wing

♂

♀

thorax brown with
pair of conspicuous
yellow spots

abdomen brown
with very small
pale spots

♂

small, inconspicuous
pale markings

♀

female cerci
usually long

some populations
have shorter cerci

Similar species: Ocellated Darner grayer in coloration, with more conspicuous pale abdominal markings; largely lacks dark markings at base of wings. **Twilight Darner** lacks pale spots on thorax and dark markings at base of wings.

Ocellated Darner • *Boyeria grafiana*

63–65 mm, 2.5–2.6 in. Grayish-brown darner found at rocky forest streams and rocky lakes. Eastern. Thorax with prominent yellow lateral spots; abdomen with small pale dorsal markings and lateral spots. Wing bases without obvious dark streaks. Often hangs in deep shade during the day, becoming most active in late afternoon into evening.

pale frontal stripes noticeable in flight

early June to mid-October

♂

wing base markings largely absent

thorax grayish-brown with pair of conspicuous yellow spots

pale green markings on top of abdomen

♀

male abdomen tip pale

♂

pale lateral markings conspicuous

♀

female cerci short

Similar species: Fawn Darner is browner with smaller abdominal markings, has dark markings at base of wings.

Harlequin Darner • *Gomphaeschna furcillata*

53–60 mm, 2–2.4 in. Small, dark darner inhabiting bog ponds and shallow swamp pools. Eastern. Overall pattern complex: thorax with series of pale stripes, abdomen with green, grayish, and dull orange markings. Male face, eyes, and markings greener, female grayer. Often perches vertically on tree trunks, occasionally flat on the ground.

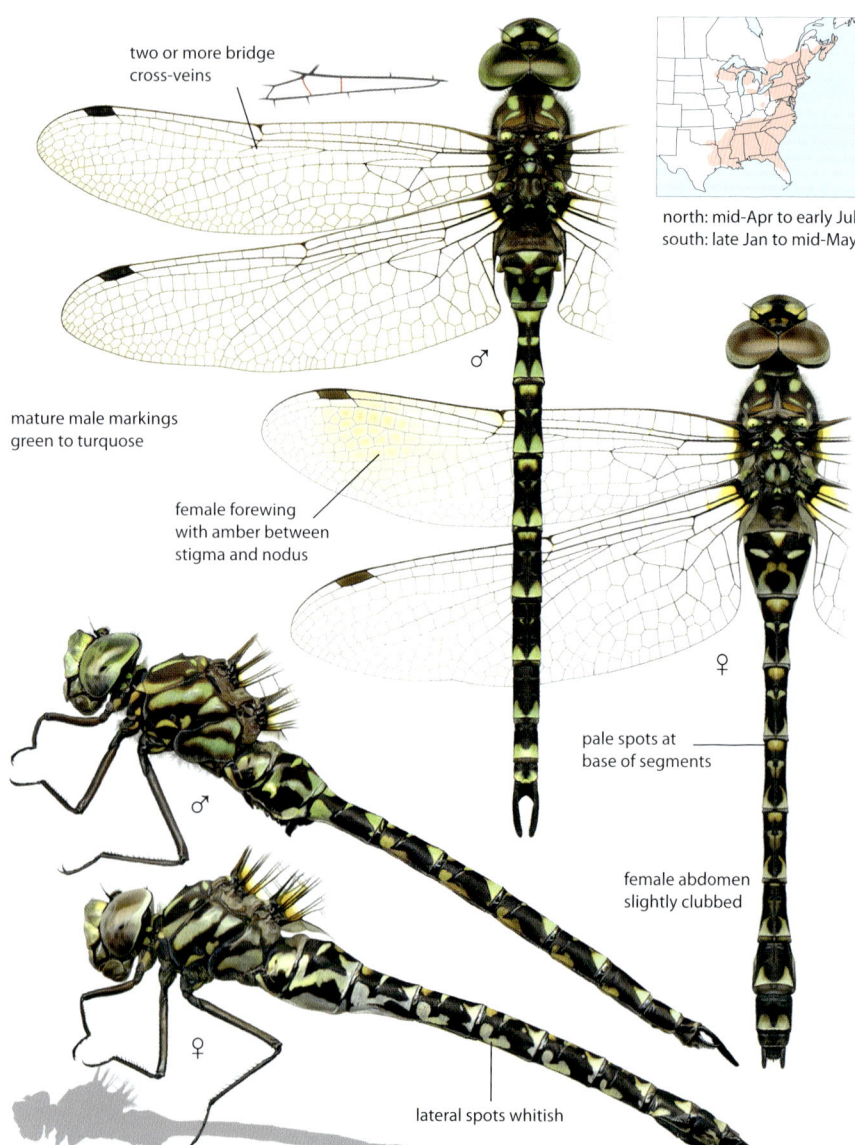

two or more bridge cross-veins

north: mid-Apr to early Jul
south: late Jan to mid-May

mature male markings green to turquose

female forewing with amber between stigma and nodus

♂

♀

pale spots at base of segments

female abdomen slightly clubbed

♂

♀

lateral spots whitish

Similar species: Taper-tailed Darner is duller in pattern, male lacking any green coloration. Middle abdominal segments are dark at the base. Female Taper-tailed has orange markings on side of abdomen and black hook-shaped mark on side of S2. Taper-tailed Darner has broader wings with single bridge cross-vein; forewing of female with amber wash at nodus.

Taper-tailed Darner • *Gomphaeschna antilope*

52–60 mm, 2–2.4 in. Small, dull-colored darner of eastern coastal plain swamps and bogs. Sexes similar with complex mottled pattern, marked in gray and orange; eyes greenish-brown. Male lacks green markings. Female abdomen with orange lateral spots. Perches on tree trunks and hangs vertically from branches.

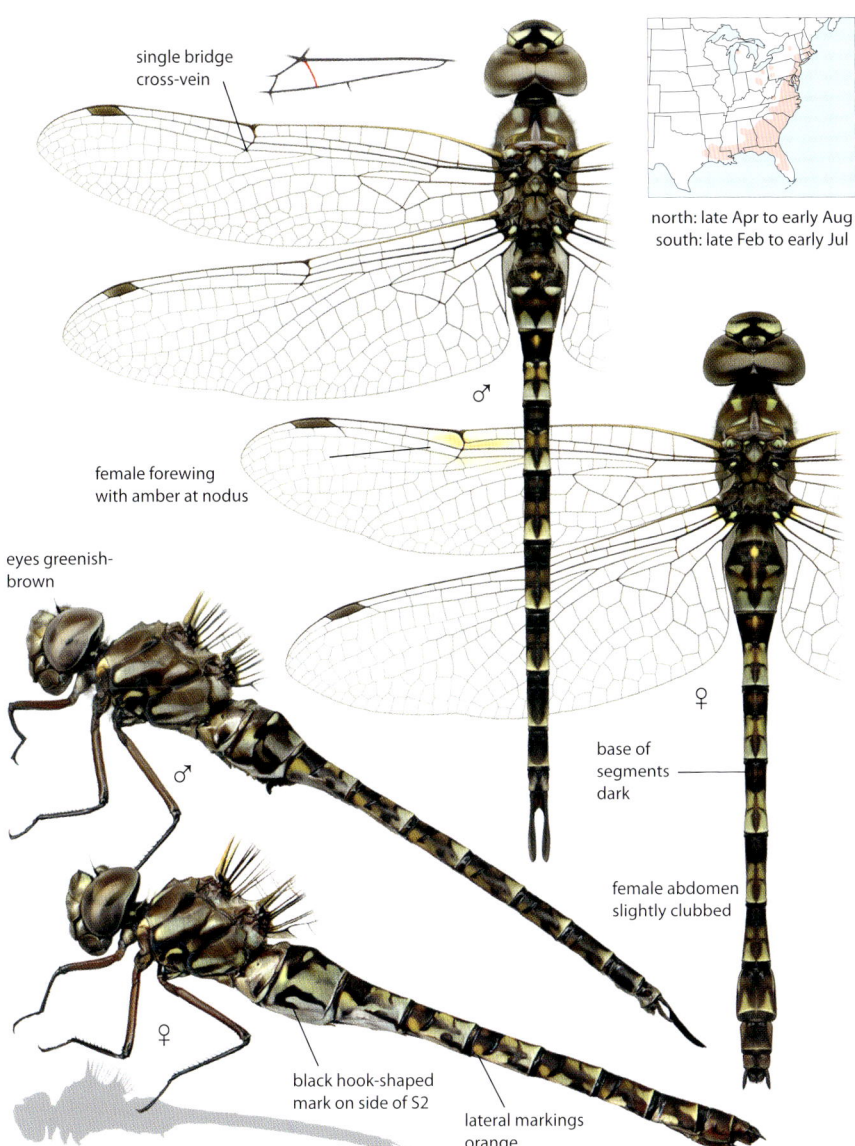

single bridge cross-vein

north: late Apr to early Aug
south: late Feb to early Jul

female forewing with amber at nodus

eyes greenish-brown

♂

♀

base of segments dark

female abdomen slightly clubbed

♂

♀

black hook-shaped mark on side of S2

lateral markings orange

Similar species: Harlequin Darner has brighter pattern; male with green markings. Middle abdominal segments have dorsal pale spots at base, which Taper-tailed lacks. Female Harlequin lacks orange markings on side of abdomen; black lateral mark on S2 triangular, not hook-shaped. Harlequin Darner has narrower wings with two bridge cross-veins; forewing of female with amber wash between nodus and stigma.

Clubtails • Family Gomphidae

The clubtails are a very large and distinctive family. As their name implies, most species, but not all, have clubbed abdomens (with S7–9 enlarged). Males are more likely to feature clubs, but many females also have them. Clubtails' eyes are widely separated, similar to the petaltails, but clubtail females lack the bladed ovipositor possessed by petaltails, most having a small subgenital plate under S8. Clubtail males have a forked epiproct, but the form of the male's appendages varies widely. The color of the appendages also varies among genera, which can aid identification. Overall, most clubtails are fairly muted in color, often predominately black or brown. Many of our species have yellow markings when immature, the yellow shifting and darkening to shades of green with age. Their wings are largely clear, with only the sanddragons having small dark marks at the base of the wings. As many species are similar in pattern, clubtails can be challenging to identify in the field.

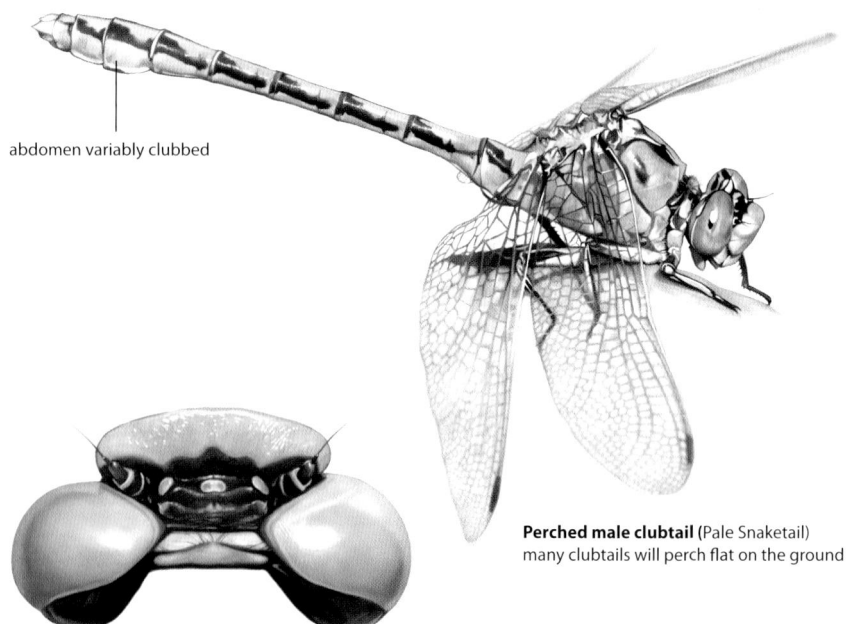

abdomen variably clubbed

Perched male clubtail (Pale Snaketail)
many clubtails will perch flat on the ground

Clubtail head: eyes widely separated

Most clubtails are found at running water. They tend to perch horizontally, sometimes with the abdomen raised. Males on territory will often settle on rocks in the middle of a stream or along the shore, on gravel banks, on sand bars, and on overhanging branches. Some elusive species perch high in trees, where they are difficult to detect and observe. Clubtails generally hunt from perches, although many males make extended patrol flights over the water. Clubtail flight is typically low. They do not soar or gather in feeding swarms. Females oviposit in flight alone, releasing their eggs by tapping the tip of their abdomens onto the water's surface. Of over 1,000 species worldwide, 103 are North American.

Greater Forceptails, genus *Aphylla,* page 87
Large slender clubtails with short legs and flanged abdomens. The name forceptail is derived from the male's large pincer-like cerci. The epiproct is largely absent, making the forceps-shaped cerci even more prominent. The lower rear corner of S10 is elongated to an acute point and takes the place of the epiproct when the male grips the female during mating.

Face is pale with brown cross-stripes. Eyes gray-blue when mature. The thorax is dark with well-developed pale stripes, yellow to green-gray in our species. The legs are notably short. The abdomen is long, slender, and slightly clubbed, both sexes having leaf-like flanges on the sides of S8 and 9, although these can be quite narrow. The abdomen is brown with pale markings that suggest weak rings, segments toward the tip distinctly more orange in color. The male's cerci are brown, the female orange. 24 species, three North America.

Lesser Forceptails, genus *Phyllocycla,* page 90
Similar to greater forceptails with the male having large pincer-like cerci and largely lacking an epiproct. Male lesser forceptails lack the pointed corners on S10 that are present in greater forceptails. Females largely lack abdominal flanges. 31 species, one North America.

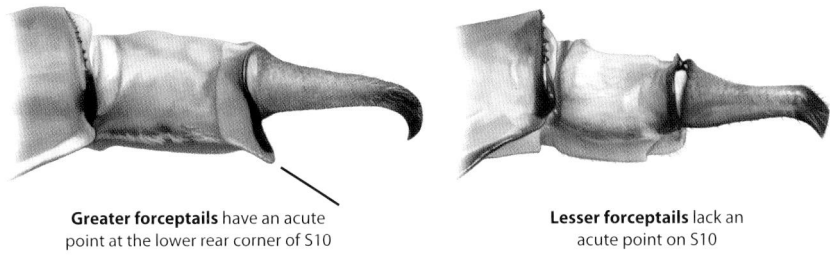

Greater forceptails have an acute point at the lower rear corner of S10

Lesser forceptails lack an acute point on S10

Leaftails, genus *Phyllogomphoides,* page 91
The leaftails are a genus of large Neotropical clubtails. Both males and females have lateral flanges on S8 and 9, with those of the male conspicuously large and leaf-like. Both sexes have pale appendages. The male cerci are long and pincer-like with blunt tips, each cercus having a small tooth dorsally and another along the inner margin. The female cerci have pointed tips, are slightly curved, and are long—longer than S10. The male's epiproct is short, divided into two thin, upcurved branches. The thorax is stout and variably striped. Legs are short, with femurs striped in some species. The abdomen is black with lateral pale markings and middorsal pale stripes. Male club often orange-brown in color with wide flanges, but the width of the flanges is variable even within the same species. Female abdominal flanges narrower and also variable in width. Of 47 species, three are North American.

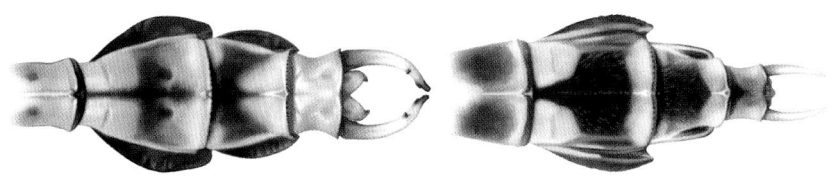

Leaftail abdomen tips: Five-striped Leaftail, male (L), female (R)

Spinylegs, genus *Dromogomphus,* page 94
The spinylegs are three eastern North American species. They are characterized by their very long hind legs, particularly the femur, which extends to the base of S3. The femur also bears a row of long prominent spines used to capture large prey. The head is proportionally small compared to the thorax, which is stout, pale, and has a strong pattern of dark stripes. The abdomen is slender, those of the males moderately clubbed. Females are only slightly clubbed.

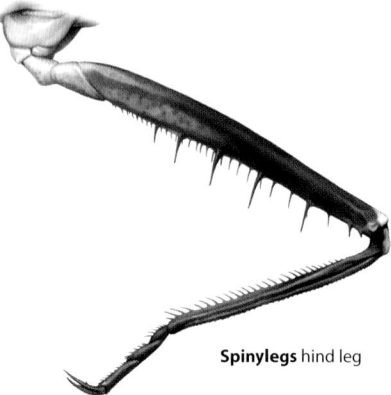

Spinylegs hind leg

Pond Clubtails, genus *Arigomphus,* page 97
A small genus comprising seven North American species. These clubtails are among the few that breed in still water. They are stout in appearance, with large heads and sturdy thoraxes. The face is pale without any dark markings, eyes green or blue upon maturity. The thorax is notably pale with dark markings largely restricted to a pair of short, separated frontal stripes and a pair of stripes at each shoulder. Pond clubtails' coloration varies from yellow-green to gray, most being dull green in appearance. The male's abdomen is only slightly clubbed, the female unclubbed. Most species have their end abdominal segments marked with orange or brown, and S10 is often pale. Both sexes have pale cerci. Females have a well-developed subgenital plate, spout-like in some species, a feature rare among clubtails.

The shape of the abdomen is rather distinctive, tapering at the tip with S9 being both narrow and long.

Male pond clubtail abdomen

Bantam Clubtails, genus *Hylogomphus,* page 104
The bantam clubtails are a genus of six species found in eastern North America. They inhabit clean rivers and streams, often perching flat on leaves. They can be recognized by their small size and stocky build. Males have large abdominal clubs that are oval in shape. The female abdomen is much less clubbed but very stout and has a rounded tip. Legs and appendages are black, and the male cercus is short and thick, with a ventral tooth near the tip. The male cerci and the branches of the epiproct spread equally wide.

The face is typically pale, but some species have black cross-stripes, which are often important for species identification. Eyes are green or blue upon maturity. The color of the thorax is muted, ranging from yellow to dull green to green-gray. The thorax is marked with black frontal and shoulder stripes, but the lateral stripe pattern varies among species. The abdomen is mostly black with small pale dorsal markings, limited to mostly small dots on the male; short streaks on the female, but both sexes have the top of S8-10 black. Pale spots mark the sides of the abdomen, and most species have large yellow spots on the sides of S8 and 9.

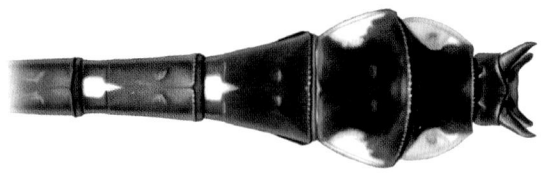

Male bantam clubtails have oval-shaped clubs

Least Clubtails, genus *Stylogomphus,* page 110

These are our smallest clubtails. Two species occur at rocky streams and small rivers in eastern North America, while nine other species are found in Asia. Although well patterned, their diminutive size makes them easy to overlook as they perch flat on rocks or among vegetation. They most resemble the pygmy clubtails (with which they were once grouped), but least clubtails have narrow pale abdominal rings and their appendages are pale. Our two species are identical in pattern and can be separated only by the structure of the male's terminal appendages and the female's subgenital plate. Intermediate forms have been found, suggesting hybridization between the species.

Male least clubtail abdomen with narrow pale rings and pale appendages

Pygmy Clubtails, genus *Lanthus,* page 112

Measuring only 1.5 in long, these dark clubtails are among our smallest. Two species are found at forested streams and rivers in the Northeast, while a third species occurs in Asia.

Both of our species are similar in appearance and are best separated by their lateral thoracic pattern. Face is yellow with black cross-stripes. Eyes green at maturity. The thorax is mostly black with yellow sides variably striped with black. Legs are black. The male abdomen is slightly clubbed and almost completely black; pale markings are restricted to the abdomen base, with small dots laterally. Yellow abdominal markings are more extensive in females, but the unclubbed abdomen is still largely black. Appendages are black.

Grappletail, genus *Octogomphus,* page 114

Grappletail is the lone species in this genus, a small, slender clubtail found at streams along the West Coast. Like its nearest relatives, the pygmy and least clubtails of the East, Grappletail has a nearly all-black abdomen, but no western clubtail is similar. The thorax is mostly pale with prominent, wide, black shoulder stripes. The pale marking at the front of the thorax is shaped like an urn. The black abdomen is barely marked with hairline dorsal pale stripes and lateral dots. The male abdomen is slightly clubbed and widest at S10. The female is unclubbed. The male cerci are forked and spread wide, and the epiproct uniquely has four branches. Correspondingly, the female occiput has a four-humped ridge to accommodate the male's epiproct.

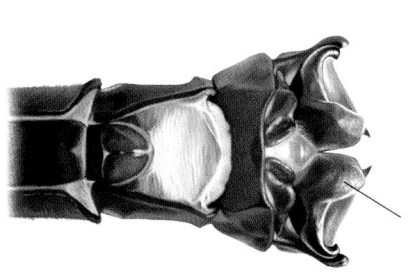

Male Grappletail ventral abdomen tip
epiproct has four prongs

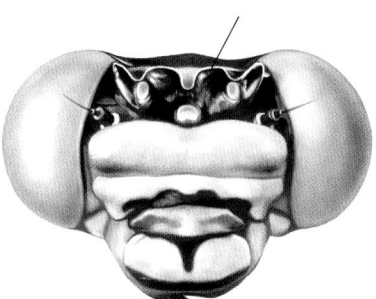

Female Grappletail head
occiput with four-humped ridge

American Clubtails, genus *Phanogomphus,* page 115

Seventeen species make up this rather diverse group of North American clubtails. Most are eastern, with only three found west of Texas. Many are widely distributed, while others are severely limited in range. They are small to medium-sized, with head, thorax, abdomen, and legs moderate in proportion. They share some traits in genital structure and venation, but their appearance varies substantially. Most are notably dull in coloration, but a few are brightly patterned. The male is weakly clubbed in the majority of species, while others have medium-sized clubs (females largely lack clubs). Most species emerge in the spring, and they are often the most common clubtails in their area. Some species are restricted to ponds and lakes, while many are able to utilize both flowing and still-water habitats.

Despite their variation, American clubtails share some pattern similarities. The face is pale, with only Clearlake Clubtail having an obvious dark facial stripe. Mature eye color is typically blue-gray. The thorax is yellow, particularly when immature, becoming dull yellow-green to green-gray with age. All species have well-developed dark black or brown stripes. The lateral thoracic pattern is often useful for species identification but can be difficult to discern, as it is often obscured by the wings. Legs vary from pale-striped to all black. The abdomen is dark with pale middorsal and lateral stripes; the abdomen pattern is helpful for identification in many cases.

Majestic Clubtails, genus *Gomphurus,* page 132

Thirteen North American species make up this genus of mostly large well-marked clubtails. Their typical habitats are rivers and large streams. Both sexes are conspicuously clubbed, the club impressively large in some species. The head appears small compared to the robust thorax, and the femur of the hind leg is distinctly long, extending close to the margin of S3. Majestic clubtails' large size and conspicuous abdominal club help separate them from most other clubtails but species in other genera are similar in size and/or have large clubs need to be considered carefully.

Pattern-wise, the face is pale, with some species having dark cross-stripes that can aid in identification. Mature eye color varies from green to blue to gray. Overall body coloration is generally muted, typically ranging from yellow to dull green. All species have well-developed dark thoracic frontal and shoulder stripes, while lateral thoracic stripes are present in most. Legs are all black in most species, so any presence of pale stripes should be noted. The abdomen is black, with pale middorsal and lateral stripes and spots. The pattern of S8 and 9 in particular is often useful to separate species.

Male majestic clubtail abdomen (Cobra Clubtail)

Dragonhunter, genus *Hagenius,* page 145

The Dragonhunter is the only species in its genus. It is our largest clubtail, brawny in structure and measuring up to 3.5 in in length, making it one of North America's largest dragonflies. Its range is eastern, and it inhabits a wide variety of rivers and streams. Southern populations are largest in size. The head is small in contrast with the massive thorax. The legs are long, particularly the hind legs, which are used to capture large prey. The wings are long and narrow, as are the stigmas. The abdomen of both sexes is slightly clubbed, the club widest at S9. Males are often seen flying with the end of the abdomen curled downward, forming a J shape.

Face is pale, with thin black cross-stripes. Eyes are green. Thorax is black, with two wide yellow lateral stripes, and very narrow yellow shoulder and frontal stripes. Legs are black. Abdomen is black with yellow middorsal and lateral stripes; S10 and top of S9 are black. Appendages are black and very short.

Appalachian Clubtails, genus *Stenogomphurus,* page 146

This genus consists of two forest-stream species found in eastern North America. They were formerly considered majestic clubtails but lack the long hind legs and large clubs that characterize that group. The head and thorax are moderate in size; the abdomen is relatively long and slightly clubbed. Wings often have darkened tips, unusual for clubtails.

The two species are similar in appearance. The face is pale with black cross-stripes, the patterns of which are useful in separating the species. The eyes are dark green when mature. The thorax is yellow-green to green-gray, with well-developed black frontal and shoulder stripes. Legs are black. The abdomen is long and black with very narrow pale middorsal stripes, often reduced to short dashes or dots, particularly on males. Both sexes have the top of S8-10 black. Side of the abdomen is marked with small pale dots, larger yellow spots on the sides of S8 and 9. Males have small clubs. Females often are identifiable to genus by their long straight abdomen with a distinctly diamond-shaped club. Appendages are black.

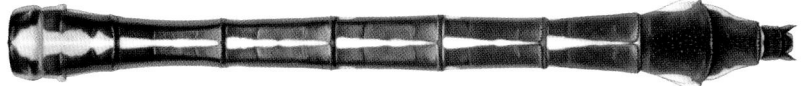

Female Appalachian clubtail abdomen: long with diamond-shaped club

Sanddragons, genus *Progomphus,* page 148

Nearly 70 species of sanddragon occur in the New World, but only four are found in our area. They breed primarily at sand-bottomed streams and lakes, their larvae adapted to burrowing. They are our only clubtails with dark wing markings, having dark streaks at the base of each wing. Our species are predominately brown and yellow, lacking any of the green coloration that marks most clubtails. The legs are short. The male abdomen is slightly clubbed, the female hardly at all. The cerci are fairly long and contrastingly pale. The male epiproct is completely divided into two movable parts, and there is an additional tongue-like structure of unknown utility above them.

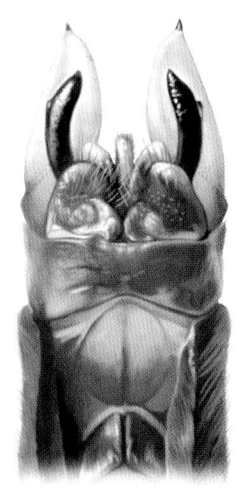

Male sanddragon abdomen tip: epiproct divided

The face is pale, often with dark cross-stripes. Eyes are mainly gray, greener in some species, browner in others. The thorax is mostly brown with brown shoulder stripes wide and often fused. Wings have small dark marking at their bases. The abdomen is black with yellow or yellow-orange middorsal pale markings, some species with pale rings; the top of S9-10 is mostly dark.

Snaketails, genus *Ophiogomphus*, page 152
Of the 28 species of snaketails, 19 are North American; the rest are found in Europe and Asia. They can be recognized by their mostly green thorax and face, bright green in most species. Ringtails may also have a bright green thorax, but their abdomens have complete pale rings that snaketails lack. Snaketail abdomens are black or brown, with pale lateral and middorsal stripes and spots. The pattern of markings repeat along the middle abdominal segments and suggest the markings of a snake. Snaketails' legs are short. Both sexes are moderately clubbed, and their appendages are pale. Most require pristine habitats, primarily clean, clear, rocky streams. Males patrol low along the water and perch flat on rocks and on shoreline vegetation.

Snaketail faces are pale yellow-green to bright green, and two of our species have black cross-stripes. Eye color ranges from green to blue to gray. The thorax varies from bright green to pale yellow-green, variably striped in black or brown, most with well-developed frontal and shoulder stripes. Legs vary from mostly pale to all black. All species have pale markings laterally and often dorsally on all abdominal segments.

Ringtails, genus *Erpetogomphus*, page 171
The ringtails are a largely Neotropical genus inhabiting rivers and streams, found in our area primarily in the Southwest. Small to medium-sized with short legs, the ringtails are among our most colorful and well-patterned dragonflies. The face is pale, eyes blue to blue-gray. The thorax is pale, yellow to bright green in most species, and variably striped with brown. The abdomen has prominent pale rings on the middle segments, typically S4–7, from which the genus gets its name. Viewed from the side, these rings are whitish, while dorsal markings are a different color in many species. The end segments may also differ in color from the rest of the abdomen, yellower in some species, rusty brown in others. Males have small abdominal clubs, while females largely lack them. Both sexes have pale appendages. The male cerci are finger-like and straight or bend downward in the middle. Of 23 species, eight are North American.

Hanging Clubtails, genus *Stylurus*, page 179
Eleven of the 30 species of hanging clubtails are found in North America, while the rest are Eurasian. They are named for their habit of hanging downward when perched on leaves and branches, but they will alight more horizontally on stouter perches. They inhabit primarily rivers and streams, and many fly late in the summer and into autumn. Males typically patrol for long periods of time, so are most often seen in flight over water.

Our species are diverse in appearance, although they share a number of anatomical characters. The male's anterior hamule is small and simple, and the terminal appendages are similar throughout the genus; the cerci are simple in form, lacking any teeth or branches. The forks of the epiproct spread about as wide as the cerci. The female subgenital plate is short. Hanging clubtails have a distinctive head shape with a relatively flat face; the projection of the frons is short compared to most other clubtails. Other characters helpful to recognize hanging clubtails in the field include short legs and long slender abdomens, with males being moderately clubbed. Pattern-wise, most of our species have a pair of isolated pale stripes at the front of the thorax.

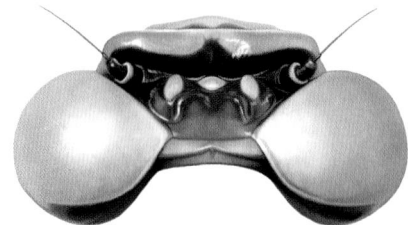

Hanging clubtails have a relatively flat face

Two-striped Forceptail • *Aphylla williamsoni*

71–76 mm, 2.8–3 in. Large, dark southeastern clubtail inhabiting mud-bottomed lakes, ponds, and slow streams. Thorax dark brown, with two wide pale lateral stripes. Flanged abdomen slightly clubbed, dark with pale markings, forming weak rings; end segments distinctly orange. Male cerci forceps-shaped, lack epiproct.

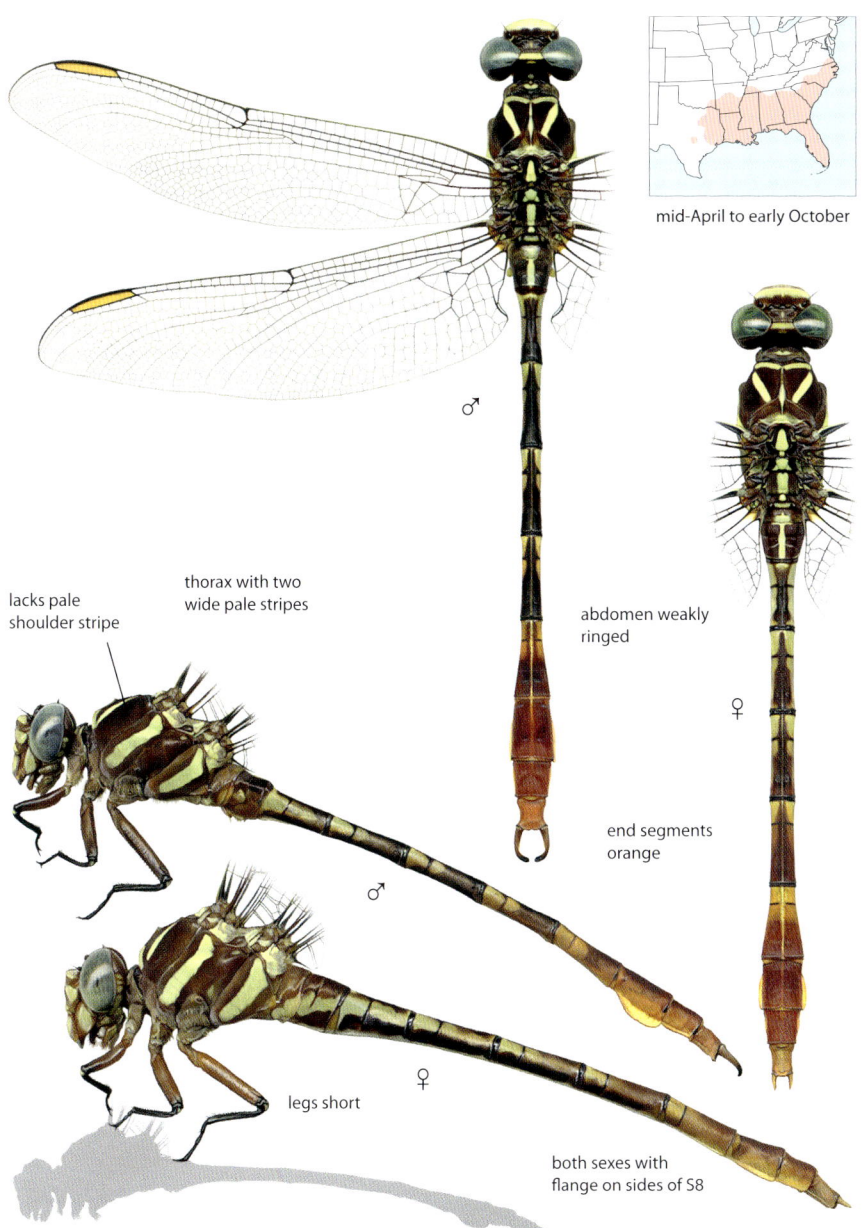

mid-April to early October

lacks pale shoulder stripe

thorax with two wide pale stripes

abdomen weakly ringed

end segments orange

legs short

both sexes with flange on sides of S8

Similar species: Broad-striped and **Narrow-striped Forceptails** have three lateral pale thoracic stripes and pale shoulder stripes.

Broad-striped Forceptail • *Aphylla angustifolia*

62–68 mm, 2.4–2.7 in. Large, slender clubtail found at ponds, slow rivers, and pools of intermittent streams, primarily in coastal Texas and Louisiana. Side of thorax with narrow pale stripe between two wider stripes. Abdomen brown, weakly ringed, orange toward tip. Both sexes with narrow abdominal flanges. Male lacks epiproct; cerci forceps-shaped.

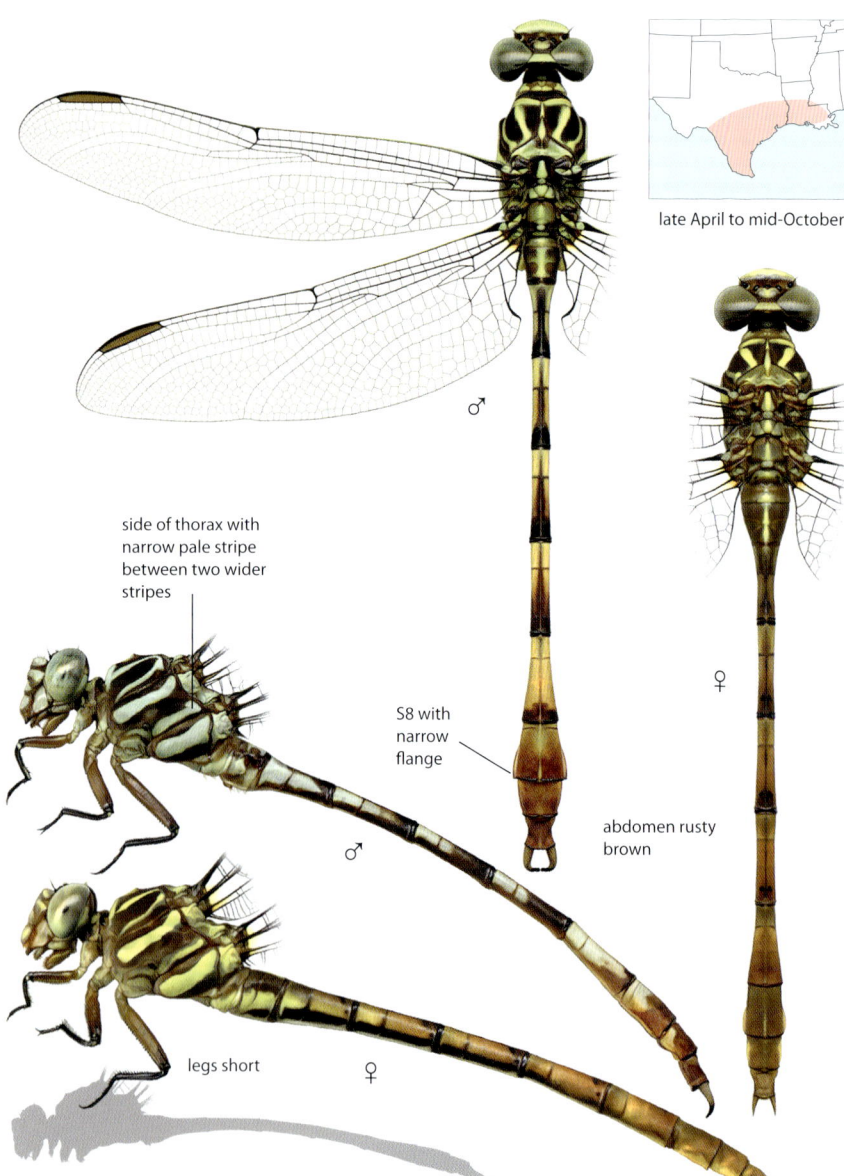

late April to mid-October

side of thorax with narrow pale stripe between two wider stripes

S8 with narrow flange

abdomen rusty brown

legs short

♂

♀

Similar species: Narrow-striped Forceptail has three narrow lateral thoracic stripes equal in width; both sexes with wider flanges on S8. **Two-striped Forceptail** has only two pale lateral thoracic stripes, lacks pale shoulder stripe. **Ringed Forceptail** and **Leaftails** have distinct pale rings on abdomen.

Narrow-striped Forceptail • *Aphylla protracta*

64–66 mm, 2.5–2.6 in. Large, slender, short-legged clubtail with flanged abdomen found at mud-bottomed ponds and lakes. Local in Texas and Arizona. Similar to Broad-striped Forceptail but has three narrow pale lateral thoracic stripes equal in width. Both sexes have wide abdominal flanges. Male lacks epiproct; cerci forceps-shaped.

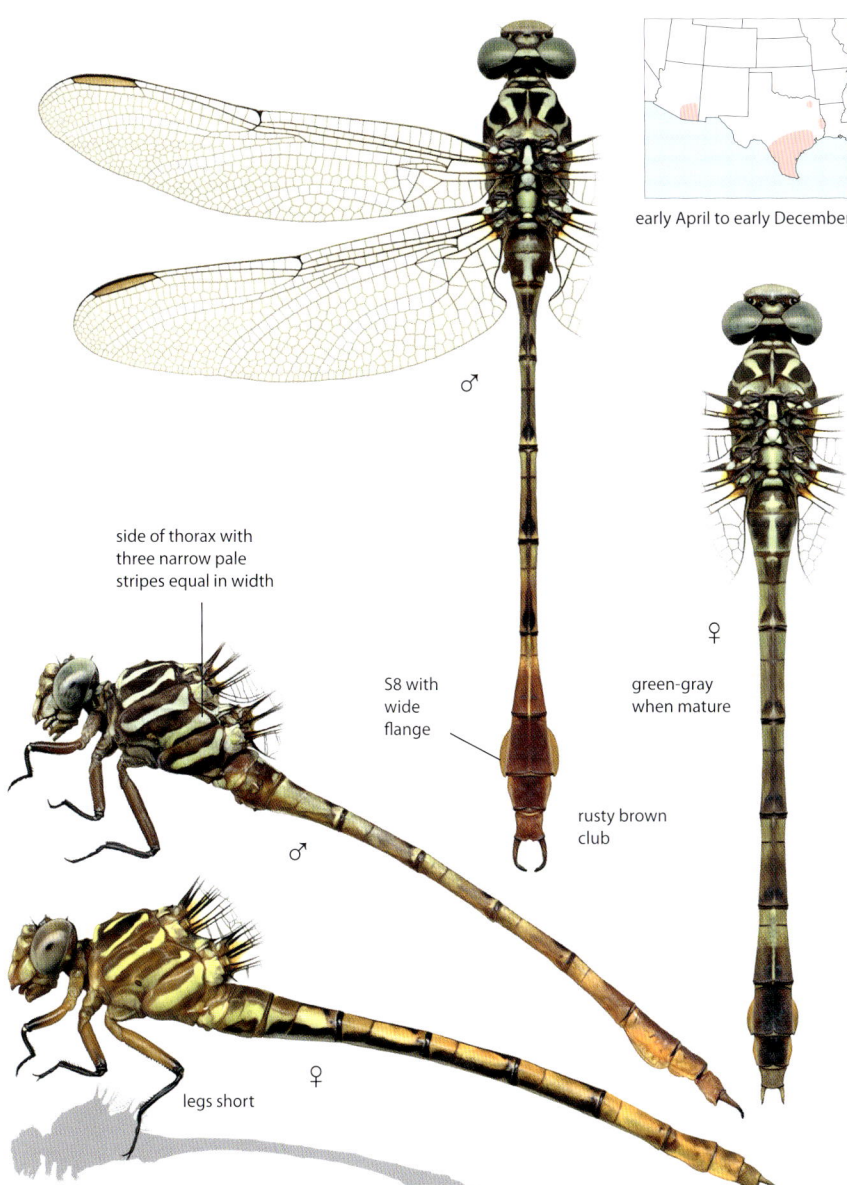

early April to early December

side of thorax with three narrow pale stripes equal in width

S8 with wide flange

♂

green-gray when mature

♀

rusty brown club

legs short

♂

♀

Similar species: Broad-striped Forceptail thorax has narrow pale lateral stripe between two wider stripes; abdominal flanges narrower. **Two-striped Forceptail** thorax lacks pale shoulder stripe, has only two wide lateral stripes. **Ringed Forceptail** and **Leaftails** have distinct pale abdominal rings.

Ringed Forceptail • *Phyllocycla breviphylla*

54–60 mm, 2.1–2.4 in. Slender, dark clubtail inhabiting streams and rivers, including the southern Rio Grande in Texas. Thorax with pale lateral stripes about equal in width. Abdomen long, black, slender with pale rings. Male abdominal flange wide; end segments brown. Tip of female abdomen mostly black. Male lacks epiproct; cerci pincer-like.

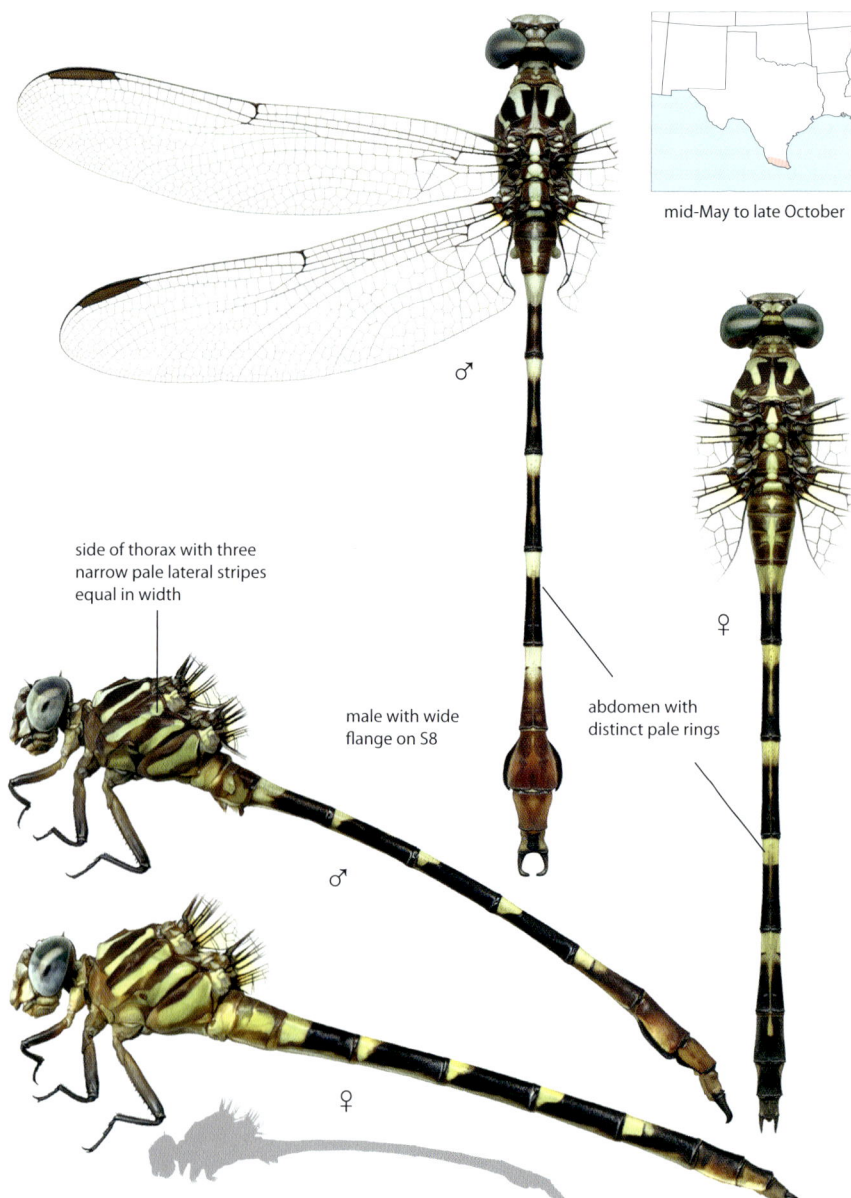

mid-May to late October

side of thorax with three narrow pale lateral stripes equal in width

male with wide flange on S8

abdomen with distinct pale rings

♂

♀

♂

♀

Similar species: Broad-striped Forceptail has narrow pale lateral thoracic stripe between two wider stripes; abdomen weakly ringed. **Narrow-striped Forceptail** has weakly ringed abdomen.

West Mexican Leaftail • *Phyllogomphoides nayaritensis*

62 mm, 2.4 in. Darkly patterned leaftail found primarily in western Mexico at rivers and large streams; single New Mexico record. Thorax mostly dark with narrow pale stripes. Abdomen dark with thin middorsal pale stripes, wide complete pale ring on S7; abdomen tip mostly black. Both sexes with abdominal flanges; pale cerci.

early August to late October

pale shoulder stripe reduced or absent

side of thorax with two narrow pale stripes

S7 with wide complete pale ring

dark short legs

♂

♀

Similar species: Five-striped and **Four-striped Leaftails** have paler thoraxes with complete pale shoulder stripes and wider pale lateral stripes; legs pale-striped. **Greater Forceptails** have less contrasting abdomen patterns. **Ringed Forceptail** lacks pale dorsal stripes on abdomen; abdomen with complete pale rings.

Four-striped Leaftail • *Phyllogomphoides stigmatus*

65–70 mm, 2.6–2.75 in. Large, brightly patterned clubtail found at lakes, large ponds, and slow-flowing rivers and streams, primarily in Texas. Thorax pale with four dark stripes visible in side view. Abdomen has middorsal pale stripes on S3-5, usually complete pale rings on S5-7. Male with wide leaf-like flanges, S8-10 orange-brown. Long pincer-like cerci pale.

early May to early October

four dark stripes visible in side view

end segments orange/brown

male cerci pincer-like; epiproct short

both sexes with long pale cerci

complete pale rings S5-7

striped legs short

Similar species: Five-striped Leaftail has additional dark stripe along rear lateral margin of thorax; complete pale ring only on S7. **Greater Forceptails** have dark stripe at rear lateral margin of thorax; abdomen weakly ringed, lacking pale dorsal stripes. Legs unstriped. **Ringtails** smaller; abdomen lacking flanges. **Flag-tailed Spinylegs** has longer hind legs.

Five-striped Leaftail • *Phyllogomphoides albrighti*

60–63 mm, 2.4–2.5 in. Large, strongly marked clubtail found at clear rivers and large streams, primarily in Texas. Similar to Four-striped Leaftail, but thorax darker, with dark stripe along lower rear margin, making five stripes visible in side view. Pale rings on abdomen broken, usually only S7 ring complete. Both sexes with wide abdominal flanges.

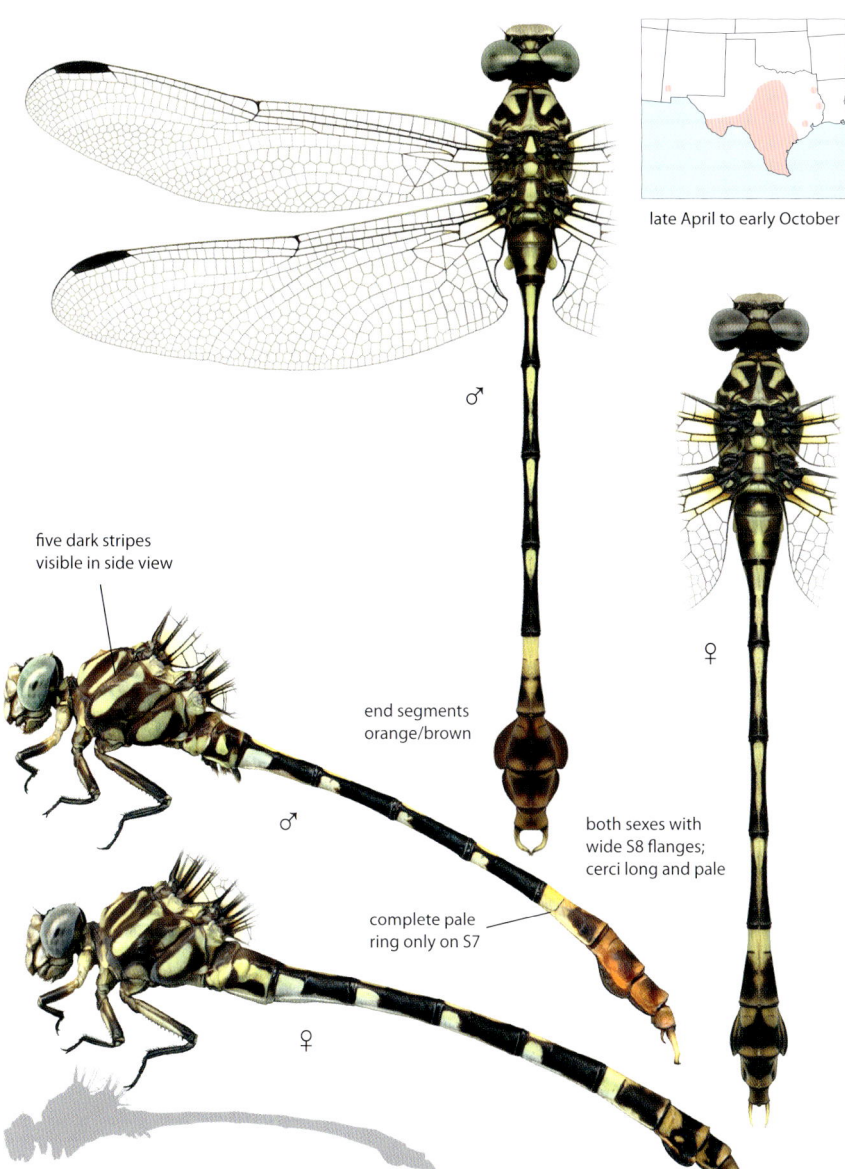

late April to early October

five dark stripes
visible in side view

end segments
orange/brown

both sexes with
wide S8 flanges;
cerci long and pale

complete pale
ring only on S7

♂

♀

♂

♀

Similar species: Four-striped Leaftail lacks dark stripe on lower rear margin of thorax; S5-7 have complete pale rings. **Greater Forceptails** have weakly ringed abdomens, lacking pale dorsal stripes; legs unstriped. **Flag-tailed Spinylegs** have longer hind legs. **Ringtails** smaller, with unflanged abdomens.

Southeastern Spinylegs • *Dromogomphus armatus*

64–74 mm, 2.5–2.9 in. Large, colorful southeastern spinylegs found at small woodland streams and seeps. Head small, face pale with dark cross-line, eyes green. Thorax yellow-green with dark stripes, including two complete lateral stripes. Abdomen dark, with dorsal pale stripes on S1-6; male moderately clubbed, wide laterally, bright orange/brown in color.

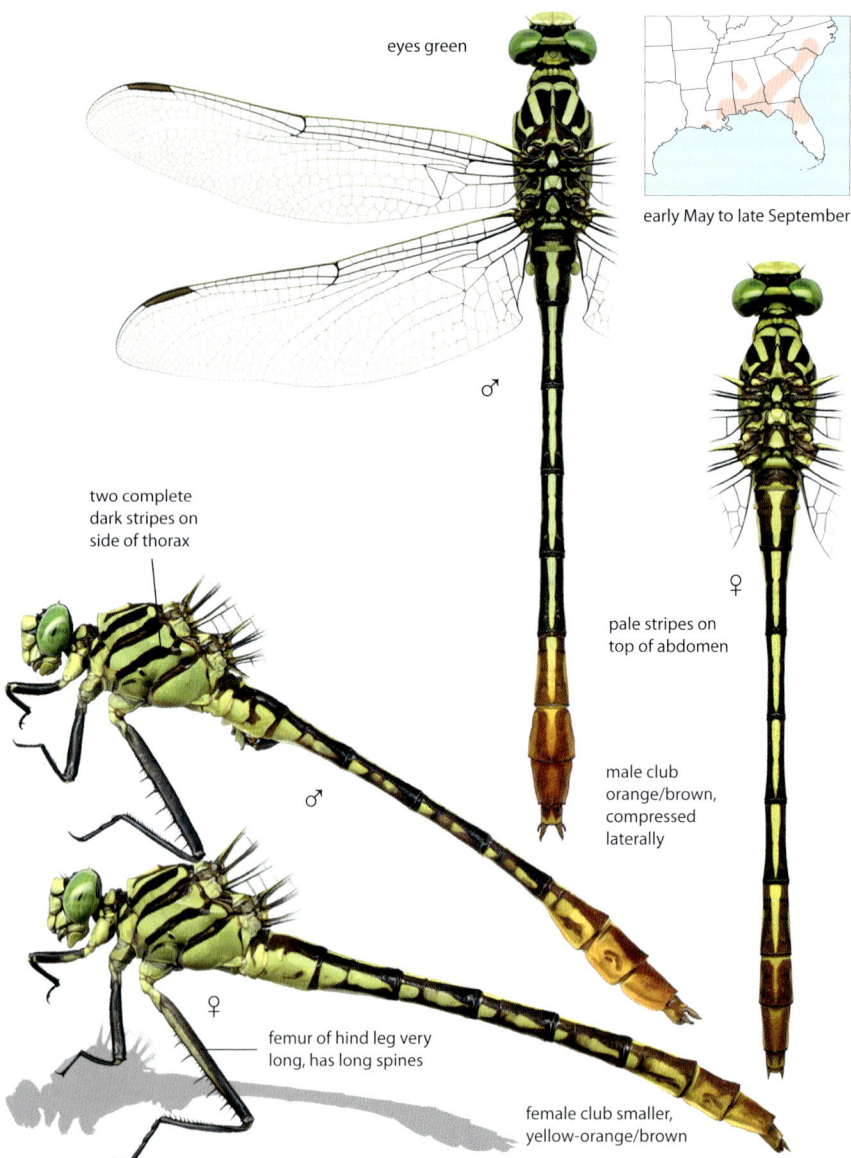

eyes green

early May to late September

two complete dark stripes on side of thorax

♂

♀

pale stripes on top of abdomen

♂

male club orange/brown, compressed laterally

♀

femur of hind leg very long, has long spines

female club smaller, yellow-orange/brown

Similar species: Flag-tailed Spinylegs smaller, with blue eyes; face unstriped. Abdomen with pale rings on S4-6. **Cocoa Clubtail** duller in pattern. Legs brown, without long spines on hind leg femur; club flatter, not compressed laterally. **Russet-tipped Clubtail** has short legs, smaller club. **Two-striped Forceptail** thorax mostly dark brown; lacks pale stripes on top of abdomen.

Flag-tailed Spinylegs • *Dromogomphus spoliatus*

56–65 mm, 2.2–2.6 in. Colorful midwestern spinylegs inhabiting slow rivers and mud-bottomed ponds and lakes. Face pale, eyes gray-blue. Thorax yellow green with dark stripes, lateral stripes often broken. Abdomen dark with pale dorsal stripes, pale rings on S4-6. Male club large, wide and flaglike in side view, orange and brown in color.

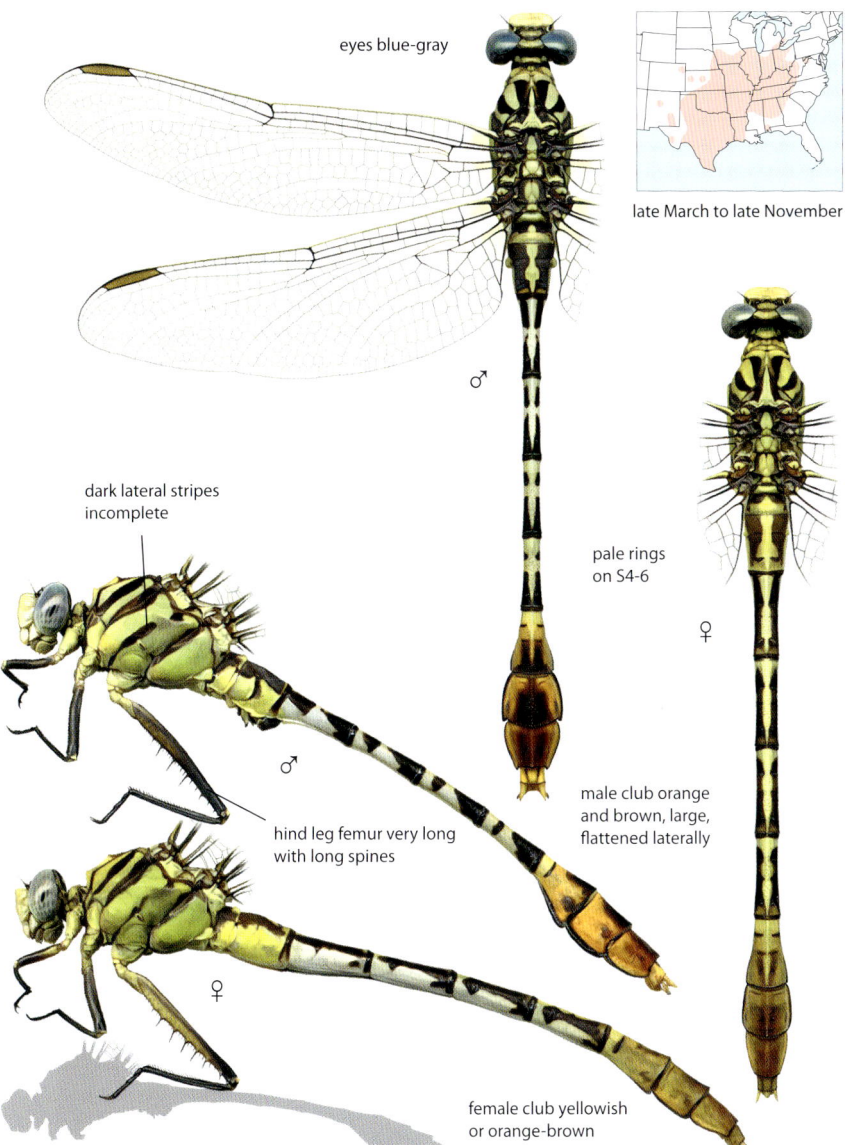

eyes blue-gray

late March to late November

dark lateral stripes incomplete

pale rings on S4-6

♂

♀

hind leg femur very long with long spines

male club orange and brown, large, flattened laterally

♂

♀

female club yellowish or orange-brown

Similar species: Southeastern Spinylegs larger, has green eyes, brown facial stripe, abdomen without pale rings. Species with orange-brown clubs, including **forceptails, leaftails, ringtails, Sulphur-tipped, Russet-tipped, Stillwater,** and **Jade Clubtails,** have shorter legs; smaller clubs not compressed laterally. **Cocoa Clubtail** duller in pattern. Legs brown, lacking long spines on hind leg femur; abdomen shorter with flatter club.

Black-shouldered Spinyleg • *Dromogomphus spinosus*

53–68 mm, 2.1–2.7 in. Common eastern spinyleg found at streams and rivers, occasionally at well-oxygenated rocky lakes. Head small, face pale, eyes green. Thorax pale with wide dark shoulder stripes. Slender abdomen black with narrow middorsal pale stripes and markings on top of all segments. Male moderately clubbed, club mostly dark.

eyes green

north: early Jun to mid-Sep
south: mid-Apr to late Nov

♂

thorax with prominent, wide, dark shoulder stripes, no lateral stripes

♂

♀

abdomen long, slender with pale dorsal stripes

♀

legs black; hind leg femur very long and armed with long spines

Similar species: Majestic clubtails (genus *Gomphurus*) are similar in size and have long legs but lack large spines on hind leg femur, and have larger abdominal clubs; most species with dark lateral thoracic stripes.

Horned Clubtail • *Arigomphus cornutus*

55–57 mm, 2.1–2.2 in. Northern pond clubtail inhabiting lakes, ponds, and slow streams. Both sexes with large plate-like occiput, female with pair of long occipital horns. Eyes blue. Thorax lightly marked with two dark frontal stripes and pair of shoulder stripes. Abdomen with dorsal pale stripes on S1-7, pale streak on S8; sides of S8-9 marked with orange.

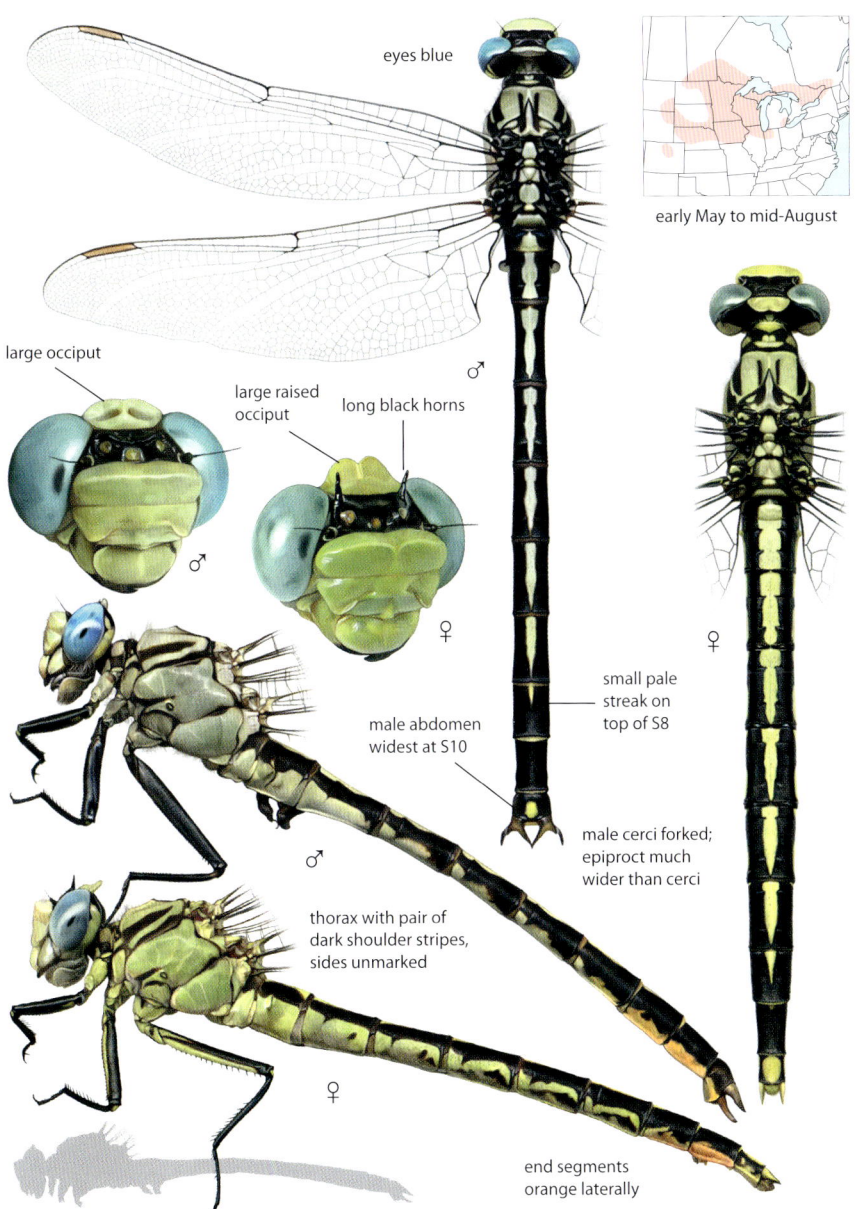

eyes blue

early May to mid-August

large occiput

large raised occiput

long black horns

♂

male abdomen widest at S10

small pale streak on top of S8

male cerci forked; epiproct much wider than cerci

♀

♀

♂

thorax with pair of dark shoulder stripes, sides unmarked

♀

end segments orange laterally

Similar species: Unicorn Clubtail has green eyes, smaller occiput, lacks pale streak on top of S8. S10 paler; male cerci unforked. **Lilypad Clubtail** has smaller occiput, lacks pale streak on top of S8; male appendages narrower and paler.

Unicorn Clubtail • *Arigomphus villosipes*

50–58 mm, 2–2.3 in. Common northeastern species found at muddy lakes, ponds, and slow streams. Thorax mostly pale, with only dark frontal and shoulder stripes. Eyes green. Abdomen mostly black, with pale middorsal stripes on S1-7. S10 mostly pale; cerci pale. Both sexes with touch of orange/brown on sides of S8-9. Both sexes have small occipital horn.

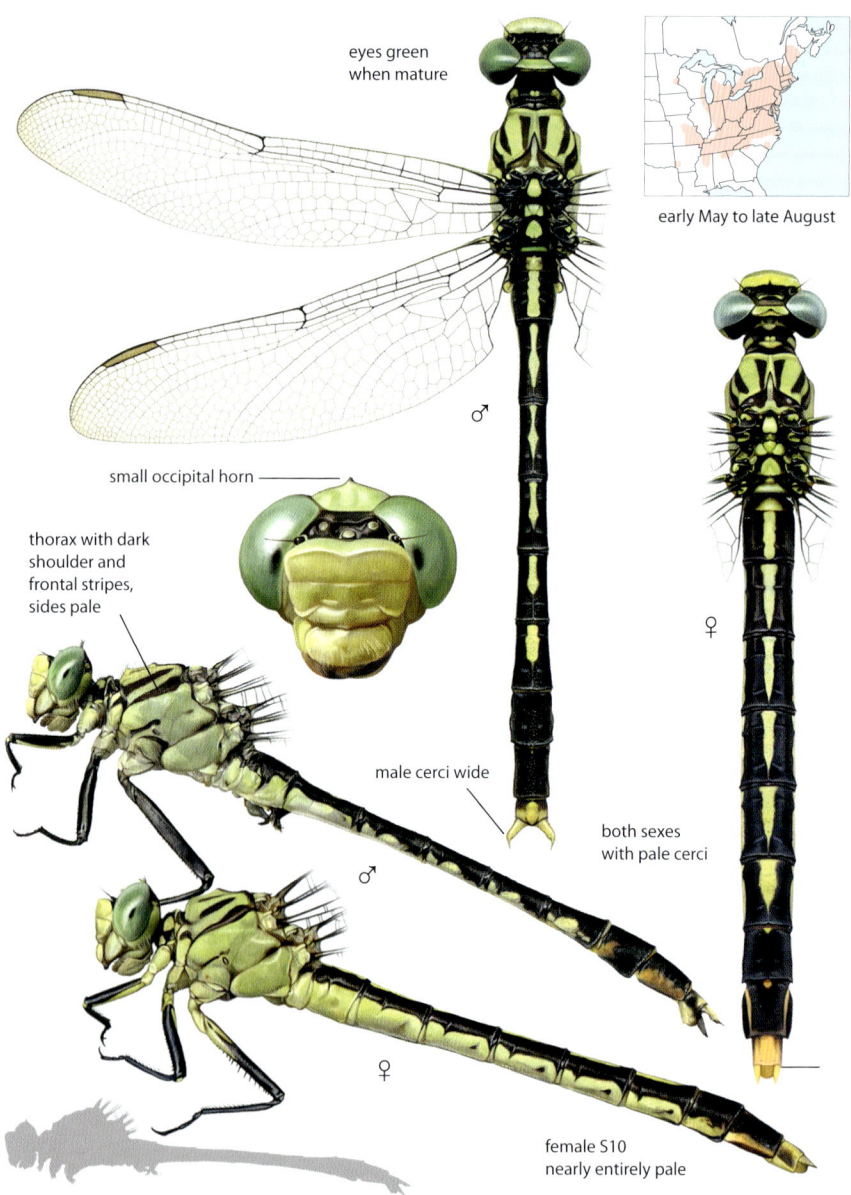

eyes green
when mature

early May to late August

small occipital horn

thorax with dark
shoulder and
frontal stripes,
sides pale

♂

male cerci wide

both sexes
with pale cerci

♀

♂

♀

female S10
nearly entirely pale

Similar species: Lilypad Clubtail most similar, has bluer eyes when mature. Females with black markings on S10; males have forked cerci. **Horned Clubtail** has blue eyes, larger occiput, often more dark markings on S10. Male cerci branched, and epiproct very wide.

Lilypad Clubtail • *Arigomphus furcifer*

42–54 mm, 1.8–2.1 in. Northeastern pond clubtail found at ponds, lakes, slow streams with lilypads and other aquatic vegetation. Thorax lightly marked, having only dark shoulder and frontal stripes. Eyes blue to turquoise. Abdomen black with pale middorsal stripes on S1-7, pale spot on top of S10; male with sides of S7-9 orange-brown. Cerci of both sexes pale.

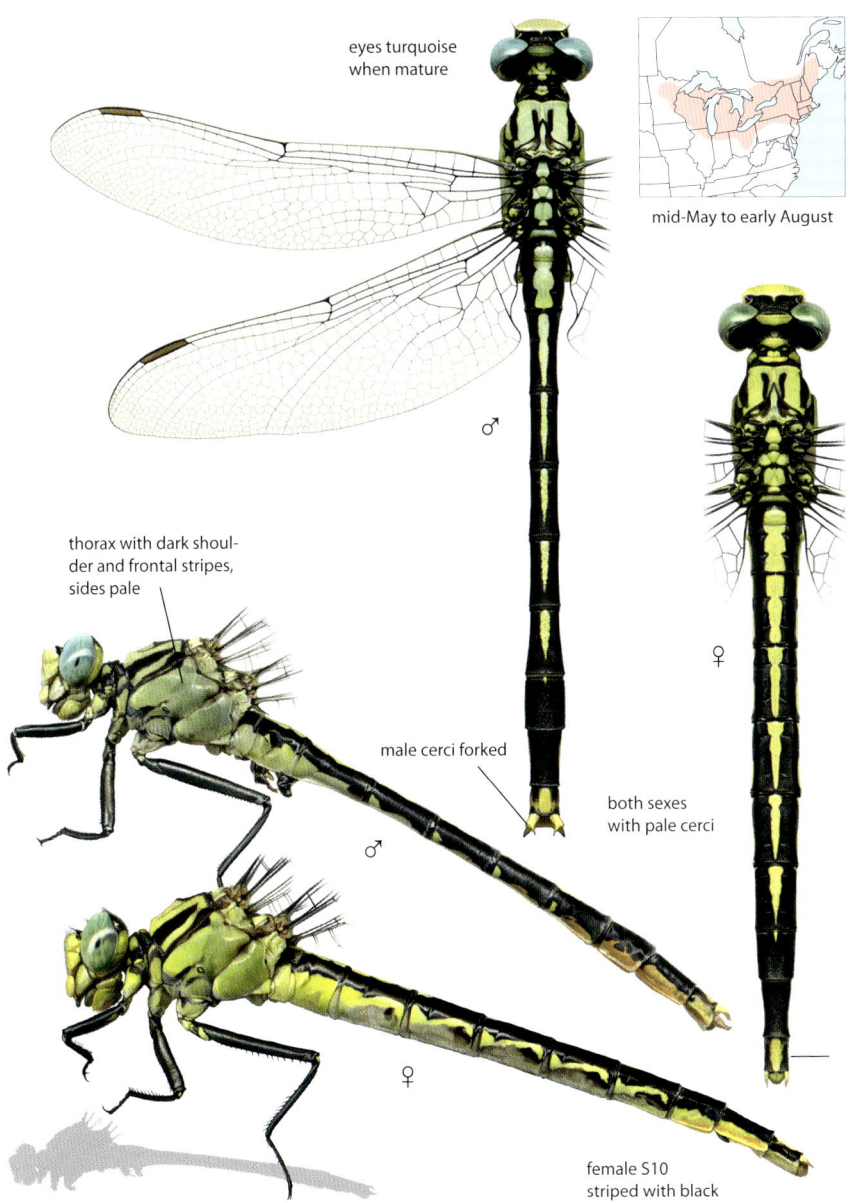

eyes turquoise
when mature

mid-May to early August

♂

thorax with dark shoul-
der and frontal stripes,
sides pale

male cerci forked

both sexes
with pale cerci

♂

♀

♀

female S10
striped with black

Similar species: Unicorn Clubtail has unforked male cerci, green eyes; female with S10 nearly entirely pale. **Horned Clubtail** male cerci wider and branched; both sexes with taller, larger occiput.

Stillwater Clubtail • *Arigomphus lentulus*

48–58 mm, 1.9–2.3 in. Pale mid-southern species with a rusty abdomen tip inhabiting open lakes and ponds. Eyes blue. Thorax pale green, lightly marked with a pair of dark brown shoulder stripes. Abdomen mostly pale, S7-10 rusty brown in color, appendages pale. Femurs mostly pale, tibiae black with pale stripes. Male epiproct longer than cerci.

eyes blue

late March to mid-August

♂

thorax with two dark
shoulder stripes

♀

♂

abdomen mostly pale,
end segments rusty
brown, cerci pale

♀

femurs mostly pale

length of S9 equal to S8

Similar species: Jade Clubtail nearly identical, has less developed dark shoulder stripes, green eyes; female S9 longer. Compare male appendages, page 396. **Bayou Clubtail** thorax more striped, and end abdominal segments dark brown in female, blackish in male.

Jade Clubtail • *Arigomphus submedianus*

51–55 mm, 2–2.2 in. Pale pond clubtail with a rusty abdomen tip inhabiting muddy ponds and lakes in the eastern Great Plains. Very similar to Stillwater Clubtail, but both sexes have only a single well-developed dark shoulder stripe. Face pale, eyes green. Male epiproct wider than cerci.

eyes green

late March to mid-August

♂

anterior shoulder stripe well developed; posterior stripe reduced or absent

♀

abdomen mostly pale, end segments rusty brown, cerci pale

♂

femurs mostly pale

♀

S9 longer than S8

Similar species: Stillwater Clubtail has two complete dark stripes at each shoulder and blue eyes. Compare male appendages, page 396. **Bayou Clubtail** thorax more extensively striped, and abdominal pattern more contrasting, with end abdominal segments darker, black in male, dark brown in female.

Bayou Clubtail • *Arigomphus maxwelli*

50–54 mm, 2–2.1 in. Mid-southern clubtail found at ponds, bayous, and slow streams. Found more often at running water than other pond clubtails. Eyes green. Thorax mostly pale with narrow black midfrontal, shoulder, and lateral stripes. Abdomen mostly pale, male ringed in appearance. S7-9 entirely dark, blackish in male. S10 and appendages pale.

eyes green

late March to early July

thorax with dark shoulder and frontal stripes, lateral stripes often broken

middle segments contrastingly pale, often ringed in appearance

S7-9 entirely dark, blackish in male, dark brown in female

cerci pale

Similar species: Jade Clubtail largely lacks posterior dark shoulder and lateral thoracic stripes; end abdominal segments rusty brown. **Stillwater Clubtail** largely lacks lateral thoracic stripes. End abdominal segments rusty brown, eyes blue. **Gray-green Clubtail** has plain, less contrasting pattern on thorax and abdomen.

Gray-green Clubtail • *Arigomphus pallidus*

60–62 mm, 2.4 in. Southeastern pond clubtail with distinctive dull pattern and coloration, found at ponds, lakes, and slow streams. Eyes bright green. Thorax gray-green, with faint brown frontal and shoulder stripes; stripes may be reduced or missing. Abdomen brown, with low-contrast pale lateral and middorsal stripes on S1-7. Cerci of both sexes pale.

eyes bright green, contrasting with dully colored body

mid-March to early October

♂

male abdomen appears ringed

thorax gray-green, dully patterned

pale mark on S7

♀

top of S10 and appendages pale

♂

♀

Similar species: Bayou and **Jade Clubtails** have more contrasting patterns; top of S7 dark. **Stillwater Clubtail** has blue eyes, more contrasting pattern.

Spine-crowned Clubtail • *Hylogomphus abbreviatus*

34–35 mm, 1.3–1.4 in. Brightly patterned bantam clubtail found at clean rivers with rock and mud, in the Northeast and Appalachia. Face unmarked. Thorax greenish-yellow with dark stripes, anterior lateral stripe incomplete, posterior lateral stripe complete. Abdomen with small dorsal yellow markings on S1-7; top of S8-10 black. Large yellow lateral markings on S8-9.

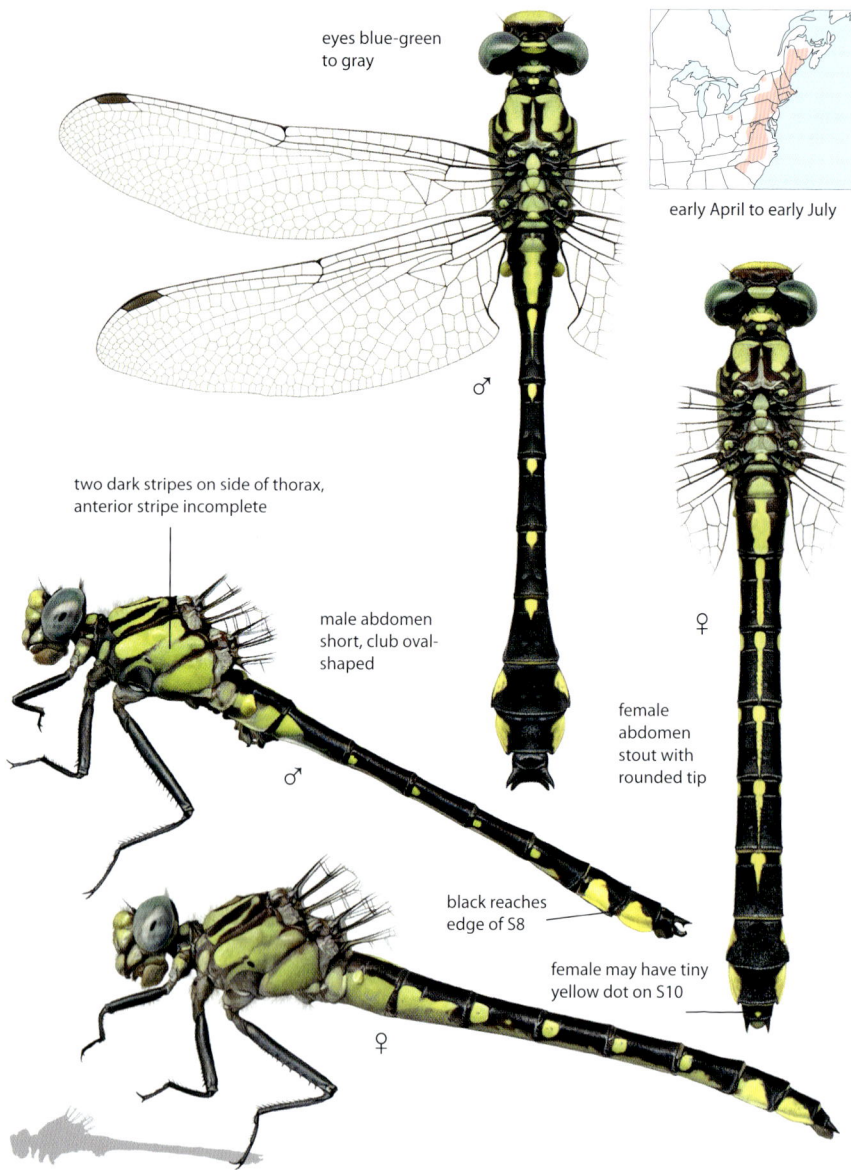

eyes blue-green
to gray

early April to early July

two dark stripes on side of thorax,
anterior stripe incomplete

male abdomen
short, club oval-
shaped

♂

♀

female
abdomen
stout with
rounded tip

black reaches
edge of S8

female may have tiny
yellow dot on S10

♀

Similar species: Banner Clubtail has darker face, two complete lateral thoracic stripes, with area between them darkened. Northern form of **Piedmont Clubtail** largely identical but usually has no black on the edge of S8; female lacks yellow spot on S10. Also compare male appendages and female subgenital plates, pages 397–99.

Banner Clubtail • *Hylogomphus apomyius*

35–37 mm, 1.4–1.5 in. Small, chunky bantam clubtail found at clean, sand-bottomed, acidic streams, primarily southeastern. Face brownish green, eyes blue. Thorax yellow-green with dark stripes, lateral stripes complete, with area between them darkened. Abdomen with small dorsal marks on S1-7, top of S8-10 black. Large yellow lateral markings on S8-9.

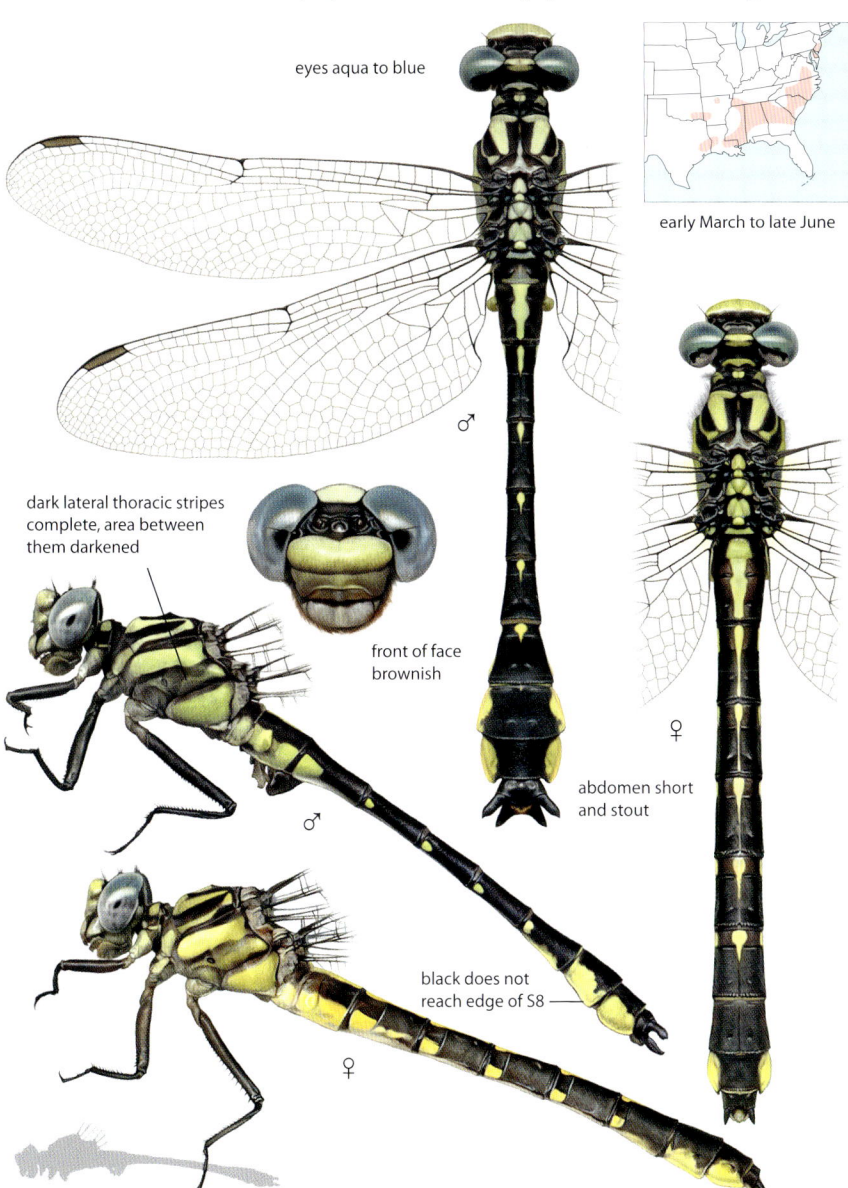

eyes aqua to blue

early March to late June

dark lateral thoracic stripes
complete, area between
them darkened

front of face
brownish

abdomen short
and stout

black does not
reach edge of S8

Similar species: Spine-crowned Clubtail has paler face, incomplete anterior lateral thoracic stripe; black reaches edge of S8. Female often with pale dot on top of S10. **Piedmont Clubtail** northern form has incomplete anterior lateral thoracic stripe. Southern form has dark facial line, area between lateral thoracic stripes not darkened.

Twin-striped Clubtail • *Hylogomphus geminatus*

39–47 mm, 1.5–1.9 in. Bantam clubtail found at sandy woodland streams in the Florida Panhandle area. Face with black cross-stripe; eyes aqua-blue to green. Thorax pale yellow-green with dark stripes, both lateral stripes complete. Male club wide. Abdomen black, with small dorsal markings on S1-7; top of S8-10 black. Large yellow lateral markings on S8-9.

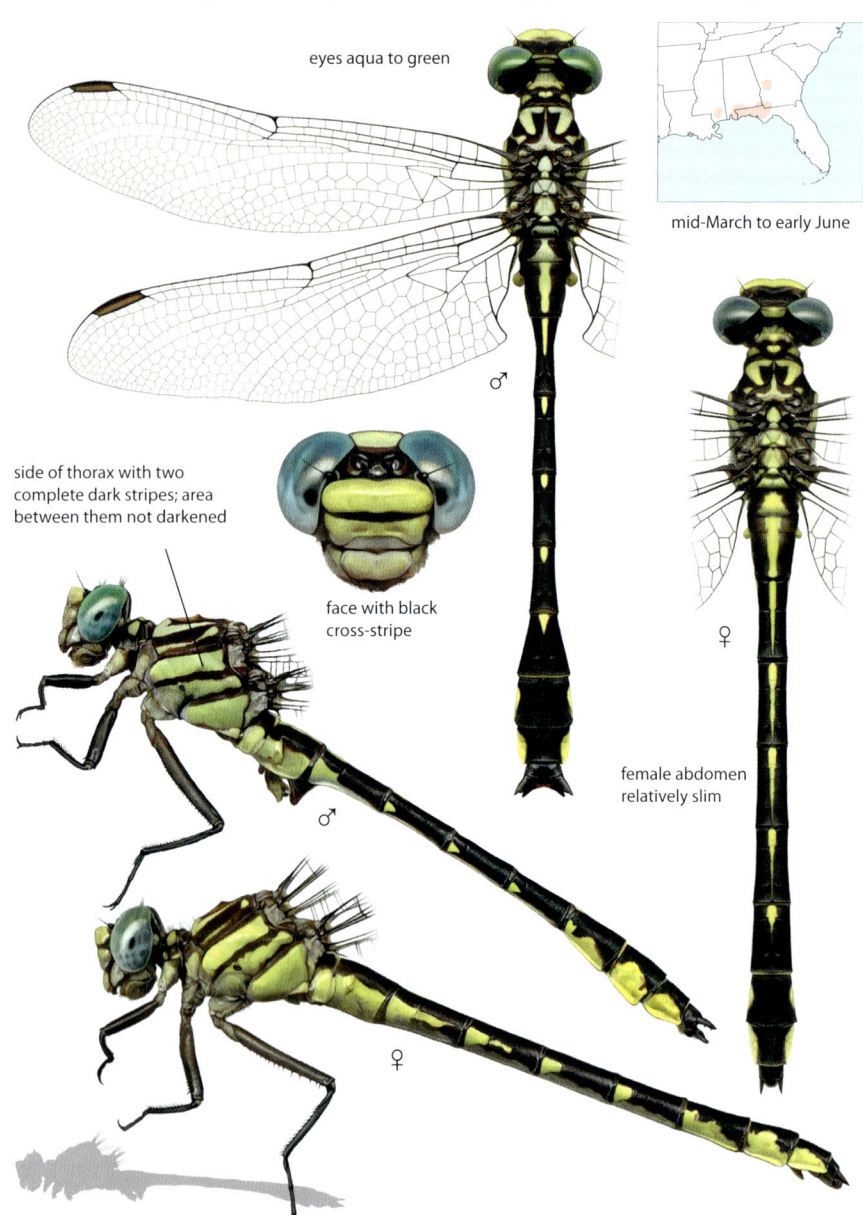

eyes aqua to green

mid-March to early June

side of thorax with two complete dark stripes; area between them not darkened

face with black cross-stripe

♂

♀

female abdomen relatively slim

♂

♀

Similar species: Piedmont Clubtail southern form identical, but range not known to overlap. Compare in-hand structures, pages 398–99. **Banner Clubtail** lacks black face stripe; dark lateral thoracic stripes complete, with area between them darkened.

Piedmont Clubtail • *Hylogomphus parvidens*

39–46 mm, 1.5–1.8 in. Southeastern bantam clubtail inhabiting clean woodland streams with sand and silt bottoms. Northern form has unstriped face, green eyes, thorax with incomplete anterior lateral stripe. Southern form has a dark facial stripe, blue-green eyes, thorax with two complete lateral stripes.

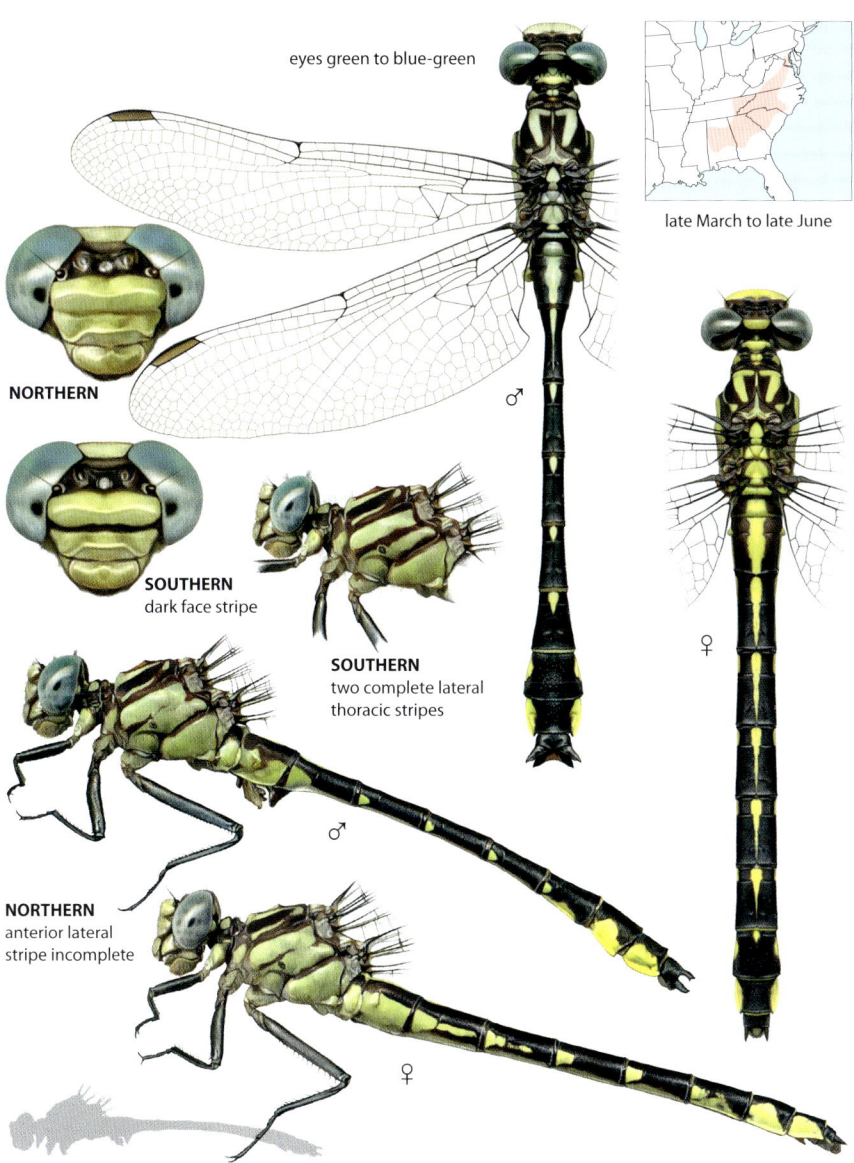

eyes green to blue-green

late March to late June

NORTHERN

SOUTHERN
dark face stripe

SOUTHERN
two complete lateral
thoracic stripes

♂

♀

♂

NORTHERN
anterior lateral
stripe incomplete

♀

Similar species: Twin-striped Clubtail identical to southern-form Piedmont, but range not known to overlap. Compare in-hand structures, pages 398–99. **Spine-crowned Club-tail** similar to northern-form Piedmont. Female often with yellow dot on S10; male occiput more convex. Compare male appendages, female subgenital plate, pages 397, 399.

Mustached Clubtail • *Hylogomphus adelphus*

44–46 mm, 1.7–1.8 in. Common dark bantam clubtail found at clean, rocky rivers and streams, in the Northeast and Appalachia. Face with black stripes, eyes green. Thorax green-gray, may have one or two complete black lateral stripes. Male abdomen almost entirely black both dorsally and laterally. Pale markings small but more extensive on female.

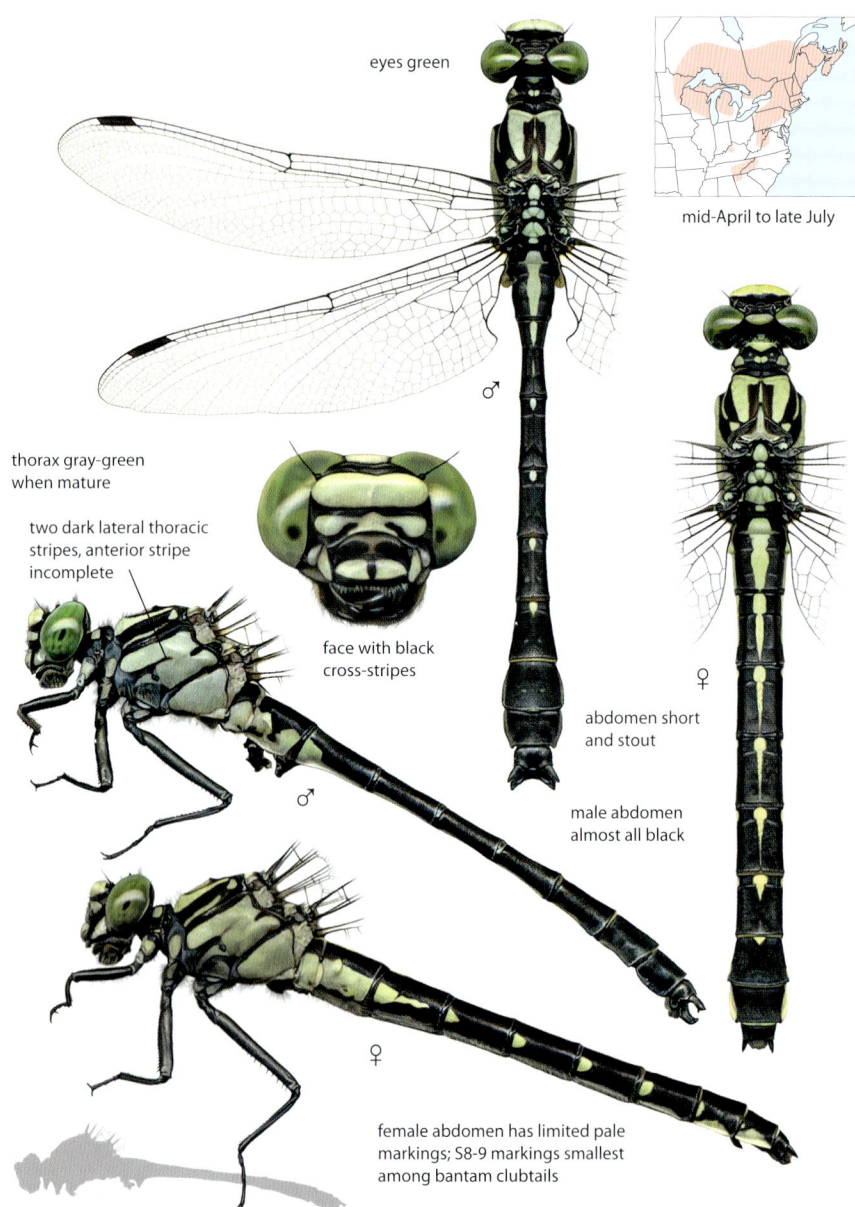

eyes green

mid-April to late July

thorax gray-green when mature

two dark lateral thoracic stripes, anterior stripe incomplete

face with black cross-stripes

♂

♀

abdomen short and stout

male abdomen almost all black

female abdomen has limited pale markings; S8-9 markings smallest among bantam clubtails

Similar species: Gray-green coloration distinctive among bantam clubtails except for **Green-faced Clubtail,** which lacks dark facial stripes and largely lacks dark lateral thoracic stripes. Male has pale lateral spots on S4-7; female with larger yellow spots on side of S8-9.

Green-faced Clubtail • *Hylogomphus viridifrons*

45–46 mm, 1.8 in. Gray-green bantam clubtail inhabiting rocky rivers and large streams in the East and Appalachia. Face pale green, eyes green. Thorax gray-green; female often yellower, dark lateral stripes reduced or absent. Male abdomen with small pale lateral markings, end segments nearly entirely black. Female with larger lateral abdominal markings.

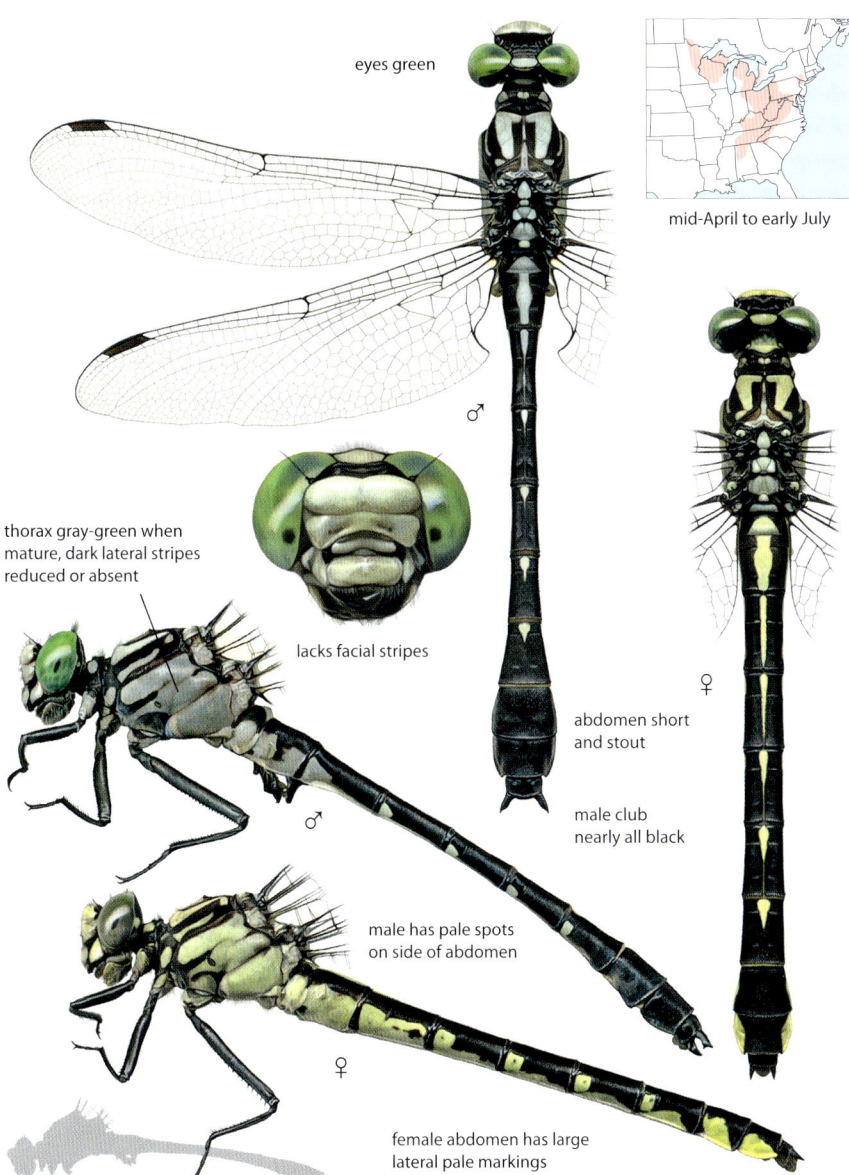

eyes green

mid-April to early July

thorax gray-green when mature, dark lateral stripes reduced or absent

lacks facial stripes

♂

♀

abdomen short and stout

male club nearly all black

male has pale spots on side of abdomen

♂

♀

female abdomen has large lateral pale markings

Similar species: Green-gray coloration distinctive among bantam clubtails except for **Mustached Clubtail,** which has dark facial and lateral thoracic stripes. Male has sides of S4-7 black; female abdomen has smaller pale lateral markings. Other similar clubtails have dark lateral thoracic stripes, males with yellow lateral markings on club.

Eastern Least Clubtail • *Stylogomphus albistylus*

31–36 mm, 1.2–1.4 in. Very small, widespread eastern clubtail found at rocky riffles of clear streams and small rivers. Thorax yellow to pale green with black stripes and isolated pale frontal stripes. Abdomen black with narrow pale yellow rings at the bases of S4-7, rings sometimes broken at top; cerci pale. Eyes green. Legs mostly black.

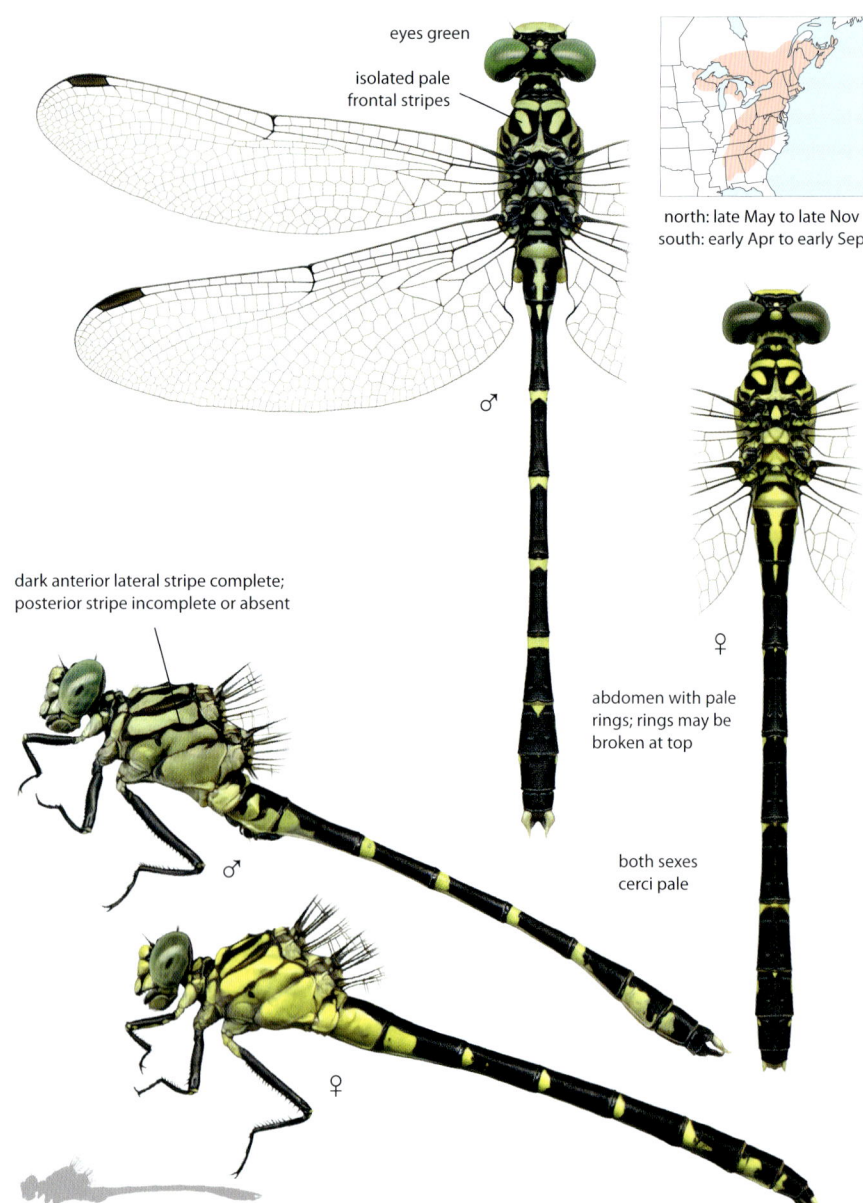

eyes green

isolated pale frontal stripes

north: late May to late Nov
south: early Apr to early Sep

♂

♀

dark anterior lateral stripe complete; posterior stripe incomplete or absent

♂

abdomen with pale rings; rings may be broken at top

both sexes cerci pale

♀

Similar species: Interior Least Clubtail identical in pattern (see male appendages, page 412). **Pygmy clubtails** similar in size, but abdomens lack pale rings and cerci are black.

Interior Least Clubtail • *Stylogomphus sigmastylus*

35–38 mm, 1.4–1.5 in. Very small, delicate clubtail inhabiting small clean streams and rivers; limited range in the central US. Pattern identical to Eastern Least Clubtail. Thorax yellow to pale green with black stripes and isolated pale frontal stripes. Abdomen with narrow pale rings; rings may be broken at top. Cerci pale. Eyes green. Legs mostly black.

mid-May to early August

Similar species: Eastern Least Clubtail identical in pattern (see male appendages, page 412). In areas of range overlap, intermediate forms have been found, suggesting the two species hybridize.

Northern Pygmy Clubtail • *Lanthus parvulus*

33–40 mm, 1.3–1.6 in. Very small, dark northeastern clubtail found at rocky forest streams, rivers in the North, mountain streams southward. Thorax mostly black above with isolated pale frontal stripes and two black lateral stripes usually joined at middle. Male abdomen nearly entirely black, female lightly marked. Eyes green. Legs and appendages black.

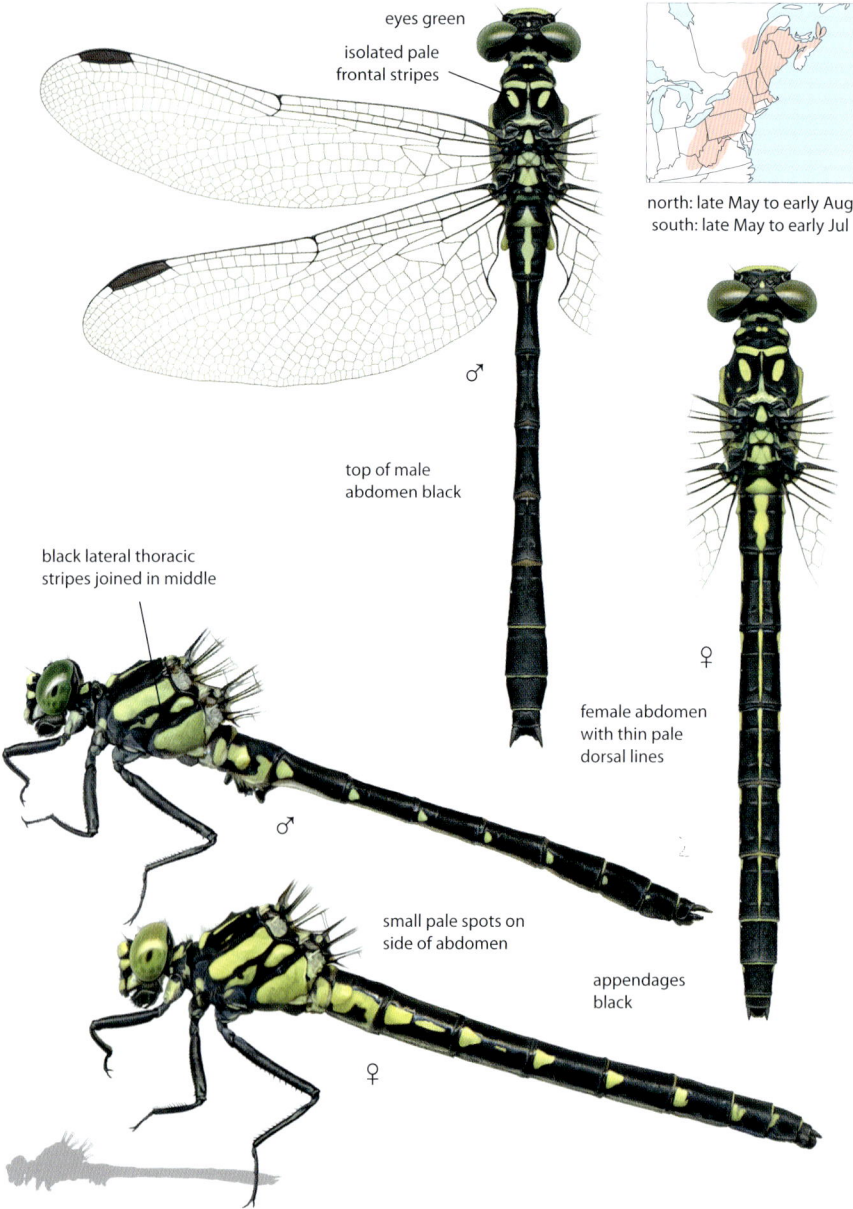

eyes green

isolated pale frontal stripes

north: late May to early Aug
south: late May to early Jul

top of male abdomen black

black lateral thoracic stripes joined in middle

♂

♂

female abdomen with thin pale dorsal lines

♀

small pale spots on side of abdomen

appendages black

♀

Similar species: Southern Pygmy Clubtail has single black lateral thoracic stripe. **Least clubtails** have pale rings on abdomen and pale appendages.

Southern Pygmy Clubtail • *Lanthus vernalis*

29–40 mm, 1.1–1.6 in. Very small, dark clubtail inhabiting clean northeastern and Appalachian forest streams and rivers. Thorax mostly black above with isolated pale frontal stripes; sides pale with single black lateral stripe. Male abdomen nearly completely black, female with pale lateral markings and dorsal lines. Eyes green. Legs and appendages black.

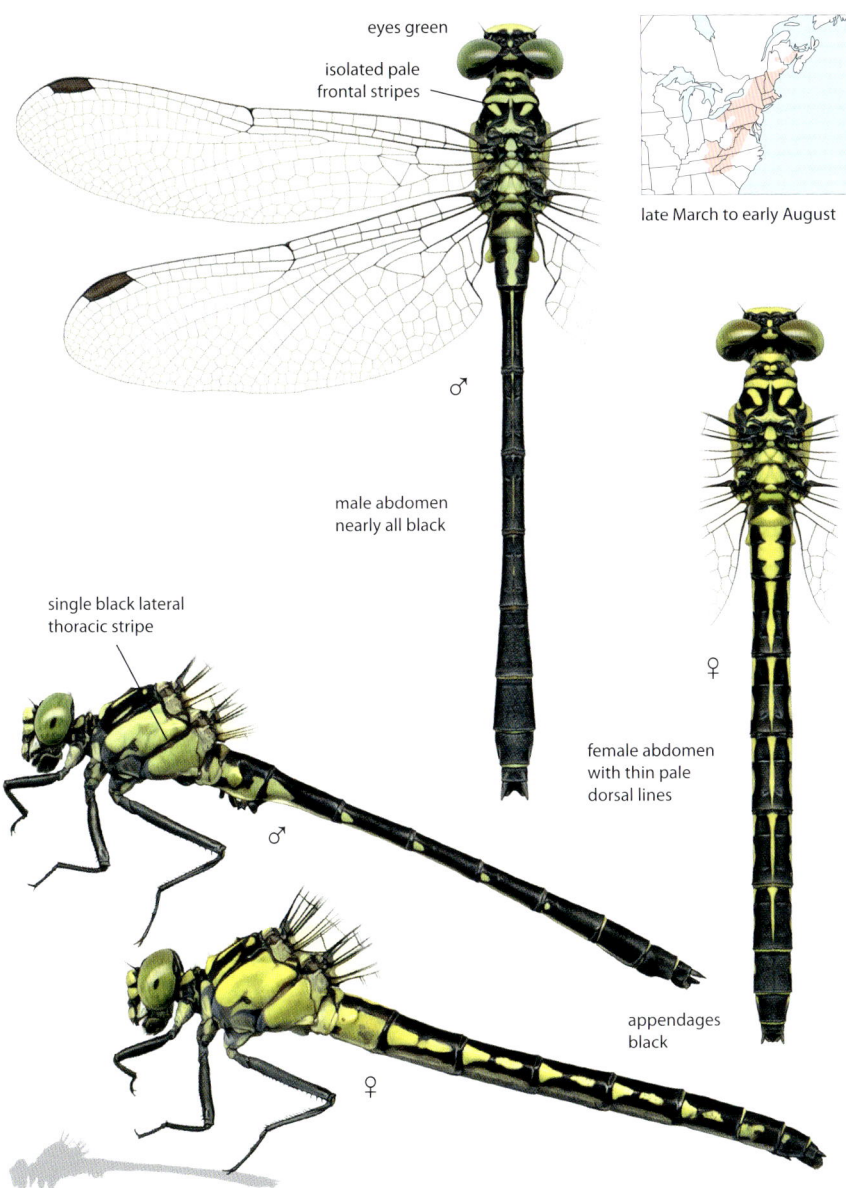

eyes green

isolated pale frontal stripes

late March to early August

♂

male abdomen nearly all black

single black lateral thoracic stripe

♂

♀

female abdomen with thin pale dorsal lines

appendages black

♀

Similar species: Northern Pygmy Clubtail has two black lateral thoracic stripes. **Least clubtails** have pale rings on abdomen and pale appendages.

Grappletail • *Octogomphus specularis*

51–53 mm, 2–2.1 in. Distinctive West Coast clubtail found at woodland, mountain, and lake outflow streams. Thorax pattern unique, with wide black shoulder stripes, pale area at front forming an urn shape. Abdomen black with pale dorsal hairline stripes. Male abdomen slightly clubbed, widest at S10, with wide cerci and four-branched epiproct.

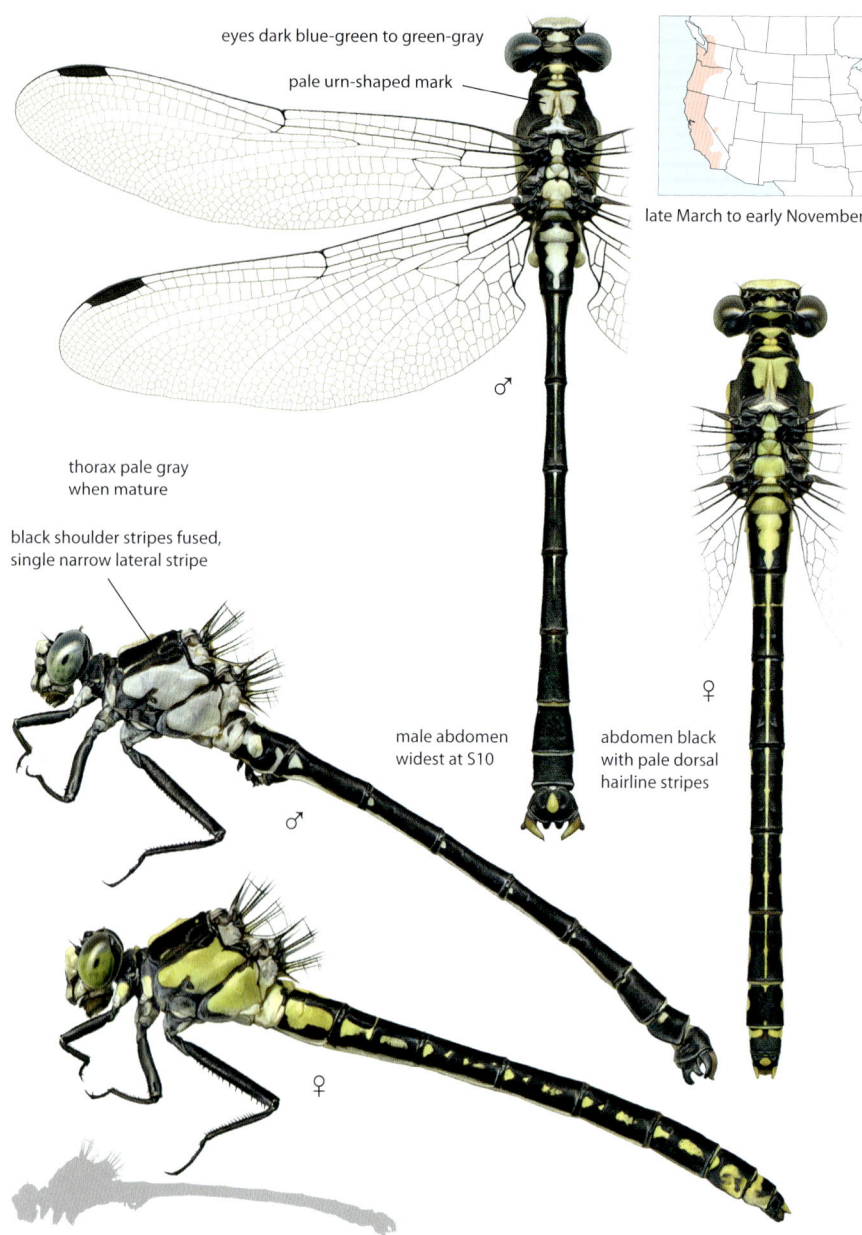

eyes dark blue-green to green-gray

pale urn-shaped mark

late March to early November

thorax pale gray when mature

black shoulder stripes fused, single narrow lateral stripe

male abdomen widest at S10

abdomen black with pale dorsal hairline stripes

♂

♀

Similar species: Pacific Clubtail has dark midfrontal thoracic stripes, abdomen with larger pale markings.

Rapids Clubtail • *Phanogomphus quadricolor*

42–45 mm, 1.7–1.8 in. Small, relatively bright-colored eastern species found at flowing rock- and mud-bottomed rivers and large streams. Area between lateral thoracic stripes usually darkened. Abdomen of both sexes mostly black with minimal pale markings: small yellow spot on side of S7, larger yellow markings on edges of S8-9. Legs black. Eyes blue.

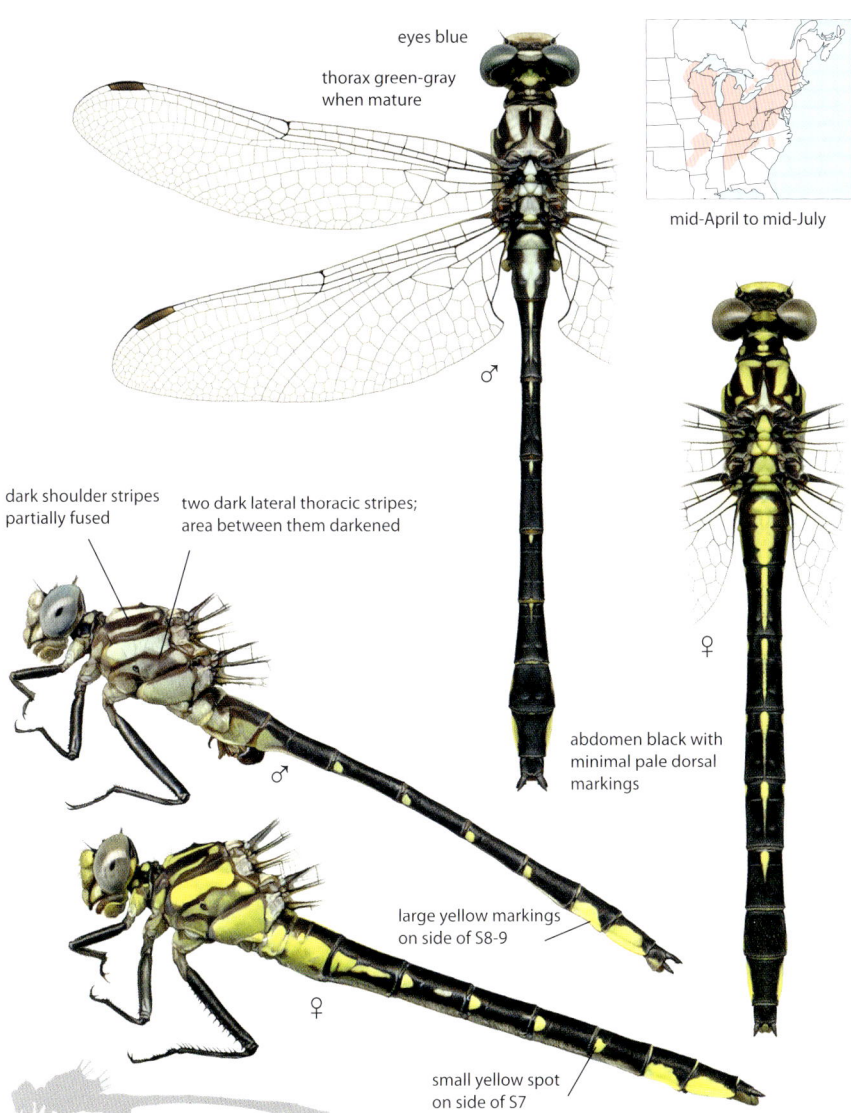

eyes blue

thorax green-gray when mature

mid-April to mid-July

♂

dark shoulder stripes partially fused

two dark lateral thoracic stripes; area between them darkened

♀

abdomen black with minimal pale dorsal markings

♂

large yellow markings on side of S8-9

♀

small yellow spot on side of S7

Similar species: Harpoon and **Beaverpond Clubtails** most similar, but larger; area between dark lateral stripes not darkened. Males with less prominent yellow on the sides of S8 and 9, females with yellow stripe on side of S7. **Lancet, Tennessee, Sulphur-tipped,** and **Pronghorn Clubtails** have yellow on top of S9. **Ashy** and **Dusky Clubtails** duller in pattern, with dark stripe at lateral rear margin of thorax. **Bantam clubtails** are stockier, with wider clubs; female abdomen rounded at tip, with S9 shorter than S8.

Beaverpond Clubtail • *Phanogomphus borealis*

44–49 mm, 1.7–1.9 in. Northeastern clubtail inhabiting mud-bottomed lakes and ponds, including beaver ponds and associated streams. Thorax with wide pale frontal stripes roughly triangular in shape; dark anterior lateral stripe narrow but complete. Abdomen with pale dorsal streaks on S1-7, sometimes small dorsal mark at base of S8. Legs black. Eyes gray-blue.

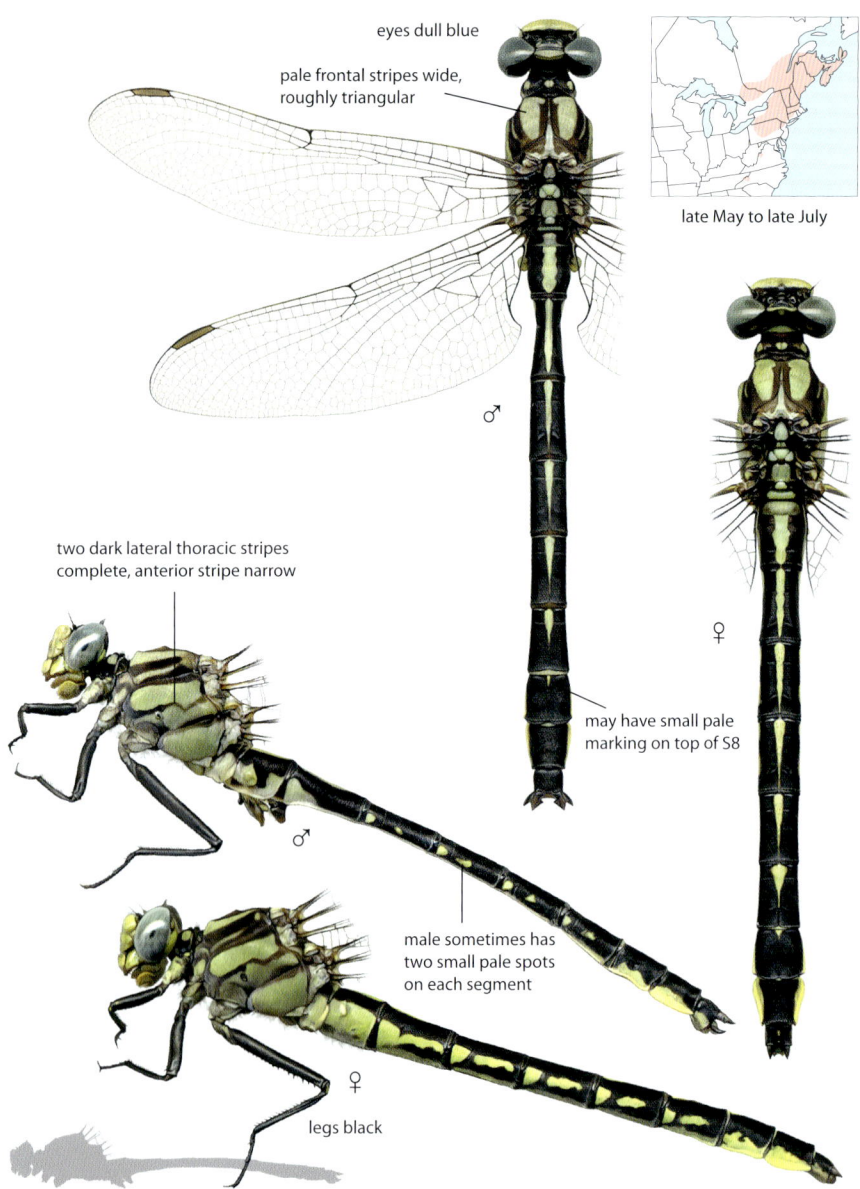

eyes dull blue

pale frontal stripes wide,
roughly triangular

late May to late July

two dark lateral thoracic stripes
complete, anterior stripe narrow

♂

♀

may have small pale
marking on top of S8

♂

male sometimes has
two small pale spots
on each segment

♀

legs black

Similar species: Harpoon Clubtail has greener eyes, pale frontal stripes with hooklike extension; female with pale hind leg femur. Compare male cerci, page 400. **Dusky** and **Ashy Clubtails** duller, have dark stripe at lower rear margin of thorax. **Rapids Clubtail** smaller; area between lateral thoracic stripes darkened. Female with only small lateral spots on S5-7.

Harpoon Clubtail • *Phanogomphus descriptus*

48–52 mm, 1.9–2 in. Northeastern and Appalachian clubtail found at running streams and rivers with silt or sand bottoms. Dark anterior lateral thoracic stripe often reduced or incomplete. Abdomen with narrow pale dorsal markings on S1-7, top of S8-9 black, male slightly clubbed. Legs black except for female's hind leg femur. Eyes gray-green.

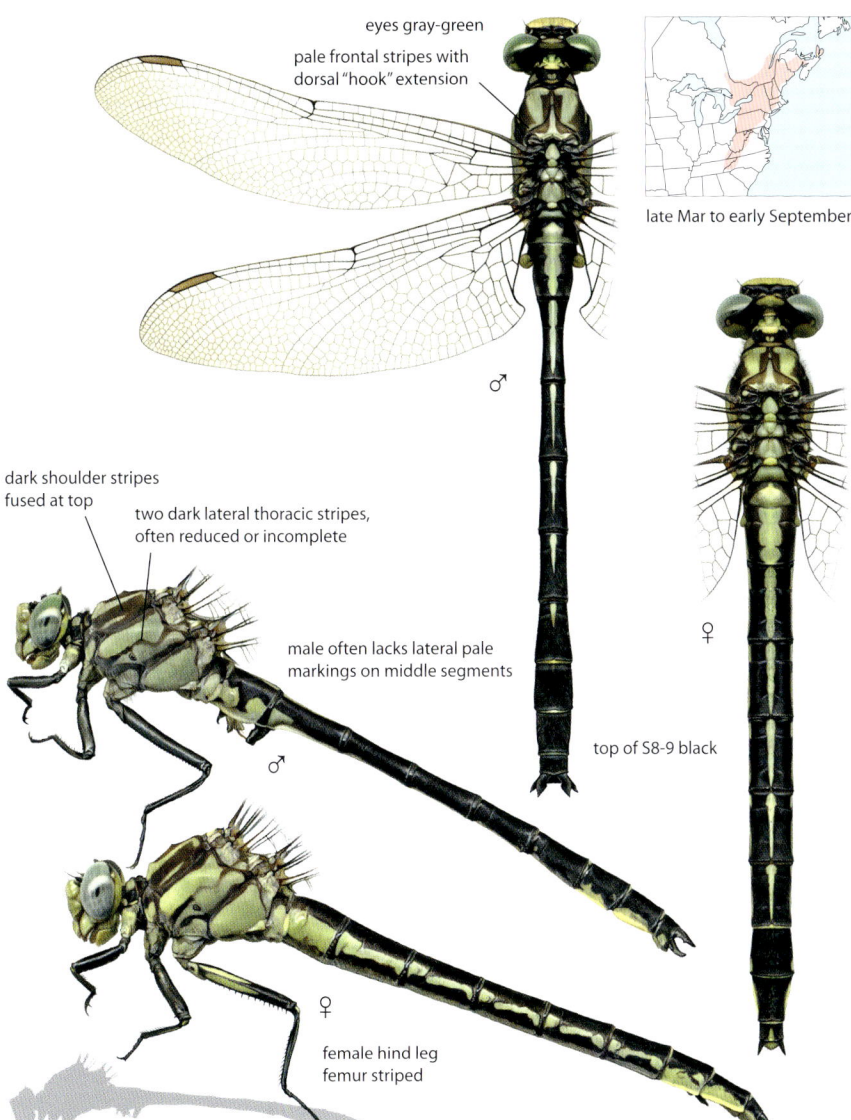

eyes gray-green

pale frontal stripes with dorsal "hook" extension

late Mar to early September

♂

dark shoulder stripes fused at top

two dark lateral thoracic stripes, often reduced or incomplete

male often lacks lateral pale markings on middle segments

♂

♀

top of S8-9 black

♀

female hind leg femur striped

Similar species: Beaverpond Clubtail has bluer eyes; pale frontal stripes largely lack dorsal "hook." Male may have double yellow lateral spots on S-7; female legs black. Compare male cerci, page 400. **Dusky Clubtail** has dark stripe on lateral rear margin of thorax, pale marking on top of S8, pale-striped tibiae. **Rapids Clubtail** smaller, area between dark lateral thoracic stripes darkened, eyes bluer. Female with only small lateral spots on S5-7. **Sable** and **Cherokee Clubtails** have black facial stripes; female with black hind leg.

Lancet Clubtail • *Phanogomphus exilis*

29–48 mm, 1.5–1.8 in. The most common eastern clubtail; widespread at slow streams, ponds, and sand-bottomed lakes. Abdomen slender, with yellow stripes on top of all segments. Side of thorax with two pale stripes, dark middle stripes fused, dark stripe along lower rear margin. Eyes dull aqua-blue in males, gray-green in females.

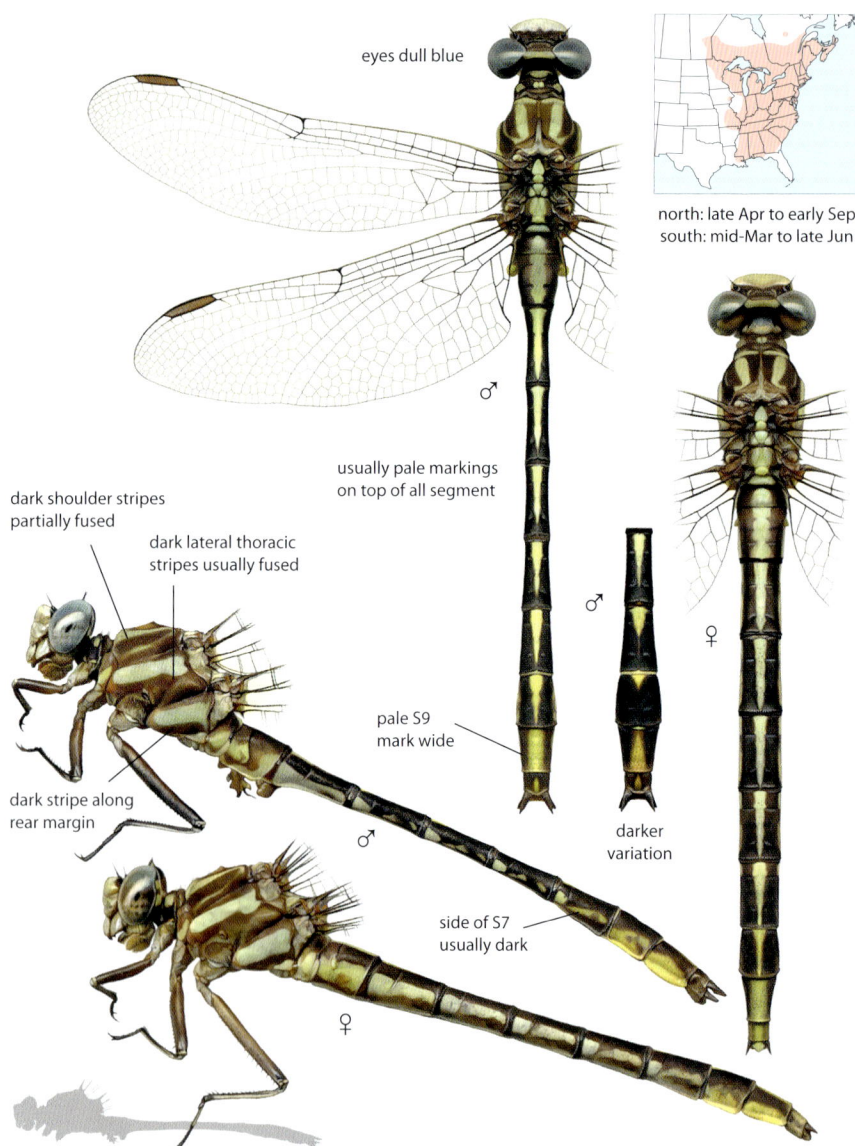

eyes dull blue

north: late Apr to early Sep
south: mid-Mar to late Jun

♂

usually pale markings
on top of all segment

dark shoulder stripes
partially fused

dark lateral thoracic
stripes usually fused

pale S9
mark wide

dark stripe along
rear margin

♂

♀

darker
variation

side of S7
usually dark

♂

♀

Similar species: Oklahoma Clubtail largely identical but not known to share range. Compare appendages, page 401. **Ashy Clubtail** browner; top of S9 usually dark. **Dusky Clubtail** has top of S9 black. **Cypress Clubtail** thorax has unfused dark shoulder and lateral stripes, lacks dark stripe at rear margin of thorax; side of S7 mostly yellow. **Harpoon** and **Beaverpond Clubtails** lack dark stripe at rear margin of thorax; lateral stripes unfused. Top of S8-9 dark.

Oklahoma Clubtail • *Phanogomphus oklahomensis*

46–49 mm, 1.8–1.9 in. Small, slender, dully colored clubtail found at mud-bottomed ponds and lakes, also slow streams. Common in Arkansas, Louisiana, eastern Texas/Oklahoma. Side of thorax with two dark stripes, the area between them darkened; dark stripe along lower rear margin. Pale markings on top of all abdominal segments.

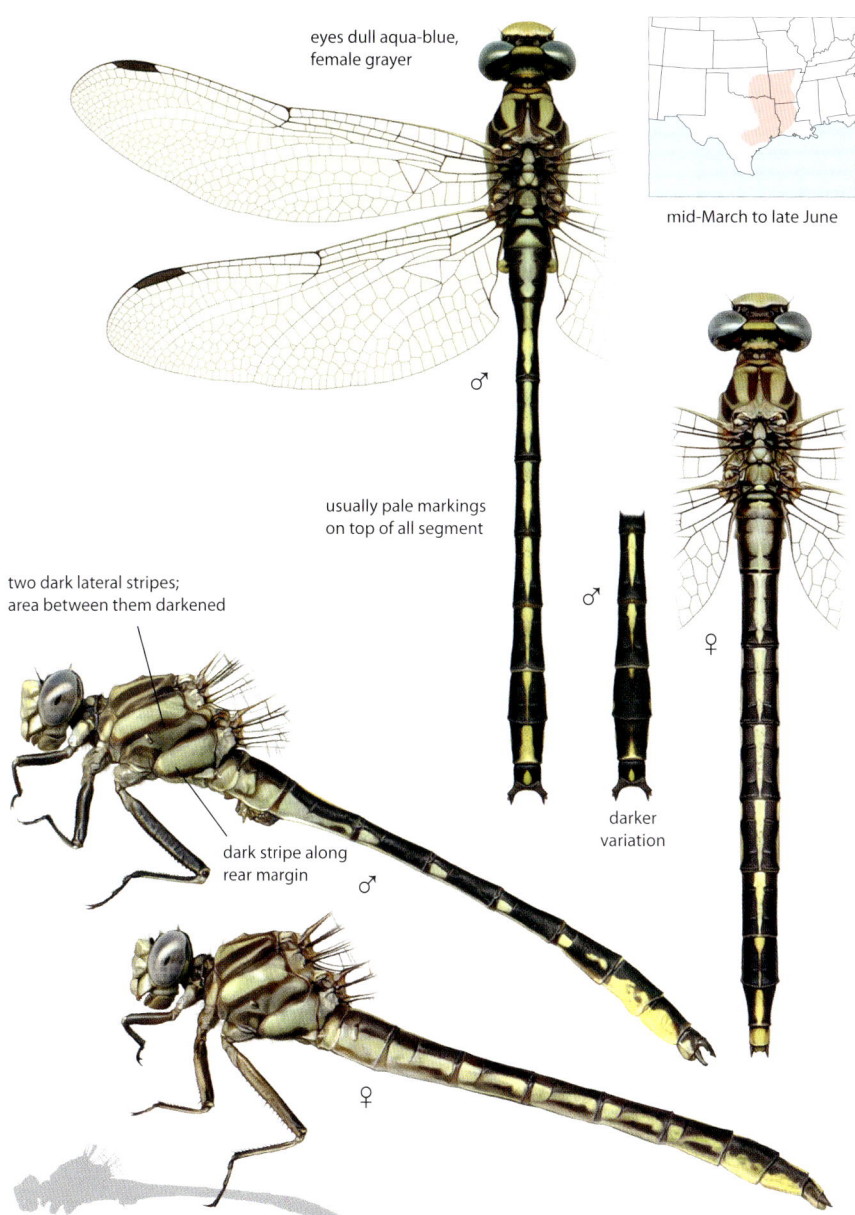

eyes dull aqua-blue, female grayer

mid-March to late June

♂

usually pale markings on top of all segment

two dark lateral stripes; area between them darkened

♂

darker variation

dark stripe along rear margin

♂

♀

♀

Similar species: Lancet Clubtail largely identical but has separate range. Compare male appendages, pages 400–401.

Ashy Clubtail • *Phanogomphus lividus*

48–56 mm, 1.9–2.2 in. Medium-sized, dull brown clubtail common and widespread in the East at slow streams, rivers, large lakes. Thorax brown. Area between dark lateral stripes darkened; dark stripe at rear margin. Abdomen with dorsal pale markings on S1-8; S9 usually dark; S10 with dorsal pale spot most often on female. Femurs brown. Eyes purple-gray.

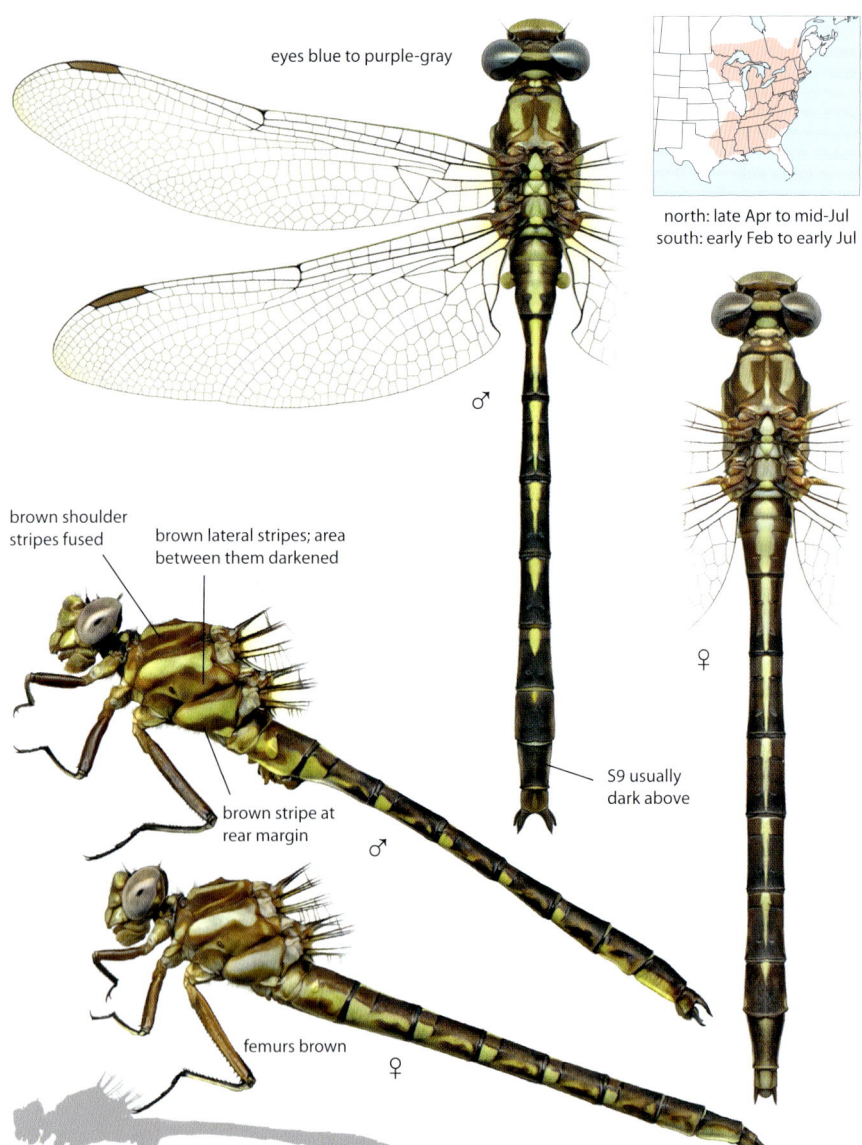

eyes blue to purple-gray

north: late Apr to mid-Jul
south: early Feb to early Jul

♂

brown shoulder stripes fused

brown lateral stripes; area between them darkened

brown stripe at rear margin

♂

S9 usually dark above

♀

femurs brown

♀

Similar species: Dusky Clubtail usually blacker; eyes bluer, femurs darker or more contrastingly striped. Compare male cerci, female subgenital plates, pages 401, 405. **Lancet Clubtail** blacker, usually has pale markings on top of all abdominal segments. **Beaverpond** and **Harpoon Clubtails** blacker, with more contrasting pattern; lack dark lateral stripe on rear margin of thorax.

Dusky Clubtail • *Phanogomphus spicatus*

46–50 mm, 1.8–2 in. Dark, dully patterned northeastern species found at ponds, lakes, and slow streams. Area between dark lateral thoracic stripes darkened, wide dark stripe at rear margin of thorax. Abdomen with dorsal pale markings on S1-8; S9 black above. Female with dorsal yellow spot on S10. Male femurs black, female striped. Eyes blue-gray.

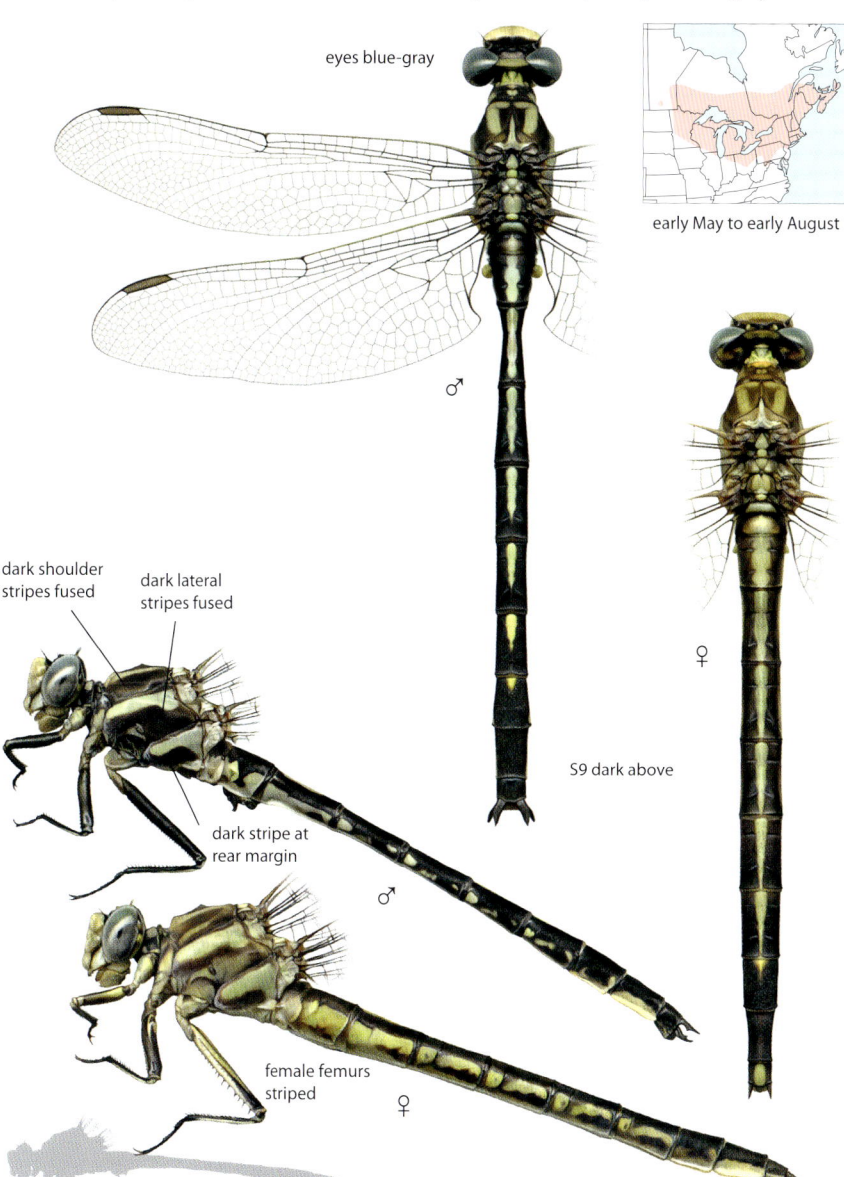

eyes blue-gray

early May to early August

♂

♀

dark shoulder stripes fused

dark lateral stripes fused

S9 dark above

dark stripe at rear margin

♂

female femurs striped

♀

Similar species: Ashy Clubtail usually browner, paler; femurs brown. Often indistinguishable in the North. Compare male cerci, female subgenital plates, pages 401, 405. **Lancet Clubtail** usually has pale markings on top of all abdominal segments. **Beaverpond** and **Harpoon Clubtails** lack dark lateral stripe on rear margin of thorax.

Cypress Clubtail • *Phanogomphus minutus*

43–50 mm, 1.7–2 in. Common far-southeastern clubtail found at slow, sandy streams and rivers, occasionally lakes and ponds. Male moderately clubbed. Abdomen with full-length dorsal yellow stripes on all segments. Thorax with two distinct dark lateral stripes. Both sexes with lobed occiput, more pronounced on female. Femurs often pale-striped.

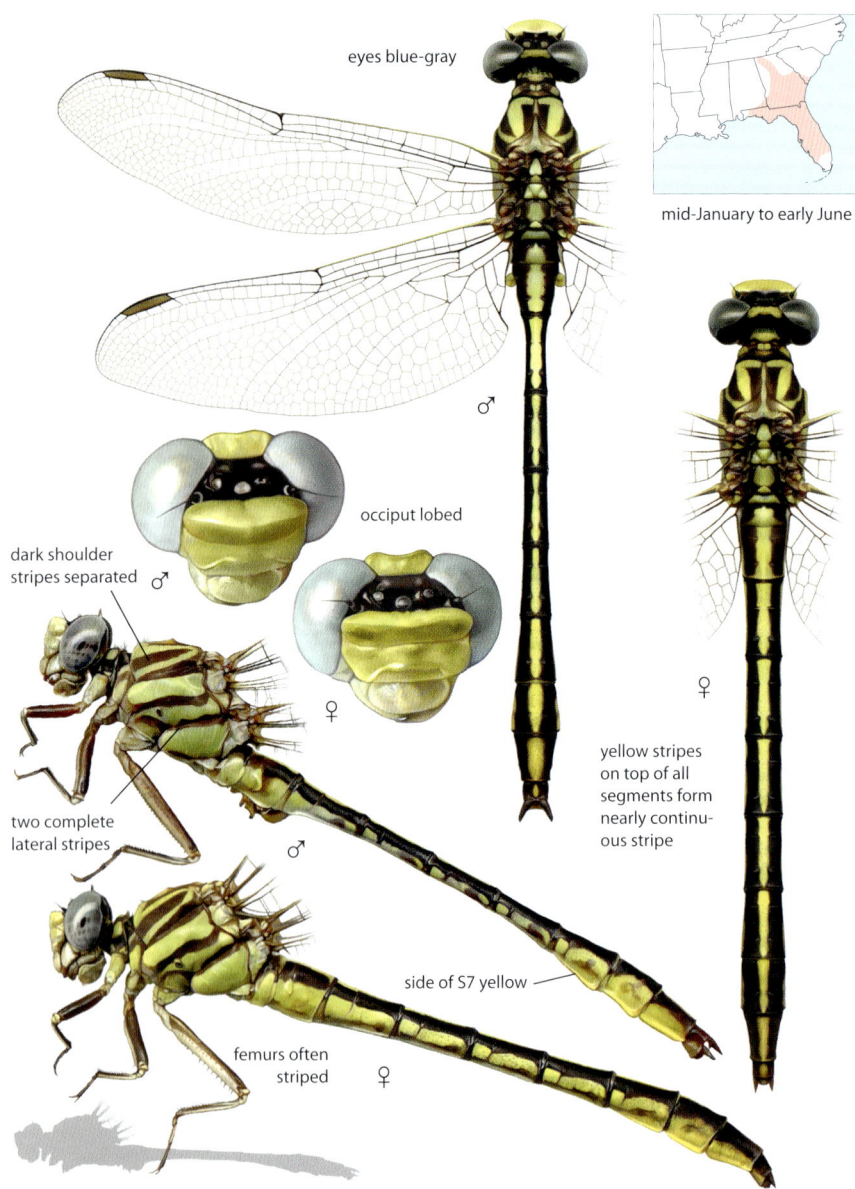

eyes blue-gray

mid-January to early June

♂

occiput lobed

dark shoulder stripes separated ♂

♀

two complete lateral stripes

♂

side of S7 yellow

yellow stripes on top of all segments form nearly continuous stripe

♀

femurs often striped ♀

Similar species: Sandhill Clubtail has brown face stripe, flatter occiput; top of S10 darker. Compare in-hand structures, pages 402, 405. **Lancet Clubtail** thorax has dark stripe at rear margin, shoulder stripes fused, dark lateral thoracic stripes fused, side of S7 mostly dark.

Sandhill Clubtail • *Phanogomphus cavillaris*

37–45 mm, 1.5–1.8 in. Small southeastern clubtail inhabiting sand-bottomed lakes. Male moderately clubbed. Abdomen with yellow dorsal stripe from S1 to S8 or 9. Face with brown facial line. Dark lateral thoracic stripes not fused. Two distinct forms: brown form (ssp. *P. c. cavillaris*) FL peninsula/GA, black form (ssp. *P. c. brimleyi*) FL Panhandle/NC.

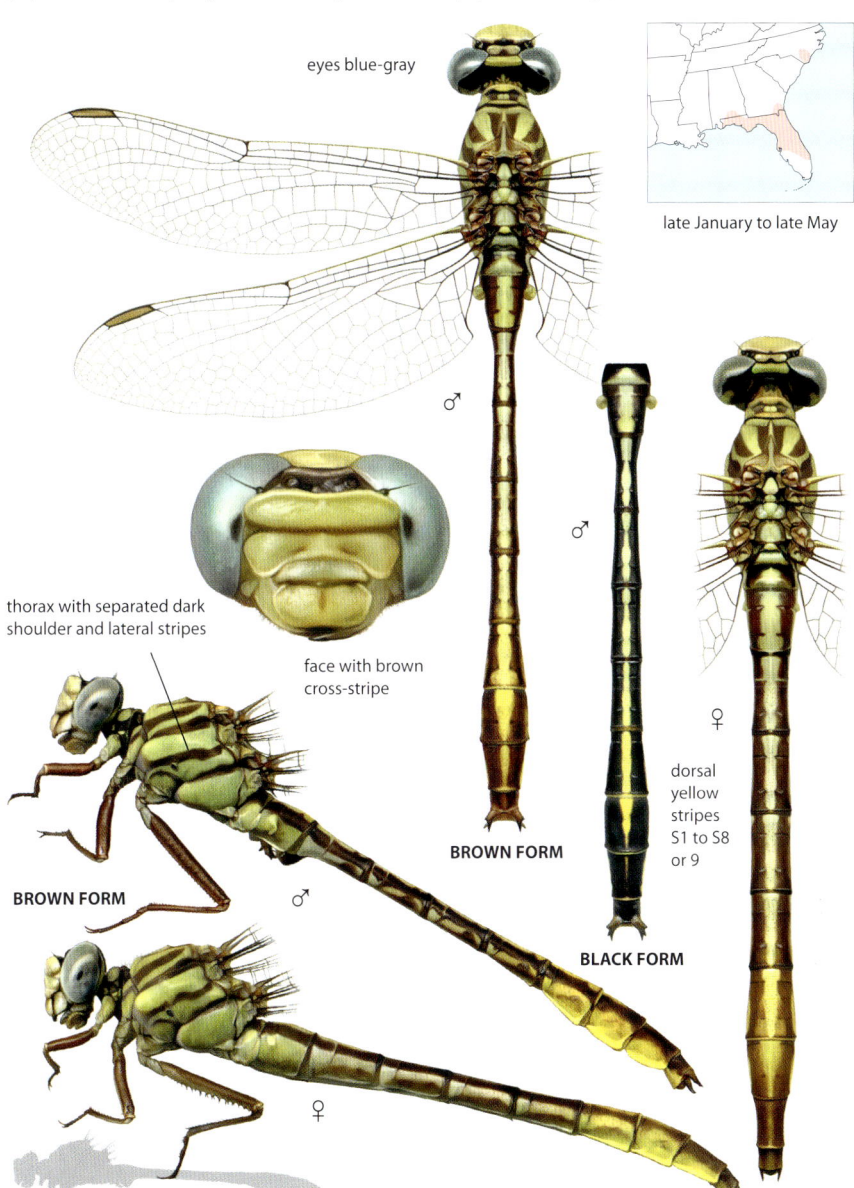

eyes blue-gray

late January to late May

thorax with separated dark shoulder and lateral stripes

face with brown cross-stripe

BROWN FORM

♂

♂

BROWN FORM

♂

dorsal yellow stripes S1 to S8 or 9

BLACK FORM

♀

♀

Similar species: Cypress Clubtail larger, lacks brown face line, has lobed occiput, often has yellow on top of S10. Also compare male cerci, female subgenital plates, pages 401, 405. **Lancet Clubtail** thorax has dark stripe at rear margin, shoulder stripes fused, dark lateral thoracic stripes fused, side of S7 mostly dark.

Westfall's Clubtail • *Phanogomphus westfalli*

39–45 mm, 1.5–1.8 in. Small, slender clubtail restricted to the Florida Panhandle at small boggy ponds and streams. S9 long. Abdomen narrow, with thin yellow dorsal stripes on S1-7; S8-9 black above. Area between dark lateral thoracic stripes not darkened. Face with dark cross-stripe. Eyes blue-gray.

eyes blue-gray

late February to late April

dark shoulder and lateral stripes complete and separated

face with black cross-stripe

abdomen slender, with thin yellow dorsal stripes on S1-7

S9 longer than S8 but shorter than S7

♂

♀

Similar species: Diminutive Clubtail nearly identical but has separate range; area between dark lateral stripes darkened. **Clearlake Clubtail** larger; S9 longer than S7. **Sandhill Clubtail** usually has larger pale marking on top of S8, S9 not elongated. **Cypress Clubtail** has yellow markings on top of S8-9, S9 not elongated. **Hodges' Clubtail** S9 not elongated.

Hodges' Clubtail • *Phanogomphus hodgesi*

41–44 mm, 1.6–1.75 in. Small, slender clubtail found at central Gulf Coast sand-bottomed rivers and streams. Abdomen narrow, with fine yellow stripes on top of S1-7; may have short pale streaks on S8-9. Female often with pale spot on top of S10. Eyes blue. Face unstriped. Thoracic stripes bold and separated. Legs with pale stripes.

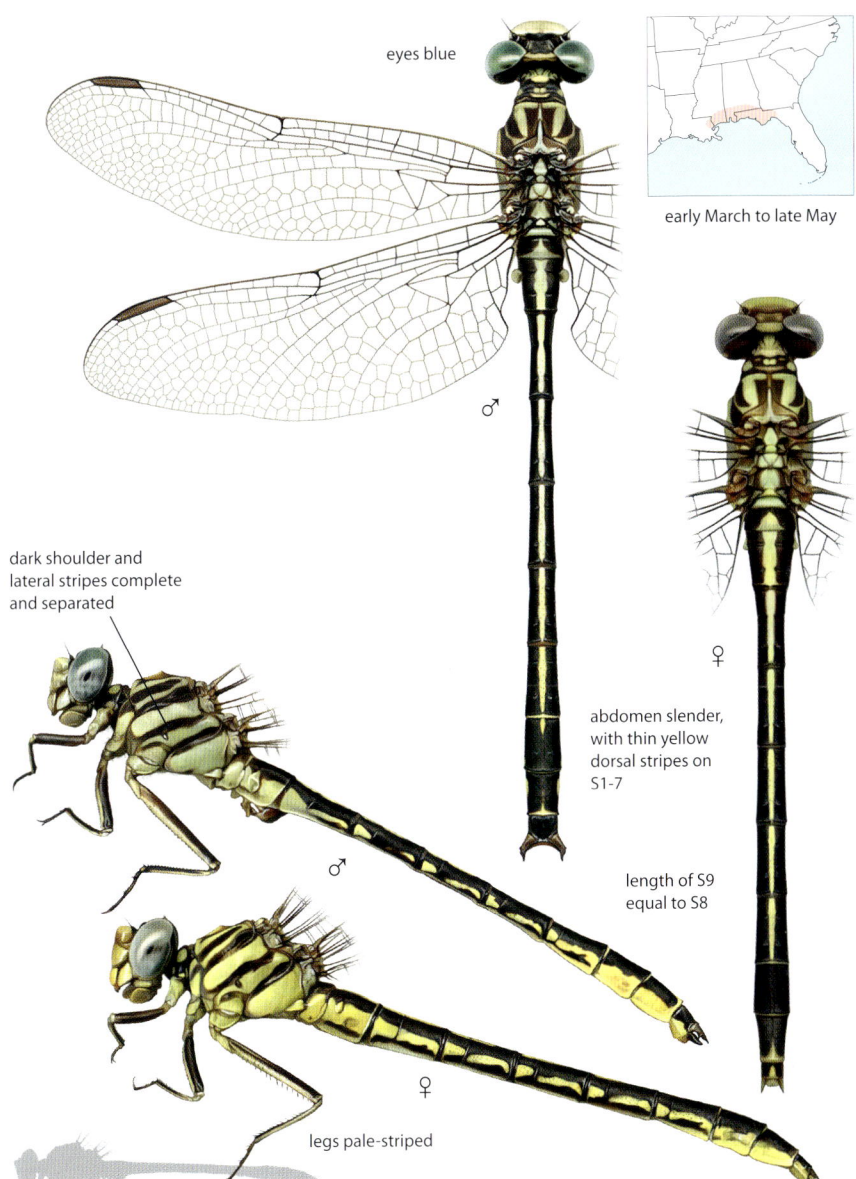

eyes blue

early March to late May

dark shoulder and lateral stripes complete and separated

♂

abdomen slender, with thin yellow dorsal stripes on S1-7

♀

♂

length of S9 equal to S8

♀

legs pale-striped

Similar species: Diminutive and **Westfall's Clubtails** have S9 longer than S8, darker legs, Westfall's with black facial stripe. **Lancet, Oklahoma, Cypress,** and **Sandhill Clubtails** have wider, full-length pale markings on top of S1-S8, and often S9.

Clearlake Clubtail • *Phanogomphus australis*

52–54 mm, 2–2.1 in. Medium-sized southeastern American clubtail found at clear, sand-bottomed lakes. Both sexes with very long S9, longer than both S7 and S8. Abdomen has fine yellow dorsal stripes; top of S9 black. Female with dorsal yellow spot on S10. Face with dark cross-line. Thorax with two dark lateral stripes. Eyes aqua-blue. Legs black.

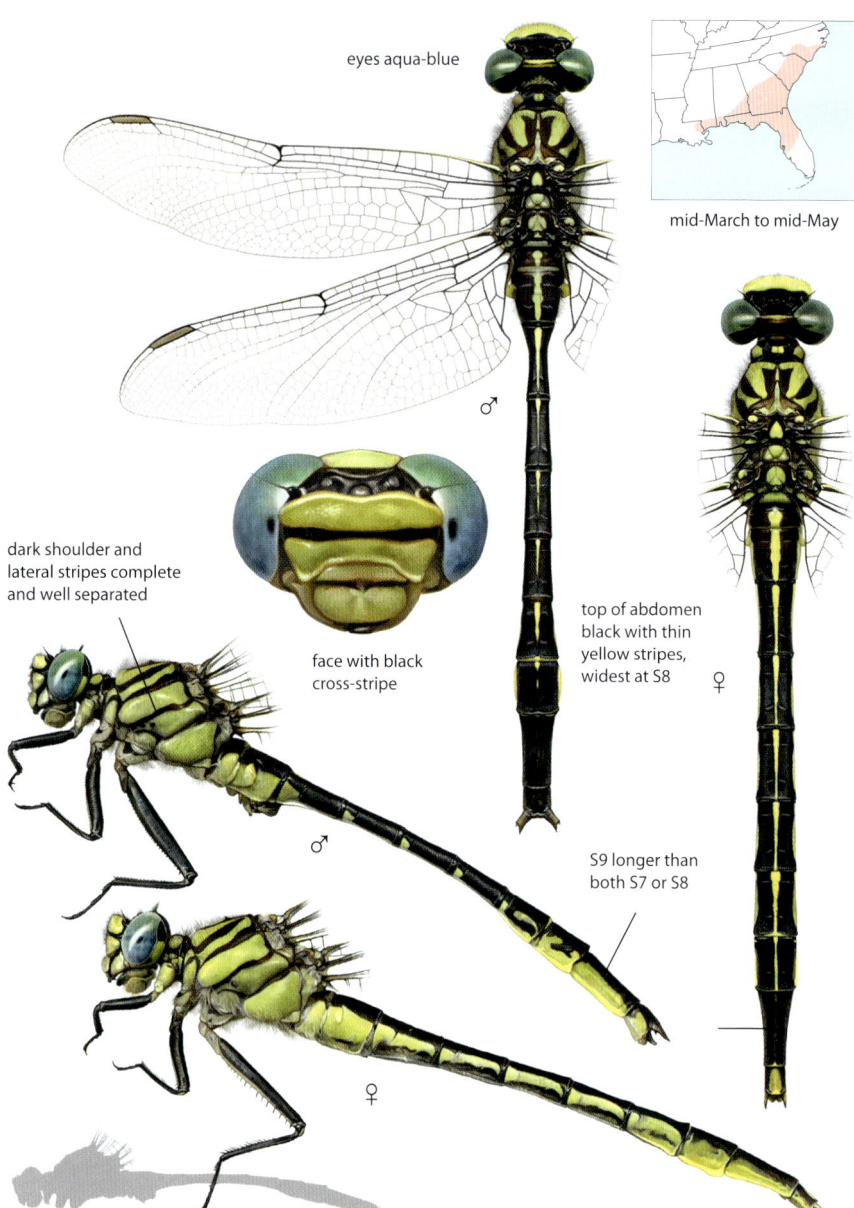

eyes aqua-blue

mid-March to mid-May

dark shoulder and lateral stripes complete and well separated

face with black cross-stripe

top of abdomen black with thin yellow stripes, widest at S8

S9 longer than both S7 or S8

Similar species: Westfall's and **Diminutive Clubtails** also have a long S9, but S9 shorter than S7. They are often noticeably smaller. **Sandhill Clubtail** lacks elongated S9. **Cypress Clubtail** lacks long S9, usually has dorsal yellow markings on all abdominal segments.

Diminutive Clubtail • *Phanogomphus diminutus*

39–43 mm, 1.5–1.7 in. Small American clubtail limited to the Carolinas at slow streams and lakes, typically boggy and acidic. S9 long, longer than S8 but shorter than S7. Abdomen with narrow yellow dorsal stripes on S1-7, S8 with yellow streak, top of S9 black. Legs brown, tibiae pale-striped. Eyes gray-blue. Area between lateral thoracic stripes darkened.

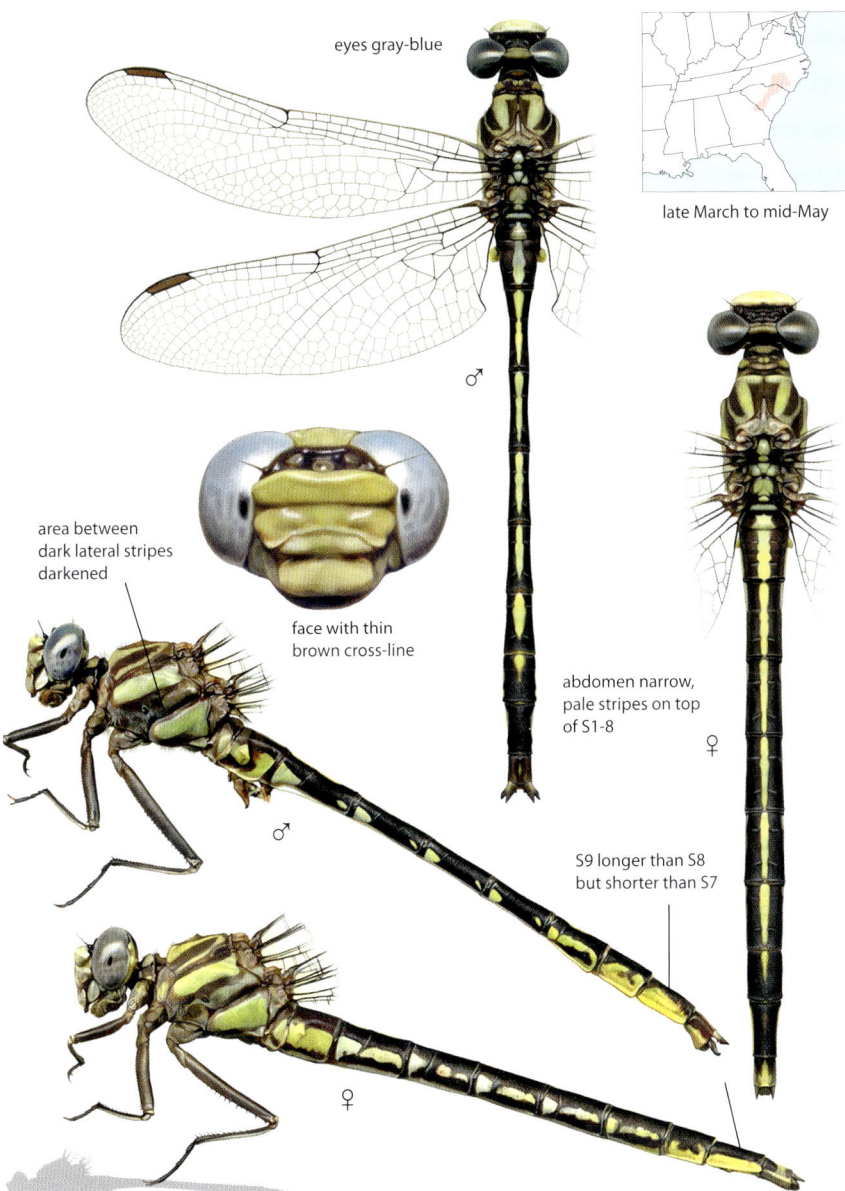

eyes gray-blue

late March to mid-May

area between dark lateral stripes darkened

face with thin brown cross-line

abdomen narrow, pale stripes on top of S1-8

♂

♀

S9 longer than S8 but shorter than S7

♀

Similar species: Clearlake Clubtail larger, S9 longer than S7, and area between dark lateral thoracic stripes not darkened; has black facial stripe. **Lancet** and **Cypress Clubtails** have abdomen more extensively marked with yellow on top of S9; S9 not elongated.

Tennessee Clubtail • *Phanogomphus sandrius*

47.5–51.5 mm, 1.9–2 in. Robust, brightly marked American clubtail inhabiting streams flowing over bedrock; very limited range in central Tennessee. Abdomen with pale markings on top of all segments, S8 with small yellow triangle, S9 with wide yellow stripe. Male moderately clubbed, female slightly. Tibiae pale-striped. Eyes blue.

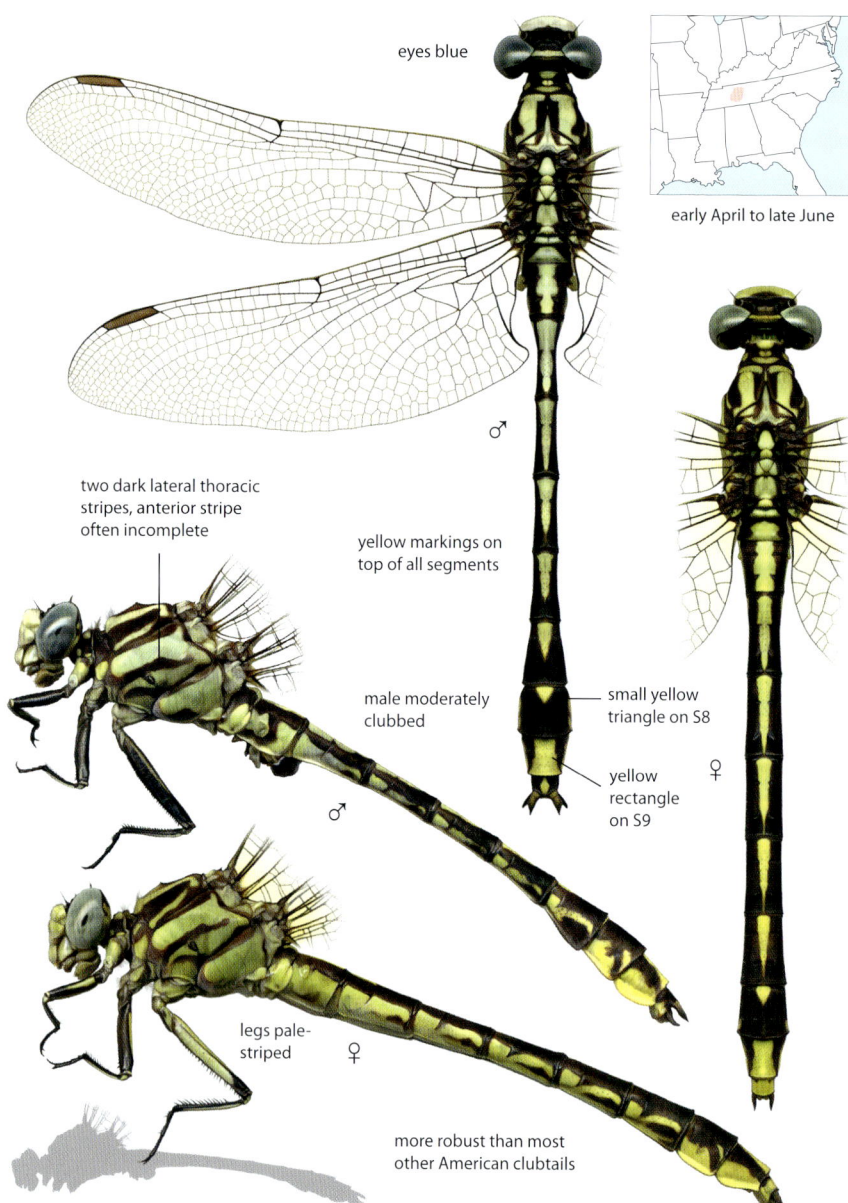

eyes blue

early April to late June

two dark lateral thoracic stripes, anterior stripe often incomplete

yellow markings on top of all segments

male moderately clubbed

small yellow triangle on S8

yellow rectangle on S9

♂

♀

legs pale-striped ♀

more robust than most other American clubtails

♂

Similar species: Handsome Clubtail has wider club and longer hind legs; tibiae black. Dark lateral thoracic stripes narrower, anterior stripe reduced. **Pronghorn Clubtail** nearly identical, but range separated; has wider dark thoracic stripes, lateral stripes often fused.

Pronghorn Clubtail • *Phanogomphus graslinellus*

47–53 mm, 1.9–2.1 in. Wide-ranging midwestern species found at slow streams, also lakes in the North. Abdomen with pale markings on top of all segments: yellow triangle on S8; S9 with broad yellow dorsal stripe. Both sexes moderately clubbed. Thorax with dark shoulder stripes often fused. Dark lateral stripes often fused; area between them darkened.

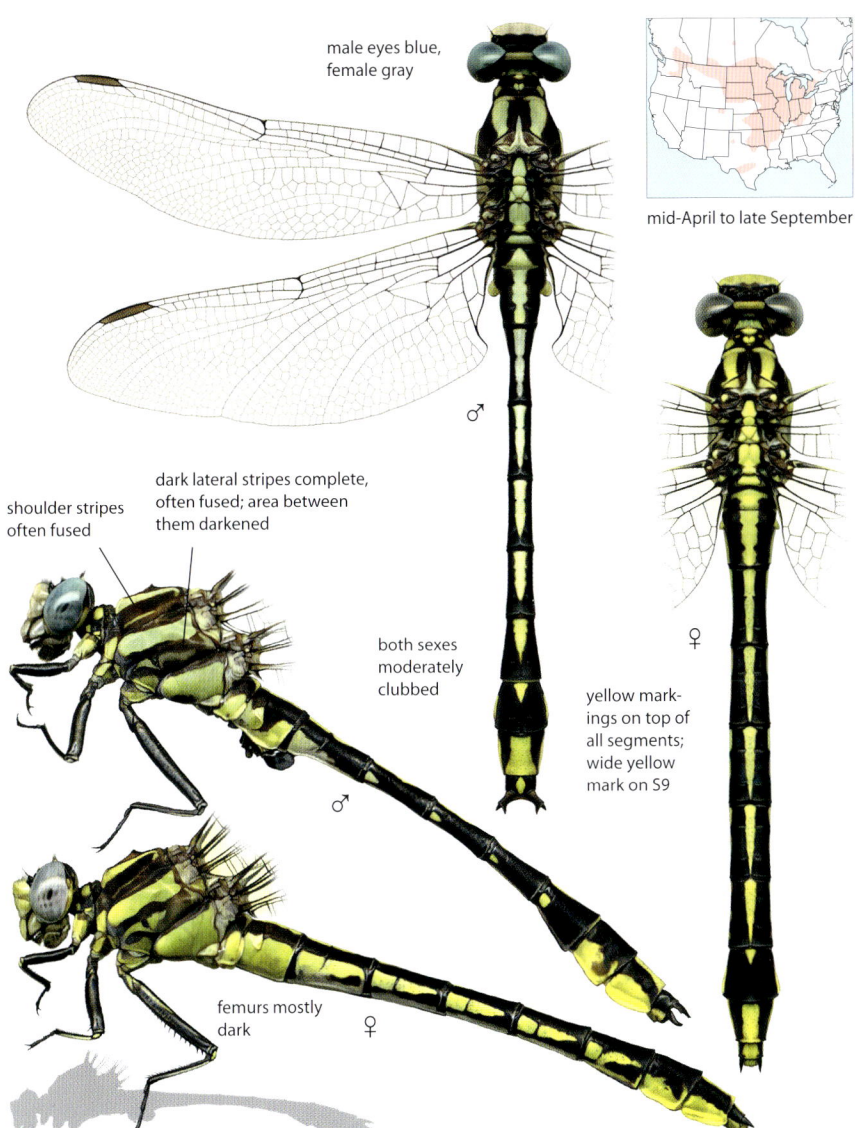

male eyes blue, female gray

mid-April to late September

♂

shoulder stripes often fused

dark lateral stripes complete, often fused; area between them darkened

both sexes moderately clubbed

yellow markings on top of all segments; wide yellow mark on S9

♀

♂

femurs mostly dark

♀

Similar species: Tennessee Clubtail similar but with separate range. **Sulphur-tipped Clubtail** paler overall. Most likely confused with **majestic clubtails,** which have longer hind leg femurs and larger clubs. In addition, **Plains Clubtail** larger; markings on sides of middle abdominal segments whitish in color. Compare male cerci, pages 403, 407. **Midland** and **Handsome Clubtails** have anterior dark lateral thoracic stripe incomplete. **Ozark Clubtail** has all-black legs, narrower yellow stripe on S9 when present.

Sulphur-tipped Clubtail • *Phanogomphus militaris*

47–53 mm, 1.9–2.1 in. Pale clubtail of open lakes, large ponds, slow streams and rivers in the south-central US. Abdomen with pale markings on top of all segments: S8 with pale dorsal triangle; S9 with wide dorsal stripe. Sides of S7-10 yellow; abdomen tip orange when mature. Thorax with well-developed pale shoulder stripe. Femurs and tibiae pale-striped.

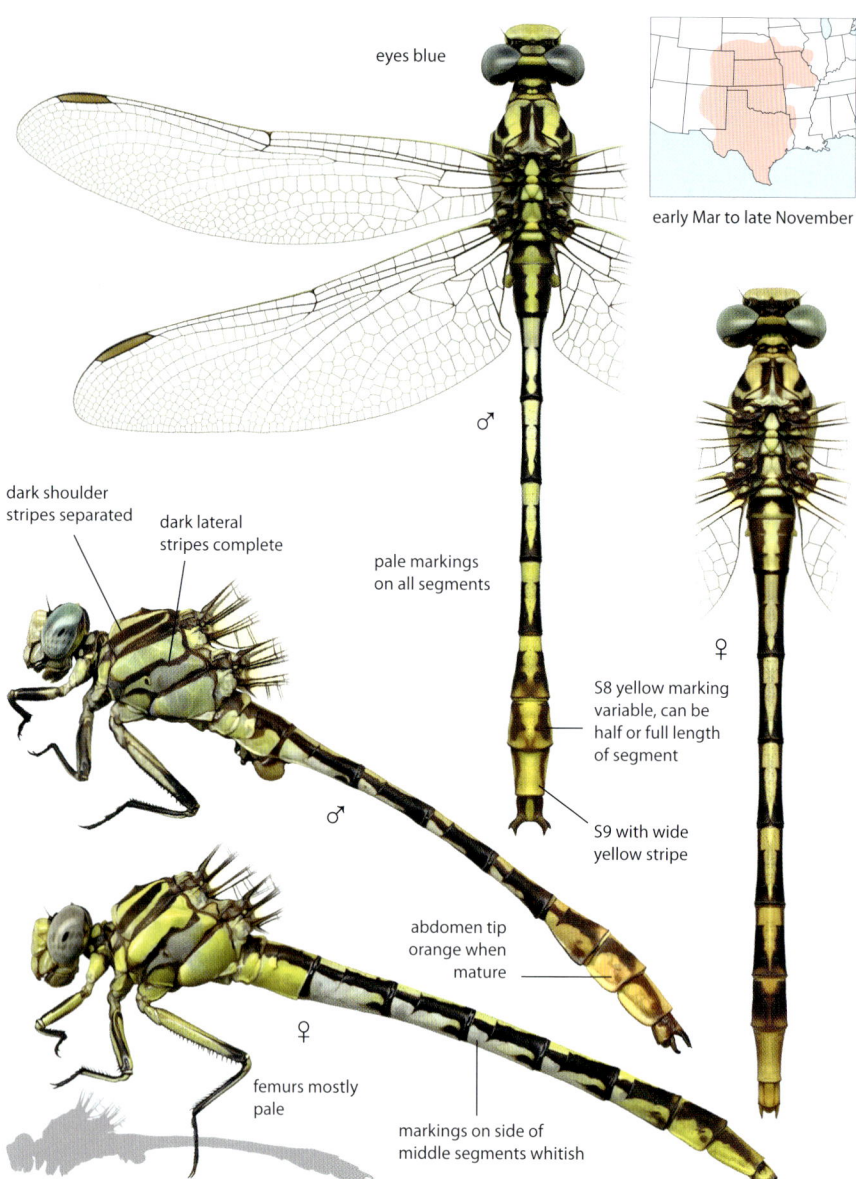

eyes blue

early Mar to late November

♂

dark shoulder stripes separated

dark lateral stripes complete

pale markings on all segments

♀

S8 yellow marking variable, can be half or full length of segment

♂

S9 with wide yellow stripe

abdomen tip orange when mature

♀

femurs mostly pale

markings on side of middle segments whitish

Similar species: Pronghorn Clubtail has darker legs, and pale shoulder stripe is reduced or lacking; lacks orange tones on abdomen. **Plains Clubtail** blacker, pale shoulder stripe reduced, legs blacker, femur of hind leg longer. **Tamaulipan Clubtail** has darker legs, with no pale stripes on tibiae.

Pacific Clubtail • *Phanogomphus kurilis*

48–57 mm, 1.9–2.1 in. Common far-western clubtail found at streams, rivers, also ponds and lakes. Thorax with fused dark shoulder stripes, single dark lateral stripe. Abdomen with narrow dorsal markings on S1-8; dorsal S9 markings variable. Large yellow marks on sides of S8-9. Eyes blue. Legs all-black.

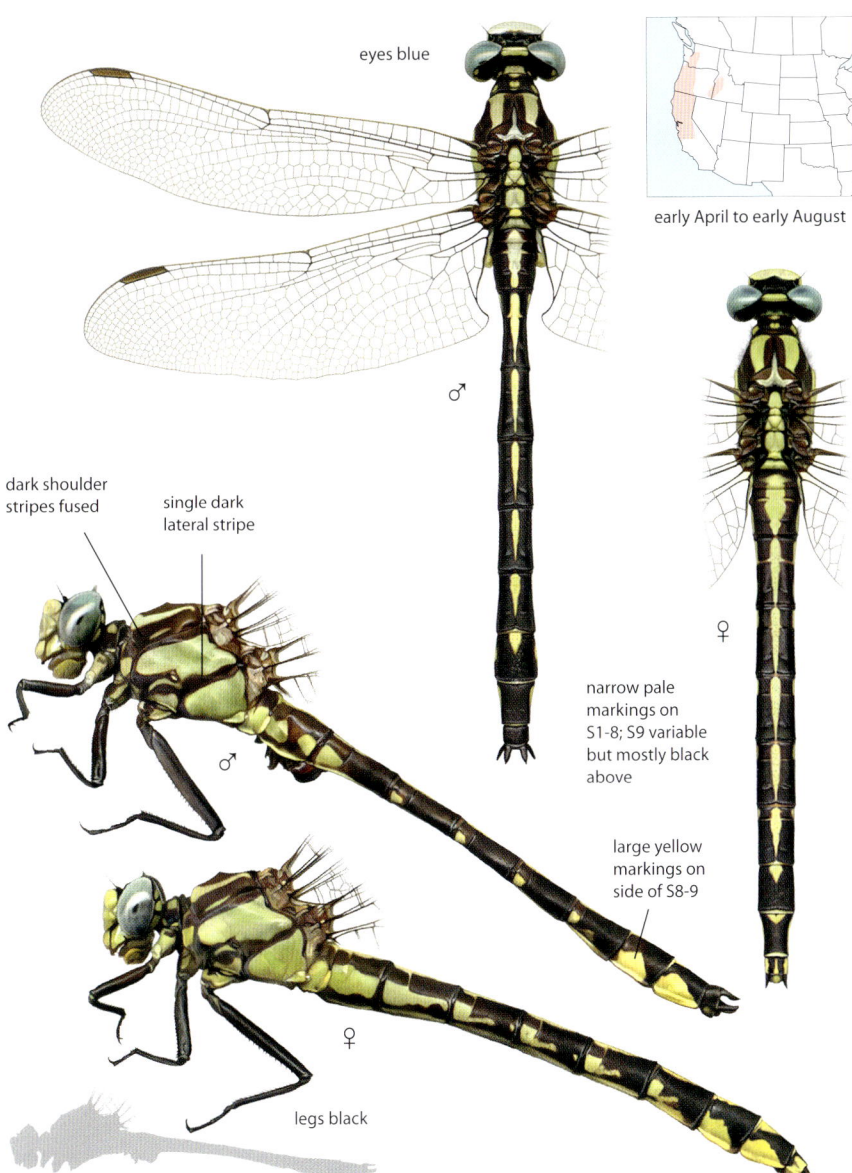

eyes blue

early April to early August

dark shoulder stripes fused

single dark lateral stripe

narrow pale markings on S1-8; S9 variable but mostly black above

large yellow markings on side of S8-9

legs black

Similar species: Only American clubtail with single lateral thoracic stripe, which also separates Pacific Clubtail from all majestic clubtails. **Grappletail** abdomen darker; thorax front lacking wide dark middorsal stripes. Co-occuring **snaketail species** have pale markings on top of all abdominal segments, paler legs.

Plains Clubtail • *Gomphurus externus*

52–59 mm, 2–2.3 in. Brightly marked Great Plains majestic clubtail inhabiting sand- and mud-bottomed rivers, and large streams. Face pale; eyes blue to gray. Thorax with complete dark brown lateral stripes; area between them gray. Dark shoulder stripes often fused at top. Abdomen with small club and yellow markings on top of all segments.

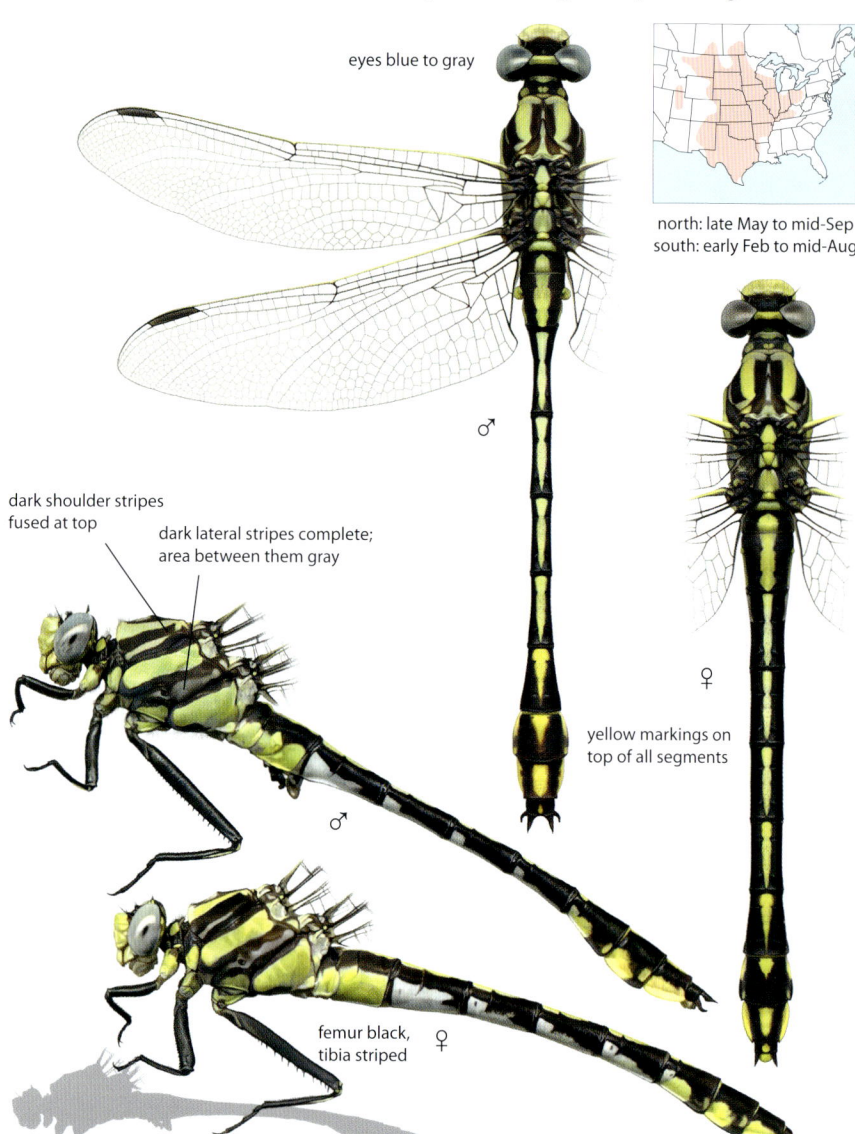

eyes blue to gray

north: late May to mid-Sep
south: early Feb to mid-Aug

♂

dark shoulder stripes
fused at top

dark lateral stripes complete;
area between them gray

♀

yellow markings on
top of all segments

♂

femur black,
tibia striped   ♀

Similar species: Pronghorn Clubtail, an American clubtail very similar in pattern, usually lacks black on edge of S8, and femur of hind leg does not reach beyond middle of S2. **Tamaulipan Clubtail** has dark shoulder stripes separated, dark stripe along rear lateral margin of thorax, pale stripe on hind leg femur; abdomen more extensively pale. **Gulf Coast Clubtail** has dark facial line; top of S9-10 black. **Midland, Handsome,** and **Splendid Clubtails** have incomplete lateral thoracic stripes.

Tamaulipan Clubtail • *Gomphurus gonzalezi*

46–50 mm, 1.8–2 in. Small, pale majestic clubtail found at rocky rivers; in South Texas, along the muddy Rio Grande. Eyes blue-gray. Side of thorax with two complete dark stripes and additional dark stripe along the rear margin. Femur of hind leg with pale stripe; tibiae dark. Abdomen with small club; club mostly yellow. Pale markings on top of all segments.

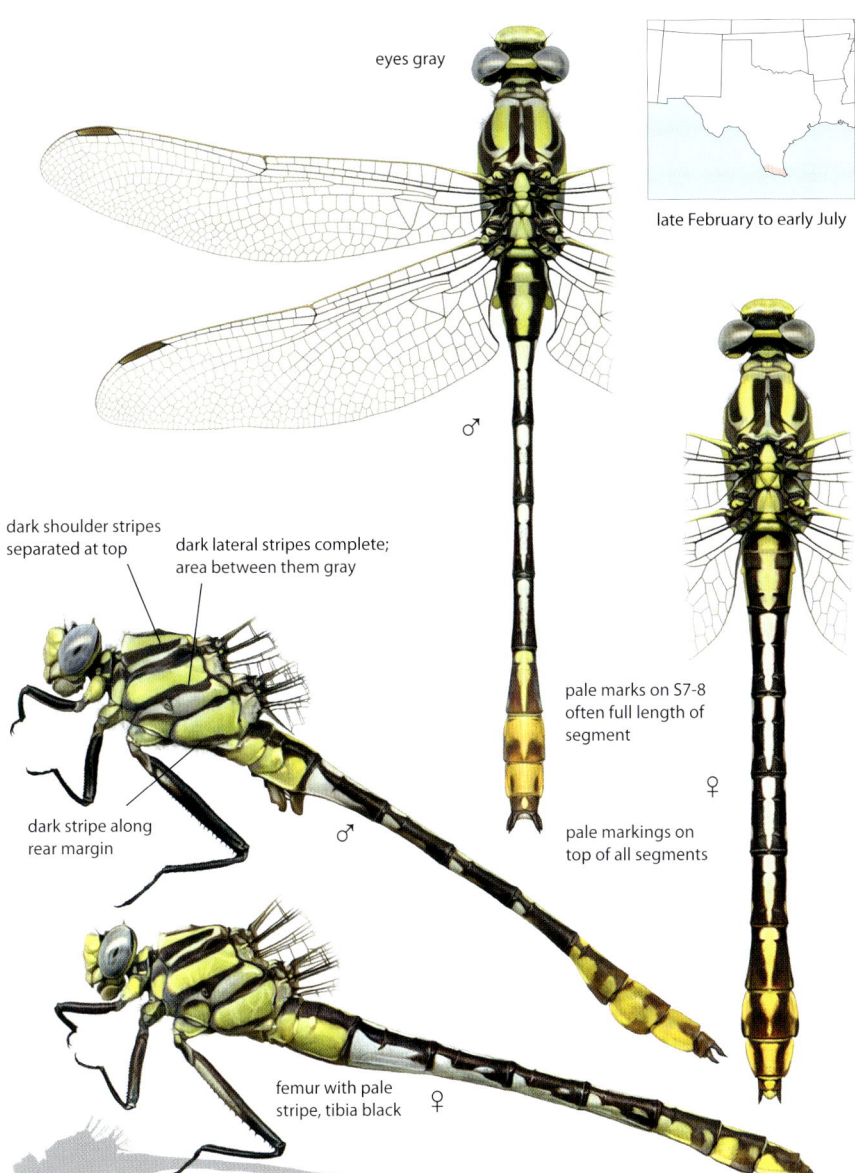

eyes gray

late February to early July

dark shoulder stripes separated at top

dark lateral stripes complete; area between them gray

dark stripe along rear margin

♂

pale marks on S7-8 often full length of segment

pale markings on top of all segments

♀

femur with pale stripe, tibia black ♀

Similar species: Sulphur-tipped Clubtail, an American clubtail, has shorter hind leg femur; femurs paler. S9 longer than S8. **Plains Clubtail** lacks dark stripe at rear lateral margin of thorax; shoulder stripes fused at top. Abdomen darker. Femurs dark; tibiae with pale stripes.

Midland Clubtail • *Gomphurus fraternus*

28–55 mm, 1.9–2.2 in. Small northeastern majestic clubtail found at flowing rivers, large streams with mud and rock bottoms, and large lakes with wave action in the North. Thorax with incomplete dark anterior lateral stripe; posterior lateral stripe usually complete. Small yellow marking on top of S8. Top of S9 usually black; some populations with yellow stripe.

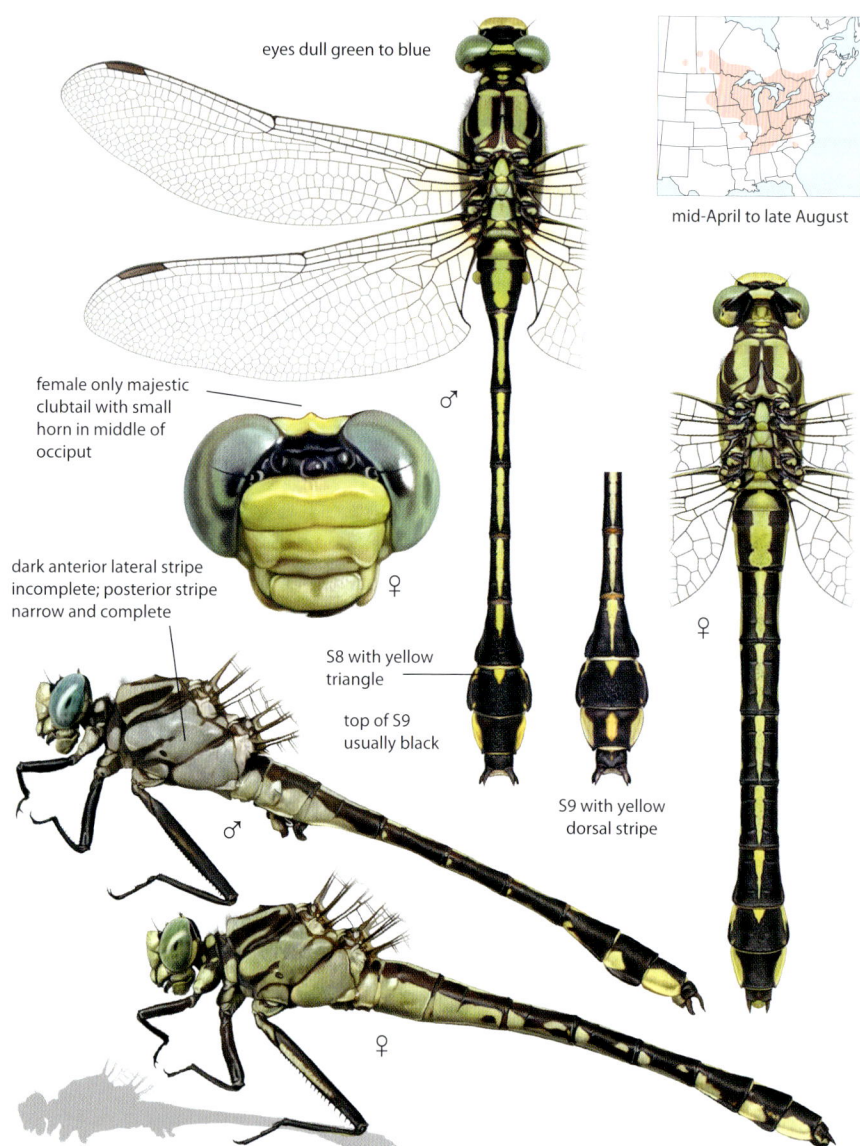

eyes dull green to blue

mid-April to late August

female only majestic clubtail with small horn in middle of occiput

dark anterior lateral stripe incomplete; posterior stripe narrow and complete

S8 with yellow triangle

top of S9 usually black

S9 with yellow dorsal stripe

♂

♀

♀

♂

♀

Similar species: Handsome Clubtail usually has yellow dorsal stripe on S9; female lacks occipital horn. Individuals of Midland Clubtail with dorsal yellow stripe on S9 best separated by the male appendages, page 407. **Plains Clubtail** also has dorsal pale marking on S9 but has two complete dark lateral thoracic stripes. **Splendid Clubtail** larger, has dark face stripe, lacks yellow triangle on top of S8.

Handsome Clubtail • *Gomphurus crassus*

54–59 mm, 2.1–2.3 in. Brightly patterned majestic clubtail inhabiting clean gravel- and rock-bottomed rivers; limited eastern interior range. Eyes blue to aqua. Thorax with incomplete dark anterior lateral stripe; posterior stripe complete. Abdomen usually with yellow markings on top of all segments; S9 dorsal marking variable but often is broad stripe.

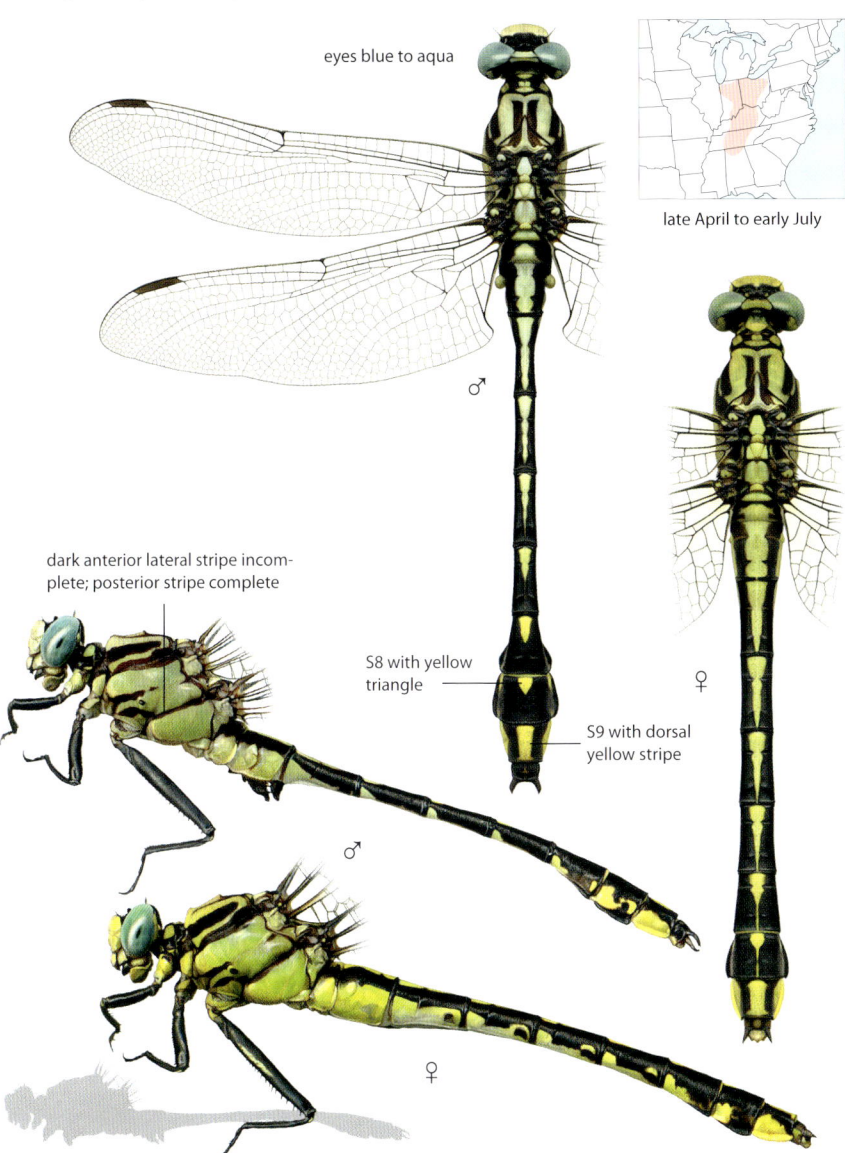

eyes blue to aqua

late April to early July

dark anterior lateral stripe incomplete; posterior stripe complete

S8 with yellow triangle

S9 with dorsal yellow stripe

♂

♀

Similar species: Midland Clubtail eyes greener. Usually lacks pale markings on top of S9, otherwise best separated by male appendages, female occiput (see pages 407, 410). **Plains Clubtail** has two complete lateral thoracic stripes. **Tennessee Clubtail** has shorter hind leg femur and smaller club, thorax with wider and more developed dark lateral stripes, tibiae with pale stripes.

Cocoa Clubtail · *Gomphurus hybridus*

50–52 mm, 2 in. Small, dull brownish majestic clubtail found at southeastern rivers with sand and silt bottoms. Eyes blue-gray. Thorax with brown stripes, lateral stripes complete. Legs brown, tibia with pale stripes. Abdomen moderately clubbed, blackish with end segments browner. Small pale lateral and dorsal markings on S8; S9 with yellow sides.

eyes blue-gray

mid-March to mid-June

♂

thoracic stripes brown, lateral stripes complete

♀

♂

end segments brown

legs brown

♀

yellow on side of S9

Similar species: Brown coloration separates Cocoa Clubtail from most other majestic clubtails. Brown-form **Cobra Clubtail** has green eyes and wider club. **Ozark Clubtail** has fused thoracic shoulder and lateral stripes, with larger yellow marking on side of S8.

Septima's Clubtail • *Gomphurus septima*

53–62 mm, 2.1–2.4 in. Dull-colored majestic clubtail found at clean, rocky rivers within its scattered eastern range. Face pale; eyes green-gray to turquoise. Thoracic stripes dark brown, shoulder stripes fused at top, lateral stripes incomplete or nearly absent. Legs dark. Abdomen black with small club. S8 and 9 with pale yellow lateral spots.

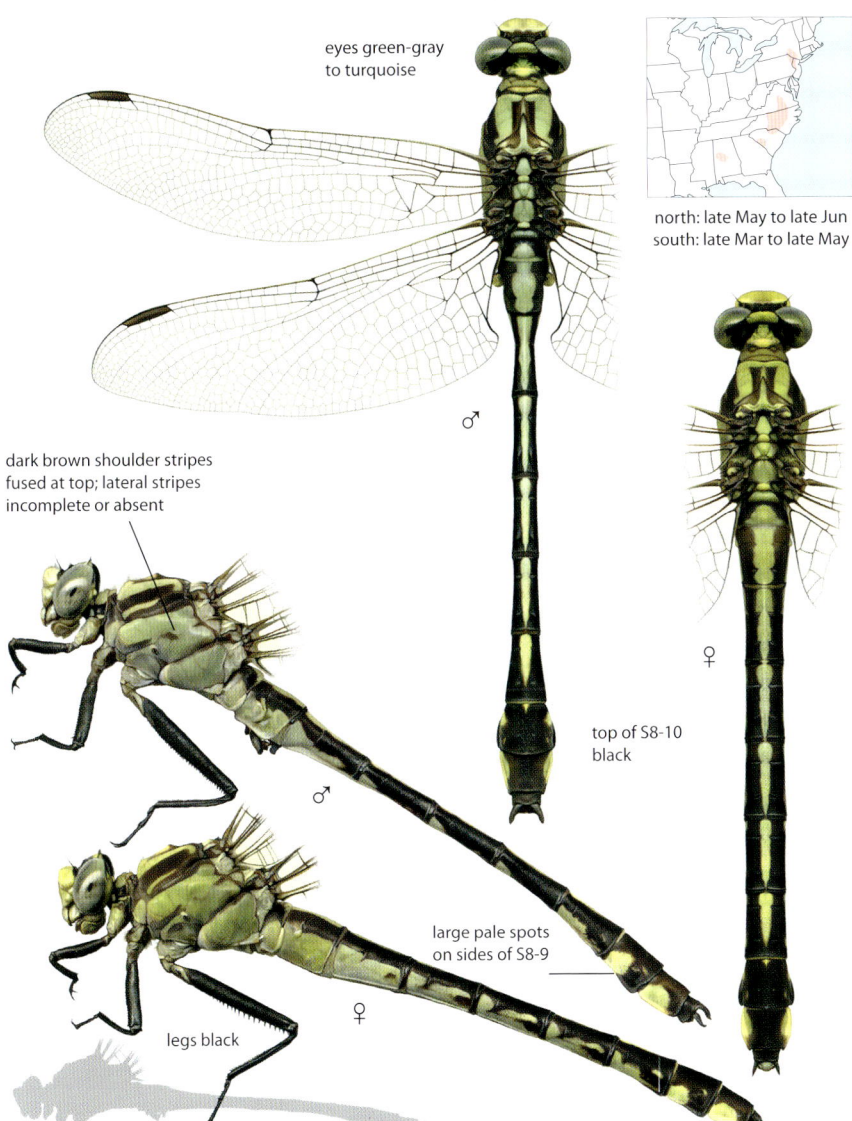

eyes green-gray to turquoise

north: late May to late Jun
south: late Mar to late May

dark brown shoulder stripes fused at top; lateral stripes incomplete or absent

♂

♀

top of S8-10 black

♂

large pale spots on sides of S8-9

legs black

♀

Similar species: Midland Clubtail has more contrasting pattern, with black thoracic stripes, brighter yellow markings on S8-9; female with occipital horn and often pale stripe on hind leg femur. **Splendid Clubtail** larger; thorax pattern well-defined with black stripes, shoulder stripes separated. **Cocoa Clubtail** browner, with complete lateral thoracic stripes and less yellow on side of brown club. **Black-shouldered Spinyleg** has narrower frontal thoracic stripes, dorsal pale marks on top of S8-10.

Ozark Clubtail • *Gomphurus ozarkensis*

50–53 mm, 2–2.1 in. Dark, dull-colored majestic clubtail found at clear rivers and streams in the Ozark Plateau. Thorax with wide, fused, dark brown shoulder stripes; lateral stripes also fused. Abdomen black; both sexes moderately clubbed, with yellow lateral spot on S8. Larger yellow marking along edge of S9; top of S9-10 black or with yellow markings.

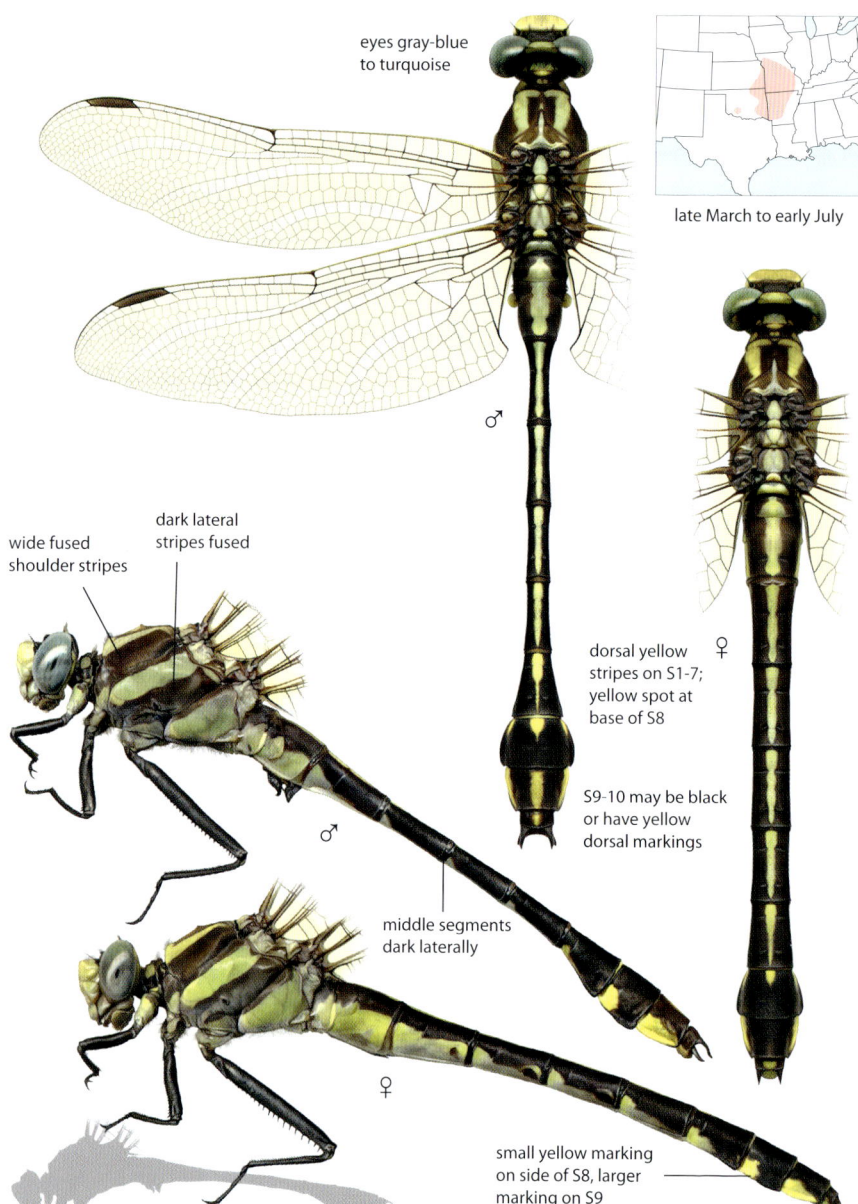

eyes gray-blue to turquoise

late March to early July

wide fused shoulder stripes

dark lateral stripes fused

♂

dorsal yellow stripes on S1-7; yellow spot at base of S8

♀

S9-10 may be black or have yellow dorsal markings

middle segments dark laterally

♂

♀

small yellow marking on side of S8, larger marking on S9

Similar species: Wide, fused thoracic stripes distinctive among majestic clubtails. **Pronghorn Clubtail** often has narrow pale stripe between its dark thoracic stripes, and has full-length yellow lateral marking on S8; both sexes with wide yellow dorsal stripe on S9.

Columbia Clubtail • *Gomphurus lynnae*

54–60 mm, 2.1–2.4 in. The only majestic clubtail found in the Northwest, inhabiting open mud- and sand-bottomed rivers. Thorax with dark shoulder stripes nearly fused, lateral stripes fused, dark stripe along lower rear margin. Thorax and S1-2 lightly pruinose upon maturity. Abdomen with large dorsal yellow markings on S1-9, lateral yellow markings on S8-9.

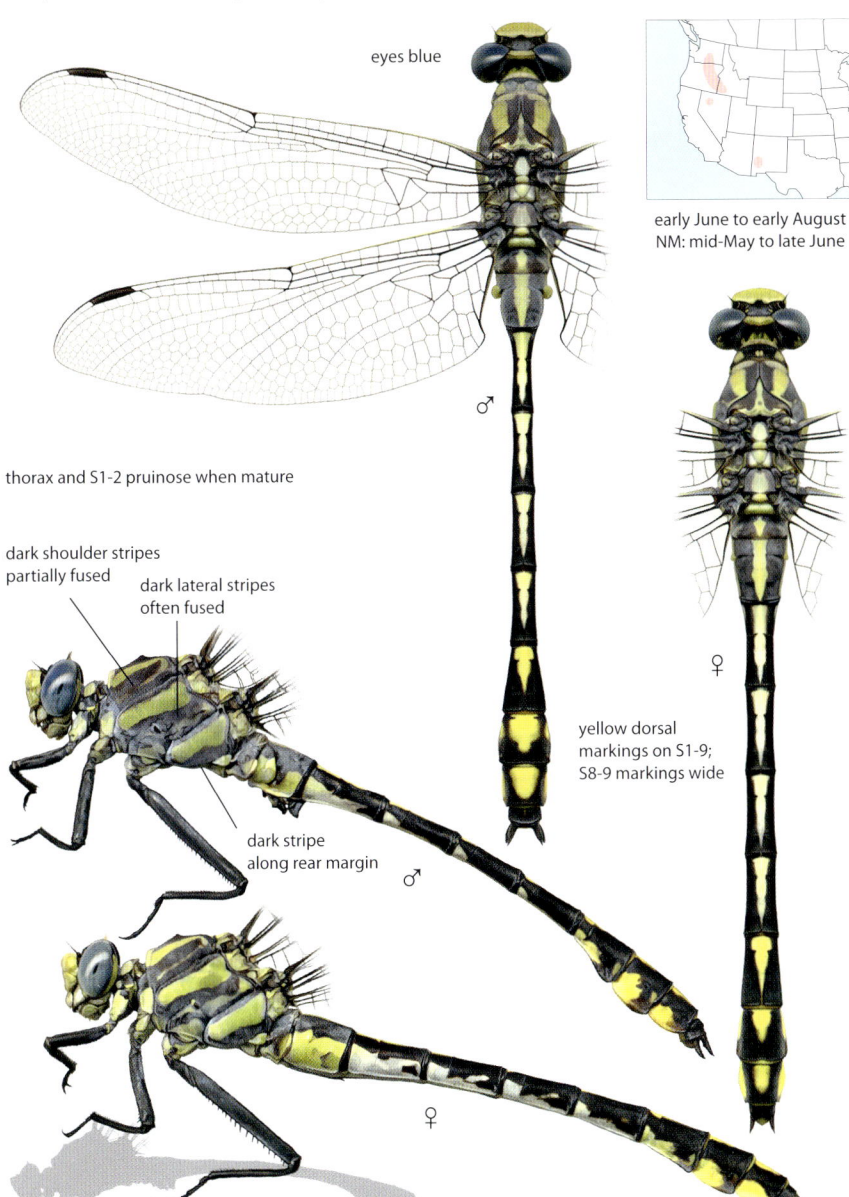

eyes blue

early June to early August
NM: mid-May to late June

thorax and S1-2 pruinose when mature

dark shoulder stripes
partially fused

dark lateral stripes
often fused

yellow dorsal
markings on S1-9;
S8-9 markings wide

dark stripe
along rear margin

♂

♀

Similar species: Pacific Clubtail has only single dark lateral thoracic stripe, lacks pruinosity on thorax; top of S9 black. **Pronghorn Clubtail** lacks pruinosity on thorax, has dark stripe along hind rear margin of thorax, and has smaller yellow marking on top of S8.

Cobra Clubtail • *Gomphurus vastus*

47–57 mm, 1.9–2.25 in. Dark majestic clubtail with very wide club; widespread in the East at flowing rivers and large streams. Face typically with wide black cross-stripe. Eyes green. Thorax usually with two complete black lateral stripes. Both sexes with very large clubs. Yellow spots on sides of S8-9; S8 spot smaller and does not touch edge of segment.

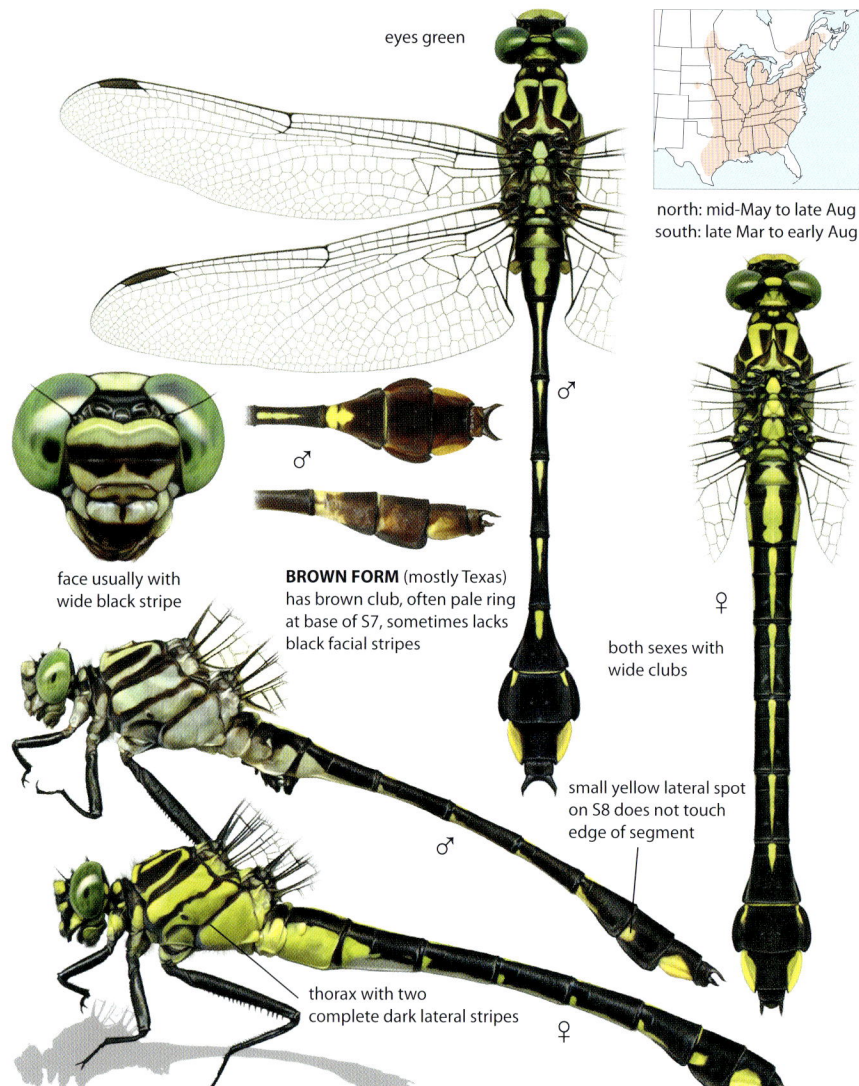

eyes green

north: mid-May to late Aug
south: late Mar to early Aug

face usually with wide black stripe

BROWN FORM (mostly Texas) has brown club, often pale ring at base of S7, sometimes lacks black facial stripes

♂

♀

both sexes with wide clubs

small yellow lateral spot on S8 does not touch edge of segment

♂

thorax with two complete dark lateral stripes

♀

Similar species: Blackwater Clubtail larger, usually with larger pale markings on S7, smaller pale spot on sides of S9. **Gulf Coast Clubtail** has narrower black face stripe, larger yellow markings on S8. **Splendid Clubtail** has narrow dark face line, incomplete anterior lateral thoracic stripe, larger yellow spot on side of S7. **Ozark Clubtail** lacks dark facial stripes, has fused thoracic stripes, S8 with dorsal yellow mark and larger yellow lateral mark. **Cocoa Clubtail** suggestive of brown-form Cobra, but lacks dark facial stripes and has brown thoracic stripes and legs and bluer eyes.

Blackwater Clubtail • *Gomphurus dilatatus*

67–72 mm, 2.6 -2.8 in. The largest majestic clubtail. Southeastern, found at sand-bottomed woodland rivers and streams often stained with tannins, occasionally lakes. Face with wide black cross-stripes. Eyes green. Thorax with two complete black lateral stripes. Both sexes with very large clubs; yellow lateral spots on S8-9 small.

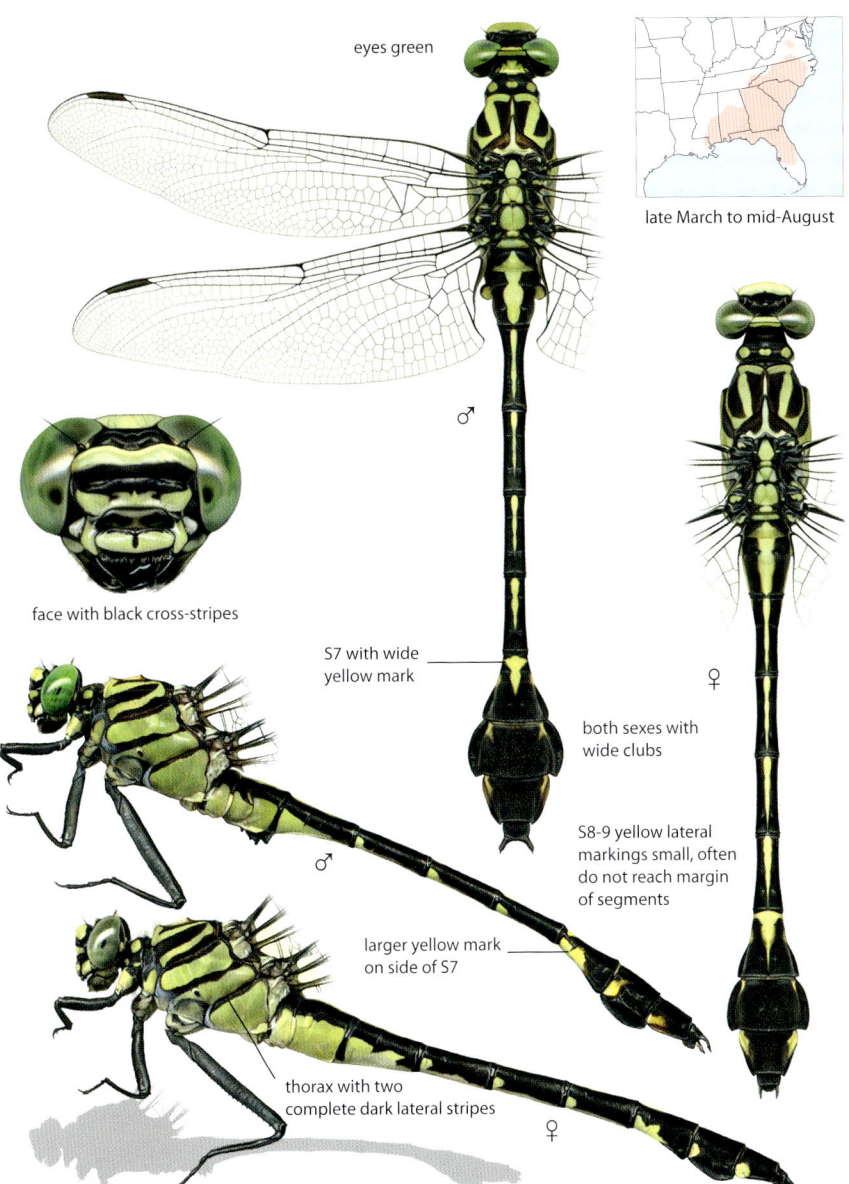

eyes green

late March to mid-August

face with black cross-stripes

S7 with wide yellow mark

♂

both sexes with wide clubs

♀

S8-9 yellow lateral markings small, often do not reach margin of segments

larger yellow mark on side of S7

thorax with two complete dark lateral stripes

♂

♀

Similar species: Cobra Clubtail smaller, with larger yellow spot on side of S9, yellow extending to edge of segment. **Gulf Coast Clubtail** with narrower dark facial stripe, larger yellow spots on S8-9. **Splendid Clubtail** with narrow black face line, incomplete anterior lateral thoracic stripe, smaller club, larger yellow spots on sides of S8-9.

Splendid Clubtail • *Gomphurus lineatifrons*

67–69 mm, 2.6–2.7 in. Large eastern upland clubtail found at flowing rivers, usually with mud and gravel amid rocks. Face with narrow black cross-stripe. Thorax with incomplete dark anterior lateral thoracic stripe. Both sexes clubbed, male club moderately wide and compressed laterally. Large lateral yellow spots on S8-9, both spots reaching segment edges.

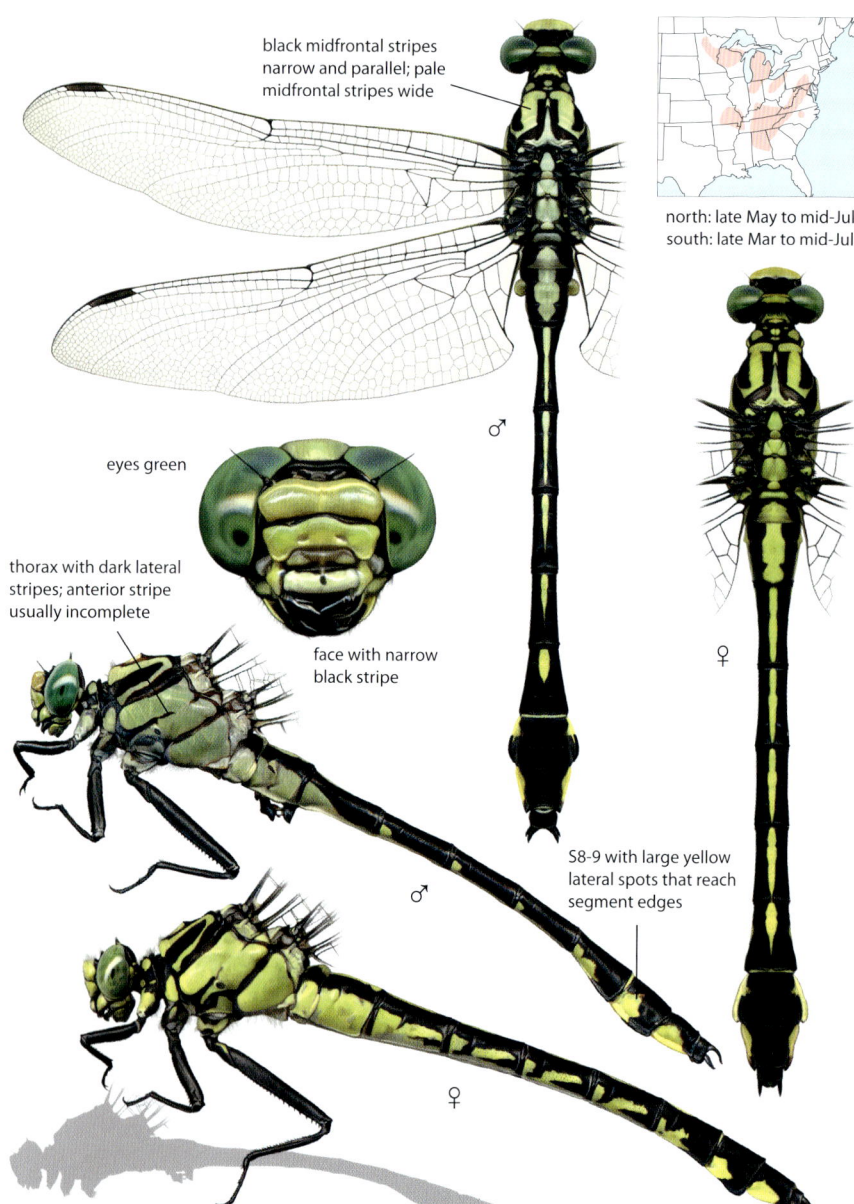

black midfrontal stripes narrow and parallel; pale midfrontal stripes wide

north: late May to mid-Jul
south: late Mar to mid-Jul

eyes green

thorax with dark lateral stripes; anterior stripe usually incomplete

face with narrow black stripe

S8-9 with large yellow lateral spots that reach segment edges

♂

♀

♂

♀

Similar species: Cobra and **Blackwater Clubtails** have wide black facial stripes, wider clubs; lateral yellow spot on S8 smaller and does not touch edge of segment. **Gulf Coast Clubtail** has complete dark lateral thoracic stripes and wider black frontal stripes.

Gulf Coast Clubtail • *Gomphurus modestus*

55–63 mm, 2.2–2.5 in. Large southern majestic clubtail found at rivers and streams with rock and sand bottoms. Face with narrow black cross-stripe. Eyes green. Thorax with complete black lateral stripes and wide frontal stripes. Abdomen of both sexes widely clubbed, with prominent lateral yellow spots on S8-9. Often has small dorsal yellow spot on S8.

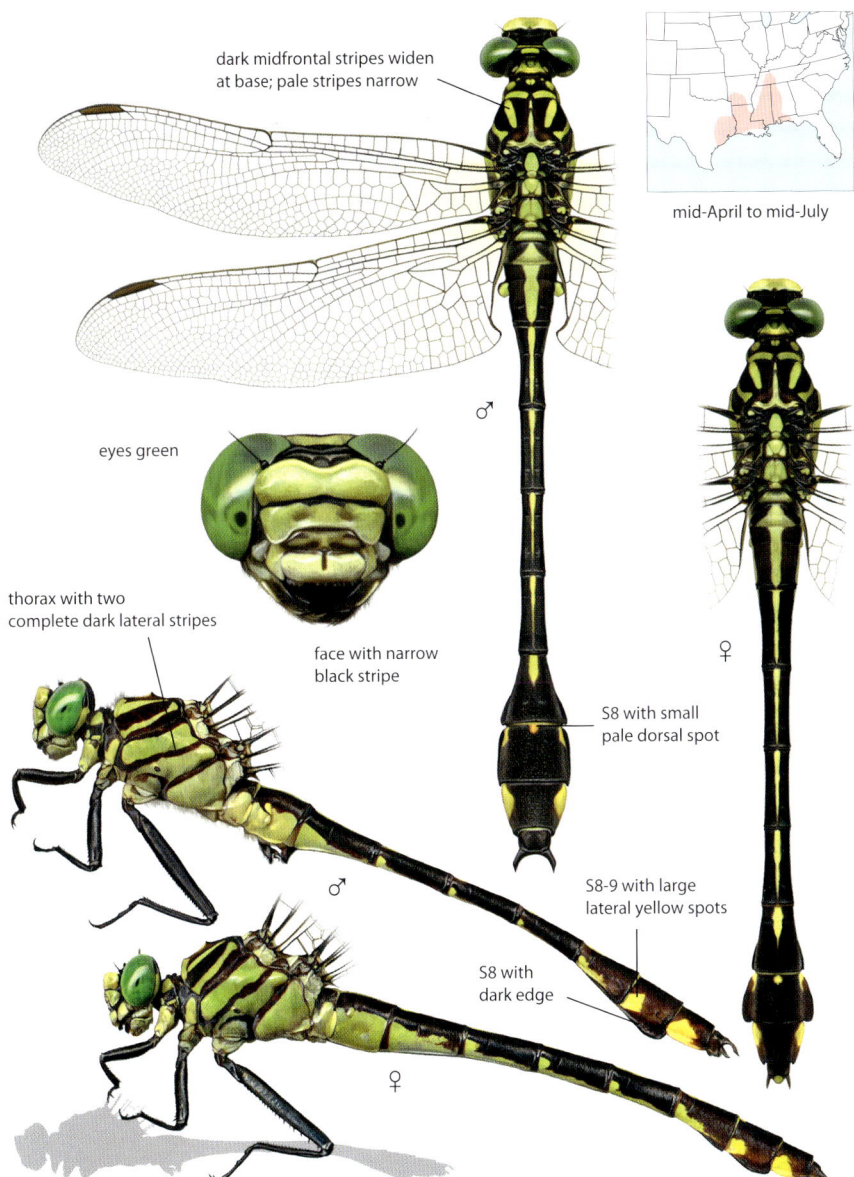

dark midfrontal stripes widen at base; pale stripes narrow

mid-April to mid-July

eyes green

thorax with two complete dark lateral stripes

face with narrow black stripe

S8 with small pale dorsal spot

♂

♀

S8-9 with large lateral yellow spots

S8 with dark edge

♂

♀

Similar species: Cobra and **Blackwater Clubtails** have wider black facial stripes and wider clubs. **Splendid Clubtail** has incomplete anterior lateral thoracic stripe, paler front of thorax, with dark frontal stripes narrower and straighter. Club smaller, with yellow lateral spot on S8 reaching edge of segment.

Skillet Clubtail • *Gomphurus ventricosus*

48–53 mm, 1.9–2.1 in. Among the smallest majestic clubtails but has the widest club. North-eastern, found at clean, slow to moderately flowing small to large rivers. Thorax with dark frontal and shoulder stripes but no obvious lateral markings. Both sexes with very large clubs, male's much wider than the thorax. Large lateral yellow spots on S8-9, edge of S8 black.

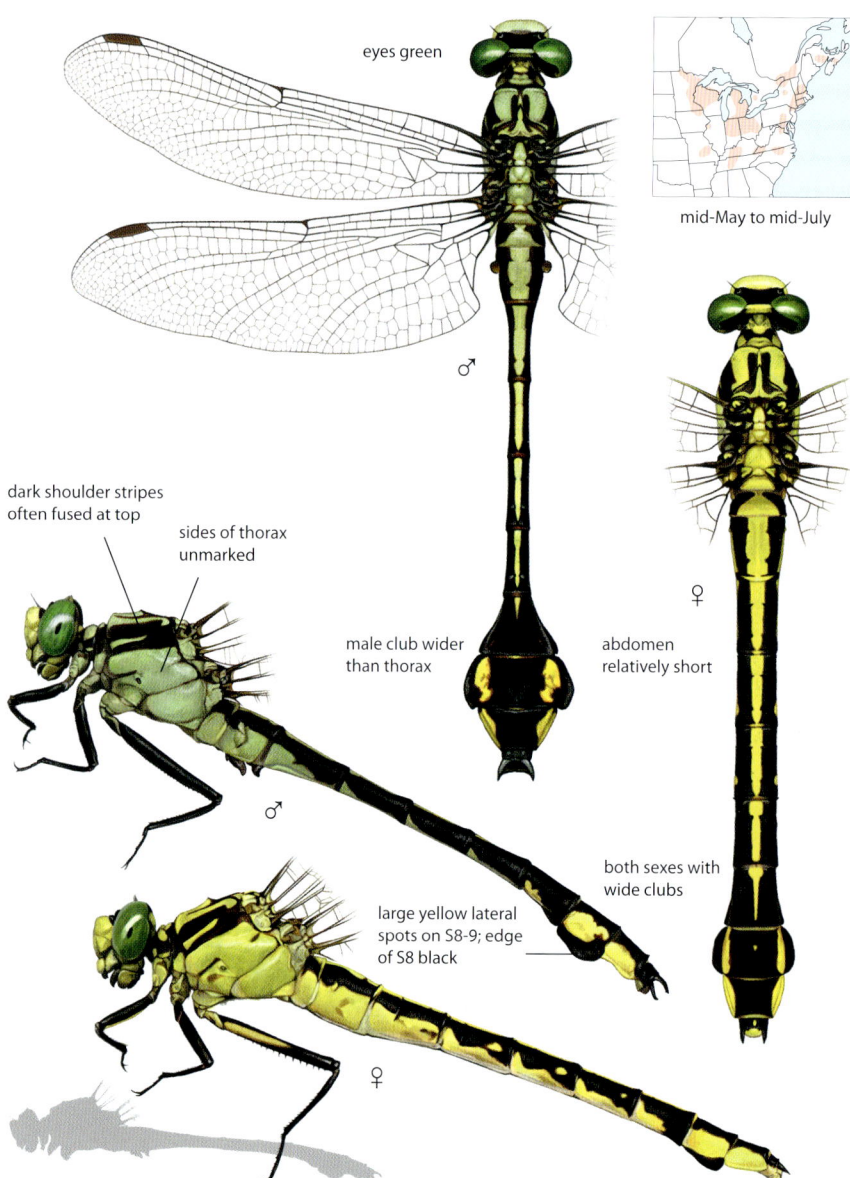

eyes green

mid-May to mid-July

♂

dark shoulder stripes often fused at top

sides of thorax unmarked

male club wider than thorax

abdomen relatively short

♀

♂

both sexes with wide clubs

large yellow lateral spots on S8-9; edge of S8 black

♀

Similar species: Combination of very large club and unmarked side of thorax separate Skillet from other majestic clubtails. **Handsome** and **Midland Clubtails** show black borders on S8 but have additional yellow dorsal marking on S8. **Black-shouldered Spinyleg** has much smaller club.

Dragonhunter • *Hagenius brevistylus*

73–90 mm, 2.9–3.5 in. The largest clubtail, common at a wide variety of eastern streams and rivers. Thorax black with two wide yellow lateral stripes, narrow yellow shoulder and frontal stripes. Abdomen slightly clubbed, black with narrow yellow middorsal stripes; S10 and top of S9 black. Appendages very short. Eyes green. Legs long and black.

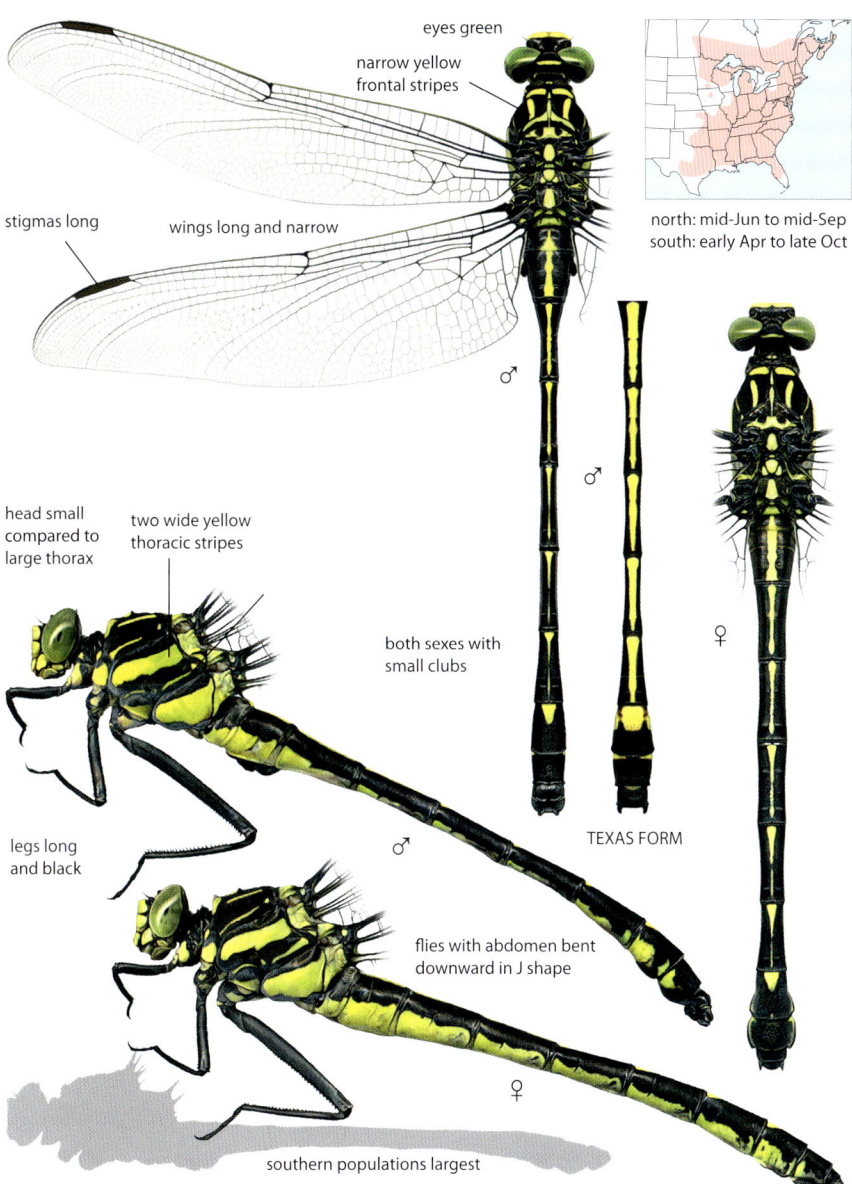

eyes green

narrow yellow frontal stripes

stigmas long

wings long and narrow

north: mid-Jun to mid-Sep
south: early Apr to late Oct

♂

♂

head small compared to large thorax

two wide yellow thoracic stripes

both sexes with small clubs

♀

legs long and black

TEXAS FORM

flies with abdomen bent downward in J shape

♂

♀

southern populations largest

Similar species: Large size distinctive among clubtails. **Majestic clubtails** smaller; thorax with wider pale frontal stripes, and abdomen usually has wider club. **River cruisers** have single yellow lateral stripe on thorax, eyes that meet at a seam. **Spiketails** have short legs and eyes that touch or nearly touch at a point.

Sable Clubtail • *Stenogomphurus rogersi*

47–50 mm, 1.9–2 in. Slender, dark clubtail of clear Appalachian forest streams. Both sexes have a long, dark abdomen with a small club; diamond-shaped club of female distinctive. Thorax with dark anterior lateral stripe widely broken. Face with two black cross-stripes; front of occiput black. Legs and appendages black. Eyes dark green.

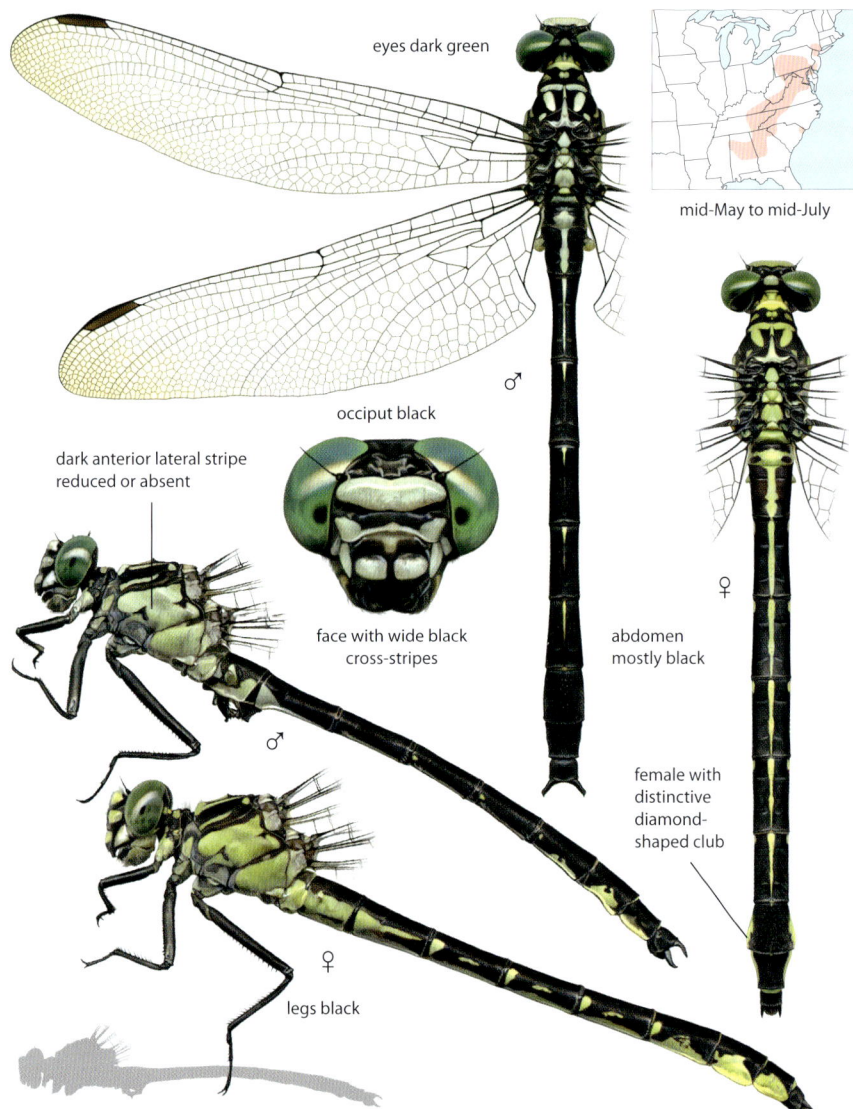

eyes dark green

mid-May to mid-July

occiput black

dark anterior lateral stripe reduced or absent

♂

face with wide black cross-stripes

abdomen mostly black

♀

female with distinctive diamond-shaped club

♀

legs black

Similar species: Cherokee Clubtail usually has complete dark anterior lateral thoracic stripe. Face with narrow cross-line; front of occiput pale. **Rapids Clubtail** smaller. Dark lateral thoracic stripes well developed; area between them darkened. Lacks dark facial stripes. **Harpoon Clubtail** has stronger dark lateral thoracic stripes. Occiput pale; lacks dark facial stripes. Female unclubbed, with pale stripe on hind leg femur. **Hanging clubtails** have isolated pale thoracic frontal stripes, dorsal pale markings on S8.

Cherokee Clubtail • *Stenogomphurus consanguis*

48–50 mm, 1.9–2 in. Slender southern Appalachian clubtail inhabiting spring-fed forest streams. Abdomen long with small club; female club diamond-shaped. Abdomen black, with thin pale dorsal streaks. Thorax usually with two narrow black lateral stripes. Face with single black line; occiput pale. Legs and appendages black. Eyes turquoise/green.

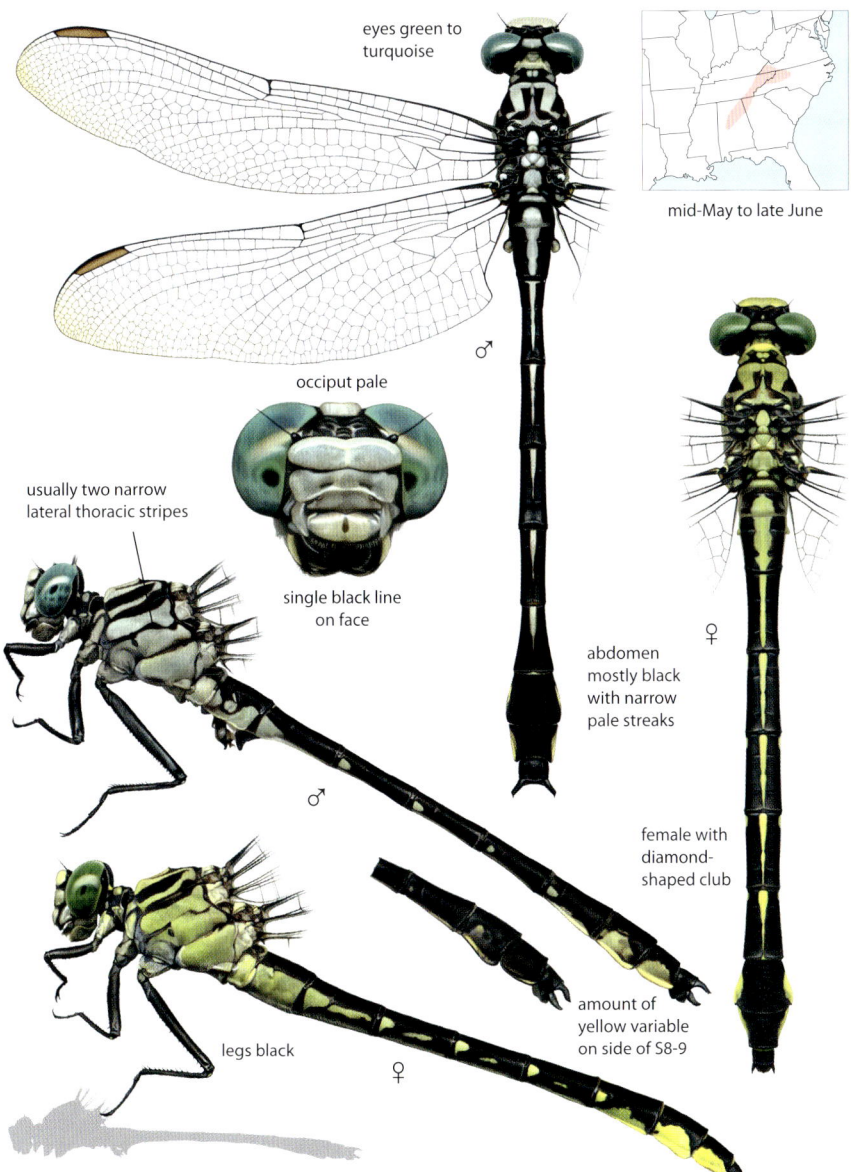

eyes green to turquoise

mid-May to late June

occiput pale

usually two narrow lateral thoracic stripes

♂

single black line on face

♂

legs black

♀

abdomen mostly black with narrow pale streaks

♀

female with diamond-shaped club

amount of yellow variable on side of S8-9

Similar species: Sable Clubtail has dark anterior lateral thoracic stripe widely broken, face with wider black facial stripes; front of occiput black. See other similar species under Sable Clubtail.

Gray Sanddragon • *Progomphus borealis*

57–60 mm, 2.2–2.4 in. Common southwestern sandddragon inhabiting sand-bottomed rocky streams and rivers. Thorax dark brown, with yellow frontal stripes, gray sides, single dark lateral stripe. Abdomen black, with yellow to orange dorsal triangles on S2-7, complete pale ring on S7, pale spots on S8-10; cerci pale. Dark mark at base of each wing.

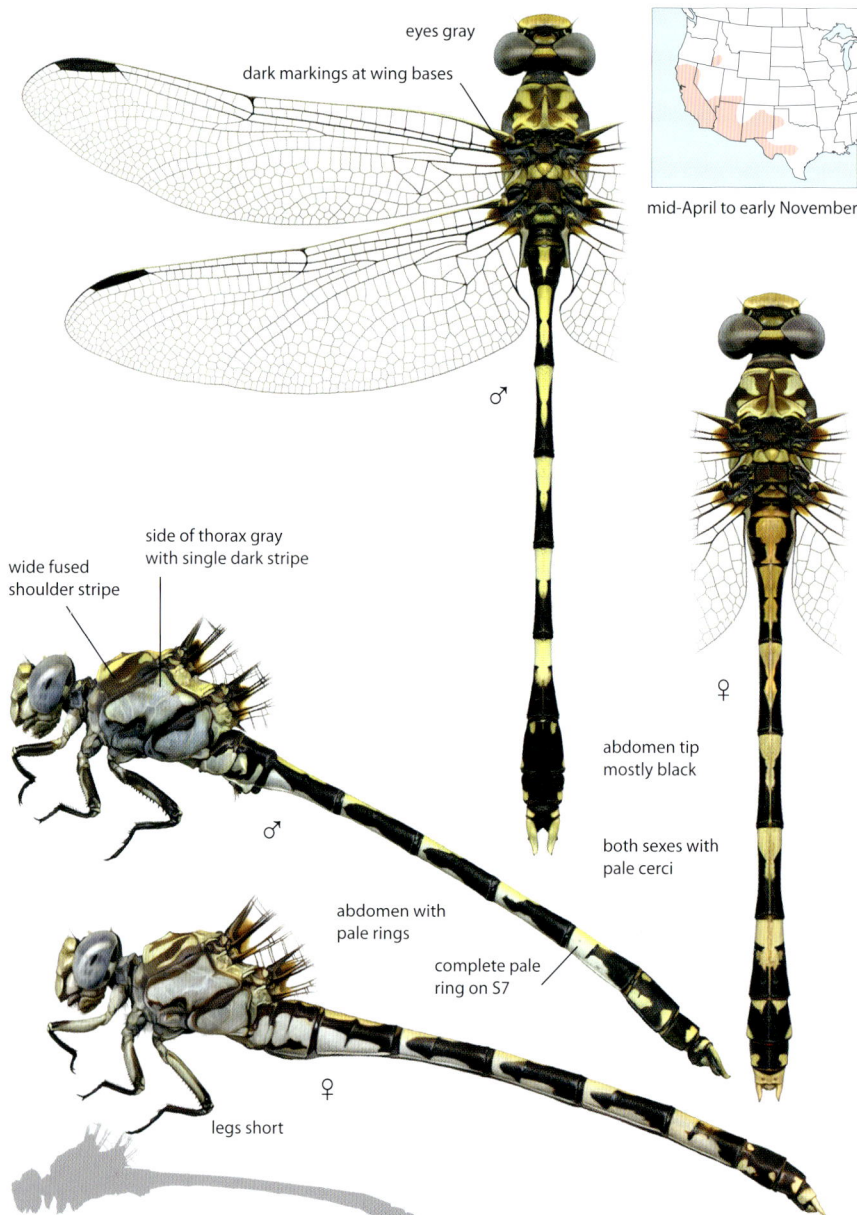

eyes gray

dark markings at wing bases

mid-April to early November

side of thorax gray with single dark stripe

wide fused shoulder stripe

♂

♀

abdomen tip mostly black

both sexes with pale cerci

abdomen with pale rings

complete pale ring on S7

♀

legs short

Similar species: Common Sanddragon has two dark lateral thoracic stripes; abdomen lacks pale rings. **Eastern** and **White-belted Ringtails** have narrow, separated shoulder stripes.

Common Sanddragon • *Progomphus obscurus*

47–53 mm, 1.9–2.1 in. Most widespread sanddragon, common at sand-bottomed streams, rivers, also sandy lakes to the north. Thorax heavily striped, including two complete lateral stripes. Abdomen black with yellow dorsal markings on S2-7. S8-10 mostly black; cerci pale. Wings with dark marks at base. Eyes olive to brown. Legs short; femurs pale-striped.

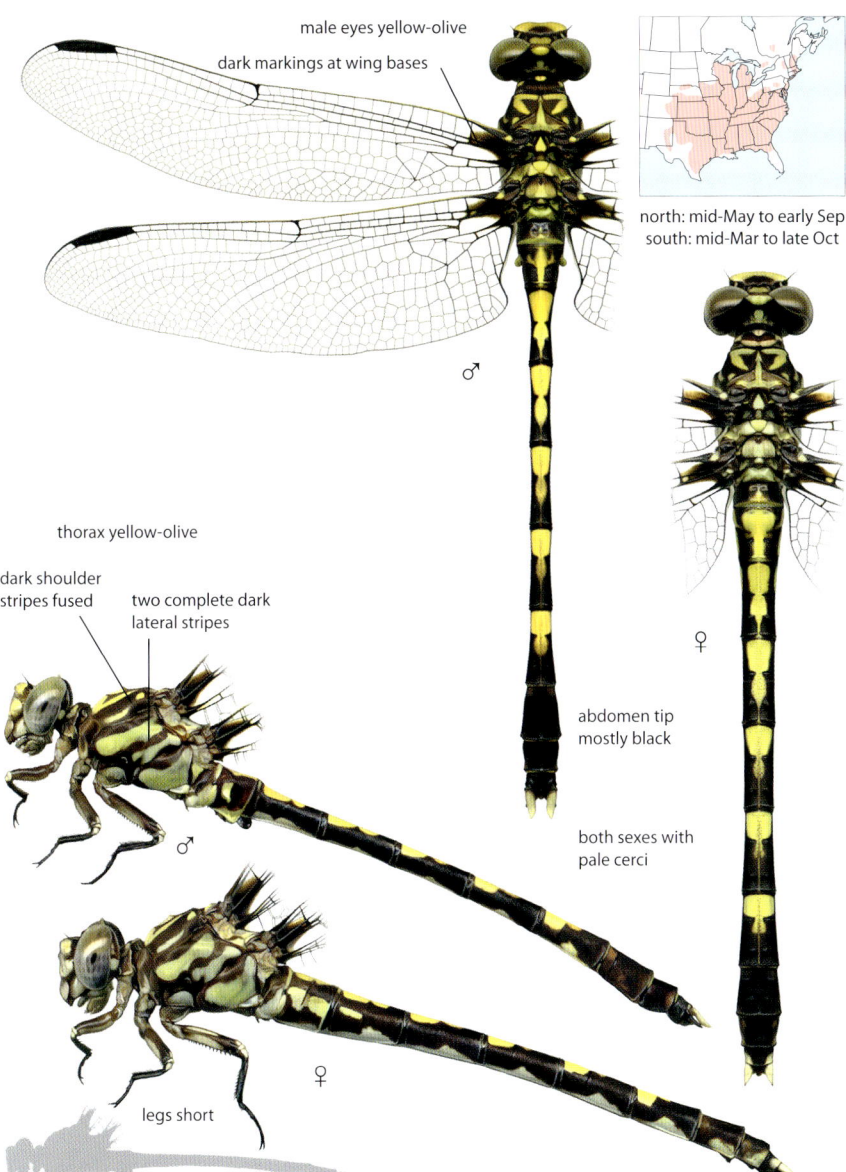

male eyes yellow-olive

dark markings at wing bases

north: mid-May to early Sep
south: mid-Mar to late Oct

♂

♀

thorax yellow-olive

dark shoulder stripes fused

two complete dark lateral stripes

♂

abdomen tip mostly black

both sexes with pale cerci

♀

legs short

Similar species: Belle's Sanddragon larger; dark lateral thoracic stripes usually fused at top. S8 with pair of small yellow dorsal rectangles; male cerci longer than S10. **Tawny Sanddragon** S8 extensively marked with orange. **Gray Sanddragon** has side of thorax gray with single dark stripe, has complete pale ring at base of S7.

Belle's Sanddragon • *Progomphus bellei*

54–60 mm, 2.1–2.4 in. Brightly patterned southeastern sanddragon found at sand-bottomed lakes and small, open, sandy streams. Thorax with fused dark brown shoulder stripe; two dark lateral stripes widely connected at bottom and often at top. Abdomen with long yellow dorsal markings on S2-7, small yellow rectangles on S8. Cerci pale, male's long.

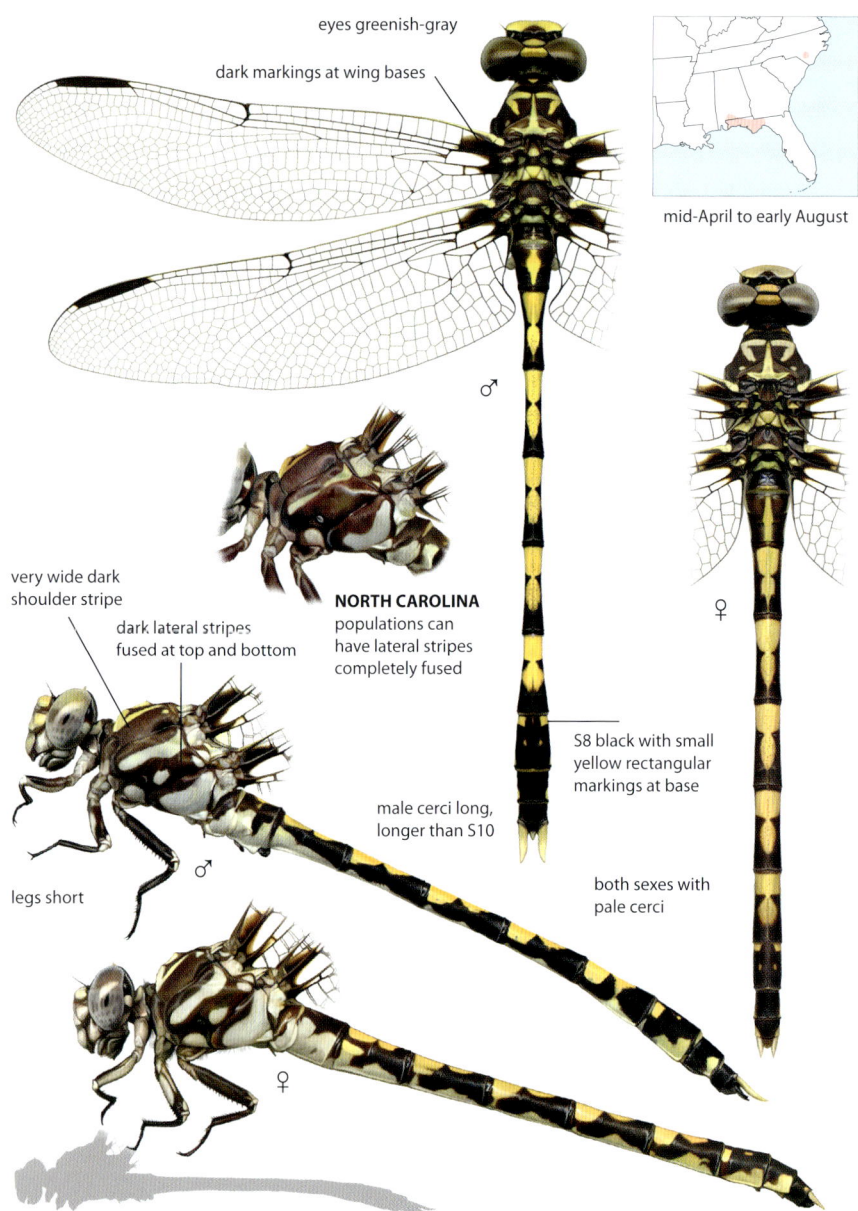

eyes greenish-gray

dark markings at wing bases

mid-April to early August

♂

NORTH CAROLINA populations can have lateral stripes completely fused

very wide dark shoulder stripe

dark lateral stripes fused at top and bottom

legs short

♂

male cerci long, longer than S10

S8 black with small yellow rectangular markings at base

both sexes with pale cerci

♀

♀

Similar species: Common Sanddragon smaller; dark lateral thoracic stripes not fused at top. S8 markings reduced or missing, and male cerci equal to S10 in length. **Tawny Sanddragon** with less contrasting pattern; S8 extensively marked in orange.

Tawny Sanddragon • *Progomphus alachuensis*

52–57 mm, 2–2.2 in. Brownish sanddragon of the Florida peninsula, found at sand-bottomed lakes, also sandy streams southward. Thorax with wide fused brown shoulder stripes, and fused dark lateral stripe with two pale spots in middle. Abdomen dark brown with orange markings on top of S2-8; S8 mostly orange. Cerci pale, male's longer than S10.

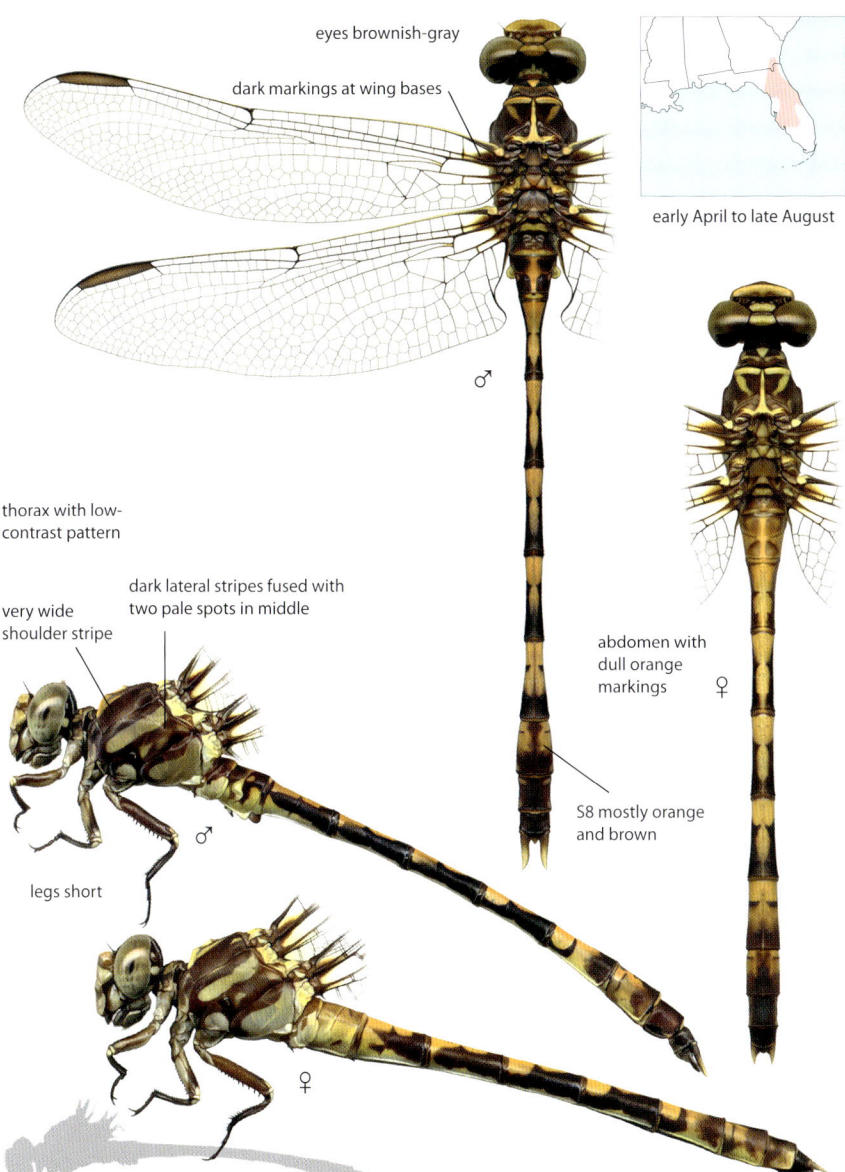

eyes brownish-gray

dark markings at wing bases

early April to late August

thorax with low-contrast pattern

dark lateral stripes fused with two pale spots in middle

very wide shoulder stripe

abdomen with dull orange markings ♀

S8 mostly orange and brown

legs short

♂

♀

Similar species: Common Sanddragon has more contrasting pattern; dark lateral thoracic stripes unfused at top, S8 mostly black, and male cerci equal to S10 in length. **Belle's Sanddragon** not known to share range; has more contrasting pattern. S8 black above with two small but well-defined rectangular pale markings.

Maine Snaketail • *Ophiogomphus mainensis*

42–46 mm, 1.7–1.8 in. Fairly common northeastern and Appalachian snaketail found at clean, rocky woodland streams and small rivers. Thorax bright green, with single black lateral stripe. Abdomen with narrow pale dorsal markings, markings on top of S8-9 reduced or absent. Eyes green. Legs black. Female with prominent forward-pointing occipital horns.

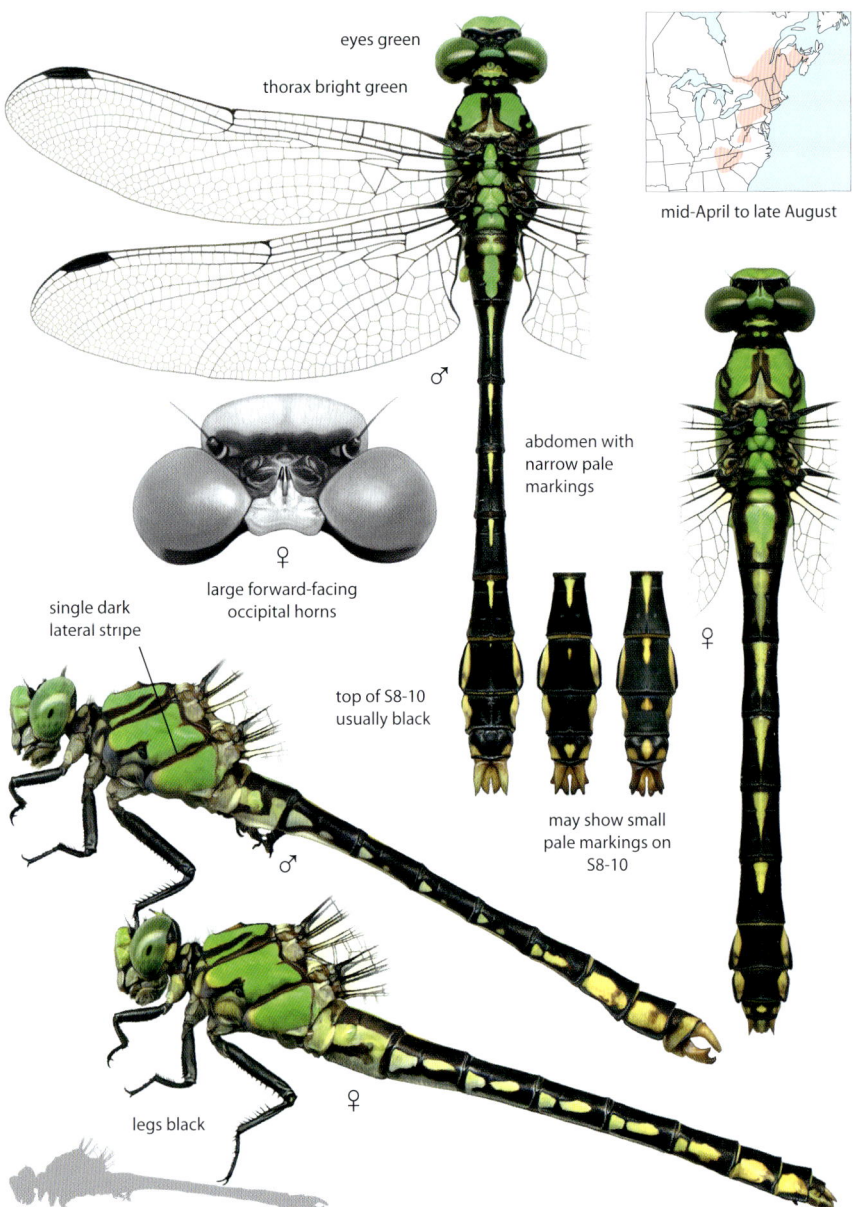

eyes green

thorax bright green

mid-April to late August

♂

abdomen with narrow pale markings

large forward-facing occipital horns

♀

single dark lateral stripe

top of S8-10 usually black

may show small pale markings on S8-10

♂

♀

legs black

♀

Similar species: Riffle Snaketail has pale markings on top of all abdominal segments; female lacks large forward-facing occipital horns. **Brook** and **Boreal Snaketails** have pale markings on top of all segments, paler femurs.

Riffle Snaketail • *Ophiogomphus carolus*

40–45 mm, 1.6–1.8 in. Northeastern snaketail inhabiting clear, sandy, and rocky streams and rivers. Thorax whitish-green with single black lateral stripe. Pale markings on all abdominal segments, small pale lengthwise rectangle on S8, pale crosswise rectangle on S9. Eyes green. Legs usually black; femur may show pale.

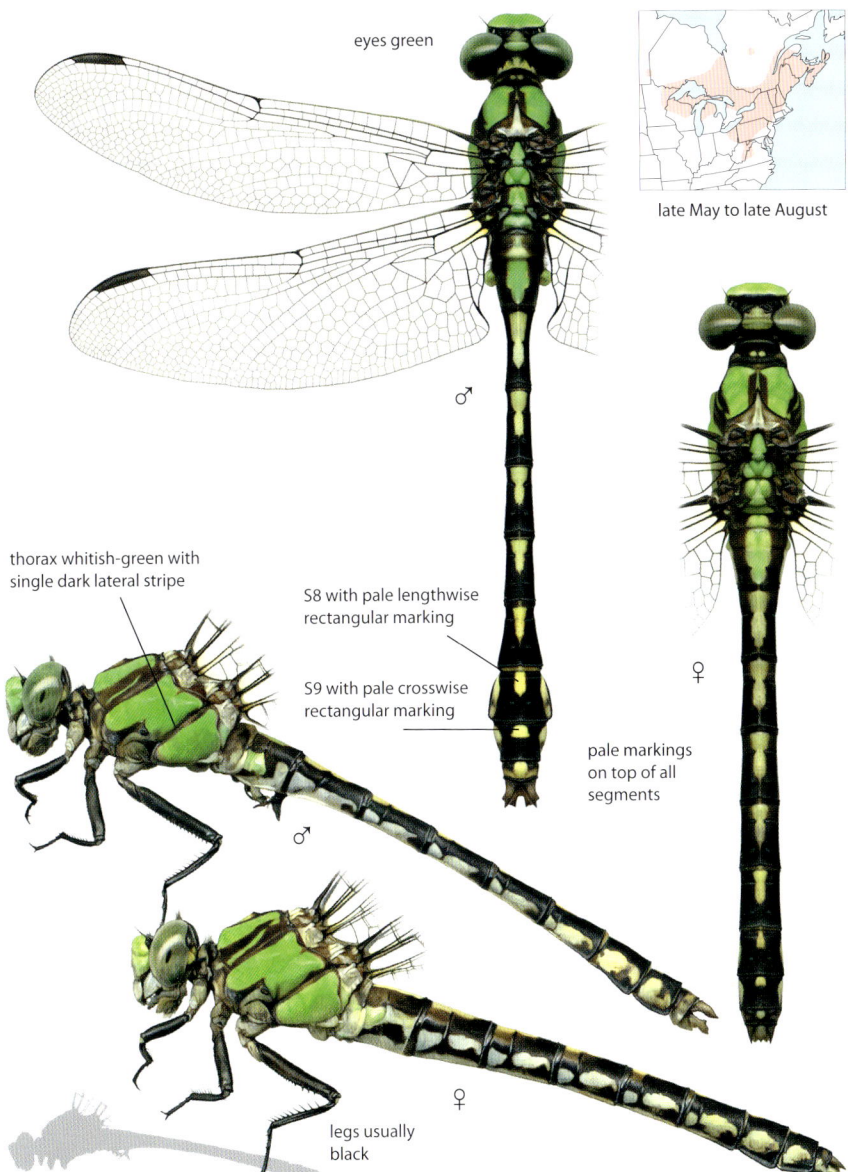

eyes green

late May to late August

♂

thorax whitish-green with single dark lateral stripe

S8 with pale lengthwise rectangular marking

S9 with pale crosswise rectangular marking

pale markings on top of all segments

♀

♂

♀

legs usually black

Similar species: Maine Snaketail usually lacks pale markings on top of S8-9. Male abdomen with very narrow dorsal stripes; female with large forward-facing occipital horns. **Brook, Boreal,** and **Sioux Snaketails** have paler legs (although Riffle may show pale on femur); pale marking on top of S8 triangular.

Brook Snaketail • *Ophiogomphus aspersus*

44–49 mm, 1.7–1.9 in. Northeastern and Appalachian snaketail found at sandy streams. Thorax grass-green, with single dark lateral stripe. Femurs pale. Abdomen with dorsal pale markings on all segments, S8 with half-length triangular mark. Eyes green. Female has postoccipital horns, lacks occipital horns; subgenital plate three-fourths the length of S9.

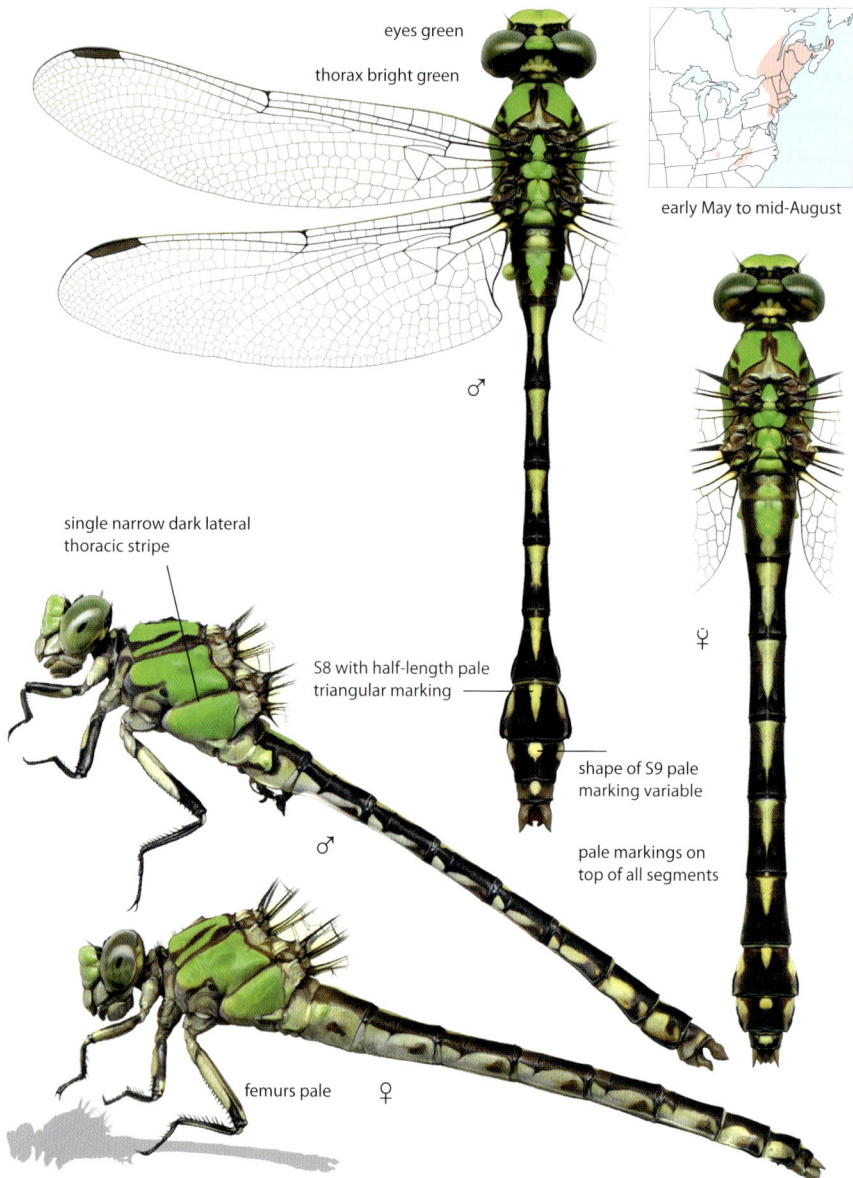

eyes green

thorax bright green

early May to mid-August

single narrow dark lateral thoracic stripe

S8 with half-length pale triangular marking

shape of S9 pale marking variable

pale markings on top of all segments

♂

♀

♂

femurs pale ♀

Similar species: Maine Snaketail usually has top of S8-9 black, black legs. **Riffle Snaketail** has rectangular pale marking on top of S8; legs usually black. **Boreal Snaketail** has black facial stripes. **Sioux Snaketail** has separate range, pale dorsal abdominal markings nearly full length of segment.

Sioux Snaketail • *Ophiogomphus smithi*

43–47 mm, 1.7–1.9 in. Brightly marked snaketail with a limited range in Wisconsin, Minnesota, Iowa, at sandy streams. Thorax green with single dark lateral stripe. Abdomen with nearly full-length pale markings on top of all segments. Legs pale-striped. Eyes pale blue-gray. Female with occipital and postoccipital horns; subgenital plate half the length of S9.

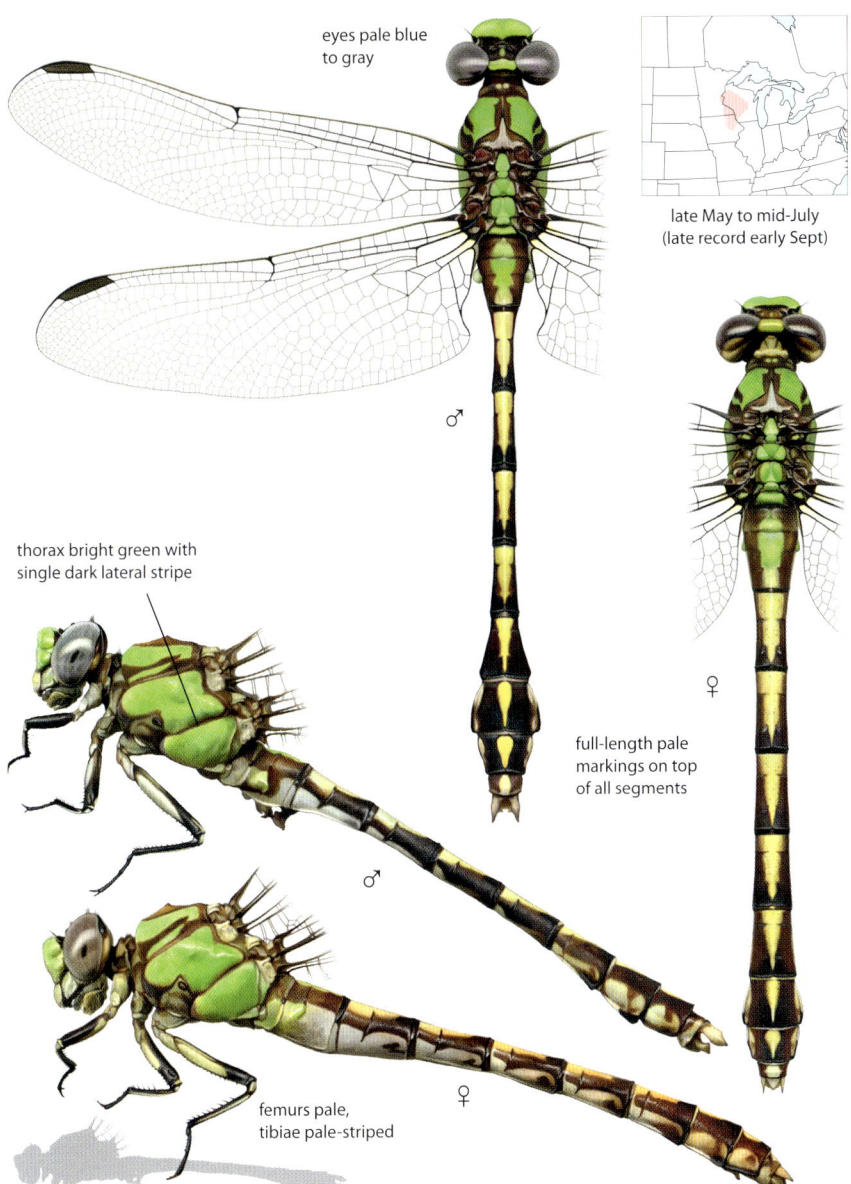

eyes pale blue
to gray

late May to mid-July
(late record early Sept)

thorax bright green with
single dark lateral stripe

full-length pale
markings on top
of all segments

femurs pale,
tibiae pale-striped

♂

♀

Similar species: Brook Snaketail most similar but with separate range. Pale dorsal S8 marking half length of segment; eyes green. **Riffle Snaketail** has smaller pale abdominal markings, S8 marking rectangular; legs usually black.

Boreal Snaketail • *Ophiogomphus colubrinus*

41–48 mm, 1.6–1.9 in. Northern snaketail inhabiting clear, running, gravel-bottomed streams and rivers. Thorax bright green with single black lateral stripe. Abdomen black with small whitish markings on top of all segments, middle segment markings T-shaped. Face with black-cross stripes. Eyes green. Femurs mostly pale.

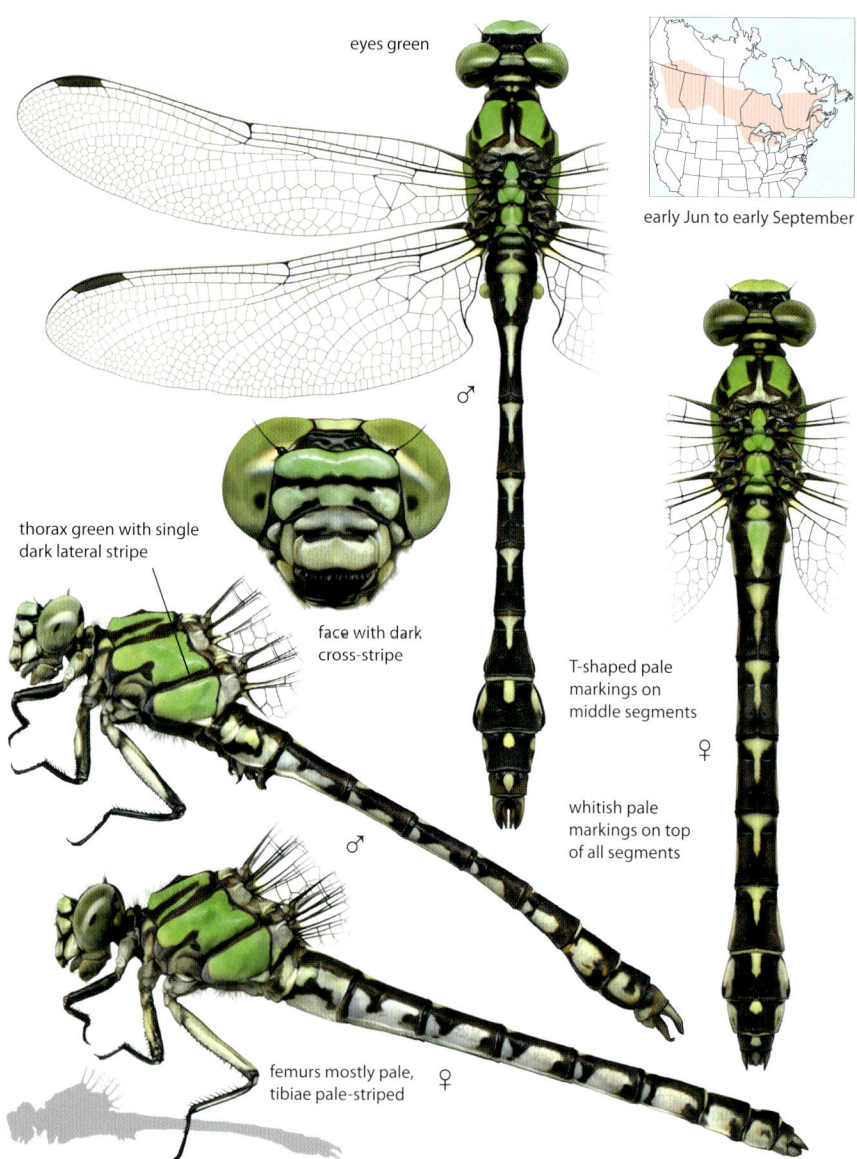

eyes green

early Jun to early September

thorax green with single dark lateral stripe

face with dark cross-stripe

♂

T-shaped pale markings on middle segments

whitish pale markings on top of all segments

♀

♂

femurs mostly pale, tibiae pale-striped ♀

Similar species: Only **Extra-striped Snaketail** has black facial stripes but has black diagonal lateral thoracic stripe; legs black. **Maine Snaketail** usually lacks dorsal pale markings on S8-9; legs black. **Riffle Snaketail** has rectangular pale marking on top of S8; legs usually black. **Brook Snaketail** has triangular dorsal pale markings on abdomen (not T-shaped).

Extra-striped Snaketail • *Ophiogomphus anomalus*

39–44 mm, 1.5–1.7 in. Small northeastern snaketail found at clear, running waters, more common at larger slow-flowing rivers. Side of thorax with partial black anterior stripe, black diagonal stripe, and complete black posterior stripe. Abdomen with narrow pale markings on top of all segments. Face with black cross-stripes. Eyes green. Legs black.

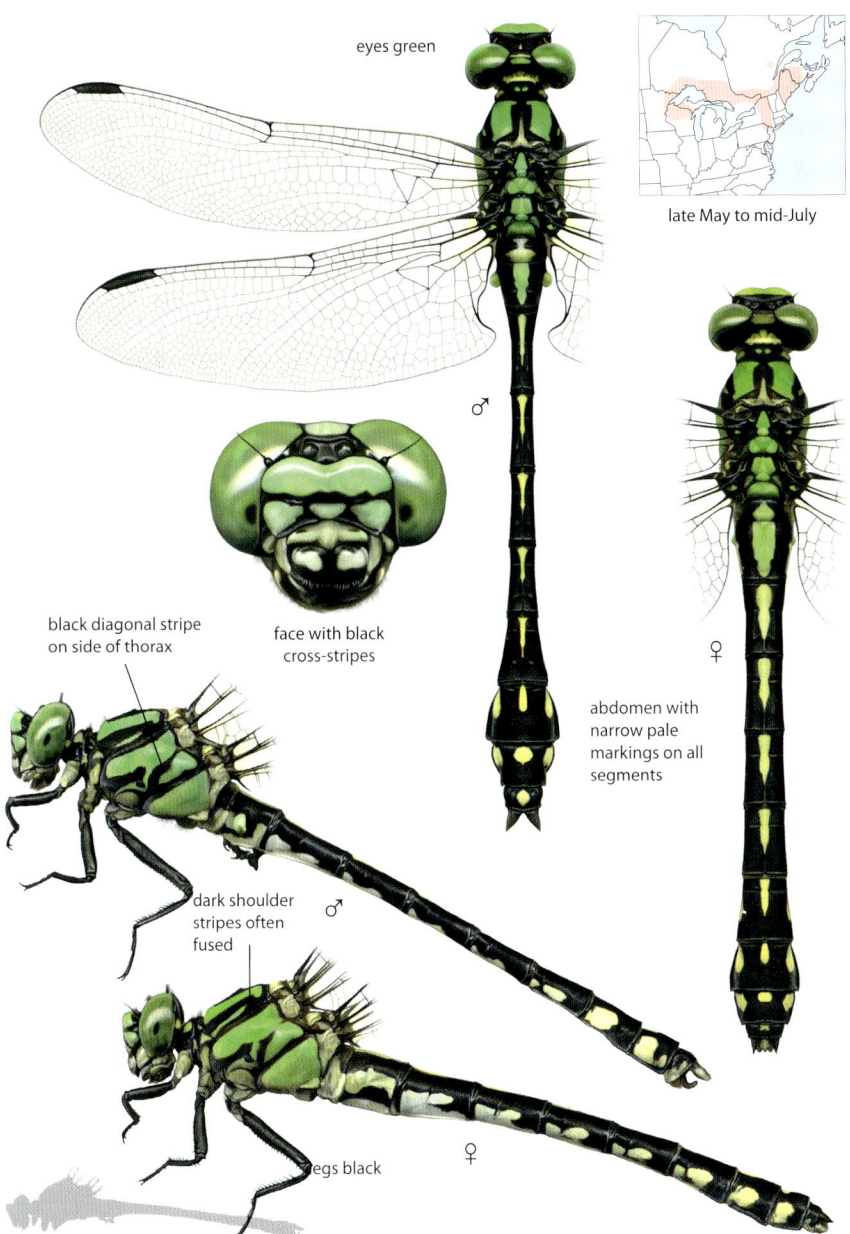

eyes green

late May to mid-July

black diagonal stripe on side of thorax

face with black cross-stripes

♂

abdomen with narrow pale markings on all segments

♀

dark shoulder stripes often fused

♂

legs black

♀

Similar species: No other snaketail has a diagonal black lateral thoracic stripe. **Boreal Snaketail** has black facial stripes, but legs are paler, thorax with single dark lateral stripe.

Edmund's Snaketail • *Ophiogomphus edmundo*

45–48 mm, 1.8–1.9 in. Rare snaketail with limited southeastern range, found at clean, rocky rivers. Thorax bright green, with two dark lateral thoracic stripes, anterior stripe often broken. Abdomen black, with pale yellow markings on top of all segments. Eyes aqua-blue to dull green. Legs mostly black.

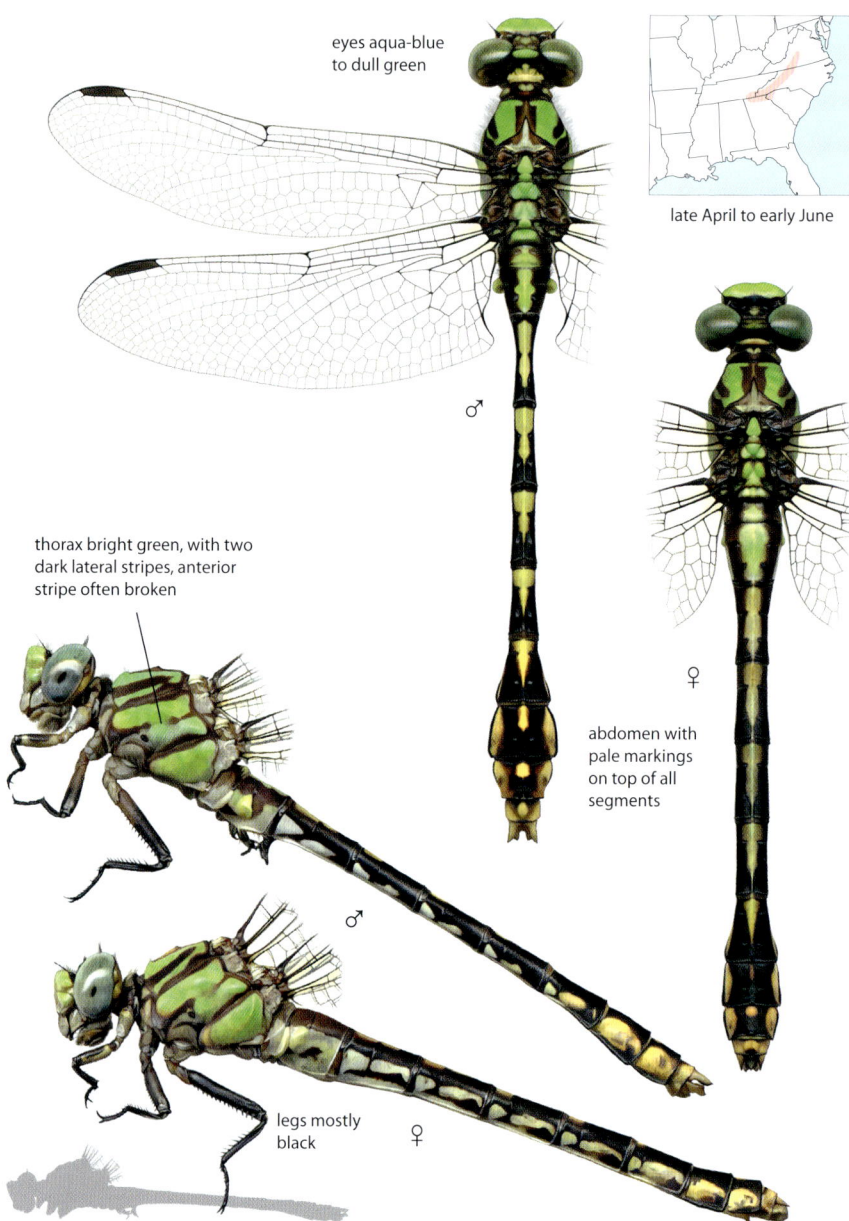

eyes aqua-blue to dull green

late April to early June

thorax bright green, with two dark lateral stripes, anterior stripe often broken

♂

♀

abdomen with pale markings on top of all segments

legs mostly black

♂

♀

Similar species: Maine Snaketail has single black lateral thoracic stripe, green eyes. **Appalachian Snaketail** lacks dark anterior lateral thoracic stripe; femurs mostly pale.

St. Croix Snaketail • *Ophiogomphus susbehcha*

50–52 mm, 2 in. Snaketail local to clean, slow rivers in the Chippewa and St. Croix watersheds, Minnesota and Wisconsin. Thorax bluish-green, with complete dark posterior lateral stripe. Abdomen black, with pale dorsal markings on all segments; S10 almost completely pale. Eyes blue. Femurs black with pale bases. Male epiproct longer than cerci.

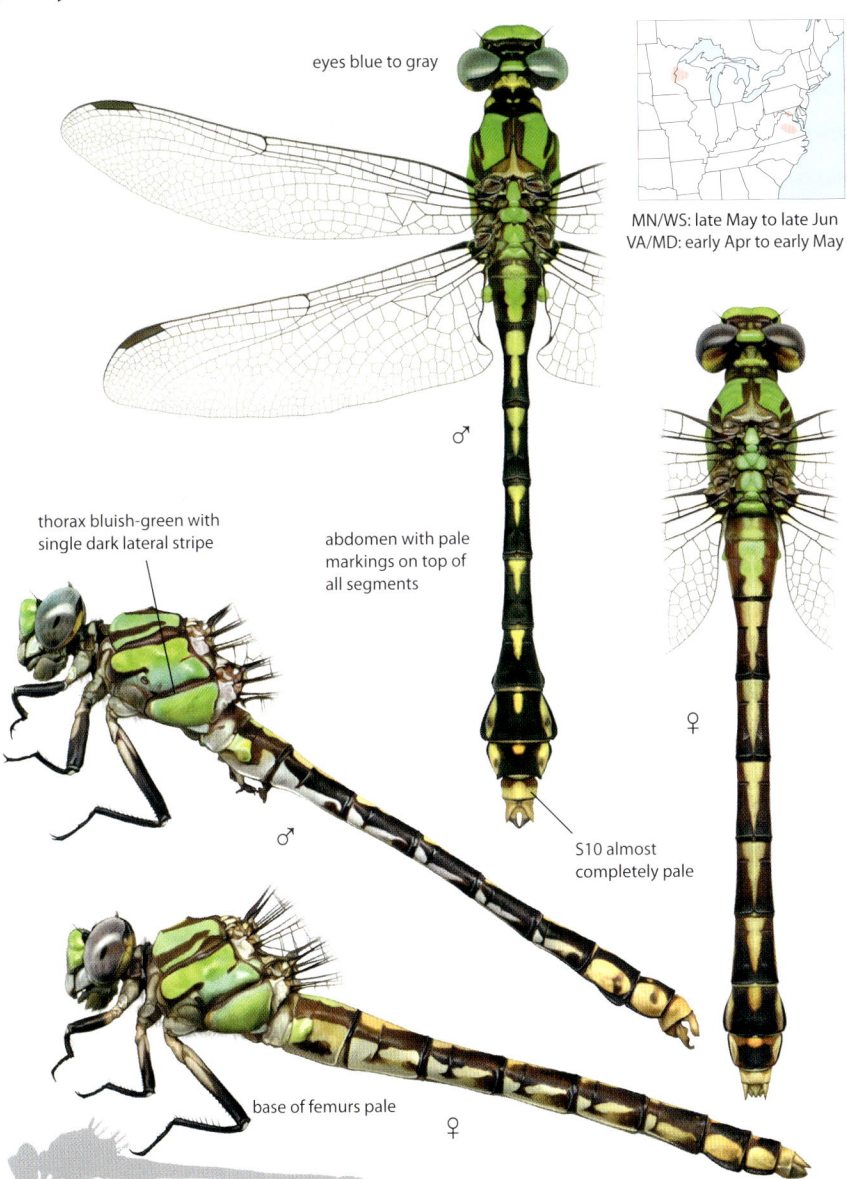

eyes blue to gray

MN/WS: late May to late Jun
VA/MD: early Apr to early May

thorax bluish-green with
single dark lateral stripe

abdomen with pale
markings on top of
all segments

♂

♀

♂

S10 almost
completely pale

base of femurs pale

♀

Similar species: Disjunct population in Virginia and Maryland similar in structure but likely an undescribed species. **Sioux, Riffle,** and **Brook Snaketails** have black markings on S10; Riffle and Brook have green eyes. **Boreal Snaketail** has black facial stripes, pale whitish abdominal markings, black on S10; femurs mostly pale.

Appalachian Snaketail • *Ophiogomphus incurvatus*

40–49 mm, 1.6–1.9 in. Small, colorful southeastern snaketail found at clear, gravel- and mud-bottomed streams. Thorax bright green with brown stripes, single lateral stripe sometimes reduced or absent. Abdomen with pale markings on all segments, club and S10 yellow to orange. Eyes aqua-blue in male, female gray. Femurs mostly pale.

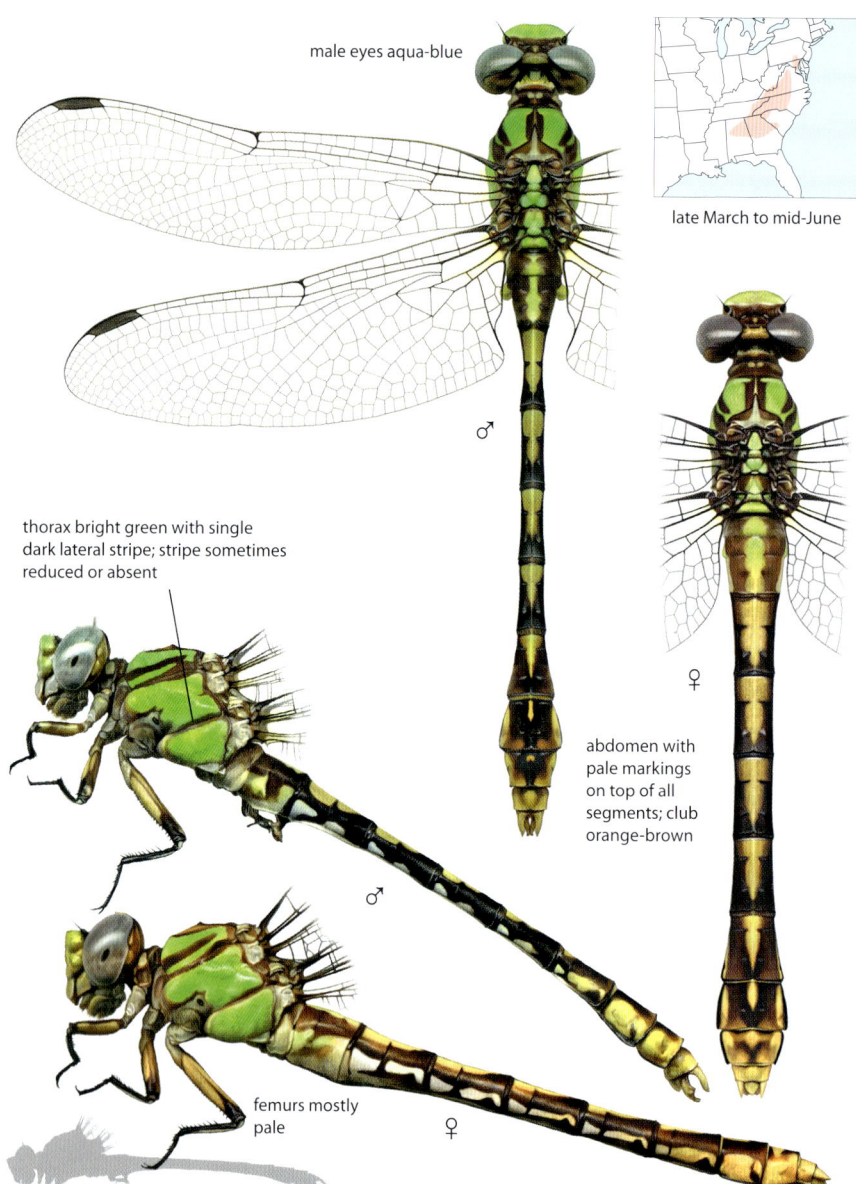

male eyes aqua-blue

late March to mid-June

thorax bright green with single dark lateral stripe; stripe sometimes reduced or absent

♂

♀

abdomen with pale markings on top of all segments; club orange-brown

♂

femurs mostly pale

♀

Similar species: Edmund's Snaketail has two dark lateral thoracic stripes, darker legs. **Maine Snaketail** usually lacks pale markings on top of S8-9, and has green eyes, black legs. **Brook Snaketail** has green eyes, more black on S10. **Rusty Snaketail** largely lacks dark frontal and lateral thoracic stripes; abdomen brown.

Southern Snaketail • *Ophiogomphus australis*

44–46 mm, 1.7–1.8 in. Small, rare snaketail with limited southern range, inhabiting small, gravel-bottomed streams. Thorax bright yellow-green with two dark brown lateral stripes, narrow but usually complete. Abdomen with pale markings on all segments; S10 mostly pale, dorsal markings orange. Eyes blue-gray. Femurs mostly pale.

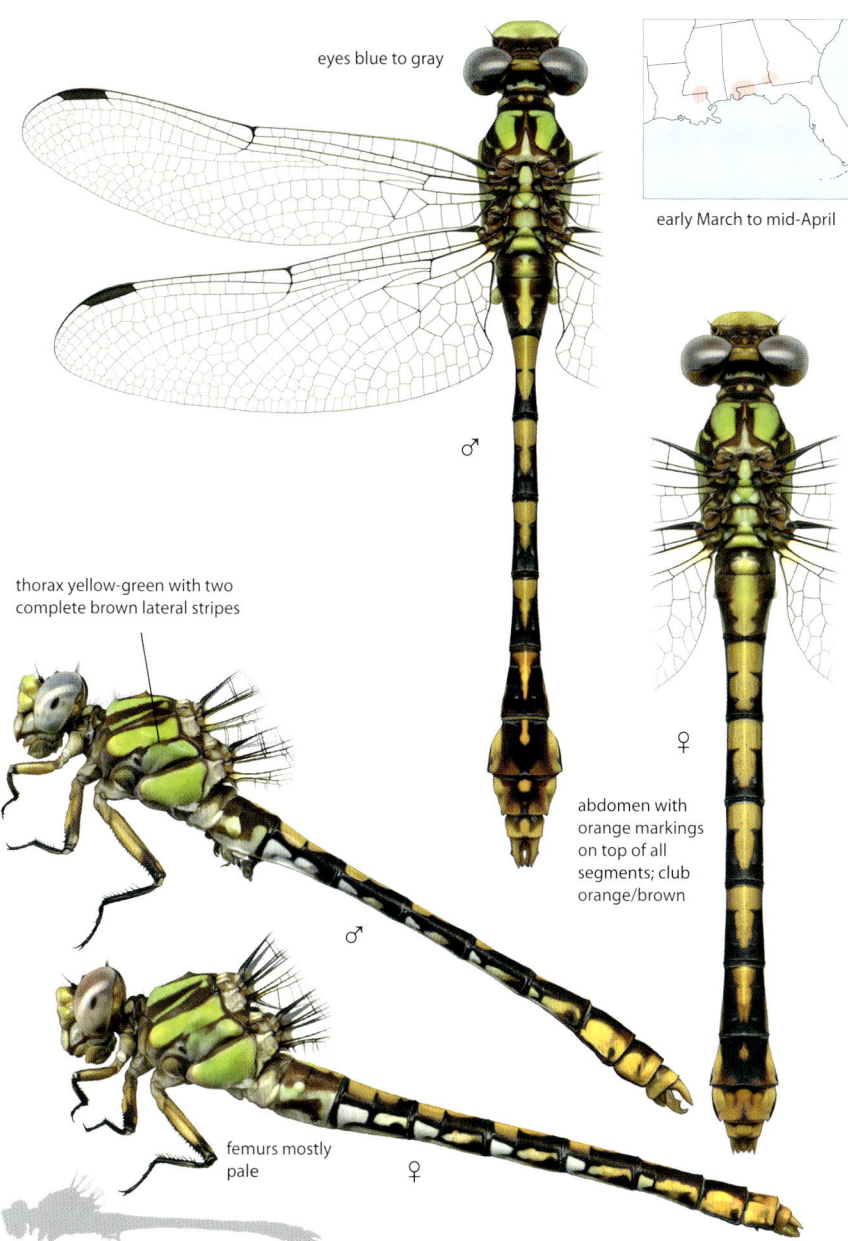

eyes blue to gray

early March to mid-April

thorax yellow-green with two complete brown lateral stripes

♂

♀

abdomen with orange markings on top of all segments; club orange/brown

femurs mostly pale

♂

♀

Similar species: Appalachian Snaketail not known to share range, lacks dark anterior lateral thoracic stripe.

Rusty Snaketail • *Ophiogomphus rupinsulensis*

45–54 mm, 1.8–2.1 in. Common, pale northeastern snaketail found at large flowing streams and rivers. Thorax bright green with reduced brown shoulder stripes, frontal and lateral stripes vestigial or absent. Abdomen mostly brown with low-contrast pattern, pale markings suggesting rings. Eyes green. Femurs mostly pale, tibiae pale-striped.

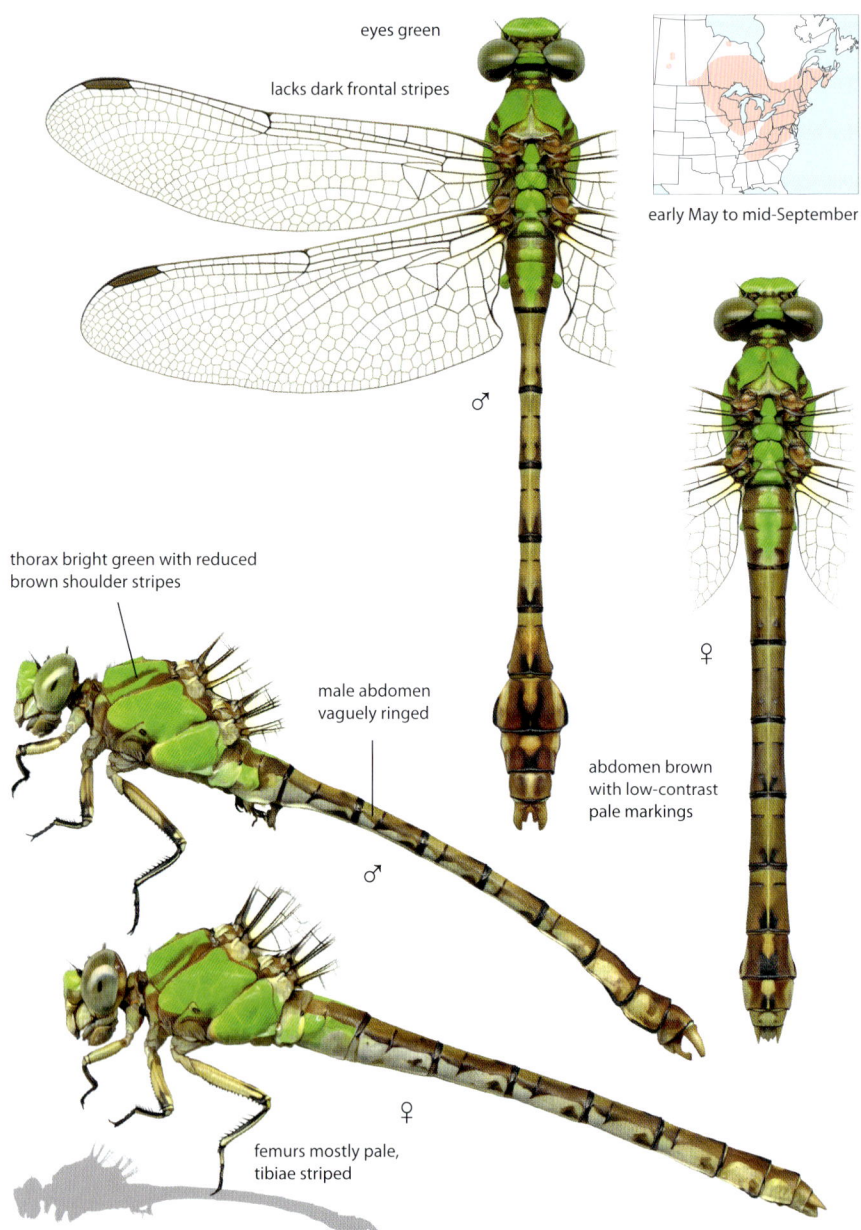

eyes green

lacks dark frontal stripes

early May to mid-September

thorax bright green with reduced brown shoulder stripes

♂

male abdomen vaguely ringed

♂

abdomen brown with low-contrast pale markings

♀

♀

femurs mostly pale, tibiae striped

Similar species: Acuminate Snaketail has bluer eyes, more contrasting thoracic and abdominal pattern. Compare male appendages and female subgenital plate, pages 416, 420.

Acuminate Snaketail • *Ophiogomphus acuminatus*

49–53 mm, 1.9–2.1 in. Rare southeastern snaketail local at clean, shaded, gravel-bottomed streams. Thorax bright yellow-green, with brown shoulder stripes; frontal and lateral stripes vestigial or absent. Abdomen mostly brown, with pale markings on all segments. Eyes blue or gray. Femurs mostly pale, tibiae with partial pale stripe.

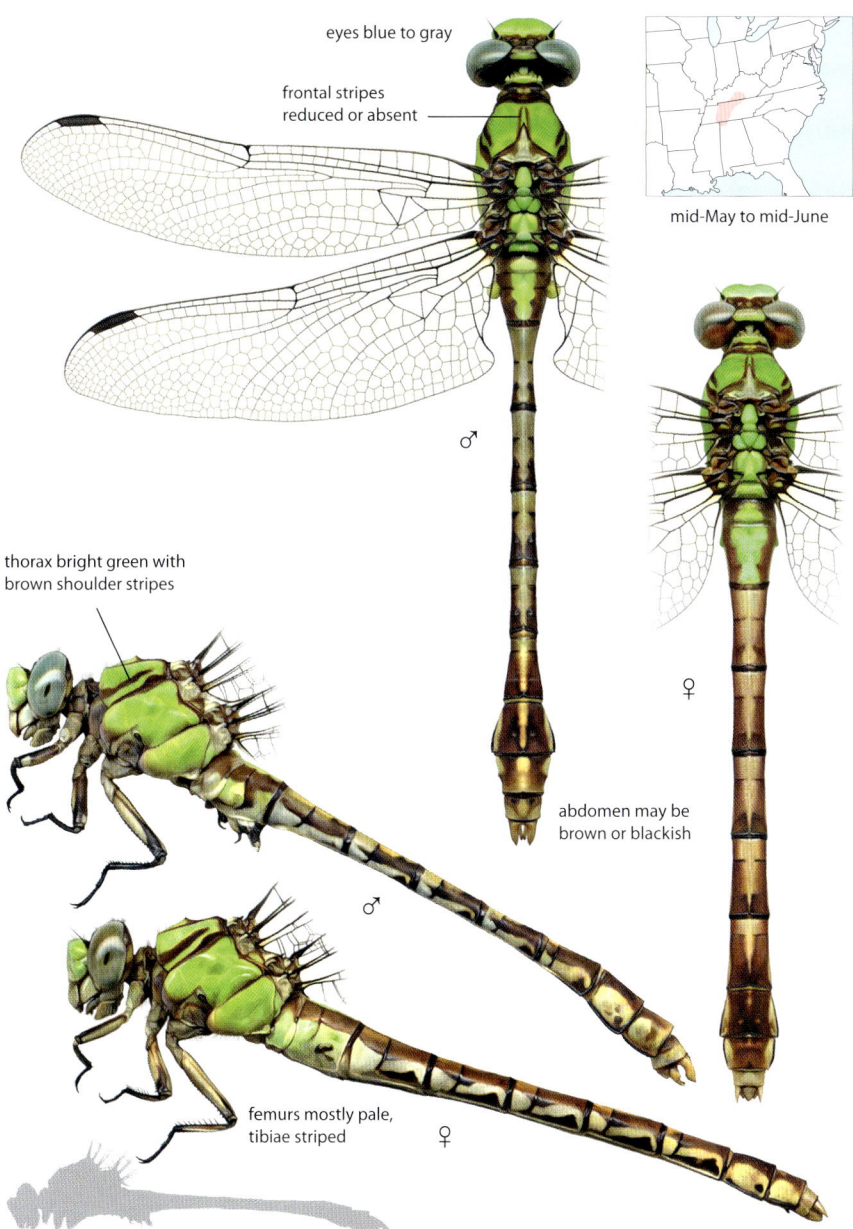

eyes blue to gray

frontal stripes reduced or absent

mid-May to mid-June

thorax bright green with brown shoulder stripes

♂

abdomen may be brown or blackish

♀

♂

femurs mostly pale, tibiae striped

♀

Similar species: Rusty Snaketail has greener eyes; thorax and abdomen pattern paler with less contrast. Compare male appendages and female subgenital plate, pages 415, 419.

Pygmy Snaketail • *Ophiogomphus howei*

31–34 mm, 1.2–1.3 in. Very small, stocky eastern snaketail inhabiting clean rivers with gravel and mud bottoms. Thorax bright green with two dark lateral stripes, anterior stripe incomplete. Abdomen distinctly short with pronounced club; dorsal pale markings minimal, top of S9 black. Basal half of hindwing amber. Eyes green, legs black.

eyes green

early May to mid-July

basal half of hindwing amber

♂

dark shoulder stripes fused at top

anterior lateral stripe broken, posterior stripe complete

♂

♀

abdomen short with wide club

♀

legs black

Similar species: Tiny size and stocky shape distinctive. No other snaketail has amber wing markings.

Westfall's Snaketail • *Ophiogomphus westfalli*

49–50 mm, 1.9–2 in. Lightly patterned snaketail found at clear, rocky rivers in the Ozark/Ouachita region, primarily in Arkansas and Missouri. Thorax bright green, largely unmarked; dark shoulder stripes vestigial or absent. Abdomen dark with pale markings on top of all segments. Eyes greenish-blue in male, female gray-green. Legs mostly pale.

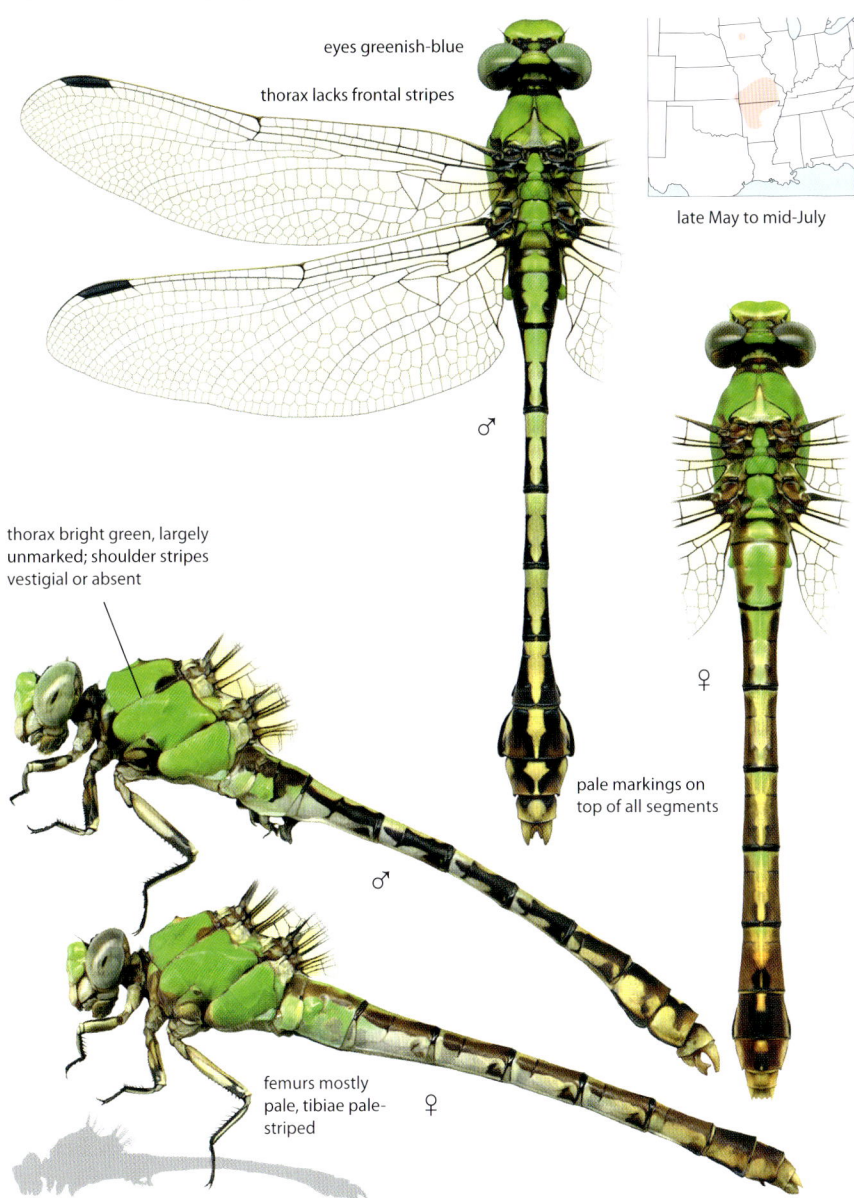

eyes greenish-blue

thorax lacks frontal stripes

late May to mid-July

thorax bright green, largely unmarked; shoulder stripes vestigial or absent

pale markings on top of all segments

♂

♀

femurs mostly pale, tibiae pale-striped

♂

♀

Similar species: Only snaketail species within its limited range. **Rusty Snaketail** has two brown shoulder stripes; abdomen brown, with less contrasting pattern. **Eastern Ringtail** has dark thoracic stripes and pale abdominal rings.

Pale Snaketail • *Ophiogomphus severus*

49–52 mm, 1.9–2 in. Common, wide-ranging western snaketail found at streams, rivers, also large lakes in the North. Thorax pale green to olive; shoulder stripes reduced, anterior shoulder stripe an isolated dark dash. Abdomen with large pale markings on all segments. Eyes pale blue-green to gray. Legs mostly pale.

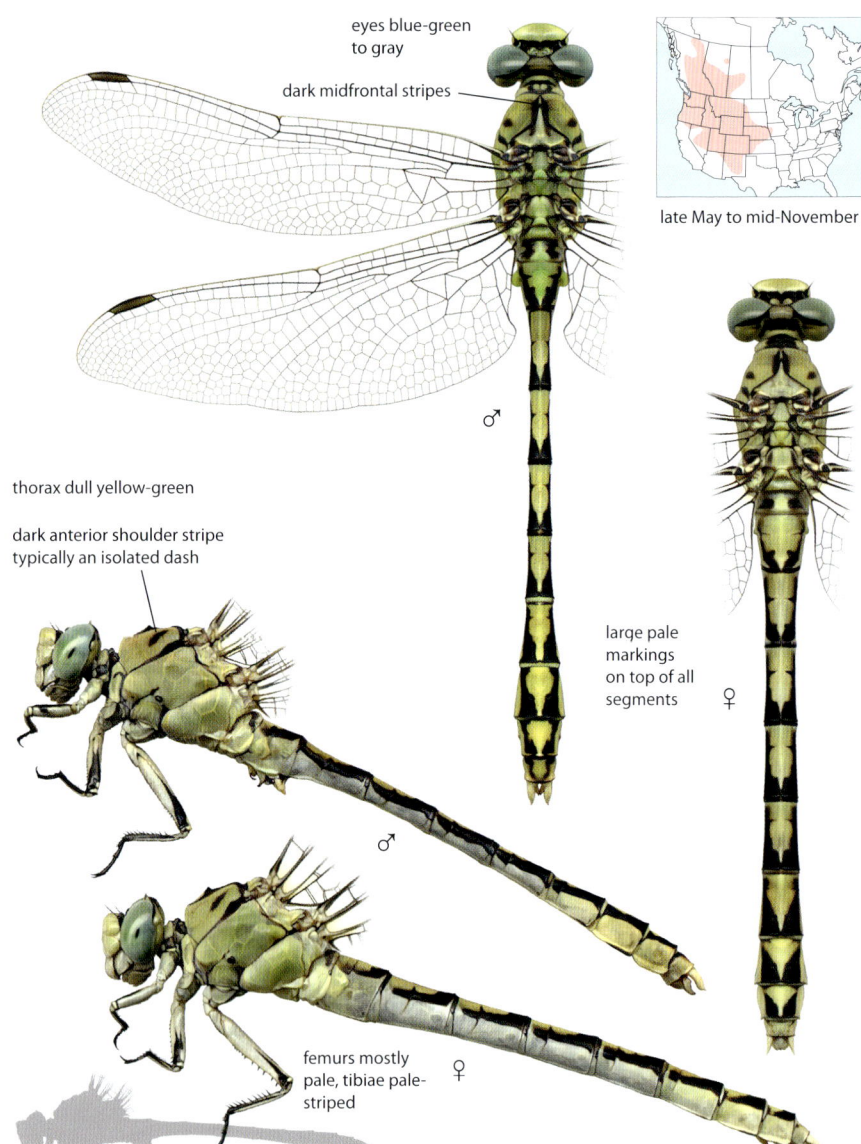

eyes blue-green to gray

dark midfrontal stripes

late May to mid-November

thorax dull yellow-green

dark anterior shoulder stripe typically an isolated dash

large pale markings on top of all segments

♂

♂

♀

femurs mostly pale, tibiae pale-striped

♀

Similar species: Arizona Snaketail nearly identical but ranges not known to overlap; dark anterior shoulder marking often more reduced. Compare male appendages, female occiput, pages 416, 420. Pale form of **Great Basin Snaketail** usually has more developed thoracic stripes. Compare appendages, page 417. Other **western snaketails** have more extensive dark thoracic stripes.

Arizona Snaketail • *Ophiogomphus arizonicus*

48–54 mm, 1.9–2.1 in. Pale-colored snaketail of southwestern mountain streams. Thorax pale green, maturing to olive, nearly unmarked; dark shoulder stripes reduced, anterior shoulder stripe a weak, isolated dash. Abdomen with pale markings on all segments. Eyes pale aqua-blue. Femurs mostly pale, tibiae pale-striped.

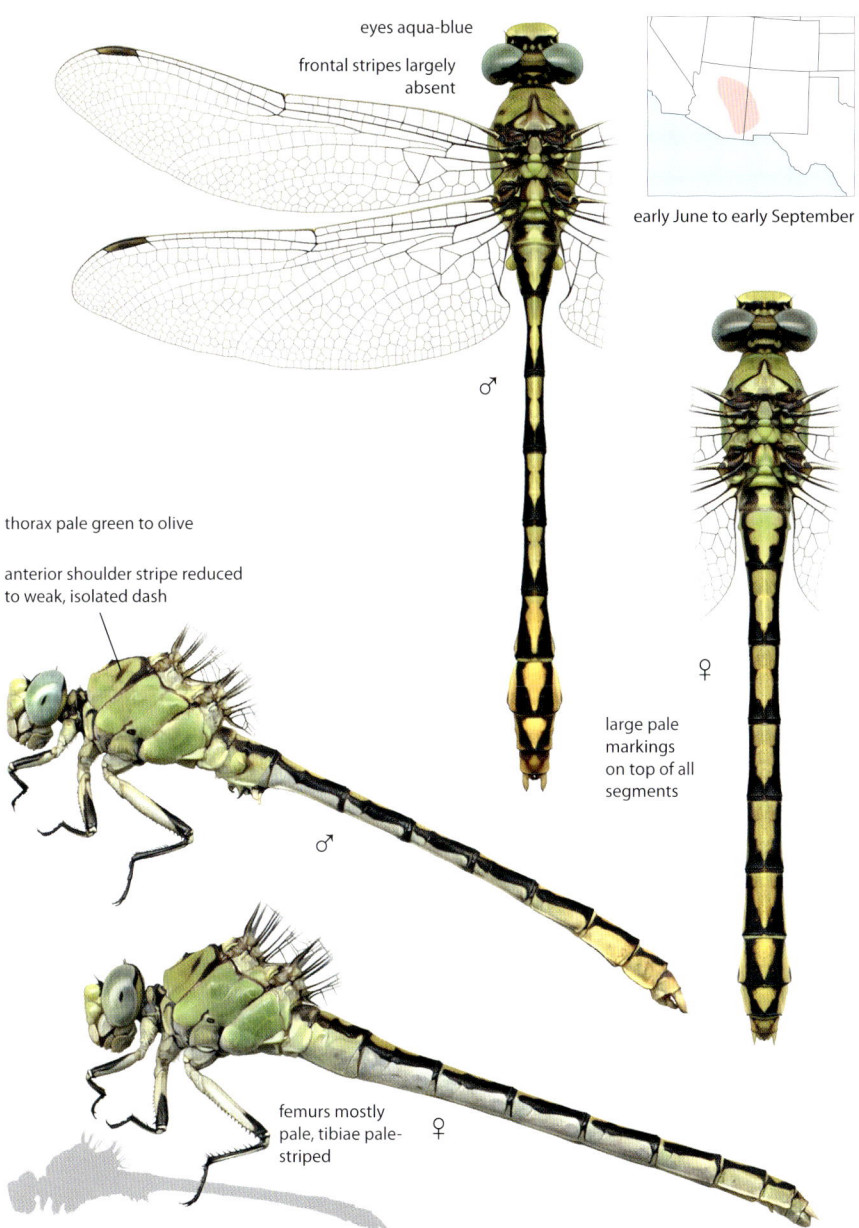

eyes aqua-blue

frontal stripes largely absent

early June to early September

thorax pale green to olive

anterior shoulder stripe reduced to weak, isolated dash

♂

large pale markings on top of all segments

♀

femurs mostly pale, tibiae pale-striped

♀

Similar species: Pale Snaketail nearly identical but not known to share range; anterior shoulder dash usually stronger. Compare male appendages, female occiput, pages 416, 420.

Sinuous Snaketail • *Ophiogomphus occidentis*

46–52 mm, 1.8–2 in. Brightly marked northwestern snaketail found at slow rivers and large streams, also lakes. Thorax bright green with well-developed dark frontal and shoulder stripes, anterior shoulder stripe narrow, subtly S-curved. Abdomen with wide pale markings atop all segments, S10 largely pale. Eyes gray. Femurs mostly pale, tibiae pale-striped.

eyes gray

early April to early October

dark anterior shoulder stripe narrow and S-curved, often fused with posterior shoulder stripe at top

♂

wide, short pale markings on top of all segments

♀

♂

femurs mostly pale, tibiae pale-striped

♀

Similar species: Great Basin Snaketail has straight anterior shoulder stripe, bluer eyes, greener abdominal markings, S10 marked with black. **Bison Snaketail** has fused dark shoulder stripes resembling single wide stripe; tibiae black.

Great Basin Snaketail • *Ophiogomphus morrisoni*

50–53 mm, 2.–2.1 in. Far-western snaketail inhabiting streams and rivers, sometimes lakes, including alkaline waters. Thorax grayish yellow-green with well-developed black stripes, anterior shoulder stripe wide and relatively straight. Abdomen black with pale markings on all segments. Eyes blue. Femurs partially pale, tibiae pale-striped.

eyes blue

early June to early September

two dark shoulder stripes, anterior stripe wide and straight

♂

abdomen with pale markings on top of all segments

♀

♂

♀

♂

GREAT BASIN FORM
larger and paler, thoracic stripes narrower; anterior shoulder stripe sometimes reduced to isolated dash; pale abdominal markings more extensive

Similar species: Sinuous Snaketail has narrow, wavy, dark anterior shoulder stripe; S10 paler. **Pale Snaketail** similar to Great Basin form of Great Basin Snaketail, but epiproct shorter than cerci (see page 416).

Bison Snaketail • *Ophiogomphus bison*

49–51 mm, 1.9–2 in. Far-western snaketail of rocky, running streams. Thorax bright green, with fused dark brown shoulder stripes appearing as single wide stripe; single narrow dark lateral stripe. Abdomen with pale markings on all segments. Eyes blue to gray. Femurs partially pale, tibiae black.

eyes blue to gray

early May to mid-October

♂

dark shoulder stripes fused, resemble single wide stripe

abdomen with pale markings on top of all segments

♀

♂

base of femurs pale ♀

Similar species: Fused shoulder stripes separate Bison Snaketail from other snaketail species. **Pacific Clubtail** has duller green thorax, narrow pale dorsal stripes on abdomen, black legs.

Eastern Ringtail • *Erpetogomphus designatus*

49–55 mm, 1.9–2.2 in. Brightly marked ringtail inhabiting gravel- and sand-bottomed rivers and streams. Only ringtail found east of Texas. Thorax yellow-green with brown stripes, posterior shoulder and posterior lateral stripes complete. Abdomen with prominent rings; dorsal markings yellow, lateral marks white, end segments rusty brown, appendages pale.

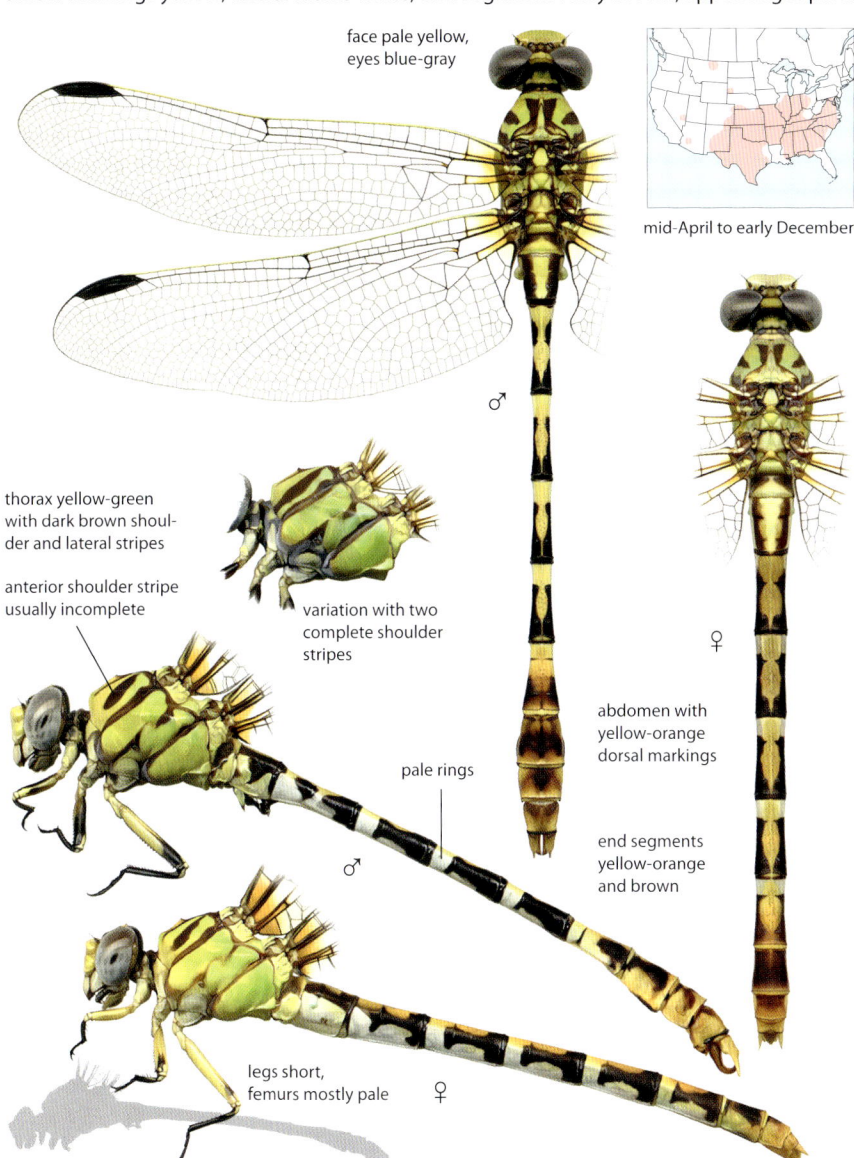

face pale yellow, eyes blue-gray

mid-April to early December

thorax yellow-green with dark brown shoulder and lateral stripes

anterior shoulder stripe usually incomplete

variation with two complete shoulder stripes

pale rings

abdomen with yellow-orange dorsal markings

end segments yellow-orange and brown

legs short, femurs mostly pale

♂

♀

Similar species: White-belted Ringtail has white stripe on side of thorax. **Serpent Ringtail** thorax dull green with complete dark lateral stripes. **Blue-faced Ringtail** has blue face, aqua-green thorax and dorsal abdominal markings, pale rings in middle of segments. **Dashed** and **Yellow-legged Ringtails** have reduced thoracic striping. **Flag-tailed Spinylegs** larger, with large club; femur of hind leg black, very long and armed with long spines.

Serpent Ringtail • *Erpetogomphus lampropeltis*

41–56 mm, 1.6–2.2 in. Well-marked southwestern ringtail found at shallow, rocky streams and rivers. Thorax with complete dark brown shoulder, frontal, and lateral stripes. Abdomen conspicuously white-ringed; dorsal markings dirty pale yellow or yellow-brown. End abdominal segments rusty brown, turning blacker with age. Appendages pale.

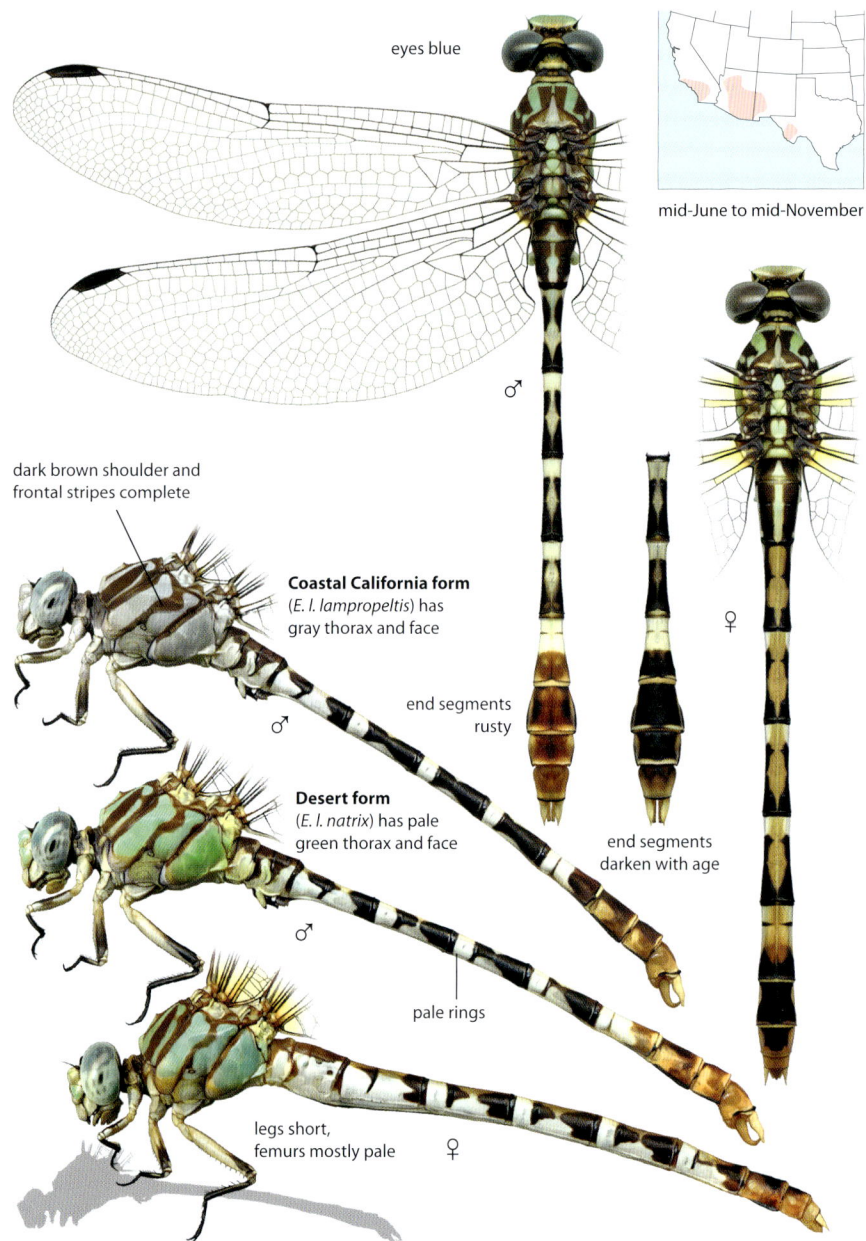

eyes blue

mid-June to mid-November

dark brown shoulder and frontal stripes complete

Coastal California form
(*E. l. lampropeltis*) has gray thorax and face

♂

end segments rusty

Desert form
(*E. l. natrix*) has pale green thorax and face

♂

end segments darken with age

♀

pale rings

legs short, femurs mostly pale

♀

Similar species: White-belted Ringtail thorax yellow with white stripe on side. **Eastern Ringtail** thorax yellow-green; dark lateral stripes reduced or broken.

White-belted Ringtail • *Erpetogomphus compositus*

46–55 mm, 1.8–2.2 in. Pale yellowish ringtail of western streams and rivers. Thorax pale yellow with dark brown stripes; white stripe (belt) between the dark lateral stripes. Abdomen dark with prominent white rings, cream-colored diamond-shaped dorsal markings. End abdominal segments rusty brown, turning blacker with age. Appendages pale.

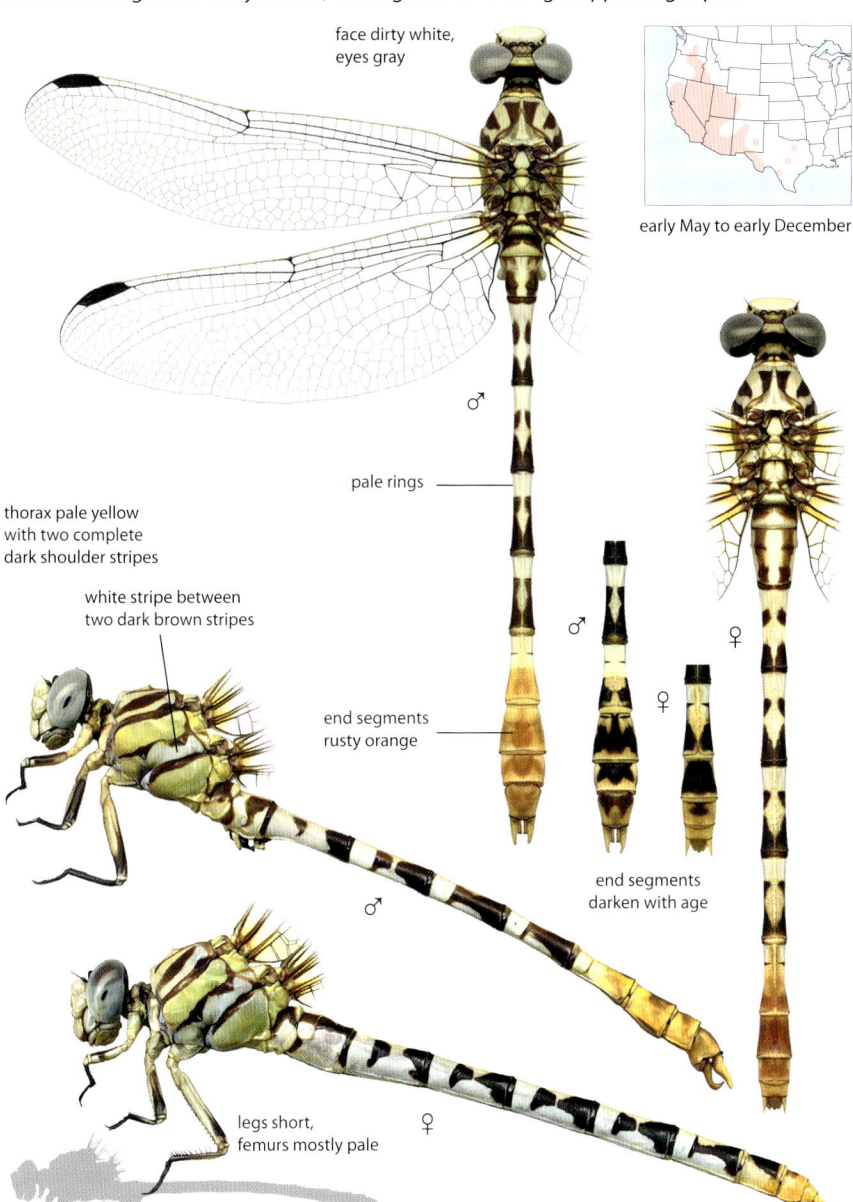

face dirty white,
eyes gray

early May to early December

♂

pale rings

thorax pale yellow
with two complete
dark shoulder stripes

white stripe between
two dark brown stripes

♂

♀

end segments
rusty orange

end segments
darken with age

♀

♂

legs short,
femurs mostly pale

♀

Similar species: Pale yellow and white thorax distinctive. **Eastern Ringtail** thorax yellow-green, lacks white stripe on side. **All other ringtails** have greener thoraxes. **Sulphur-tipped** and **Plains Clubtails** lack pale rings.

Dashed Ringtail • *Erpetogomphus heterodon*

50–55 mm, 2–2.2 in. Pale ringtail inhabiting clear, rocky mountain streams; local in New Mexico and West Texas. Face yellow-green, eyes gray. Thorax bright yellow-green with reduced brown striping. Abdomen with prominent white rings. Dorsal markings wide, dull yellow; end segment markings more yellow-orange. Appendages orange-yellow.

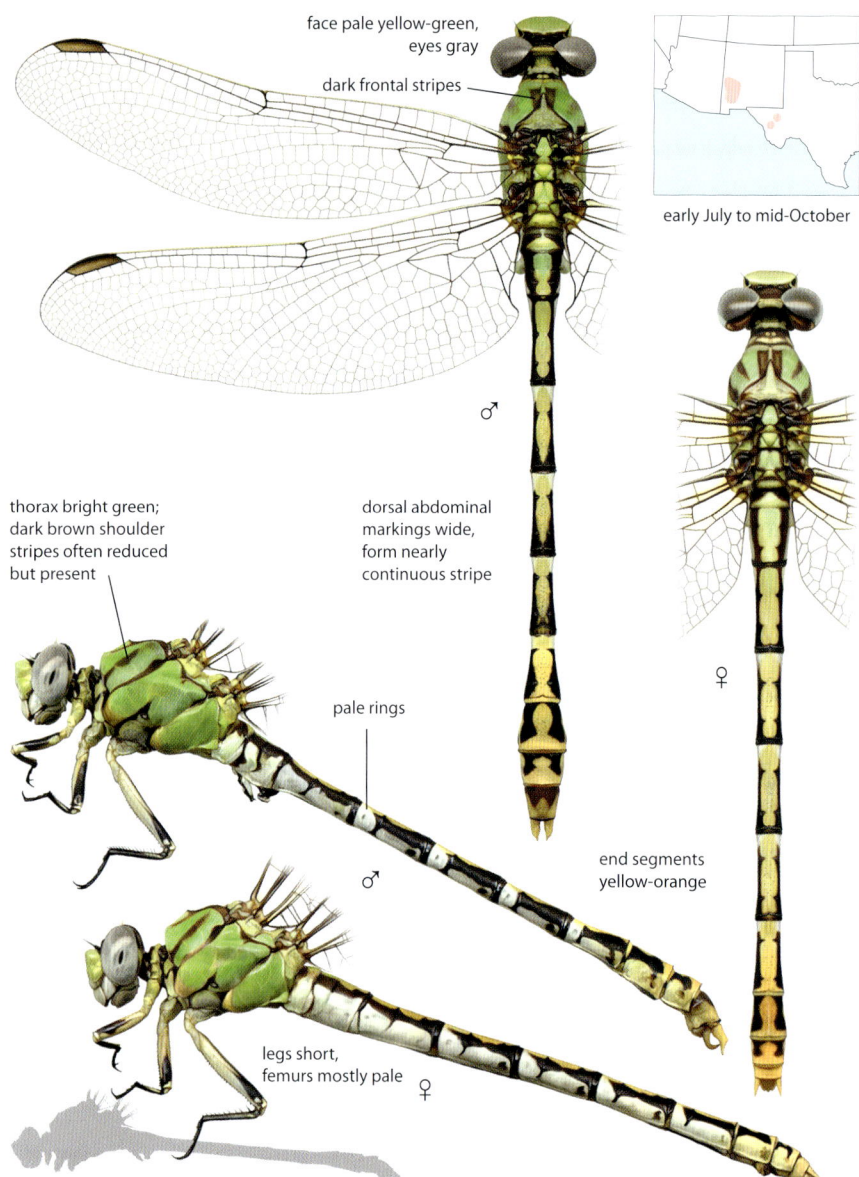

face pale yellow-green, eyes gray

dark frontal stripes

early July to mid-October

thorax bright green; dark brown shoulder stripes often reduced but present

dorsal abdominal markings wide, form nearly continuous stripe

pale rings

♂

end segments yellow-orange

♀

legs short, femurs mostly pale ♀

Similar species: Yellow-legged Ringtail most similar, but dark frontal and shoulder stripes vestigial or absent, wing base structures green, S1 completely pale, S2 with reduced dark markings. **Arizona Snaketail** duller in color. Pale abdominal markings shorter and more separated; male lacks pale abdominal rings.

Yellow-legged Ringtail • *Erpetogomphus crotalinus*

45–49 mm, 1.8–1.9 in. Bright green ringtail found at rocky mountain streams; local in southern Arizona and New Mexico. Thorax largely unmarked green. Abdomen with whitish rings; S1-2 almost entirely pale. Green-yellow dorsal markings wide and nearly full length of segments; end segments orange-yellow. Appendages orange-yellow.

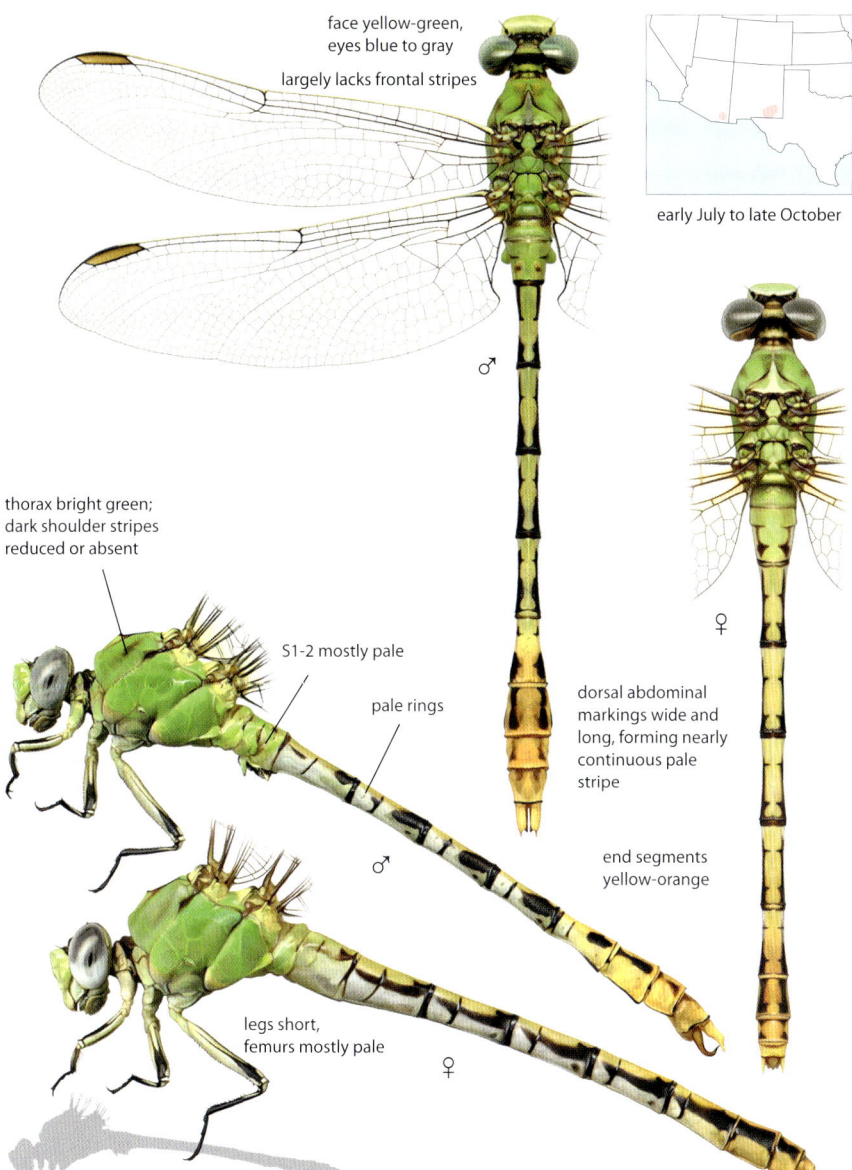

face yellow-green, eyes blue to gray

largely lacks frontal stripes

early July to late October

thorax bright green; dark shoulder stripes reduced or absent

S1-2 mostly pale

pale rings

dorsal abdominal markings wide and long, forming nearly continuous pale stripe

end segments yellow-orange

legs short, femurs mostly pale

Similar species: Dashed Ringtail thorax has more developed frontal and dark shoulder stripes, abdomen has dark markings on S1 and dark stripes on S2. **Arizona Snaketail** duller green, and pale dorsal abdominal markings are shorter and more separated; male lacks pale abdominal rings.

Straight-tipped Ringtail • *Erpetogomphus elaps*

39–52 mm, 1.5–2 in. Brightly colored ringtail of mountain streams, rare in southeastern Arizona. Face aqua-blue. Thorax bright green with only reduced brown shoulder stripe. Abdomen with white lateral rings; dorsal markings aqua in male, green to rusty in females. End segments rusty brown, appendages pale, male cerci long and straight.

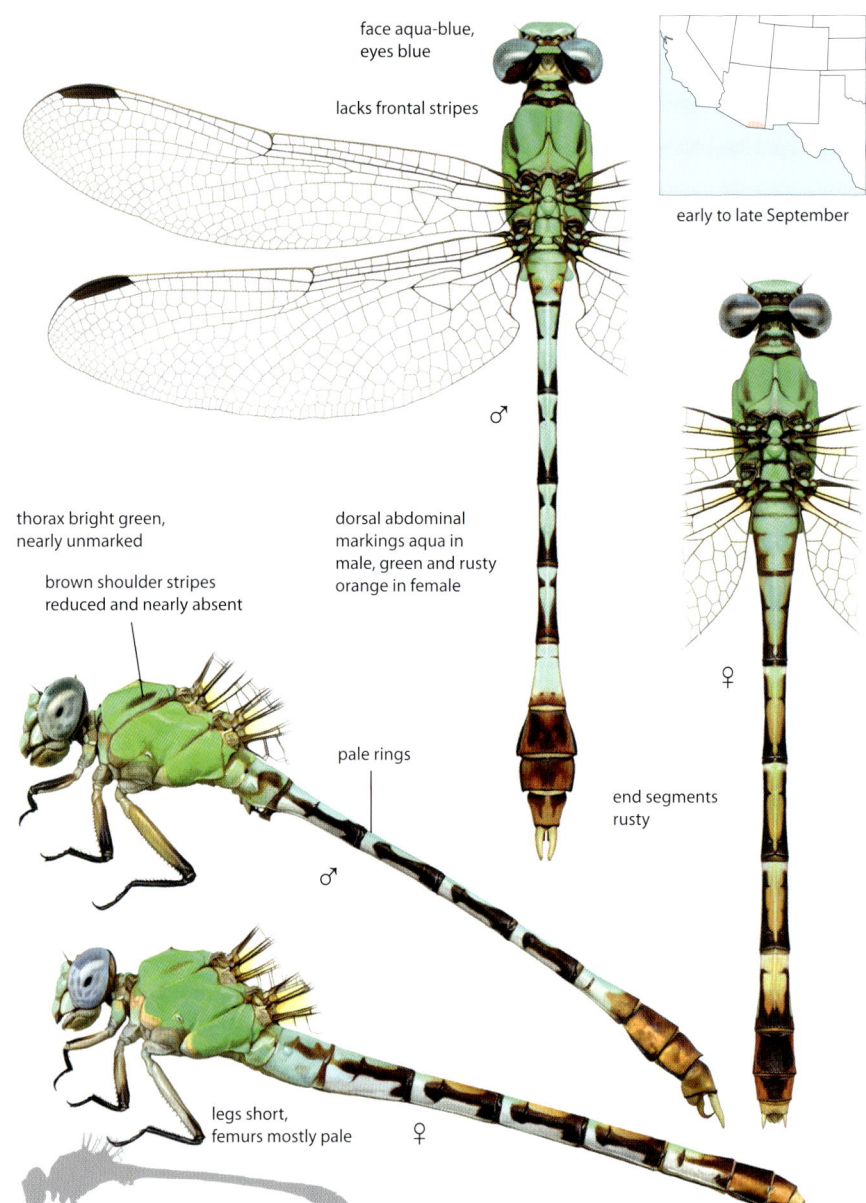

face aqua-blue, eyes blue

lacks frontal stripes

early to late September

thorax bright green, nearly unmarked

brown shoulder stripes reduced and nearly absent

dorsal abdominal markings aqua in male, green and rusty orange in female

pale rings

end segments rusty

legs short, femurs mostly pale

♂

♀

Similar species: Dashed and **Yellow-legged Ringtails** have yellow-green thoraxes, lack blue in face and abdomen. End abdominal segments paler; male cerci bends downward. **Blue-faced** and **Black-tailed Ringtails** have well-developed dark brown thoracic stripes.

Blue-faced Ringtail • *Erpetogomphus eutainia*

38–44 mm, 1.5–1.7 in. Small, colorful ringtail found at clear, spring-fed rivers and streams; local in central Texas. Face aqua-blue. Thorax bright green with complete brown shoulder and lateral stripes. Abdomen with white rings in the middle of segments; dorsal markings aqua to pale green. End segments rusty brown, appendages pale brown.

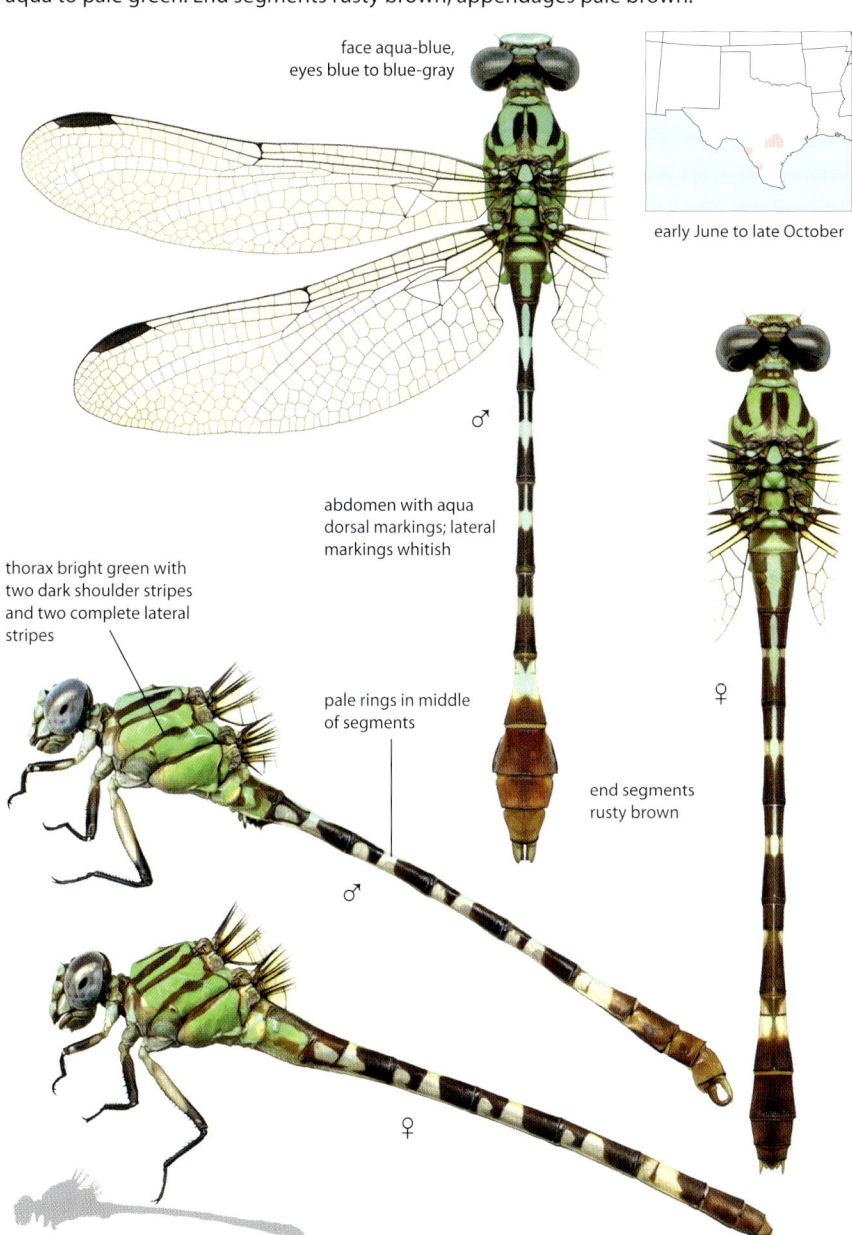

face aqua-blue, eyes blue to blue-gray

early June to late October

abdomen with aqua dorsal markings; lateral markings whitish

thorax bright green with two dark shoulder stripes and two complete lateral stripes

pale rings in middle of segments

end segments rusty brown

♂

♀

Similar species: Pale abdominal rings in the middle of segments distinctive. All other ringtails have pale abdominal rings at the base of segments.

Black-tailed Ringtail • *Erpetogomphus molossus*

47–48 mm, 1.85–1.9 in. Brightly patterned ringtail of rocky, spring-fed, mountain streams; recorded in southern Arizona. Face and eyes turquoise. Thorax bright aqua with well-developed frontal and shoulder stripes. Abdomen with narrow white rings and thin pale middorsal stripes. End segments mostly black above, appendages pale, male cerci long.

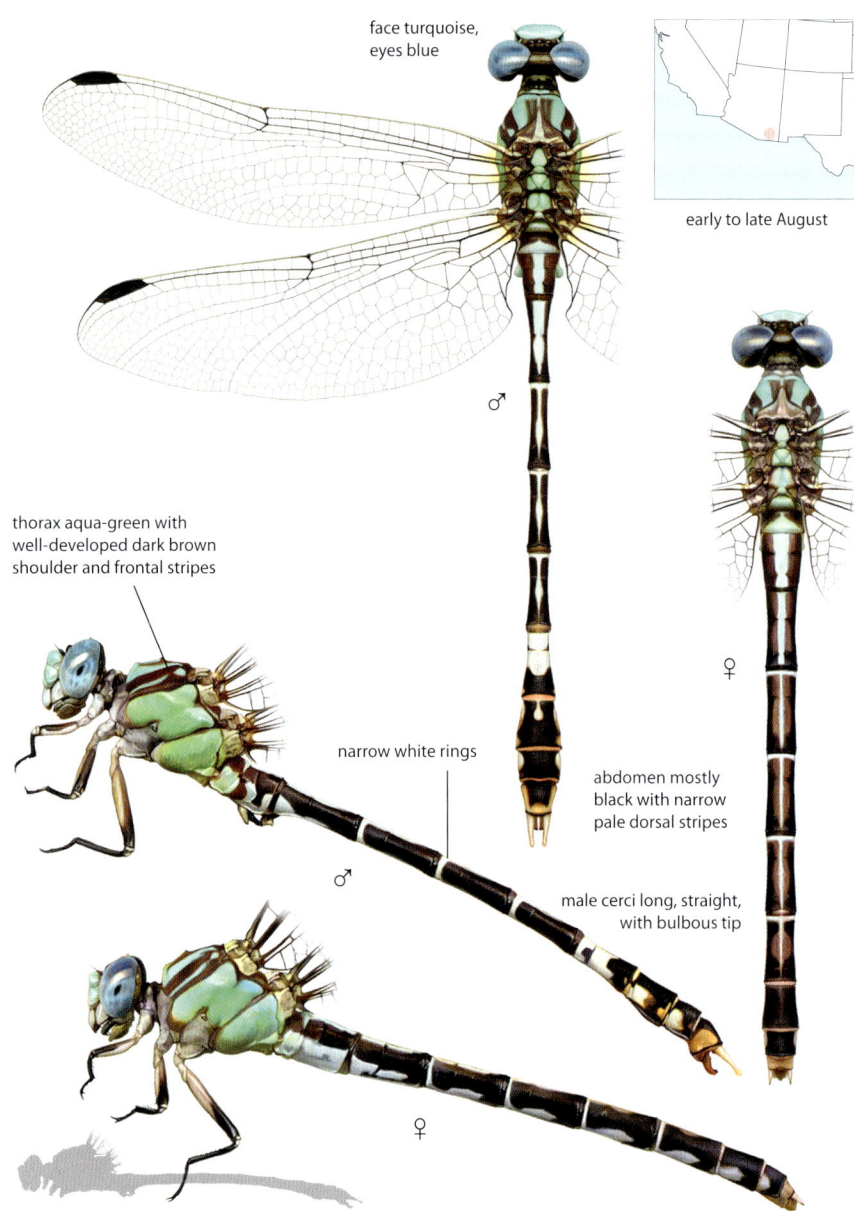

face turquoise, eyes blue

early to late August

♂

thorax aqua-green with well-developed dark brown shoulder and frontal stripes

narrow white rings

♂

♀

abdomen mostly black with narrow pale dorsal stripes

male cerci long, straight, with bulbous tip

♀

Similar species: Serpent and **Blue-faced Ringtails** have brown lateral thoracic stripes, more pale markings on side of abdomen. **Straight-tipped Ringtail** lacks dark thoracic stripes.

Brimstone Clubtail • *Stylurus intricatus*

41–55 mm, 1.6–2.2 in. Pale western hanging clubtail inhabiting rivers in open country, sometimes irrigation canals. Thorax pale yellow-green with limited dark brown stripes, sides unmarked. Abdomen mostly yellow with pale rings on S3-7, S8-10 almost completely pale; appendages pale. Eyes blue. Femurs mostly yellow, tibiae yellow-striped.

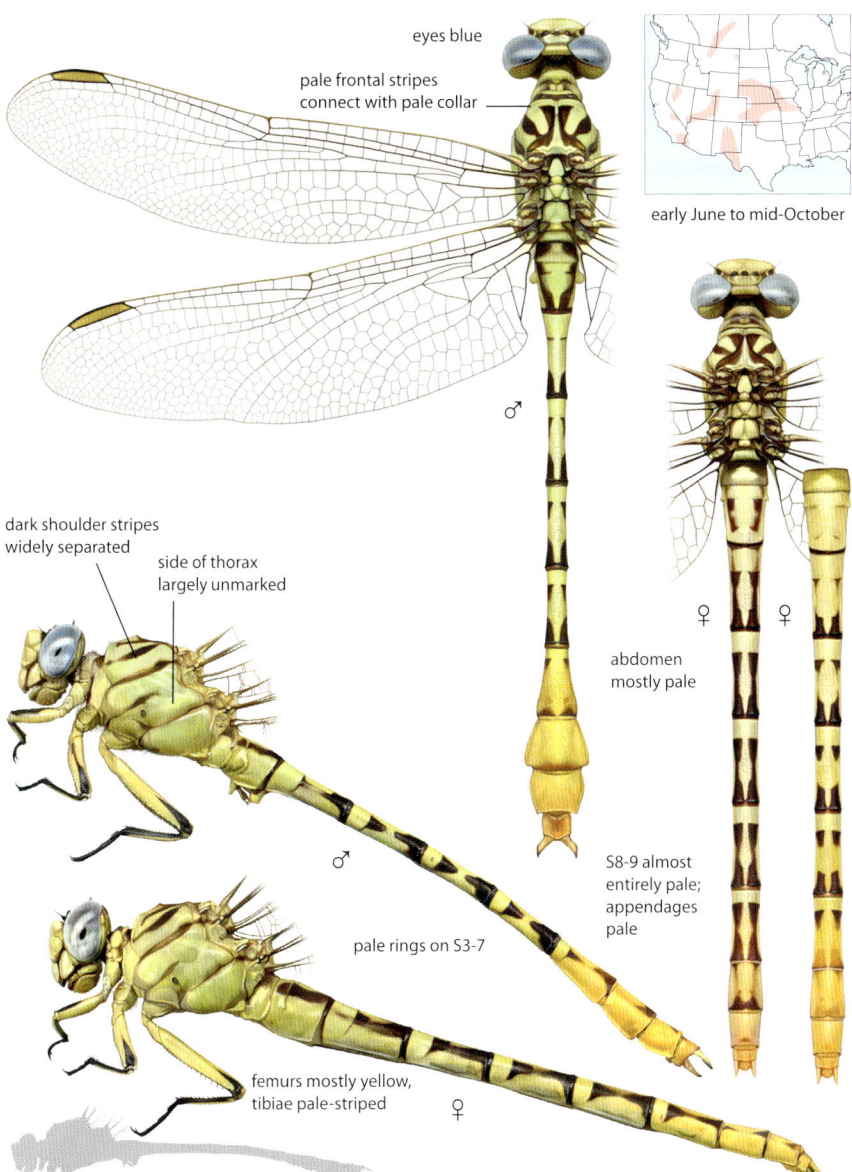

eyes blue

pale frontal stripes connect with pale collar

♂

early June to mid-October

dark shoulder stripes widely separated

side of thorax largely unmarked

♀

♀

abdomen mostly pale

♂

S8-9 almost entirely pale; appendages pale

pale rings on S3-7

femurs mostly yellow, tibiae pale-striped

♀

Similar species: Russet-tipped Clubtail has two dark lateral thoracic stripes, lacks complete pale abdominal rings; end segments and appendages darker. **Eastern Ringtail** has brighter green thorax, end abdominal segments darker, stigmas black. **White-belted Ringtail** has two dark lateral thoracic stripes with area between them whitish.

Olive Clubtail · *Stylurus olivaceus*

56–60 mm, 2.2–2.4 in. Robust, dully colored western clubtail found at mud- and sand-bottomed rivers and streams. Thorax green-gray, with bracket-shaped pale frontal stripes, prominent black shoulder stripes, lateral stripes incomplete or absent. Abdomen stout, with pale dorsal T-shaped markings on S4-7, rounded dorsal spot at base of S8. Eyes blue.

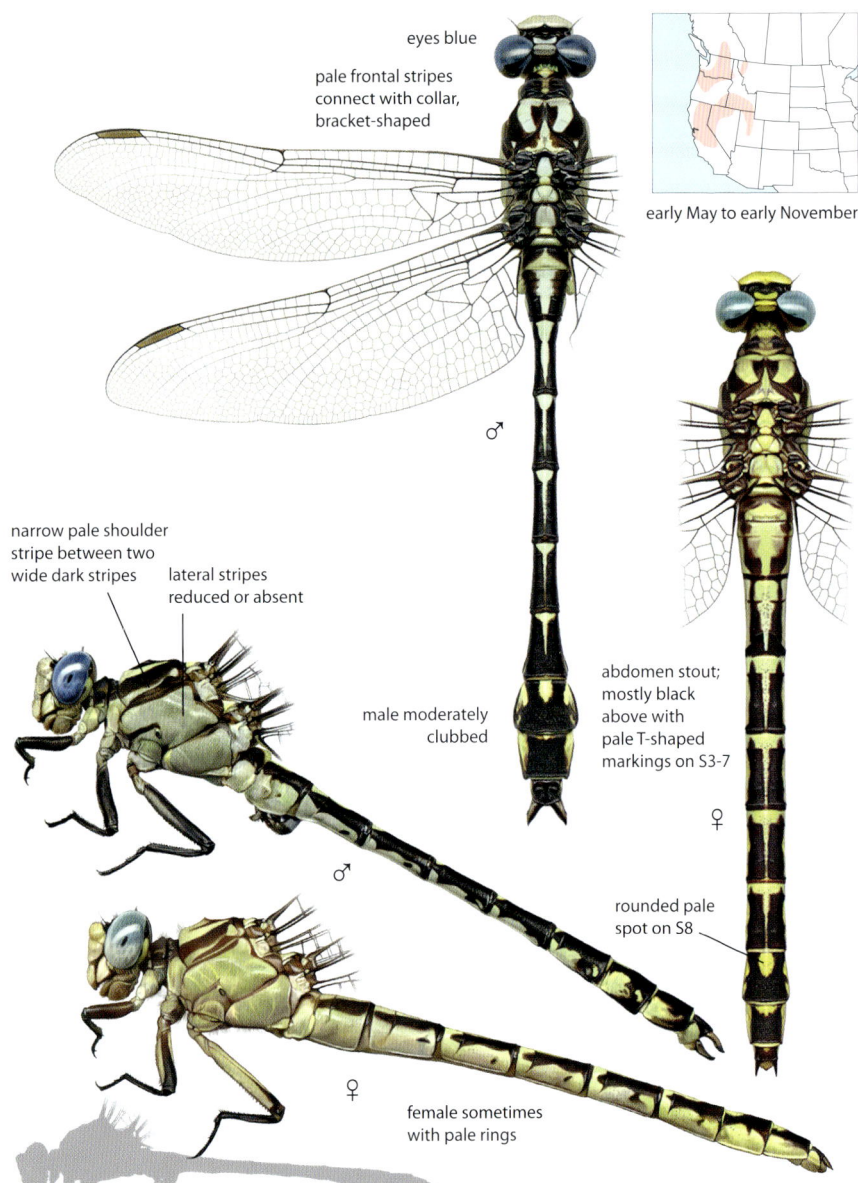

eyes blue

pale frontal stripes connect with collar, bracket-shaped

early May to early November

♂

narrow pale shoulder stripe between two wide dark stripes

lateral stripes reduced or absent

male moderately clubbed

abdomen stout; mostly black above with pale T-shaped markings on S3-7

♀

rounded pale spot on S8

♂

♀

female sometimes with pale rings

Similar species: Brimstone Clubtail brighter in color, pattern lighter. **Pacific Clubtail** has black posterior lateral thoracic stripe. **Columbia Clubtail** has two dark lateral thoracic stripes, large yellow marking on top of S9.

Zebra Clubtail • *Stylurus scudderi*

57–58 mm, 2.2–2.3 in. Robust northeastern hanging clubtail found at clean woodland streams and rivers. Thorax with isolated pale frontal stripes, dark shoulder stripes fused, two wide dark lateral stripes. Abdomen stout, black with pale rings, male widely clubbed. Face with black cross-stripes, eyes green. Legs black.

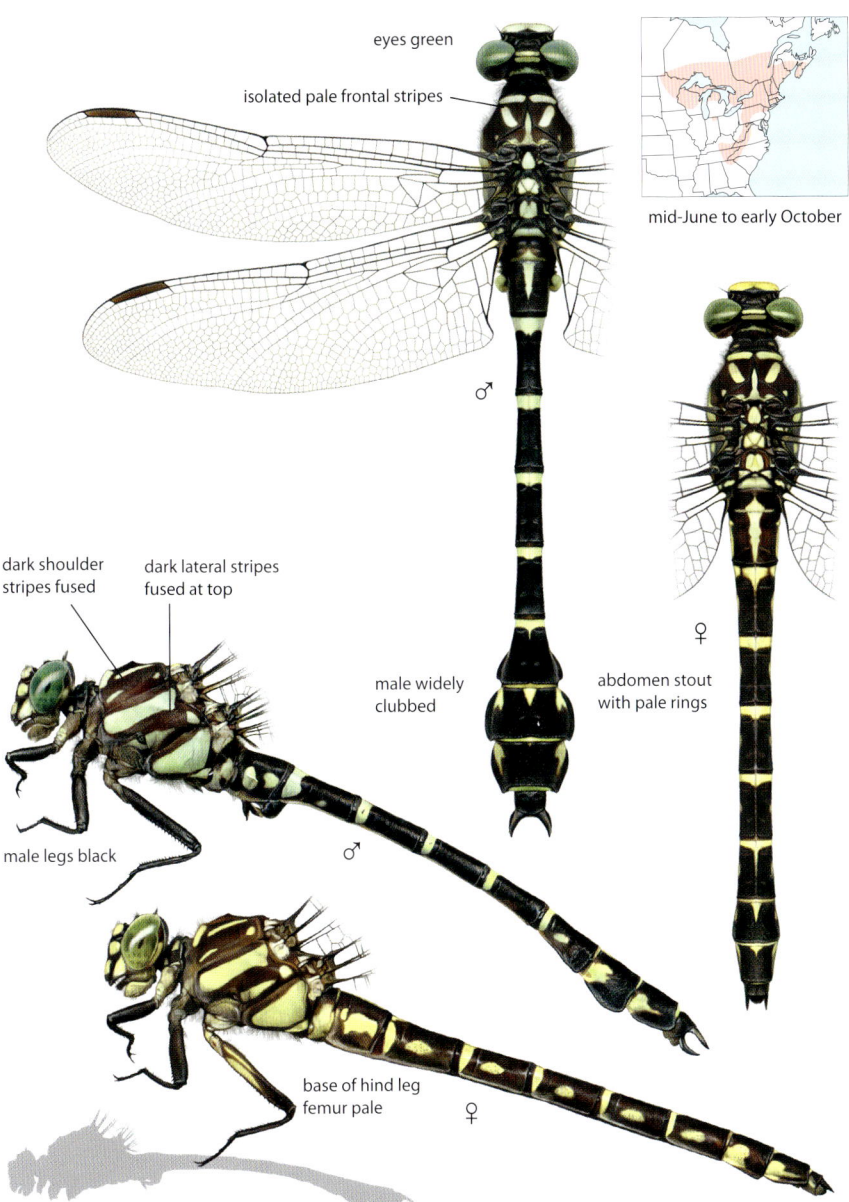

eyes green

isolated pale frontal stripes

mid-June to early October

♂

dark shoulder stripes fused

dark lateral stripes fused at top

male widely clubbed

abdomen stout with pale rings

♀

male legs black

♂

base of hind leg femur pale

♀

Similar species: Least clubtails are much smaller; dark lateral thoracic stripes narrow, appendages pale. **Tiger Spiketail** has eyes that touch. Male abdomen unclubbed, female with long horizontal subgenital plate.

Arrow Clubtail • *Stylurus spiniceps*

57–68 mm, 2.2–2.7 in. The largest hanging clubtail; inhabits large rivers, from the Northeast south to Arkansas. Thorax with isolated pale frontal stripes, wide dark shoulder and lateral stripes. Abdomen long; both sexes slightly clubbed, with small pale spots on top of S3-8. S9 distinctly long (longer than S8). Eyes green, face dark. Legs black.

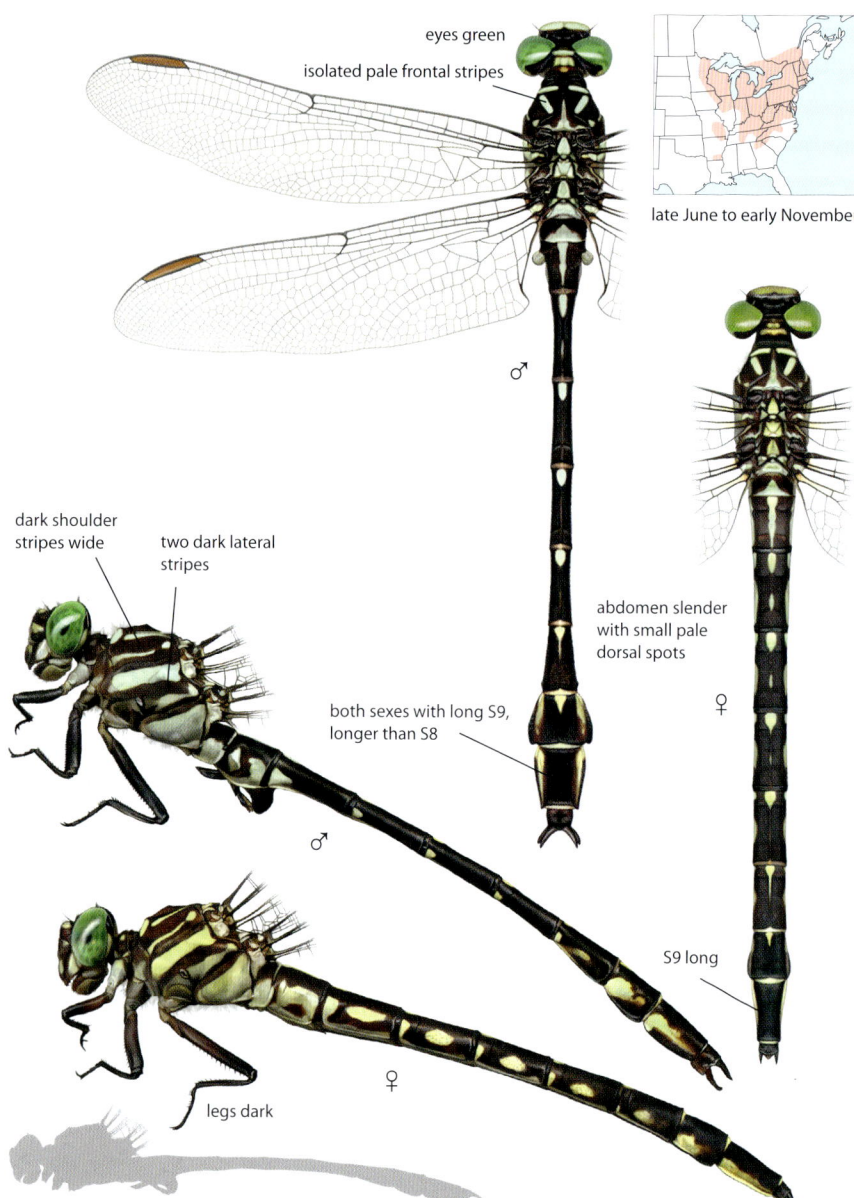

eyes green

isolated pale frontal stripes

late June to early November

♂

dark shoulder stripes wide

two dark lateral stripes

abdomen slender with small pale dorsal spots

both sexes with long S9, longer than S8

♀

♂

S9 long

legs dark

♀

Similar species: Elusive Clubtail has narrower dark lateral thoracic stripes, blue eyes, larger pale marking on side of S8, shorter S9. **Laura's Clubtail** has brighter coloration, narrower dark lateral thoracic stripes, pale dorsal abdominal stripes, shorter S9.

Riverine Clubtail • *Stylurus amnicola*

47–49 mm, 1.8–1.9 in. Eastern hanging clubtail favoring medium to large rivers. Thorax with pair of nearly parallel pale frontal stripes flanking pale three-pointed star, wide black shoulder stripes, dark lateral stripes incomplete. Abdomen with full-length dorsal pale stripes on S1-7, pale triangle at base of S8. Male club wide, with large yellow lateral spots.

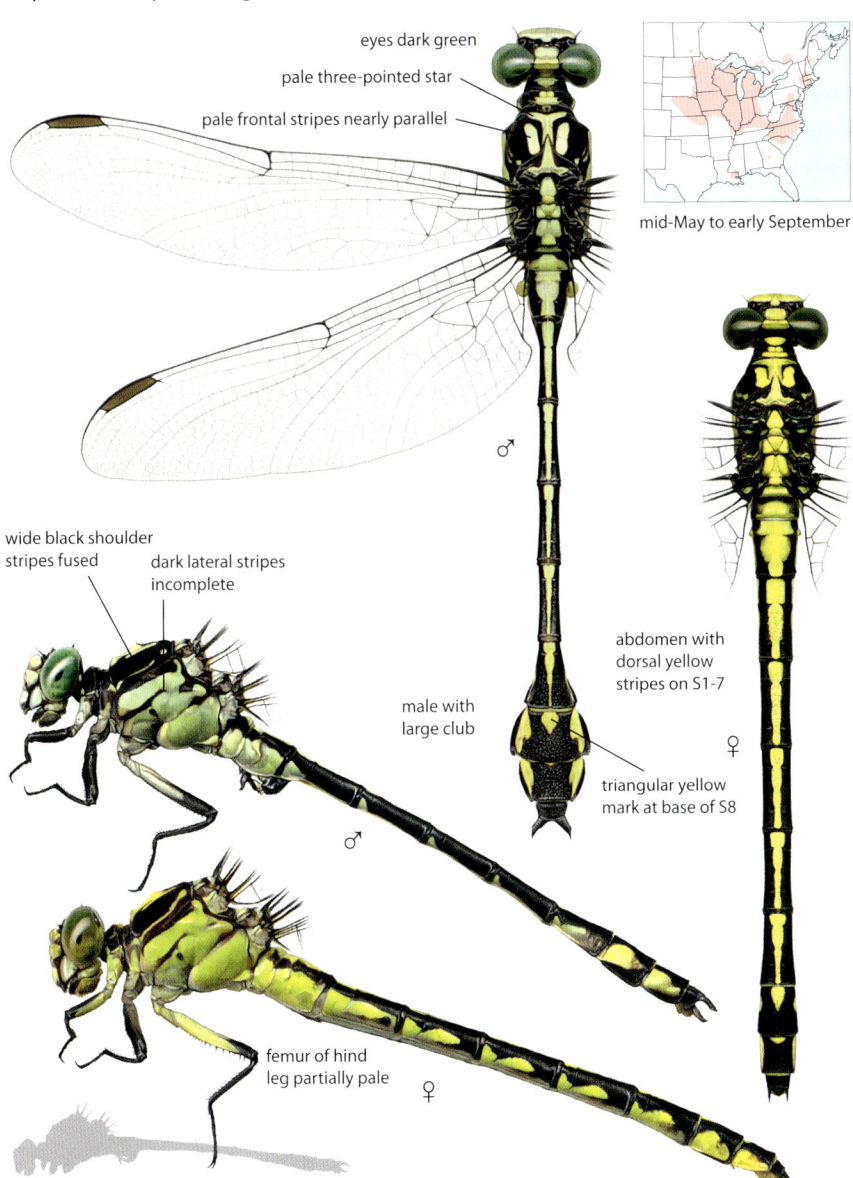

eyes dark green

pale three-pointed star

pale frontal stripes nearly parallel

mid-May to early September

♂

wide black shoulder stripes fused

dark lateral stripes incomplete

male with large club

abdomen with dorsal yellow stripes on S1-7

♀

triangular yellow mark at base of S8

♂

femur of hind leg partially pale

♀

Similar species: Townes' and **Elusive Clubtails** lack three-pointed pale frontal star; pale frontal stripes divergent, black shoulder stripes narrower and more separated, lateral stripes more developed. **Black-shouldered Spinyleg** has similar thoracic pattern but longer, all-black hind legs, yellow markings on top of S9-10, smaller lateral markings on S8-9.

Townes' Clubtail • *Stylurus townesi*

52–54 mm, 2–2.1 in. Dark, slender southeastern clubtail inhabiting sand-bottomed streams and rivers. Thorax with isolated pale frontal stripes, well-defined black shoulder stripes, and two dark lateral stripes, with anterior stripe stronger. Abdomen mostly dark, with narrow dorsal pale markings on S1-8, pale lateral markings on S7-9. Both sexes clubbed.

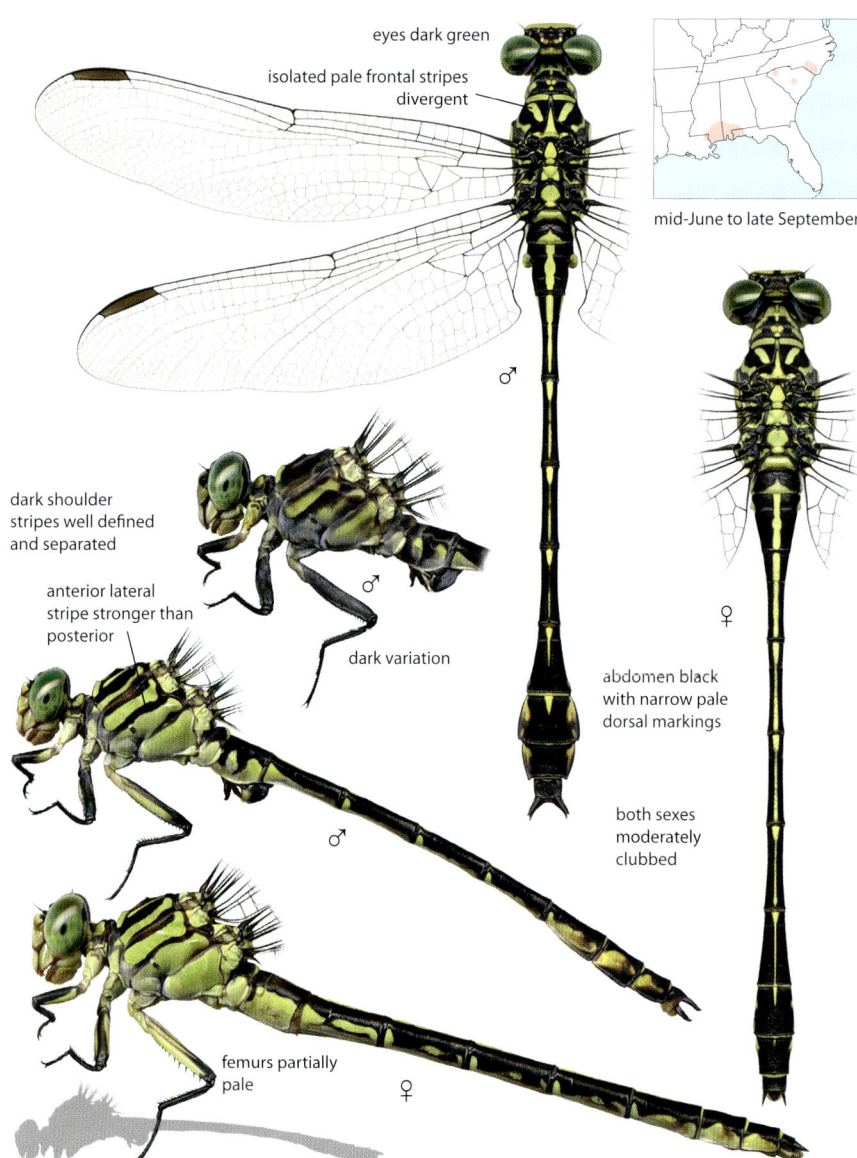

eyes dark green

isolated pale frontal stripes divergent

mid-June to late September

dark shoulder stripes well defined and separated

anterior lateral stripe stronger than posterior

dark variation

♂

♂

♂

abdomen black with narrow pale dorsal markings

both sexes moderately clubbed

♀

femurs partially pale

♀

Similar species: Riverine Clubtail has three-pointed pale star at front of thorax, wider black shoulder stripes, incomplete dark lateral stripes. **Elusive Clubtail** has two well-defined dark lateral thoracic stripes, blue eyes, blacker legs. **Laura's Clubtail** has browner club and abdomen tip; area between lateral thoracic stripes often darkened. **Yellow-sided Clubtail** has fused, wide dark shoulder stripes, lacks dark lateral thoracic stripes; eyes blue.

Elusive Clubtail • *Stylurus notatus*

52–64 mm, 2–2.5 in. Dark eastern hanging clubtail sought at slow, usually large, rivers, also large lakes. Thorax gray at maturity, with isolated frontal stripes, well-defined black shoulder stripes, two dark lateral stripes. Abdomen with pale dorsal spots on S3-8, pale lateral markings on S7-9. Eyes blue. Legs black; female hind leg femur pale at base.

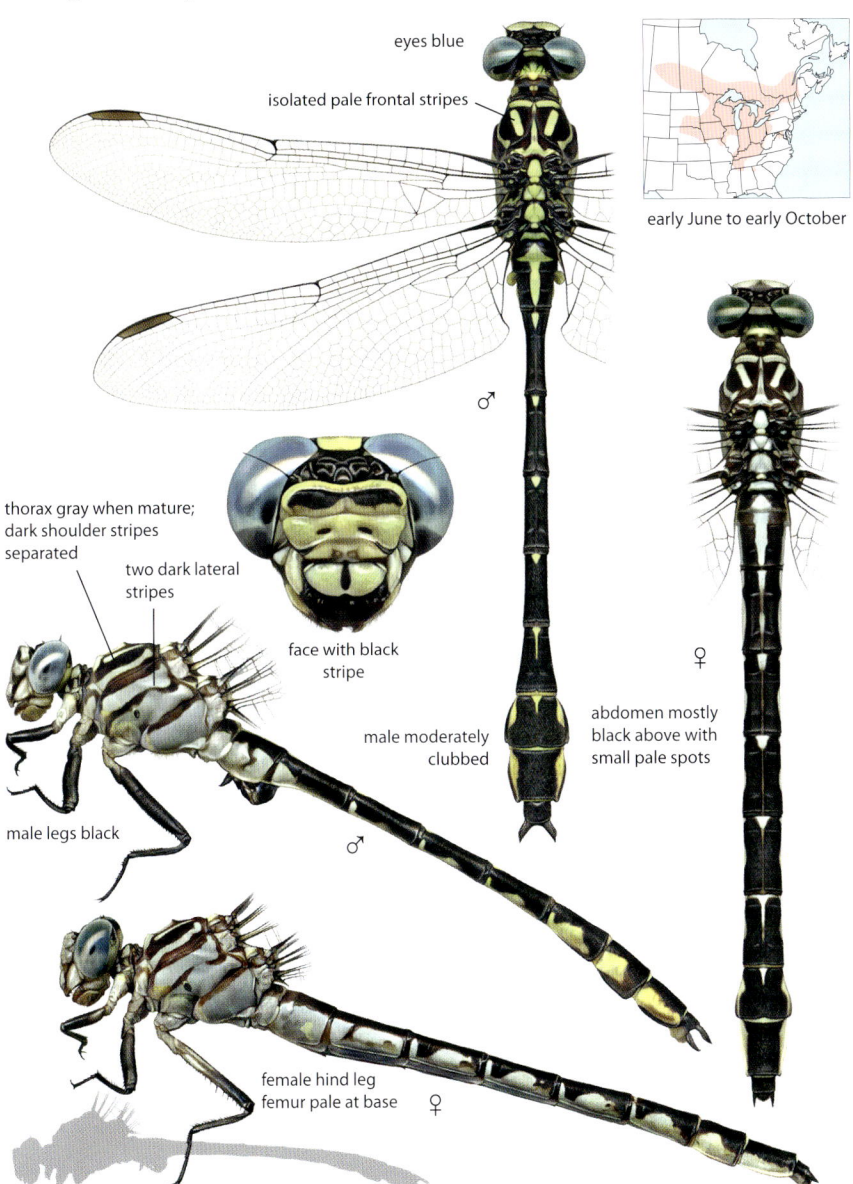

eyes blue

isolated pale frontal stripes

early June to early October

thorax gray when mature;
dark shoulder stripes
separated

two dark lateral
stripes

face with black
stripe

male moderately
clubbed

abdomen mostly
black above with
small pale spots

male legs black

♂

♀

female hind leg
femur pale at base ♀

Similar species: Riverine Clubtail has green eyes and wider dark shoulder stripes, dark lateral thoracic stripes incomplete or absent; abdomen with dorsal yellow markings more extensive. **Townes' Clubtail** has green eyes, pale-striped femurs. **Arrow Clubtail** has green eyes. Wider dark lateral thoracic stripes sometimes fused. S9 much longer than S8.

Russet-tipped Clubtail • Stylurus plagiatus

57–66 mm, 2.2–2.6 in. Wide-ranging hanging clubtail found at streams, rivers, canals, and lakes. **Eastern form:** Thorax gray-green; pale frontal stripes usually isolated, two dark lateral stripes complete. Abdomen dark with rusty club, and pale dorsal streaks but no distinct pale lateral markings. Eyes green. Femurs mostly brown or pale.

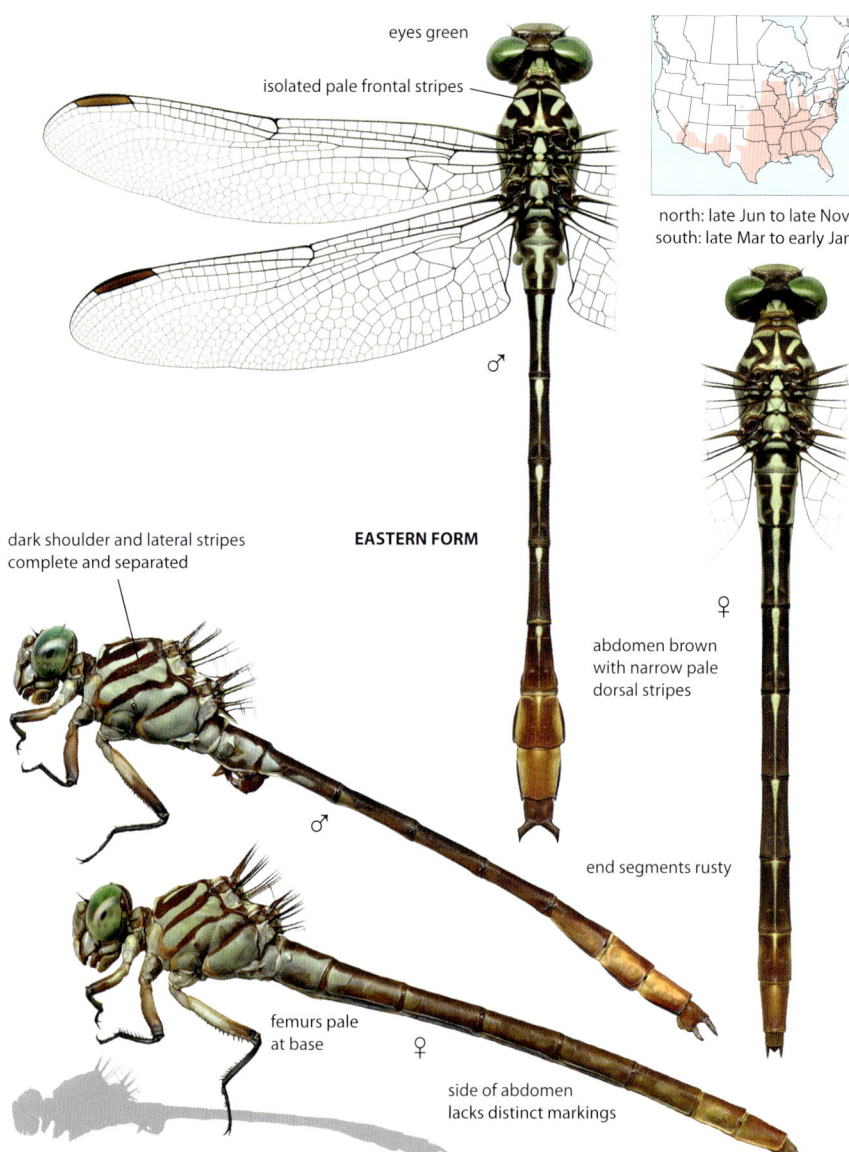

eyes green

isolated pale frontal stripes

north: late Jun to late Nov
south: late Mar to early Jan

♂

dark shoulder and lateral stripes
complete and separated

EASTERN FORM

♂

♀

abdomen brown
with narrow pale
dorsal stripes

end segments rusty

femurs pale
at base

♀

side of abdomen
lacks distinct markings

Similar species: Only a few clubtails in the East have rusty clubs. **Shining Clubtail** brighter yellow-green, with wide, fused dark shoulder stripes, dark lateral stripes often incomplete. **Southeastern** and **Flag-tailed Spinylegs** thorax also brighter in color; abdomen has pale lateral markings. Hind legs much longer. **Two-striped Forceptail** thorax mostly dark brown, with two pale lateral stripes.

Western form: Thorax yellow-green. Pale frontal stripes not isolated, dark stripes well separated. Abdomen with pale lateral markings, yellow spots or bands on S8-9; club wide and slightly rusty. Eyes blue. Femurs partially pale. Intermediate forms have blue eyes and abdomen with rusty club, but extent of lateral pale markings variable.

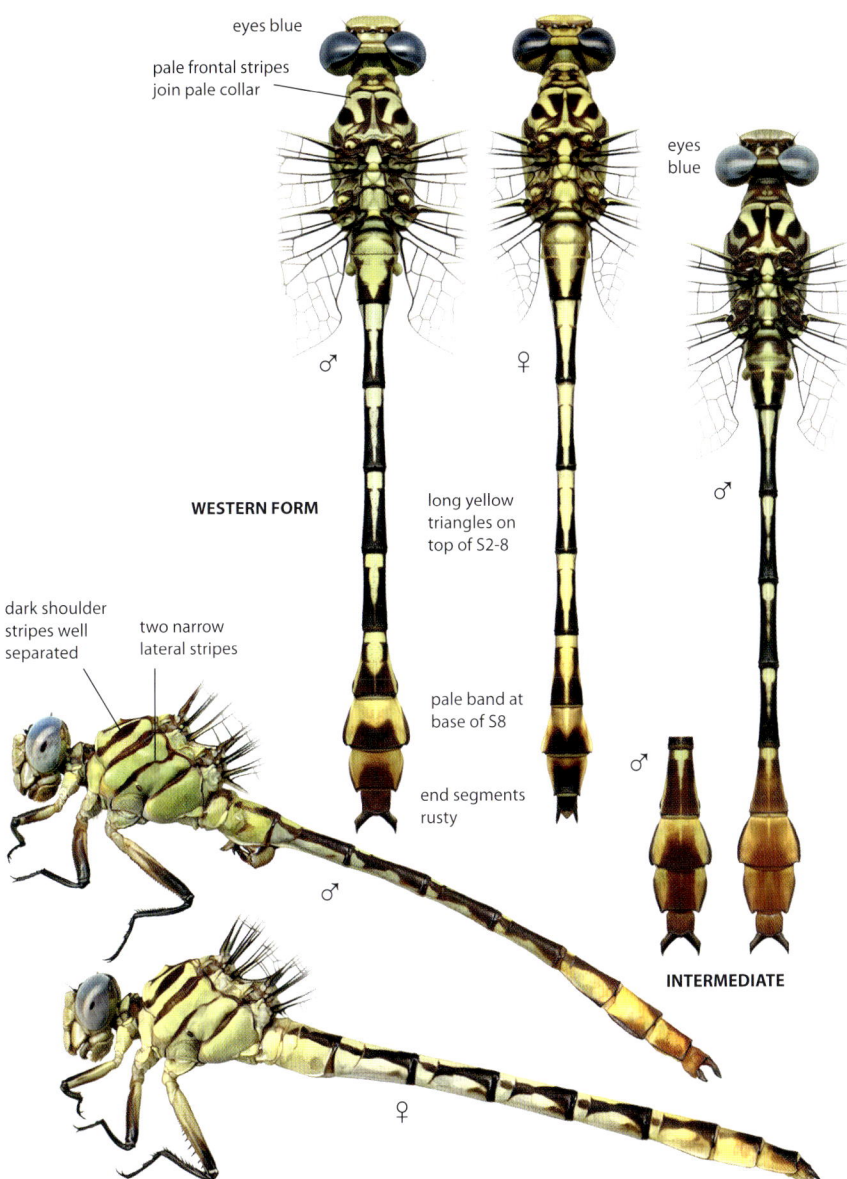

eyes blue

pale frontal stripes
join pale collar

eyes
blue

WESTERN FORM

long yellow
triangles on
top of S2-8

dark shoulder
stripes well
separated

two narrow
lateral stripes

pale band at
base of S8

end segments
rusty

♂

♀

♂

♂

♂

INTERMEDIATE

♂

♀

Similar species: Sulphur-tipped Clubtail has more contrasting abdomen pattern, S9 with wide yellow stripe. **Brimstone Clubtail** dark lateral thoracic stripes reduced or missing; S8-10 completely or nearly completely yellow. **Forceptails** have darker thoraxes with narrower yellow lateral stripes. **Ringtails** have pale abdominal rings.

Laura's Clubtail • *Stylurus laurae*

60–64 mm, 2.4–2.5 in. Uncommon eastern species found at clean woodland streams with mud and sand bottoms. Thorax with isolated pale frontal stripes, dark shoulder stripes separated by narrow pale stripe, two dark lateral stripes. Abdomen with narrow yellow dorsal stripes, S8-9 with large yellow lateral spots; end segments rusty. Eyes green.

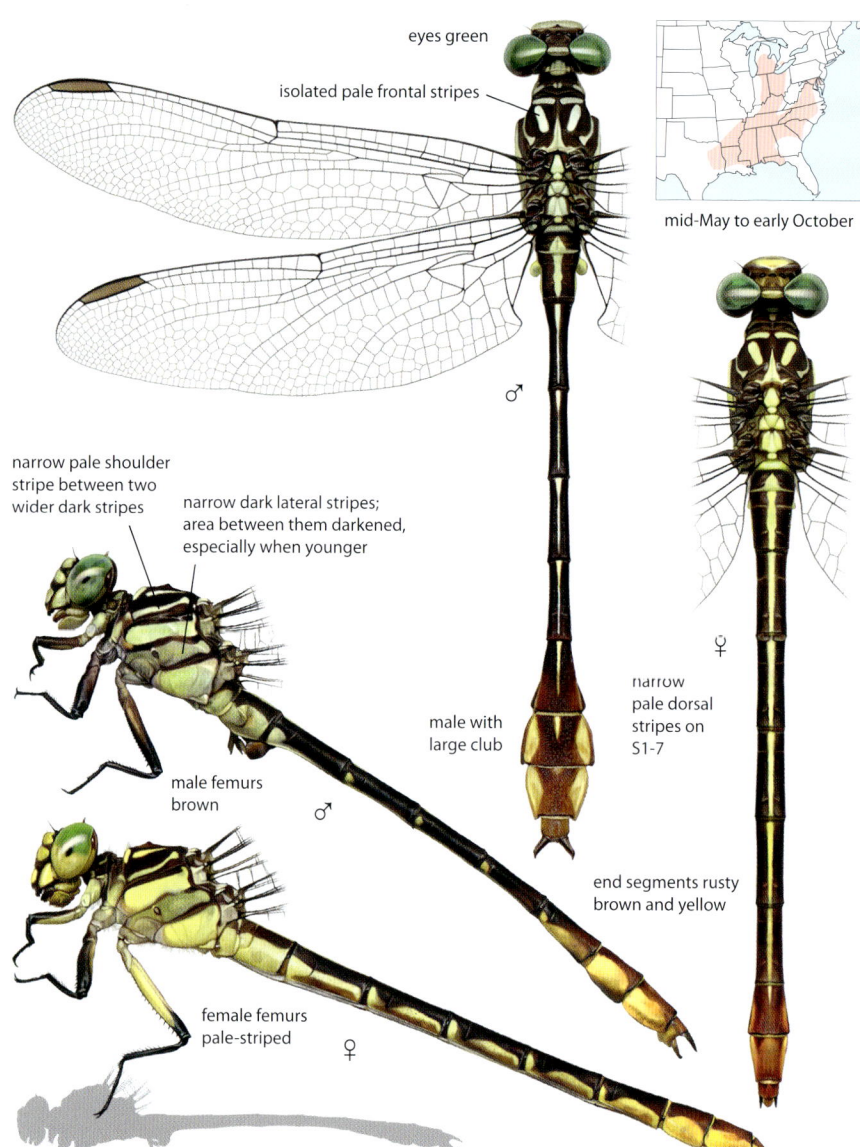

eyes green

isolated pale frontal stripes

mid-May to early October

♂

narrow pale shoulder stripe between two wider dark stripes

narrow dark lateral stripes; area between them darkened, especially when younger

male with large club

♀

narrow pale dorsal stripes on S1-7

male femurs brown

♂

end segments rusty brown and yellow

female femurs pale-striped

♀

Similar species: Russet-tipped Clubtail has pale and dark shoulder stripes about equal in width, and abdomen with less contrasting pattern; lacks pale markings on sides of abdomen. **Shining Clubtail** has fused dark shoulder stripes and largely lacks a pale shoulder stripe; abdomen pale-ringed. **Southeastern Spinyleg** larger; pale frontal stripes not isolated, hind leg much longer.

Shining Clubtail • *Stylurus ivae*

58–61 mm, 2.3–2.4 in. Brightly marked southeastern hanging clubtail inhabiting sand-bottomed woodland streams. Thorax yellow-green, with isolated pale frontal stripes, wide dark shoulder stripe, narrow dark lateral stripes. Abdomen with dorsal yellow stripes on S2-7, pale rings at base of middle segments; S7-10 mostly yellow and orange. Eyes green.

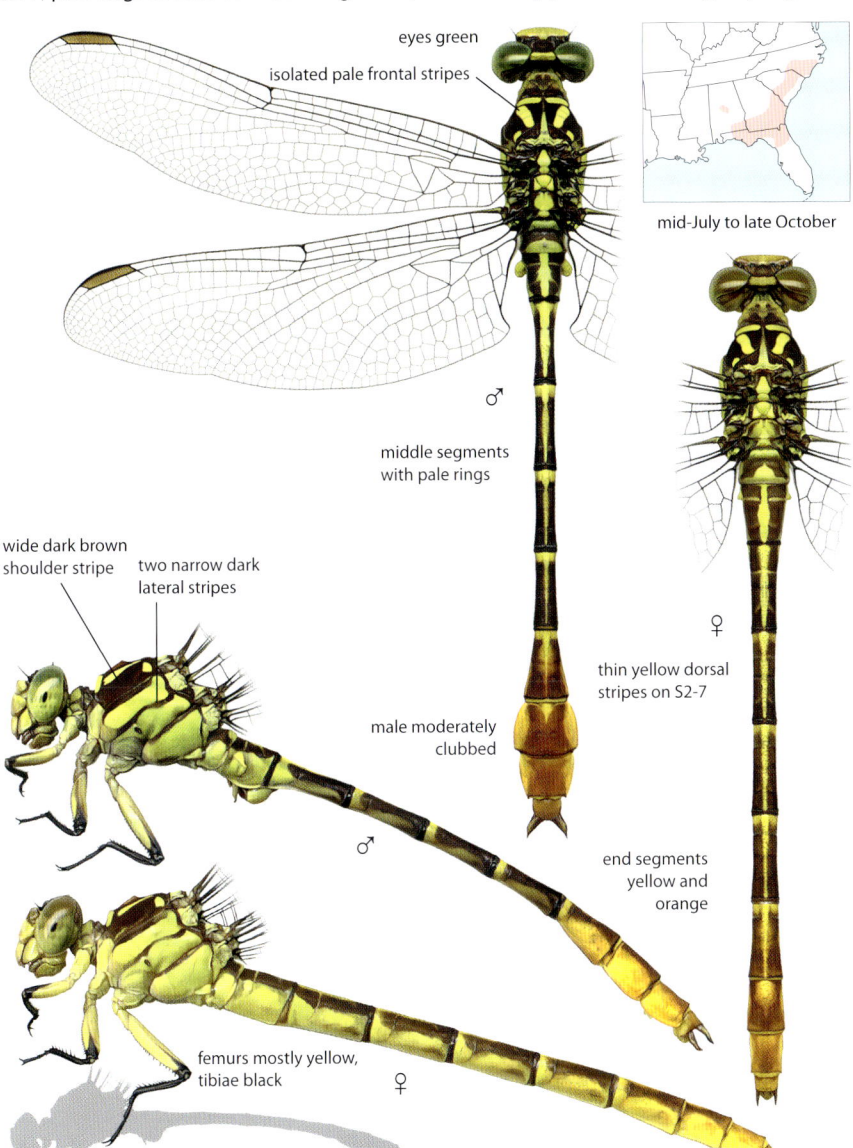

eyes green

isolated pale frontal stripes

mid-July to late October

♂

middle segments with pale rings

♀

thin yellow dorsal stripes on S2-7

wide dark brown shoulder stripe

two narrow dark lateral stripes

male moderately clubbed

♂

end segments yellow and orange

femurs mostly yellow, tibiae black

♀

Similar species: Russet-tipped Clubtail much duller, with gray-green thorax; lacks obvious pale markings on side of abdomen. **Southeastern Spinylegs** has separated dark shoulder stripes and abdomen lacks pale rings; hind leg much longer and black. **Yellow-sided Clubtail** has blue eyes; side of thorax unmarked yellow, sides of middle abdominal segments largely without pale markings.

Yellow-sided Clubtail • *Stylurus potulentus*

48–52 mm, 1.9–2 in. Scarce, slender mid-Gulf Coast species found at clean, sand-bottomed woodland streams and rivers. Thorax with unmarked yellow sides, isolated pale frontal stripes, wide dark shoulder stripes fused. Abdomen with narrow dorsal pale stripes on S1-8, sides of middle segments mostly dark. Yellow lateral spots S8-9. Face dark, eyes blue.

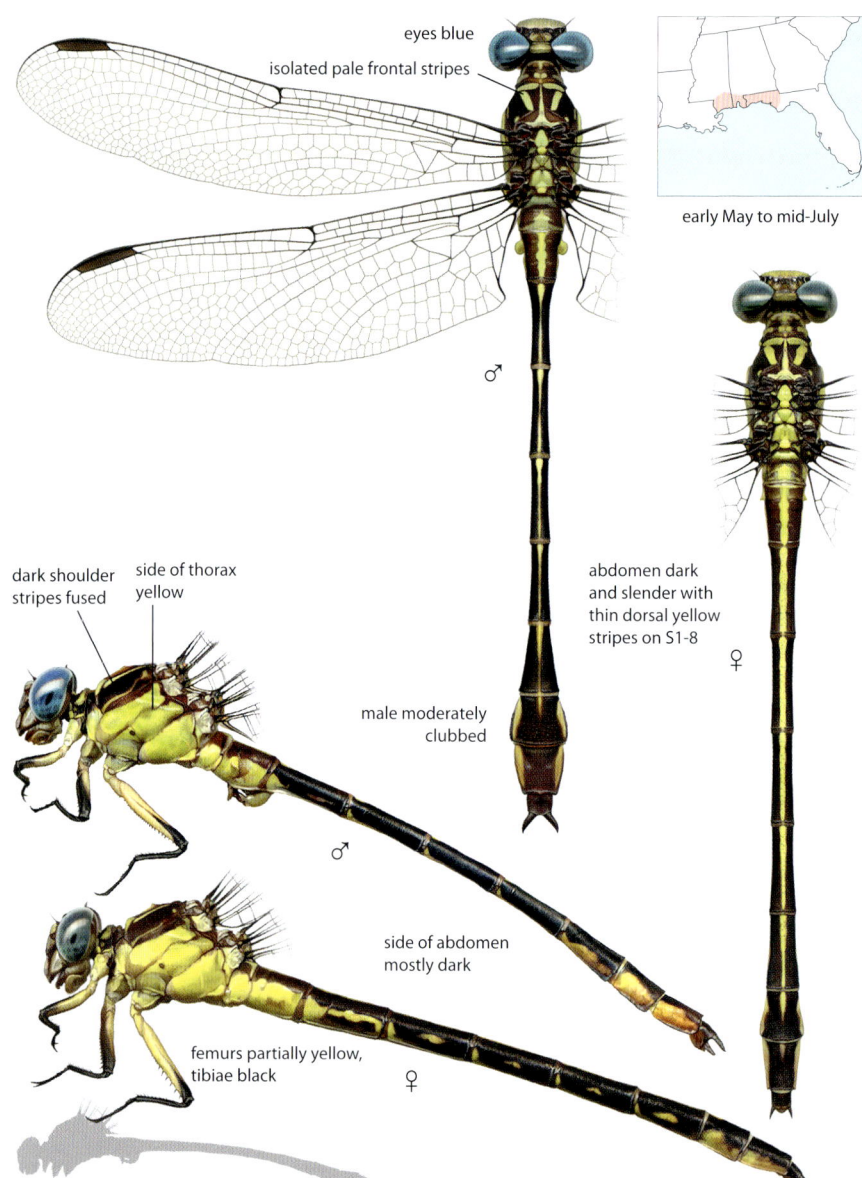

eyes blue

isolated pale frontal stripes

early May to mid-July

♂

dark shoulder stripes fused

side of thorax yellow

abdomen dark and slender with thin dorsal yellow stripes on S1-8

♀

male moderately clubbed

♂

side of abdomen mostly dark

femurs partially yellow, tibiae black

♀

Similar species: Shining Clubtail has green eyes, dark but narrow lateral thoracic stripes, yellow abdominal rings and lateral markings. **Townes' Clubtail** has green eyes, dark lateral thoracic stripes, and separated dark shoulder stripes. **Laura's Clubtail** has green eyes, dark lateral thoracic stripes, more pale markings on side of abdomen, browner end segments.

Spiketails • Family Cordulegastridae

Spiketails are large brown or black dragonflies, most with contrasting yellow markings. The thorax features two broad pale lateral stripes along with two pale frontal stripes. The legs are uniformly dark and relatively short. Eyes are blue or green when mature and meet or nearly meet at a single point on top of the head. The abdomen is long and slender. In the Old World, spiketails are called "goldenrings," named for their yellow-ringed abdomens. North American species are more varied. Their abdominal patterns may include rings but also spots or triangular and arrow-shaped markings.

Females have a large conspicuous subgenital plate in the form of a long spike, from which the family gets its name. Females deposit eggs in a manner similar to a sewing machine needle, bobbing up and down as they drive the eggs into the bottom substrate of the streams and seeps they inhabit. Males are usually best observed as they patrol up and down streams, their flight low and slow as they follow the water's course. Both sexes often feed in nearby clearings and fields. They perch on low branches and vertical stems, their bodies usually held at a 45-degree angle, although spiketails will also hang straight downward. Of 53 species worldwide, 10 are North American, all in the genus *Cordulegaster*.

Spiketails are usually easily separable from other dragonfly families. Cruisers are the most similar in size, pattern, and coloration but have only a single pale stripe on the side of the thorax. Their legs are longer, and females lack long subgenital plates. Many clubtails are yellow and black but have widely separated eyes. Darners have eyes that meet broadly along a seam, with most species having blue or green markings.

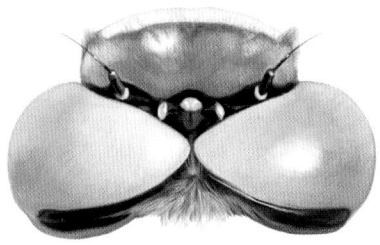

Spiketail head
eyes meet at a single point

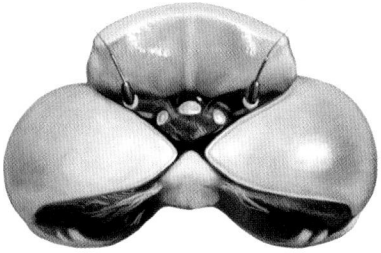

Some species have slightly separated eyes and are sometimes classified as a subgenus (*Zoraena*)

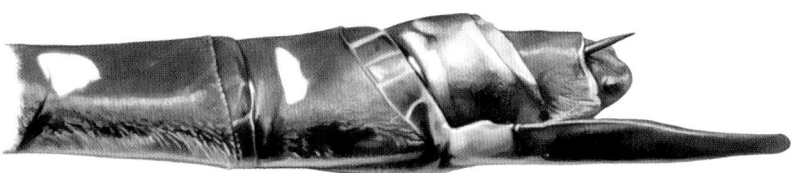

Female spiketail abdomen tip (Twin-spotted Spiketail)
horizontal subgenital plate long and spike-shaped

Arrowhead Spiketail • *Cordulegaster obliqua*

72–81 mm, 2.8–3.2 in. Large eastern spiketail found at spring-fed, mucky-bottomed forest streams. Abdomen with single row of arrow-shaped markings on top. Thorax with yellow markings between wings. Eyes green. Southern populations larger, with bluer eyes (subspecies *fasciata*). Midwestern prairie form with larger yellow abdominal markings.

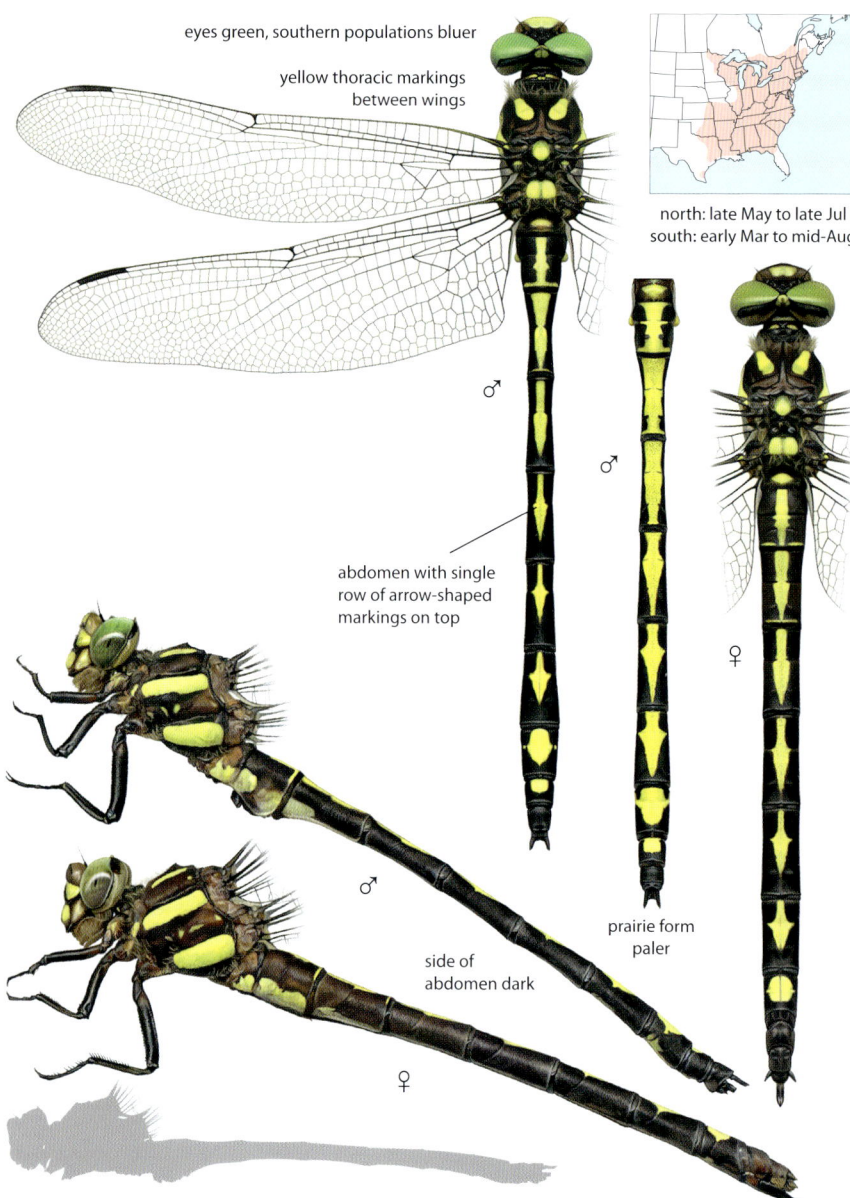

eyes green, southern populations bluer

yellow thoracic markings between wings

north: late May to late Jul
south: early Mar to mid-Aug

♂

♂

abdomen with single row of arrow-shaped markings on top

♀

prairie form paler

side of abdomen dark

♂

♀

Similar species: Single row of arrow-shaped yellow markings on top of abdomen unique among spiketails. **Cruisers** have only a single pale stripe on side of thorax. **Clubtails** have widely separated eyes.

Twin-spotted Spiketail • *Cordulegaster maculata*

64–76 mm, 2.5–3 in. Common eastern spiketail inhabiting clean forest streams and small rivers. Eyes green. Abdomen spotted with small, paired, dorsal yellow markings. Female subgenital plate very long, extending well beyond abdomen tip. Southern populations browner, with blue eyes.

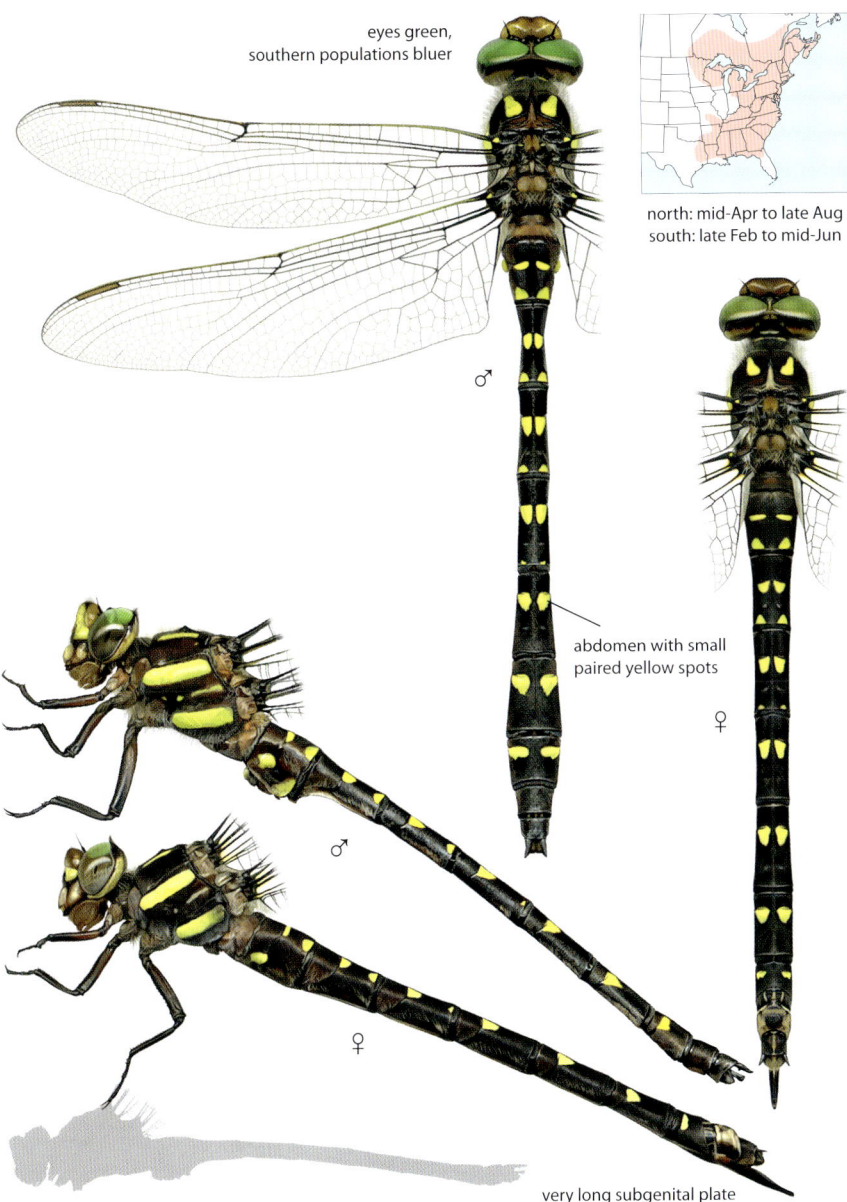

eyes green, southern populations bluer

north: mid-Apr to late Aug
south: late Feb to mid-Jun

abdomen with small paired yellow spots

♂

♀

♂

♀

very long subgenital plate

Similar species: Delta-spotted, Brown, and **Ouachita Spiketails** have yellow stripes at base of abdomen, shorter subgenital plates. **Sarracenia Spiketail** pale abdominal markings ring-like, extending farther down sides; lateral thoracic stripes gray, subgenital plate short.

Delta-spotted Spiketail • *Cordulegaster diastatops*

59–62 mm, 2.3–2.4 in. Small northeastern spiketail found at sunny small streams and seeps. Body black. Abdomen with paired triangular-shaped yellow markings, yellow lateral stripes along base segments. Thorax usually has three lateral stripes, middle one very narrow. Eyes green.

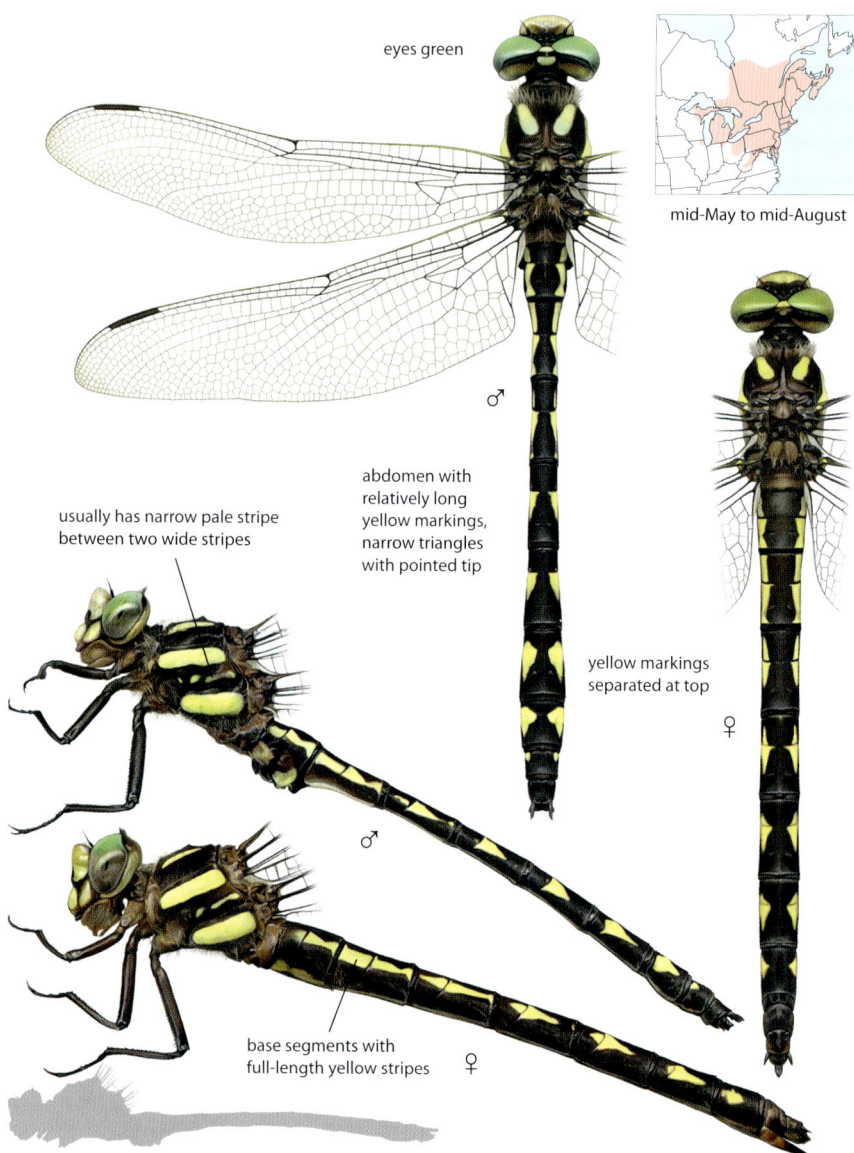

eyes green

mid-May to mid-August

abdomen with relatively long yellow markings, narrow triangles with pointed tip

♂

usually has narrow pale stripe between two wide stripes

yellow markings separated at top

♀

♂

base segments with full-length yellow stripes ♀

Similar species: Brown Spiketail legs and body brown; triangular abdominal markings blunter and less separated at top of abdomen; usually with only two lateral thoracic stripes (missing narrow middle stripe), thoracic stripes slightly narrower. **Twin-spotted Spiketail** has smaller, blunter, and more dorsally located yellow spots on abdomen; lacks stripes at abdomen base. **Ouachita Spiketail** nearly identical. Compare male appendages, page 422.

Brown Spiketail • *Cordulegaster bilineata*

55–68 mm, 2.2–2.7 in. Brown-colored spiketail found at small, often sunny, woodland streams; primarily southeastern. Legs and body brown. Thorax usually with only two yellow lateral stripes. Abdomen with paired yellow triangular markings, more separated and blunter than those of Delta-spotted Spiketail. Eyes dull green to green-gray.

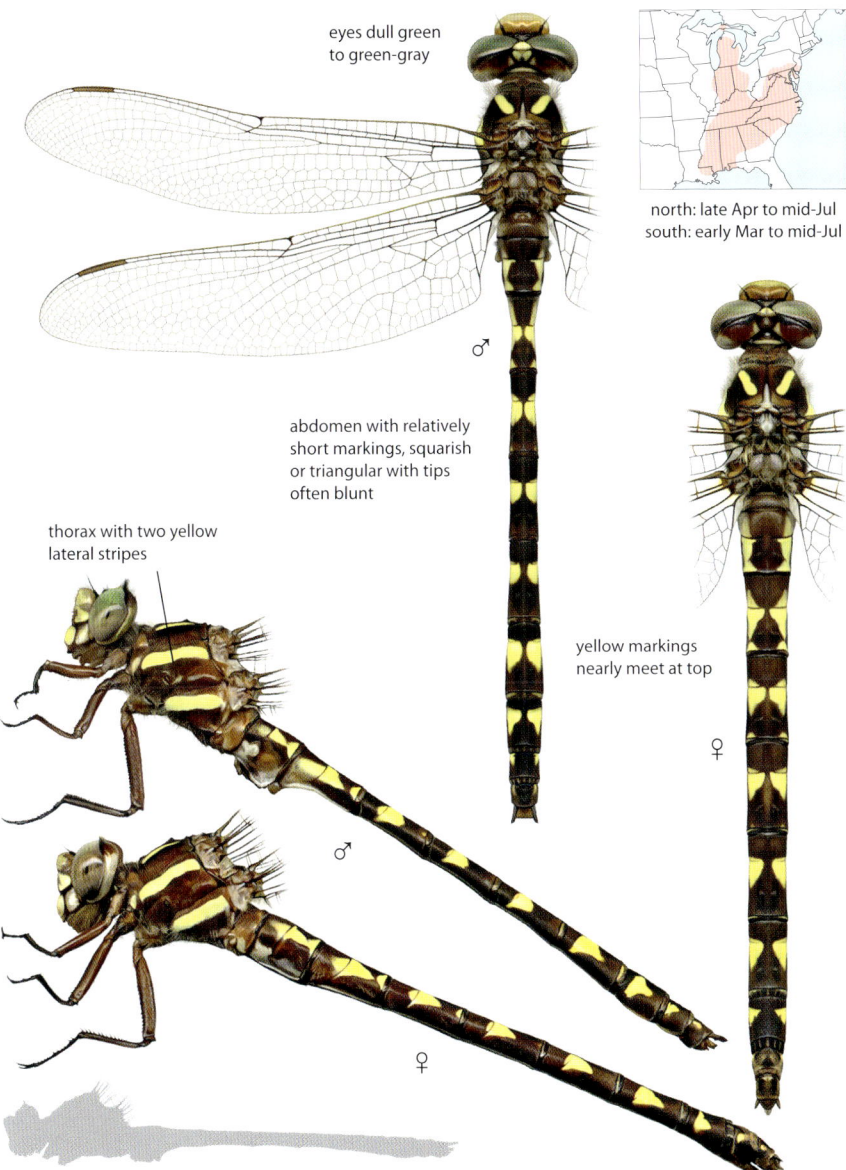

eyes dull green to green-gray

north: late Apr to mid-Jul
south: early Mar to mid-Jul

♂

abdomen with relatively short markings, squarish or triangular with tips often blunt

thorax with two yellow lateral stripes

yellow markings nearly meet at top

♀

♂

♀

Similar species: Delta-spotted Spiketail blacker in color; eyes brighter green, yellow abdominal triangles longer and more sharply pointed, thorax often has third narrow lateral stripe between two wider stripes. **Twin-spotted Spiketail** has smaller, more dorsally located yellow abdominal spots; lacks yellow stripes along base abdominal segments.

Tiger Spiketail • *Cordulegaster erronea*

65–76 mm, 2.6–3 in. Rare and local eastern spiketail found at spring-fed trickles and small forest streams. Eyes bright green; face with black cross-stripe. Thorax has yellow markings between wings. Abdomen distinctly yellow-ringed. Subgenital plate extends well beyond abdomen tip.

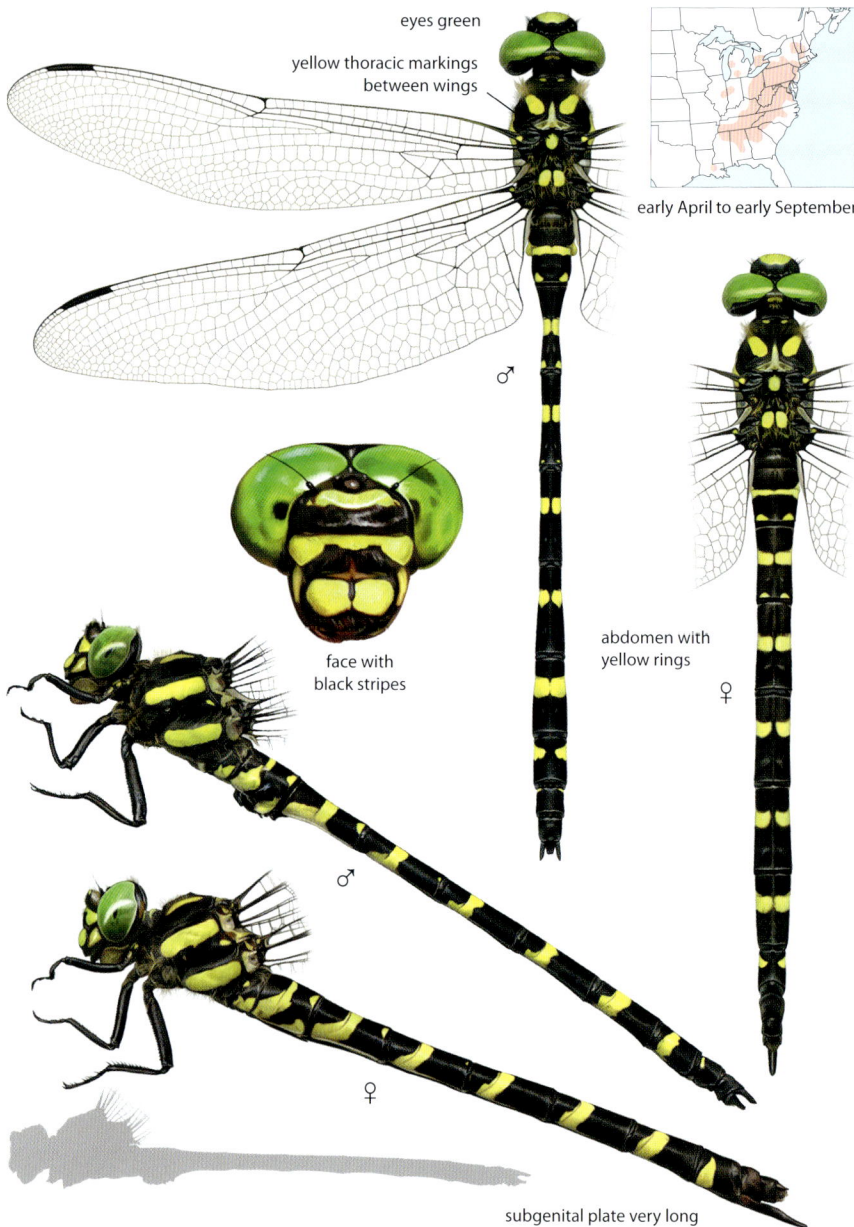

eyes green

yellow thoracic markings between wings

early April to early September

♂

face with black stripes

abdomen with yellow rings

♀

♂

♀

subgenital plate very long

Similar species: No other spiketail within its range has a ringed abdomen. Flight season begins somewhat later than other spiketails in its range.

Say's Spiketail • *Cordulegaster sayi*

60–69 mm, 2.4–2.7 in. Local southeastern spiketail inhabiting small, forested, hillside seeps, often adjacent to weedy fields or open woodland. Eyes gray-green. Thorax with magenta markings, including lateral magenta stripe between two pale yellow stripes. Abdomen with pale yellow rings.

eyes gray-green

late February to mid-July

thorax with magenta markings

magenta stripe between two pale yellow stripes

♂

abdomen with yellow rings

♀

♂

♀

subgenital plate short

Similar species: Tiger Spiketail not known to share range, lacks magenta thoracic markings, has brighter green eyes, black facial stripe; female with longer subgenital plate.

Ouachita Spiketail • *Cordulegaster talaria*

59–67 mm, 2.3–2.6 in. Spiketail with very limited range in Arkansas and Oklahoma, inhabiting small woodland streams. Similar to Delta-spotted Spiketail but does not share range; best separated by structure of male's cerci. Abdomen with paired triangular yellow markings, yellow stripes on sides of base segments. Eyes green.

eyes green

mid-April to late May

♂

abdomen with
yellow triangular
markings

yellow markings
separated at top

♀

♂

base segments with full-
length yellow stripes

♀

Similar species: Delta-spotted Spiketail largely identical, but no known range overlap. Compare male cerci, page 422. **Brown Spiketail** has separate range. Browner; yellow abdominal markings larger, blunter, less separated at top of abdomen. **Twin-spotted Spiketail** abdomen lacks yellow stripes at base, has small yellow spots, longer subgenital plate.

Sarracenia Spiketail • *Cordulegaster sarracenia*

56–64 mm, 2.2–2.5 in. Spiketail with small range in Texas and Louisiana, found at bogs with small spring-fed seeps supporting the pitcher plant, *Sarracenia alata*. Body and legs brown; thorax with two pale gray lateral stripes, pale yellow frontal stripes. Abdomen with paired yellow markings appears ringed, but markings narrowly separated at top. Eyes blue.

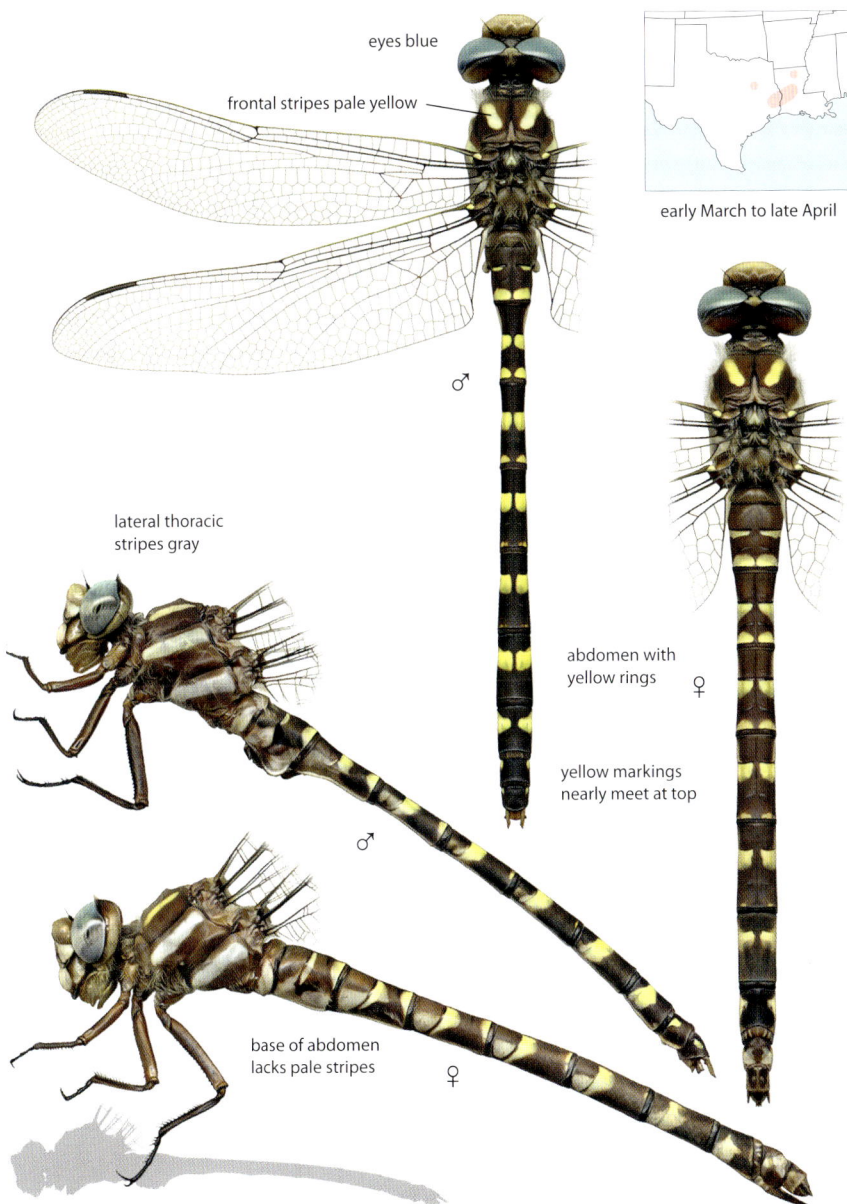

eyes blue

frontal stripes pale yellow

early March to late April

♂

lateral thoracic
stripes gray

abdomen with
yellow rings   ♀

yellow markings
nearly meet at top

♂

base of abdomen
lacks pale stripes   ♀

Similar species: Twin-spotted Spiketail has unringed abdomen marked with small yellow spots, and lateral thoracic stripes are yellow; female subgenital plate extends well beyond abdomen tip.

Apache Spiketail • *Cordulegaster diadema*

74–88 mm, 2.9–3.5 in. Large southwestern spiketail, local to small, clear, forested mountain streams. Only western spiketail with ringed abdomen. Mature eyes blue, yellow-green when juvenile. Face with wide black cross-stripes. Thorax with dorsal yellow markings between wings. Abdomen with complete yellow rings.

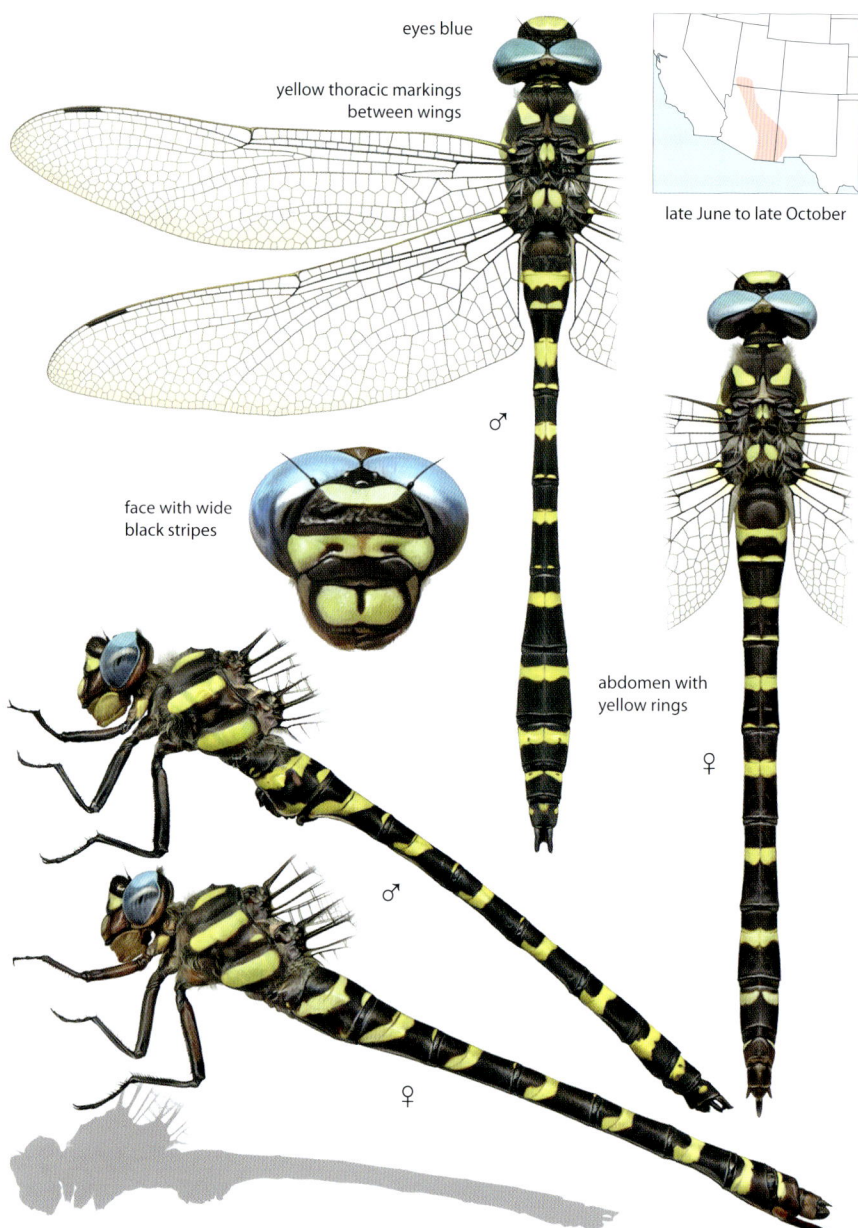

eyes blue

yellow thoracic markings between wings

late June to late October

face with wide black stripes

♂

abdomen with yellow rings

♀

♂

♀

Similar species: Pacific Spiketail has yellow spots on top of abdomen, lacks yellow rings. **Western River Cruiser** has single pale stripe on side of thorax, lacks pale abdominal rings.

Pacific Spiketail • *Cordulegaster dorsalis*

70–85 mm, 2.8–3.3 in. Fairly common western spiketail inhabiting small, clear, forested streams, widely ranging from lowlands to foothills and mountains. Eyes blue. Thorax with dorsal yellow markings between wings. Abdomen with dorsal yellow spots fused at top. Subgenital plate extends well beyond abdomen tip.

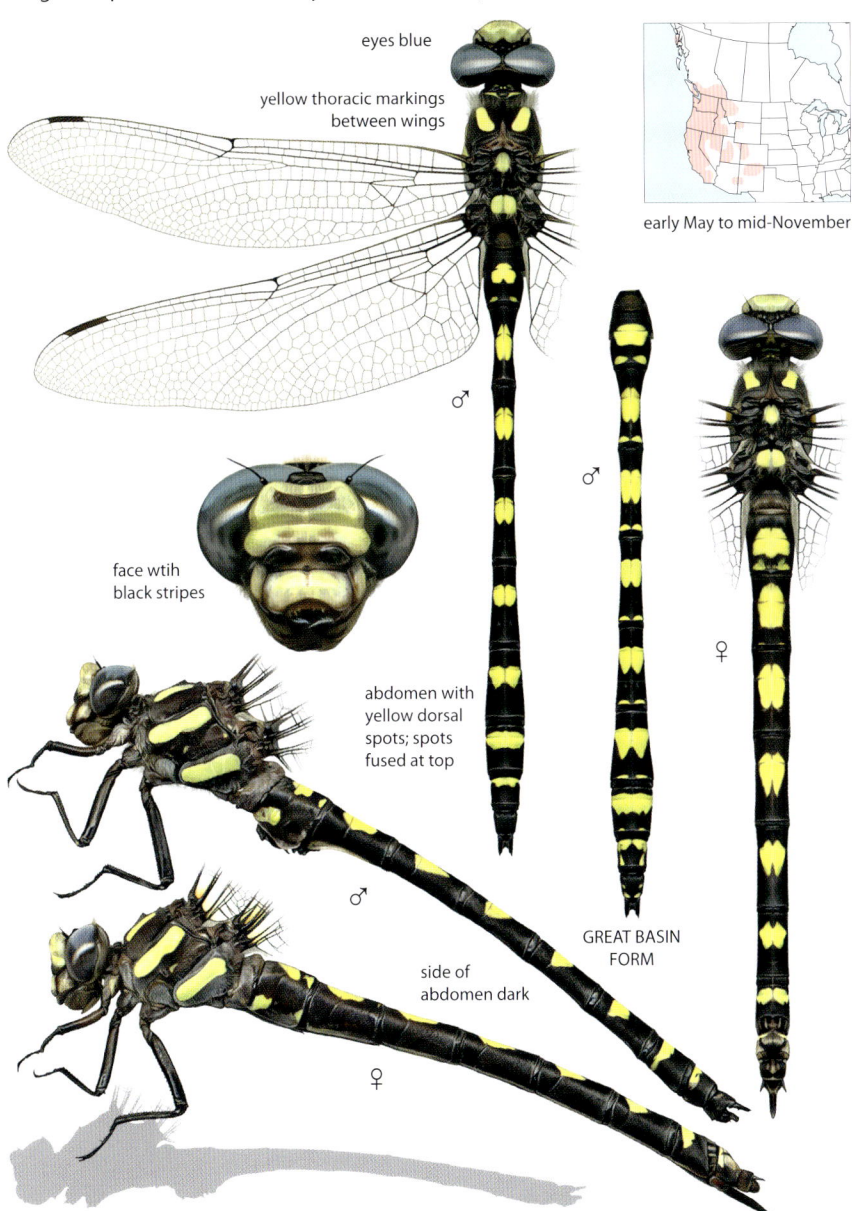

eyes blue

yellow thoracic markings between wings

early May to mid-November

face wtih black stripes

abdomen with yellow dorsal spots; spots fused at top

side of abdomen dark

GREAT BASIN FORM

♂

♂

♀

♂

♀

Great Basin form more extensively marked with yellow; found at spring-fed desert streams.

Similar species: Apache Spiketail has ringed abdomen. **Western River Cruiser** has single pale stripe on side of thorax, gray eyes.

Cruisers • Family Macromiidae

Cruisers are medium to large dragonflies typically seen in flight over streams, rivers, and lakeshores. Away from water, they feed and patrol along roads and woodland corridors and in clearings. At rest, they hang vertically or perch obliquely from branches.

Cruisers are similar in general appearance: black or brown dragonflies with pale yellow markings. All cruisers have a single pale stripe on each side of a dark thorax. Their large eyes meet at a seam and are green upon maturity in most species. The face is dark, with a pale cross-stripe. The legs are very long and uniformly dark. Cruiser wings are long, narrow, and largely clear but may be tinted. Any dark wing markings are small and restricted to the wing bases. The abdomen is variably marked with pale spots, dashes, or rings. Males of most species are moderately clubbed. Females lack an ovipositor and lay their eggs by dipping the tip of the abdomen into the water while flying at high speed. Of 126 species worldwide, nine are North American in two genera.

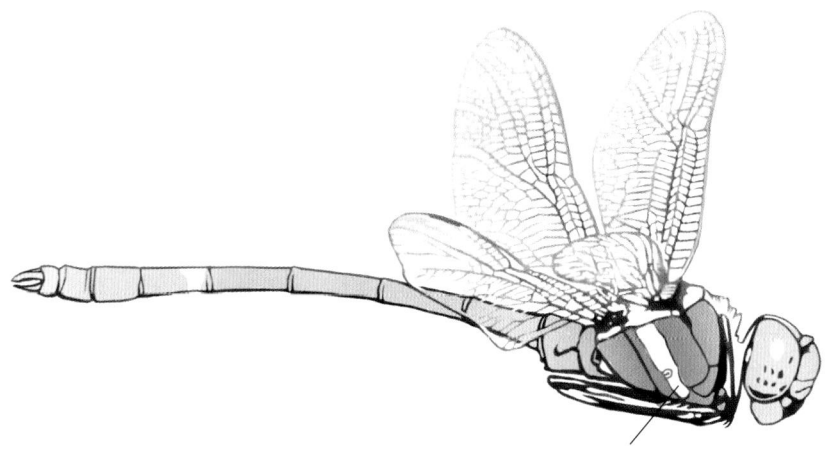

Cruiser in flight: identifiable by a single pale stripe on the side of the thorax

The single pale stripe on each side of the thorax distinguishes cruisers from other large fliers. Many spiketails are similarly dark with yellow markings, but they feature two wide lateral thoracic stripes and have shorter legs. Darners with pale thoracic stripes have two or more on each side, and spotted darners have a pair of pale lateral thoracic spots. Male river cruisers flying overhead may be separated from darners by their clubbed abdomens. A large clubtail, the Dragonhunter, has two wide yellow lateral thoracic stripes, a proportionally small head with separated eyes, and a striped abdomen.

Brown Cruisers, genus *Didymops,* page 204
This genus consists of a pair of North American species: one widespread throughout the East at rivers and streams, the other found at sand-bottomed lakes within a limited southeastern range. Both are dark brown with dull yellow markings and have dull green eyes. They fly early in the spring, their flight season largely completed before river cruisers emerge.

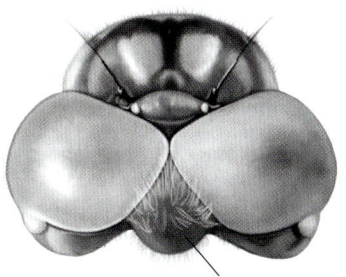

Brown cruiser head
has bulbous occiput

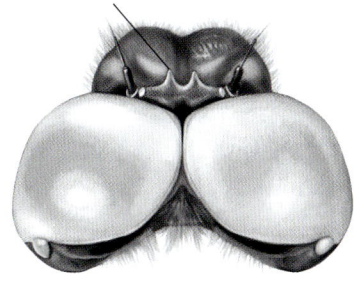

River cruiser head
has pair of cones on vertex

River Cruisers, genus *Macromia*, page 206

Of 81 species of river cruisers worldwide, seven are North American. Larger than brown cruisers, they inhabit rivers and streams, where they are characterized by their rapid flight. The majority are dark brown and black with strongly contrasting yellow markings and brilliant emerald-green eyes. Two western species are paler overall, with cream-colored markings and grayer eyes.

Most species of river cruisers can be identified by their thoracic and abdominal patterns. They are often best observed while perched or in the hand, as pattern details can be difficult to observe while the dragonfly is in swift and constant flight. Note the details of its abdominal markings, in particular the pale ring on S2, which may be complete or broken laterally and/or dorsally. Another clue is the presence and length of pale stripes at the front of the thorax, but these can be obscured by the forelegs while tucked up in flight.

Complicating river cruiser identification, some species apparently hybridize, with individuals exhibiting characters that are atypical of any recognized species. The situation remains poorly understood.

Once regarded as a separate species, the "**Wabash River Cruiser**" (*Macromia wabashensis)* is now considered an aberrant form of Royal River Cruiser or a hybrid of Royal River and Gilded River Cruisers (or possibly Bronzed River Cruiser in some locations).

It differs from Royal River Cruiser by having yellow costal veins (like Gilded River Cruiser) and having small pale spots on top of the frons. The pale abdominal markings are also larger and more fused at the top than typically on Royal River Cruiser.

Stream Cruiser • *Didymops transversa*

56–60 mm, 2.2–2.4 in. Common eastern cruiser found at rivers, streams, and occasionally large lakes. Body brown, thorax with single pale yellow lateral stripe. Immature eyes brown, dull green when mature. Rear of head brown. Abdomen dark brown with low-contrast pale markings. Small dark brown marks at base of wings.

mature eyes dull green

dark marks at base of wings

north: early Apr to early Aug
south: early Feb to mid-Jun

♂

back of head brown

thorax with single pale lateral stripe

♀

♂

abdomen brown; pale markings subtle

♀

legs very long

cerci pale

Similar species: Florida Cruiser grayer, abdomen blacker with pale rings; lacks dark wing markings. Back of head pale yellow.

Florida Cruiser • *Didymops floridensis*

65–68 mm, 2.6–2.7 in. Early-season southeastern cruiser found at sand-bottomed lakes in Florida and adjacent Alabama. Similar to Stream Cruiser but grayer, abdomen blacker. Abdomen ringed in appearance, with pale yellow-gray markings meeting at top. Mature eyes dull green. Back of head pale yellow. Lacks dark wing markings.

eyes dull green

late February to late April

back of head
pale yellow

abdomen black,
pale markings
forming rings

legs very long

cerci pale

Similar species: Stream Cruiser browner, abdomen unringed with more isolated pale markings; dark markings at wing bases. Back of head brown.

Swift River Cruiser • *Macromia illinoiensis*

65–76 mm, 2.6–3 in. Most common and widespread river cruiser; found at rivers and large streams, also open lakes, particularly to the north. Northern form, **"Illinois River Cruiser"** *M. i. illinoiensis*: Abdomen nearly all black, with prominent large yellow spot on S7. Yellow band on S2 broken both dorsally and laterally.

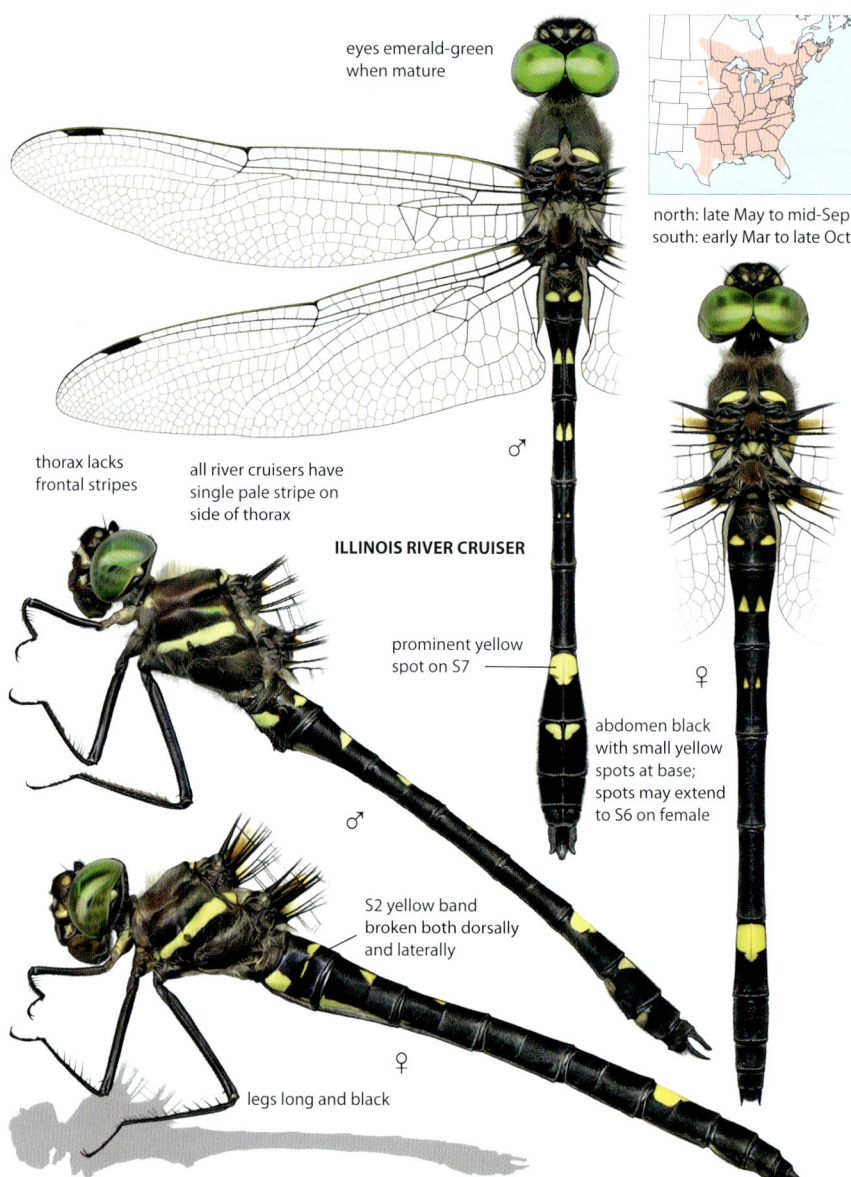

eyes emerald-green when mature

north: late May to mid-Sep
south: early Mar to late Oct

thorax lacks frontal stripes

all river cruisers have single pale stripe on side of thorax

ILLINOIS RIVER CRUISER

♂

prominent yellow spot on S7

♀

abdomen black with small yellow spots at base; spots may extend to S6 on female

♂

S2 yellow band broken both dorsally and laterally

♀

legs long and black

Similar species: Nearly all-black abdomen with prominent yellow spot on S7 often makes Illinois River Cruiser identifiable in flight. More extensively spotted individuals similar to other species, but only **female Mountain River Cruiser** has yellow band on S2 broken both dorsally and laterally.

Southern form, **"Georgia River Cruiser"** *M. i. georgina*: Only cruiser with complete yellow band on S2. Abdomen with paired yellow markings on S3-6, large dorsal yellow spot on S7, dorsal yellow mark on S8. Thorax with half-length yellow frontal stripes.

GEORGIA RIVER CRUISER

half-length yellow frontal stripes

S3 dorsal yellow marking often joins lateral stripe

S2 yellow band complete

♂

♀

abdomen with paired yellow markings on S3-6

yellow spots on top of S7-8

Similar species: All other river cruisers have yellow band on S2 broken dorsally. **Allegheny River Cruiser** also has reduced or absent frontal stripes, and S3 yellow dorsal and lateral markings are well separated; male has complete yellow ring on S7. **Mountain River Cruiser** lacks frontal stripes; S3 yellow markings well separated. **Royal River Cruiser** has smaller yellow abdominal markings; male abdomen not clubbed.

Allegheny River Cruiser • *Macromia alleghaniensis*

65–72 mm, 2.6–2.8 in. Widespread southeastern cruiser found at slow streams and rivers. Frontal stripe on thorax usually reduced or absent. Yellow ring on S2 broken at top. Paired yellow markings on S3-6. Male usually with complete yellow ring on S7, and keel of middle leg 15–20 percent of length of tibia.

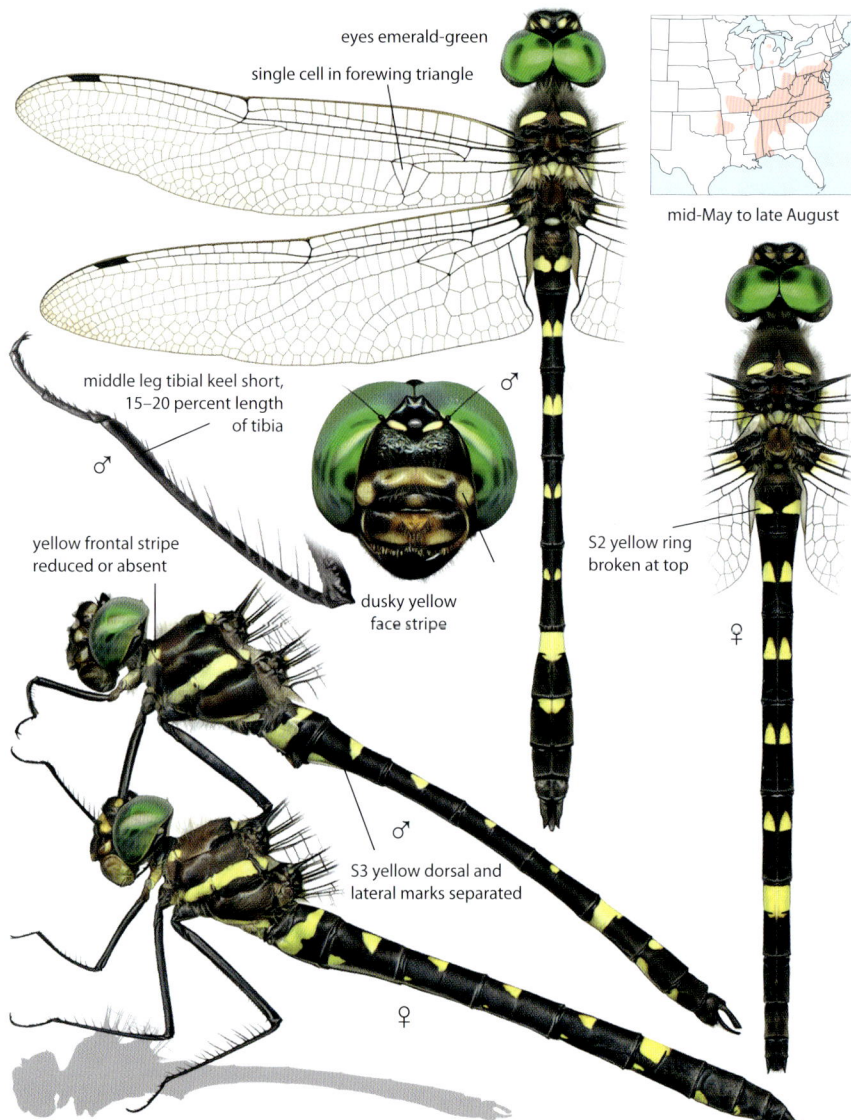

eyes emerald-green

single cell in forewing triangle

mid-May to late August

middle leg tibial keel short, 15–20 percent length of tibia

♂

yellow frontal stripe reduced or absent

dusky yellow face stripe

♂

S2 yellow ring broken at top

♀

S3 yellow dorsal and lateral marks separated

♂

♀

Similar species: Georgia River Cruiser thorax has half-length yellow frontal stripes. S2 with complete yellow ring; S3 dorsal and lateral yellow markings often joined. **Mountain River Cruiser** nearly identical. Usually two cells in forewing triangle; has bright yellow facial stripe. Male with tibial keel of middle leg half length or more; female with S2 ring broken laterally. **Royal River Cruiser** thorax with half-length yellow frontal stripes, smaller abdominal spots; male abdomen unclubbed.

Mountain River Cruiser • *Macromia margarita*

72–78 mm, 2.8–3.1 in. Southern Appalachian cruiser, local at clean, running mountain streams and rivers. Male largely identical to Allegheny River Cruiser. Usually has two cells in forewing triangle; middle leg tibial keel more than half length of tibia. Face with bright yellow cross-stripe. Female has yellow ring on S2 broken both laterally and dorsally.

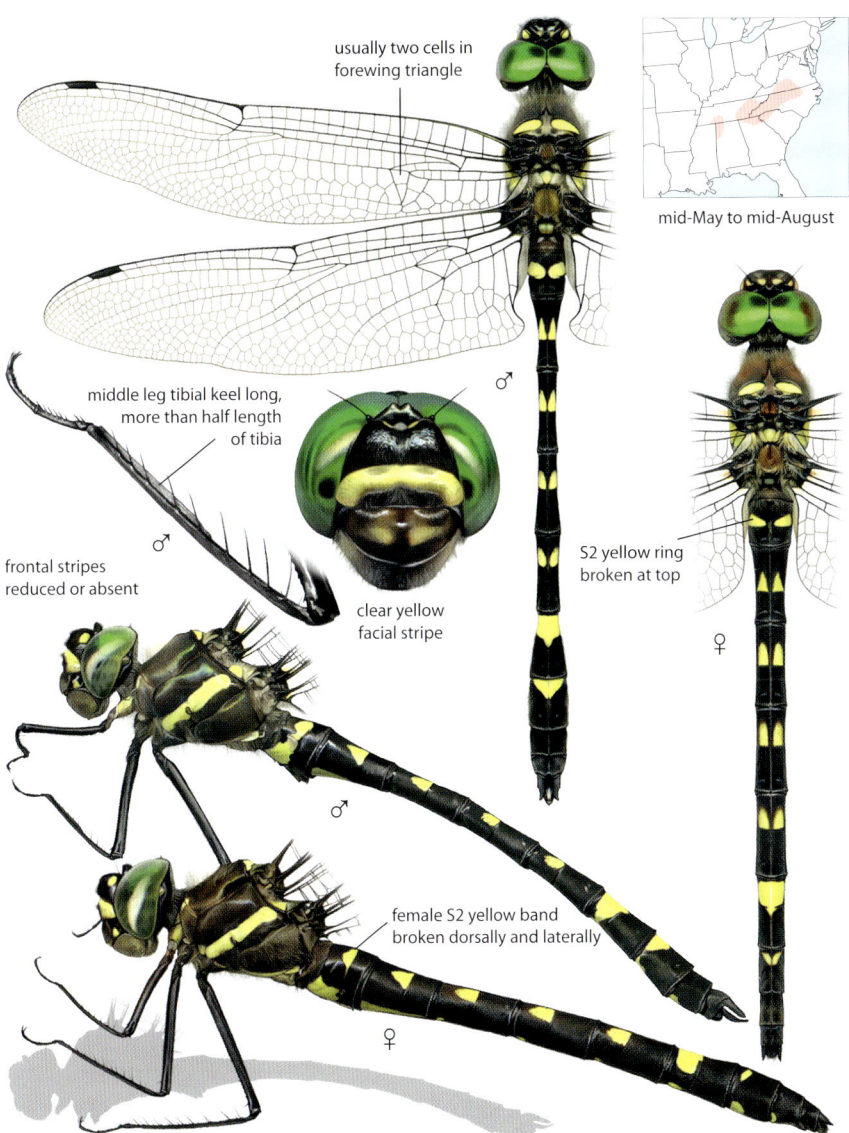

usually two cells in forewing triangle

mid-May to mid-August

middle leg tibial keel long, more than half length of tibia

♂

frontal stripes reduced or absent

clear yellow facial stripe

♂

S2 yellow ring broken at top

♀

female S2 yellow band broken dorsally and laterally

♂

♀

Similar species: Allegheny River Cruiser has dusky yellow face stripe, usually has single cell in forewing triangle. Male with shorter tibial keel of middle leg; female S2 yellow band not broken laterally. **Female Illinois River Cruiser** nearly identical to female Mountain but smaller, has dusky face stripe. **Georgia River Cruiser** has half-length frontal thoracic stripes, complete yellow ring on S2. **Royal River Cruiser** has half-length frontal thoracic stripes, smaller abdominal spots; male abdomen lacks club.

Gilded River Cruiser • *Macromia pacifica*

62–74 mm, 2.4–2.8 in. Brightly marked river cruiser inhabiting clear, flowing rivers and streams in the central US. Thorax with full-length yellow frontal stripes; S3-8 with paired, half-length yellow markings. Costal vein yellow. Apparently hybridizes with Royal River Cruiser; hybrid named Wabash River Cruiser (*M. wabashensis*) found primarily in Ohio.

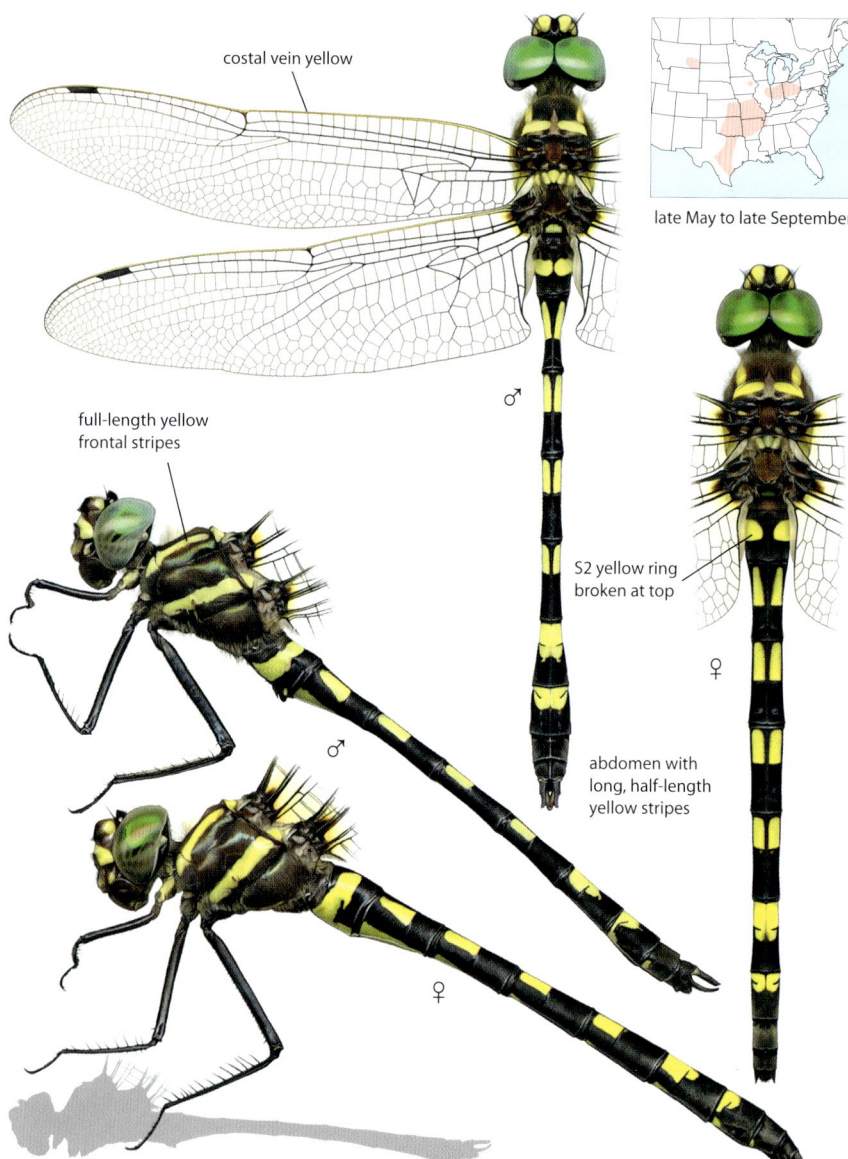

costal vein yellow

late May to late September

full-length yellow frontal stripes

♂

S2 yellow ring broken at top

♀

abdomen with long, half-length yellow stripes

♂

♀

Apparently hybridizes with **Royal River Cruiser**. See page 203.
Similar species: No other eastern species so extensively marked with yellow. **Bronze River Cruiser** paler overall, with green-gray eyes and cream-colored markings. **Western River Cruiser** has cream-colored markings, half-length frontal thoracic stripes; S2 pale ring broken laterally.

Royal River Cruiser • *Macromia taeniolata*

75–91 mm, 3–3.6 in. Very large eastern river cruiser found at clean rivers and large streams, occasionally lakes, more common southward. Top of frons dark. Thorax with half-length frontal stripes. Abdomen with small paired yellow markings, S2 yellow ring broken dorsally, S7 with pair of separated spots. Male abdomen lacks club.

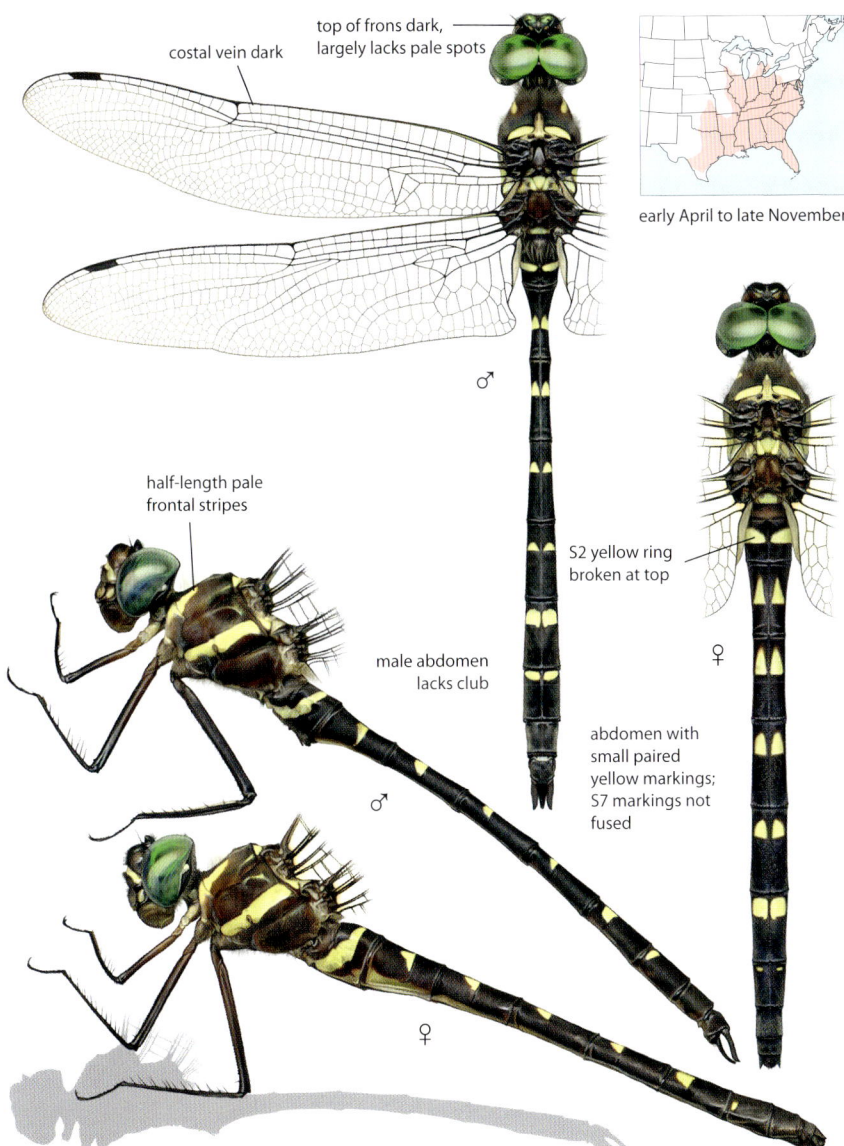

top of frons dark, largely lacks pale spots

costal vein dark

early April to late November

♂

half-length pale frontal stripes

S2 yellow ring broken at top

♀

male abdomen lacks club

♂

abdomen with small paired yellow markings; S7 markings not fused

♀

Apparently hybridizes with **Gilded River Cruiser**. See page 203.
Similar species: Georgia River Cruiser has larger yellow abdominal markings, yellow ring on S2 complete, S7 with undivided large yellow spot; male abdomen more clubbed.
Illinois River Cruiser lacks frontal stripes; yellow band on S2 broken laterally, S7 with large undivided yellow spot. Male abdomen clubbed.

Bronzed River Cruiser • *Macromia annulata*

68–73 mm, 2.7–2.9 in. Pale, dull-colored river cruiser found at rivers and large streams in dry country, primarily in Texas. Body dull brown with cream-colored markings. Thorax pruinose gray-brown, with full-length frontal stripes. Abdomen with large pale dorsal spots on S2-8, pale ring on S2 interrupted slightly at top. Eyes green-gray.

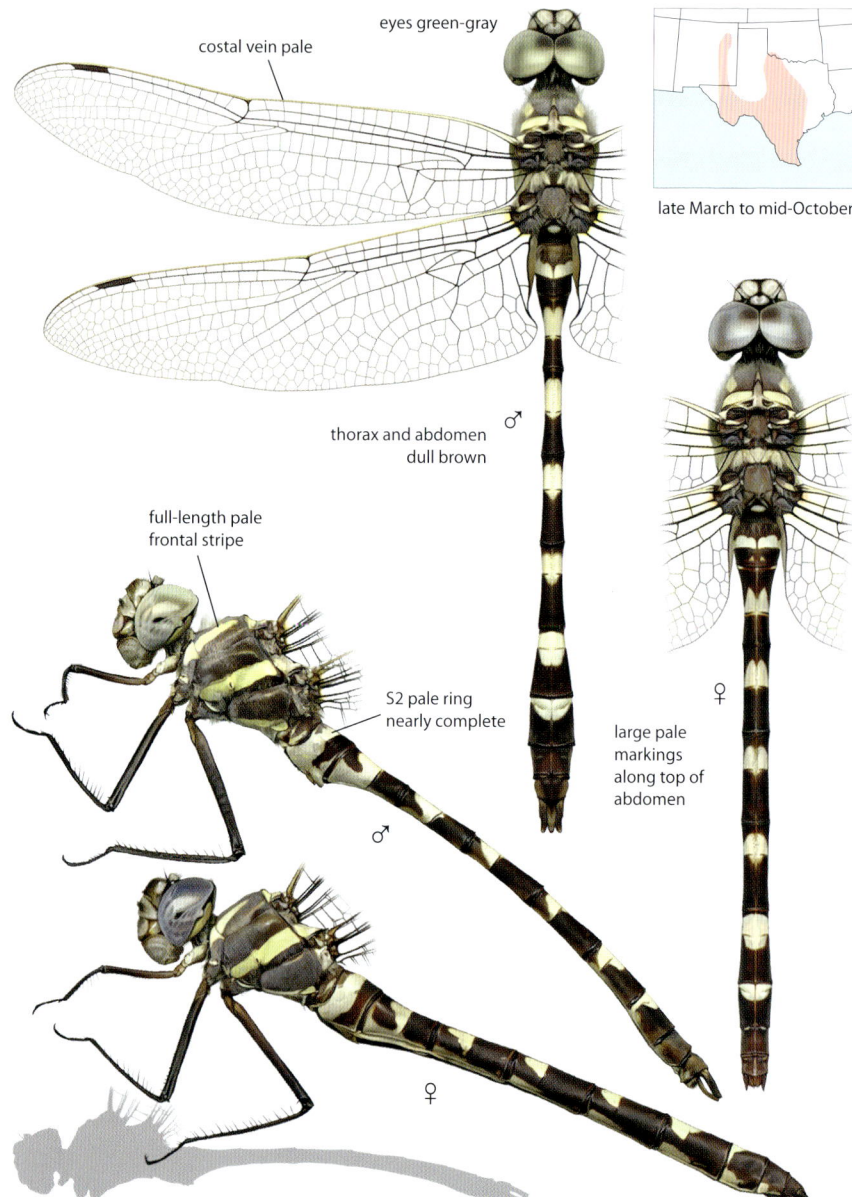

costal vein pale

eyes green-gray

late March to mid-October

thorax and abdomen dull brown

♂

full-length pale frontal stripe

S2 pale ring nearly complete

♂

large pale markings along top of abdomen

♀

♀

Similar species: Other river cruisers within its range have bright green eyes, more contrasting patterns with yellow markings. **Stream Cruiser** lacks pale frontal stripes, has low-contrast abdomen pattern and dark marks at base of each wing.

Western River Cruiser • *Macromia magnifica*

69–74 mm, 2.7–2.9 in. Dull-colored river cruiser found at western streams, rivers, irrigation canals, and large open lakes. Body grayish with cream-colored markings. Thorax lightly pruinose, has half-length pale frontal stripes. Abdomen with large pale dorsal markings on S2-9, S2 pale ring broken laterally and dorsally. Eyes gray-brown.

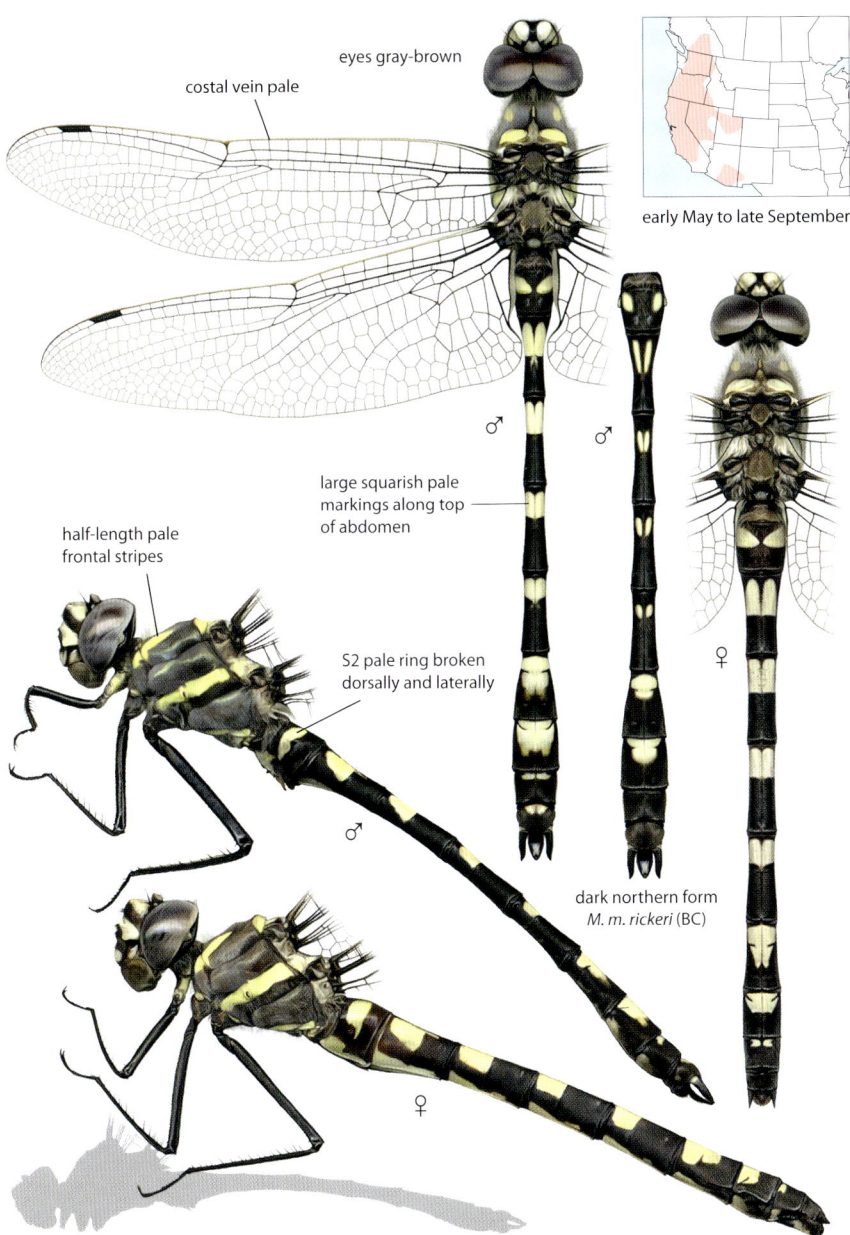

eyes gray-brown

costal vein pale

early May to late September

♂

♂

large squarish pale markings along top of abdomen

half-length pale frontal stripes

S2 pale ring broken dorsally and laterally

♂

♀

dark northern form
M. m. rickeri (BC)

♀

Similar species: Only river cruiser in its range. **Bronzed River Cruiser** has full-length frontal thoracic stripes; S2 pale ring not laterally broken.

Emeralds • Family Corduliidae

Emeralds are named for having jewel-like green eyes, a characteristic of many but not all species. Almost all emeralds are fliers, so they are most often observed as they feed and patrol on the wing. At rest, they hang vertically or perch obliquely from branches, while smaller species may rest flat on leaves. Most are dark, primarily brown and black, with some exhibiting metallic reflections on the thorax. Pale markings are typically small and inconspicuous, but these markings are often important to differentiate species. Emeralds' relatively small size and predominately dark coloration help separate them from other fliers such as darners and the similarly green-eyed cruisers. It is often possible to identify free-flying emeralds to their genus, but species identification is usually much more difficult.

Structurally, emeralds are most similar to the skimmers, but the rear margin of the emerald eye is more sinuous, and the hindwing anal loop is club-like in shape (the anal loop of skimmers is more foot- or boot-shaped). Male emeralds also have small lateral auricles on S2 that skimmers lack. The base of their hindwings has a corresponding angle and includes an anal triangle.

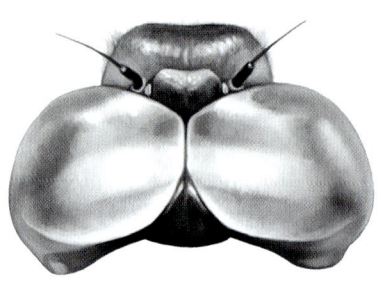

Emerald head dorsal view
eyes meet at seam

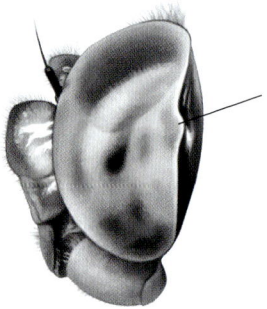

Emerald head lateral view
rear margin of eye sinuous

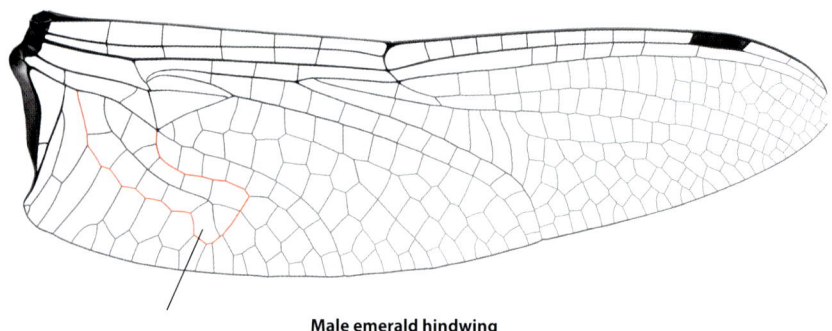

Male emerald hindwing
anal loop club-like in shape

Common Emeralds, genus *Cordulia*, page 217
This genus is composed of a single Eurasian species and a common North American one, the American Emerald. Males have a forked epiproct, unlike the similar little and striped emeralds. Their eyes are emerald-green upon maturity. The thorax is dark with metallic reflections but lacks any pale markings. The abdomen is stout and black, the male's slightly clubbed.

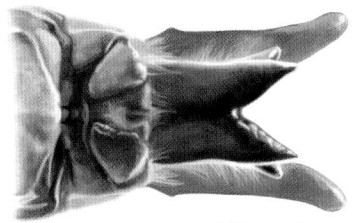

Male common emerald (ventral)
epiproct forked

Little Emeralds, genus *Dorocordulia*, page 218
Genus consists of two North American species, both northeastern. Eyes emerald-green upon maturity. Small size and unmarked thorax help separate little emeralds from the similar striped emeralds (*Somatochlora*). American Emerald is larger, marked with a whitish ring at the base of the abdomen. Little emeralds tend to perch lower than other emeralds, often landing on top of leaves.

Striped Emeralds, genus *Somatochlora*, page 220
A large and diverse group of dark, slender dragonflies, most exhibiting pale lateral thoracic stripes or spots. Of the 42 species worldwide, 26 are North American. Their highest diversity is to the north, where they breed at bog pools and fens. Some are lake species, while others utilize forest streams, particularly the few species that range into the South. Some are quite rare and range-restricted; the only dragonfly on the federal endangered species list is a striped emerald. As a group, striped emeralds are often recognizable in overhead flight by their dark and slender silhouettes, but unless found perched, most require capture for species confirmation. The eyes are brilliant green in adults of almost all species, though reddish brown in immatures. The thorax is dark brown with metallic green reflections and the abdomen mostly black (the reddish-brown Coppery Emerald an exception). All our species show pale markings on the top of the thorax between the wings, although these may become obscured with age. All have a pale ring between S2 and 3.

The presence of pale markings is useful for species identification, particularly the pattern of stripes on the side of the thorax. However, many species share similar patterns and are best separated by their reproductive structures.

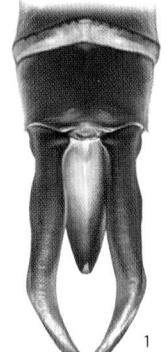
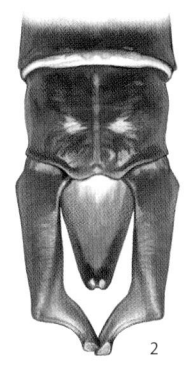
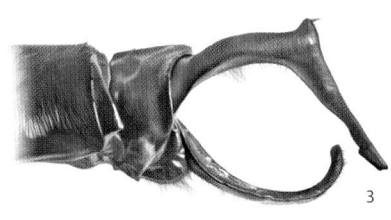

Striped emerald male cerci
come in different forms, including: 1) pincer-type, 2) bent-curled, 3) downward bent

Sundragons, genus *Helocordulia*, page 246
The sundragons are a pair of eastern North American emeralds found at running waters. The two species are similar in appearance, with differences in eye color and wing mark-

ings. The thorax is uniformly brown. The abdomen is black, with contrasting orange-yellow markings, males slightly clubbed and widest at S8. Wings have dark markings at their bases and small dark spots along the front edge of the wings.

Baskettails, genus *Epitheca*, page 248
The baskettails are a group of 13 mostly brown emeralds, 10 of which are North American. They are highly aerial, commonly seen in groups and swarms while feeding over meadows, lawns, and other open areas. Most are early-season species, breeding at ponds and slow streams, with males patrolling short beats over the water. They are dully colored, and most species do not develop green eyes. The hairy thorax is mainly brown and less metallic than most other emeralds, almost all having small yellow lateral spots. All of our species have dark wing markings, some conspicuously, others only barely. The abdomen is black with yellow lateral stripes. The male abdomen is spindle-shaped, the female more stout. Females have a very long, double-lobed subgenital plate, which they use to carry their eggs gathered in a mass (like eggs in a basket). When the egg mass is released into the water, it unfurls into a gelatinous string of several hundred eggs.

Nine of the 10 species of baskettails are classified in the subgenus *Tetragoneuria*. These are small, mostly brown, with relatively short abdomens. They are often recognizable in flight as baskettails, but identification to species is considerably more challenging. The Prince Baskettail is the lone species in the subgenus *Epicordulia*. It is larger with a long slender abdomen. In flight, it resembles a darner, but its wings have a pattern of dark markings.

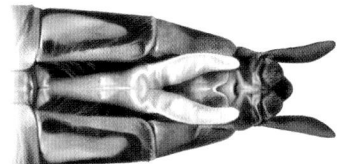

Baskettail female abdomen tip (ventral)
subgenital plate flat with long lobes

Shadowdragons, genus *Neurocordulia*, page 259
The shadowdragons are a group of seven North American species. They are crepuscular, becoming active for only short periods after dusk and, if warm enough, at dawn. Because of their habits, they are rarely encountered without effort. Identification usually requires capture, but netting them during the short window of their evening flights is very challenging. In near darkness, they often appear only as silhouettes against rippling water, their flight low, fast, and erratic. With luck, they may occasionally be found hanging in the deep shade of trees and foliage during the day.

Shadowdragons are mostly brown. Their eyes are large and dully colored. The brown thorax usually has a single pale yellow lateral spot. The abdomen is brown, weakly patterned in most species, some with blackish markings and/or yellow lateral stripes. The wings are broad, densely veined, and variably marked, most species having small brown or amber dots along the leading edge of the wings. Legs are uniformly pale brown or tan. The baskettails are most similar, but their abdominal patterns are often more contrasting and their legs blacker.

Boghaunters, genus *Williamsonia*, page 266
The boghaunters are a pair of North American species found at bogs and fens in the Northeast. Unlike other emeralds, they are perchers, most often found on the ground, on low leaves, and on tree trunks. This behavior likens them to some skimmers, but their color patterns are distinctive. Both are very small, with eyes relatively dull in color. Both species have an unmarked dark brown thorax and a black abdomen with pale rings, but abdomen pattern easily separates the two species from each other.

American Emerald • *Cordulia shurtleffii*

43–50 mm, 1.7–2 in. Common, widespread emerald found at a wide variety of ponds and lakes, including beaver, bog, and forest ponds. Eyes emerald. Side of face yellow. Thorax dark brown with green reflections. Abdomen black with whitish ring between S2 and 3. Male abdomen clubbed, epiproct forked. Female base of S3 white, cerci longer than S9.

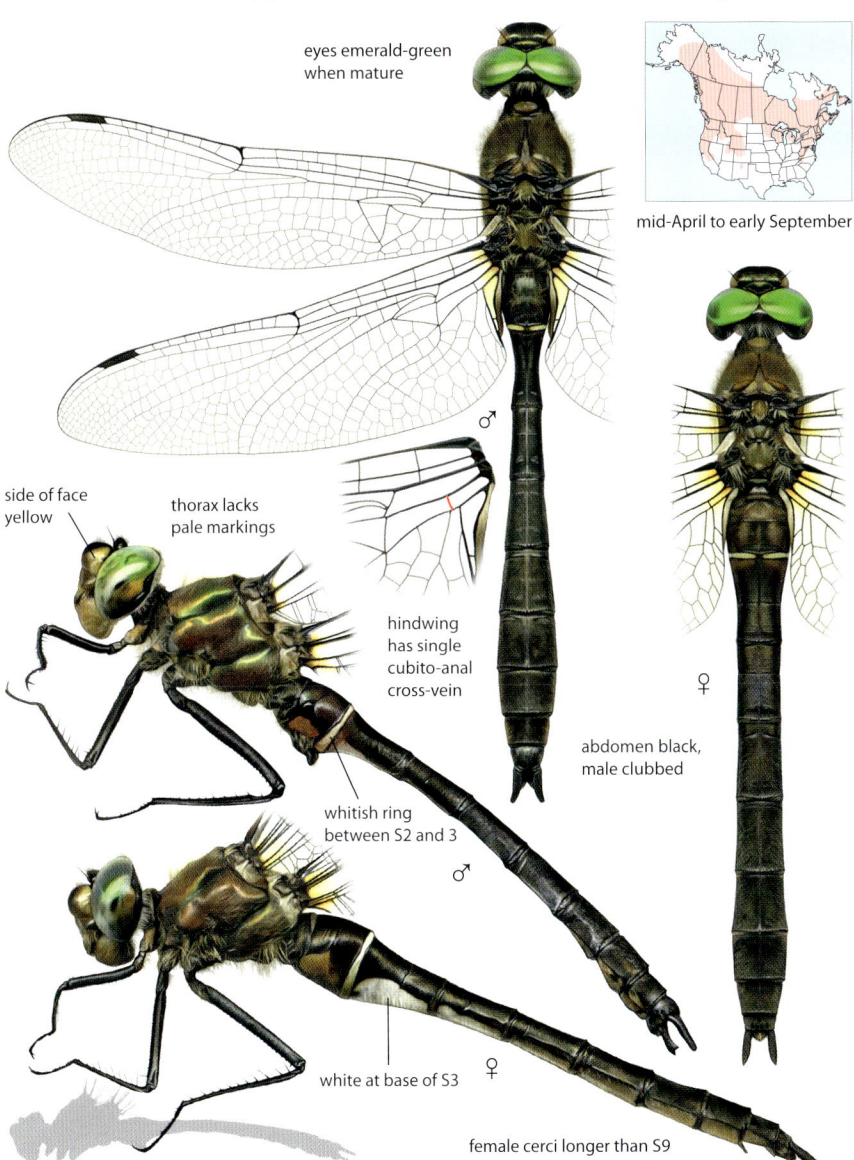

eyes emerald-green when mature

mid-April to early September

side of face yellow

thorax lacks pale markings

hindwing has single cubito-anal cross-vein

abdomen black, male clubbed

♂

whitish ring between S2 and 3

♂

white at base of S3

♀

female cerci longer than S9

♀

Similar species: Racket-tailed and **Petite Emeralds** smaller, have yellow-orange markings on S3, lack white ring between S2 and 3. **Striped emeralds** lack forked epiproct, have two cubito-anal cross-veins on hindwing; most have pale lateral thoracic markings. **Mocha Emerald** slenderer; abdomen has pale lateral spots, female with perpendicular subgenital plate. **Sundragons** have dark wing markings, abdomen with orange-yellow markings.

Racket-tailed Emerald • *Dorocordulia libera*

37–44 mm, 1.5–1.7 in. Small northeastern little emerald inhabiting bog ponds, lakes, and slow streams. Eyes emerald-green. Thorax unmarked dark brown with green reflections. Abdomen black with yellow-orange lateral markings on S2-3. Abdomen of both sexes clubbed, male club flattened and paddle-shaped. Face dark.

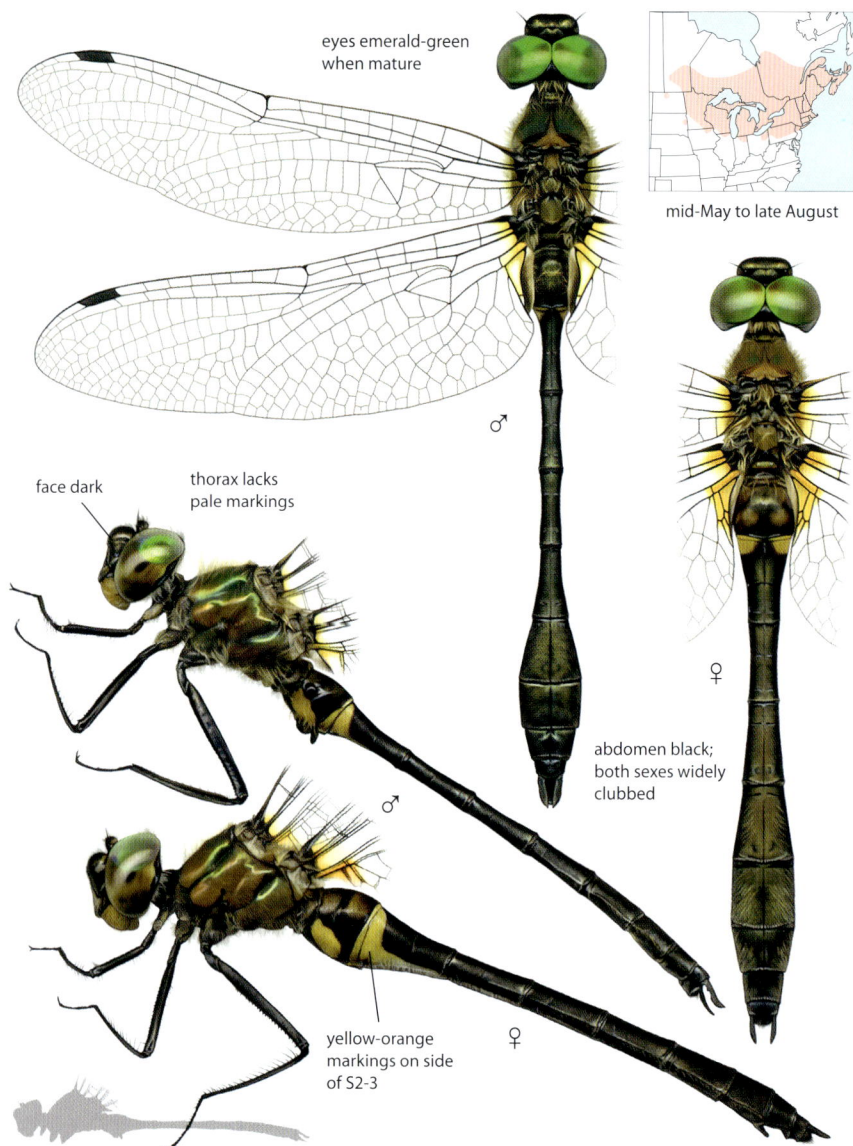

eyes emerald-green when mature

mid-May to late August

♂

thorax lacks pale markings

face dark

♂

abdomen black; both sexes widely clubbed

♀

yellow-orange markings on side of S2-3

♀

Similar species: Petite Emerald male only slightly clubbed; female abdomen unclubbed with pale lateral markings on S2-7. **American Emerald** larger, stouter; abdomen with white ring between S2 and 3. Male cerci with blunt tips, epiproct forked. **Striped emeralds** larger, with pale ring between S2 and 3; most have pale lateral thoracic markings. **Sundragons** have dark wing markings.

Petite Emerald • *Dorocordulia lepida*

37–43 mm, 1.5–1.7 in. Small, slender northeastern little emerald found at lakes, marshes, and bog ponds. Eyes emerald-green. Thorax unmarked bronze with green reflections. Male only slightly clubbed. Abdomen black with orange-yellow markings on sides of S2-3, female with additional lateral markings on S4-7. Yellow spots on sides of face.

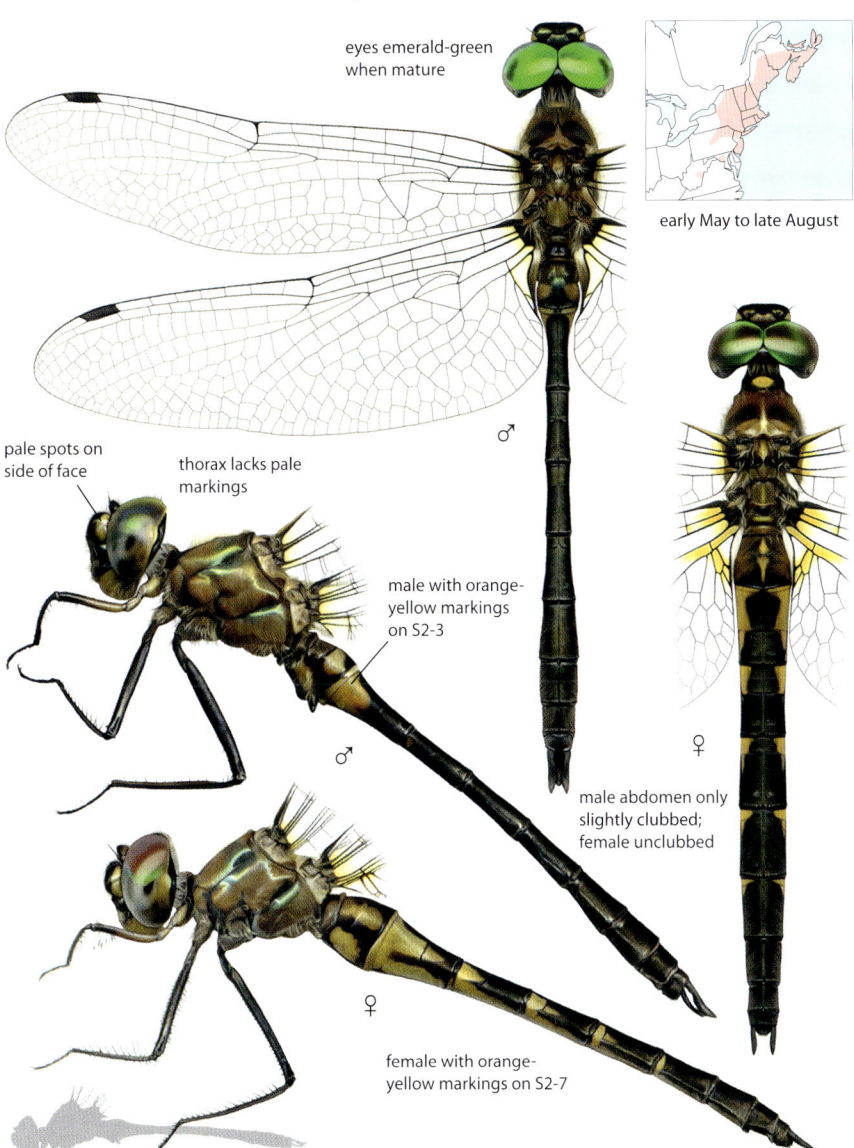

eyes emerald-green when mature

early May to late August

pale spots on side of face

thorax lacks pale markings

male with orange-yellow markings on S2-3

♂

♂

♀

male abdomen only slightly clubbed; female unclubbed

♀

female with orange-yellow markings on S2-7

Similar species: Racket-tailed Emerald has widely clubbed abdomen; female lacks yellow-orange markings on sides of S4-7. **American Emerald** larger, has whitish ring between S2 and 3, lacks orange markings on S3; male cerci with blunt tips, epiproct forked. **Striped emeralds** larger, have pale ring between S2 and 3; most have pale lateral thoracic markings. **Sundragons** have dark wing markings.

Ocellated Emerald • *Somatochlora minor*

42–50 mm, 1.7–2 in. Small striped emerald widespread in the North at clean, flowing, partly open forest streams. Eyes emerald-green. Thorax metallic brown and green with two lateral yellow oval spots. Abdomen black with pair of yellow marks on top of S3. Male abdomen short, spindle-shaped. Female subgenital plate long and perpendicular.

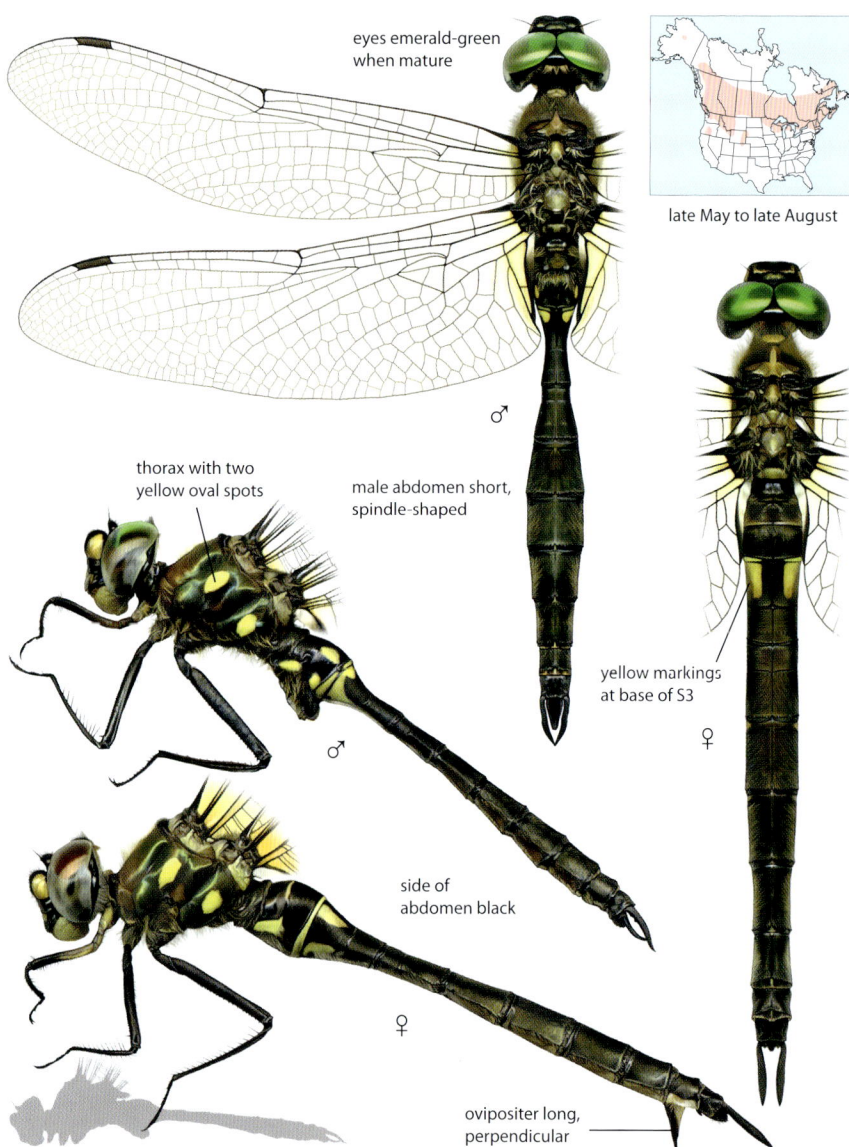

eyes emerald-green when mature

late May to late August

thorax with two yellow oval spots

male abdomen short, spindle-shaped

♂

yellow markings at base of S3

♂

side of abdomen black

♀

♀

ovipositer long, perpendicular

Similar species: Brush-tipped Emerald has pale anterior thoracic stripe; abdomen usually with small pale lateral spots. Compare male cerci, female subgenital plate, pages 422, 427. **Ski-tipped Emerald** more elongated, has pale anterior lateral thoracic stripe. **Forcipate** and **Mountain Emeralds** more elongated; **Forcipate** has pale spots on side of abdomen. Compare male cerci and female subgenital plates, pages 423, 428.

Brush-tipped Emerald • *Somatochlora walshii*

41–52 mm, 1.6–2 in. Small northern emerald found at small, slow streams through open bogs, marshes, sedge meadows. Side of thorax with pale anterior stripe and posterior oval spot. Abdomen short, spindle-shaped, S3 with yellow dorsal markings, small lateral marks on S5-7. Male cerci enlarged at tip and hairy. Subgenital plate short, obliquely angled.

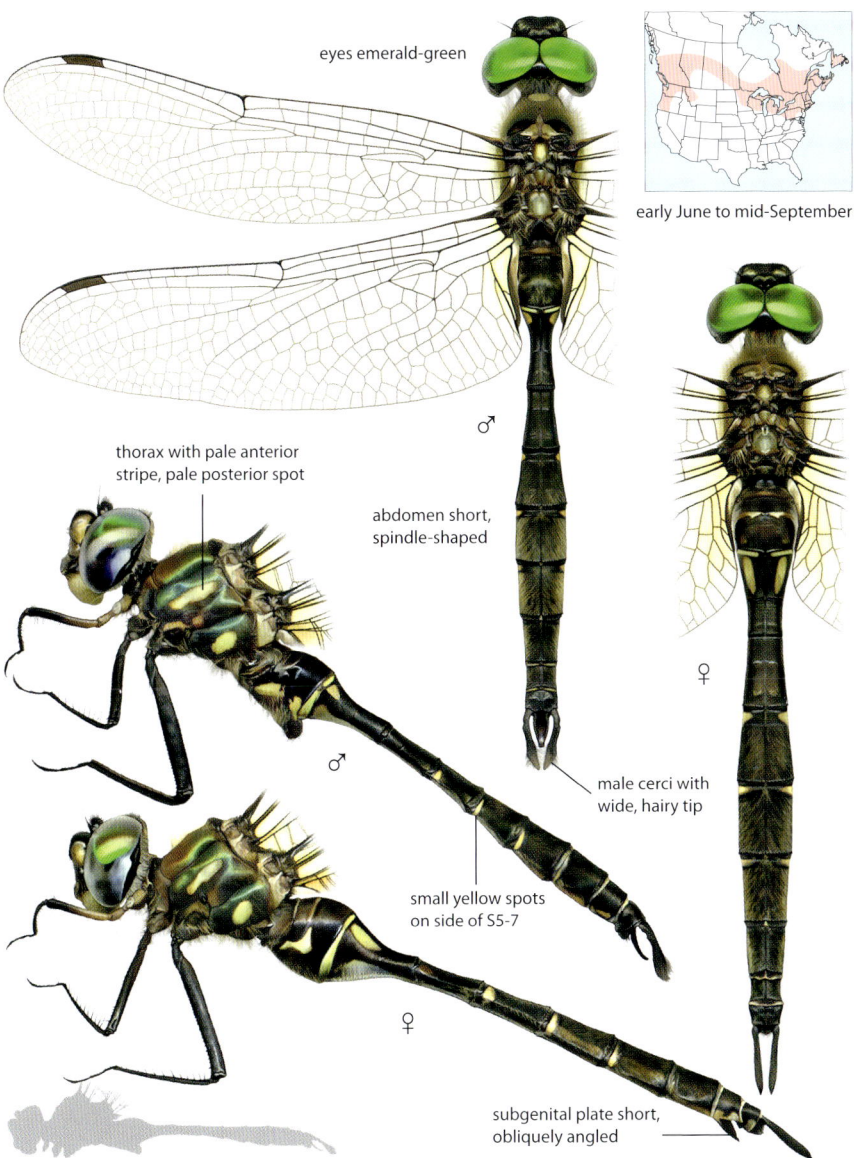

eyes emerald-green

early June to mid-September

thorax with pale anterior stripe, pale posterior spot

abdomen short, spindle-shaped

♂

♀

♂

male cerci with wide, hairy tip

small yellow spots on side of S5-7

♀

subgenital plate short, obliquely angled

Similar species: Ocellated Emerald most similar in size and shape. Side of thorax with two pale oval spots. Abdomen lacks lateral pale spots on middle segments. Compare male cerci, female subgenital plate, pages 422, 427. **Clamp-tipped** and **Ski-tipped Emeralds** larger, more elongated. Abdomen lacks pale lateral spots on middle segments. Male cerci and female subgenital plates quite different; see pages 423–24, 428–29.

Williamson's Emerald • *Somatochlora williamsoni*

53–59 mm, 2.1–2.3 in. Dark, elongated emerald found at slow streams and clear lakes from the Northeast west to Saskatchewan. Side of thorax with pale anterior stripe and posterior oval spot; marks darken with age. Dull pale spots on side of S5-8 of male, S3-8 of female. Male cerci ski-shaped, hairy in middle. Subgenital plate long and thin, perpendicular to abdomen.

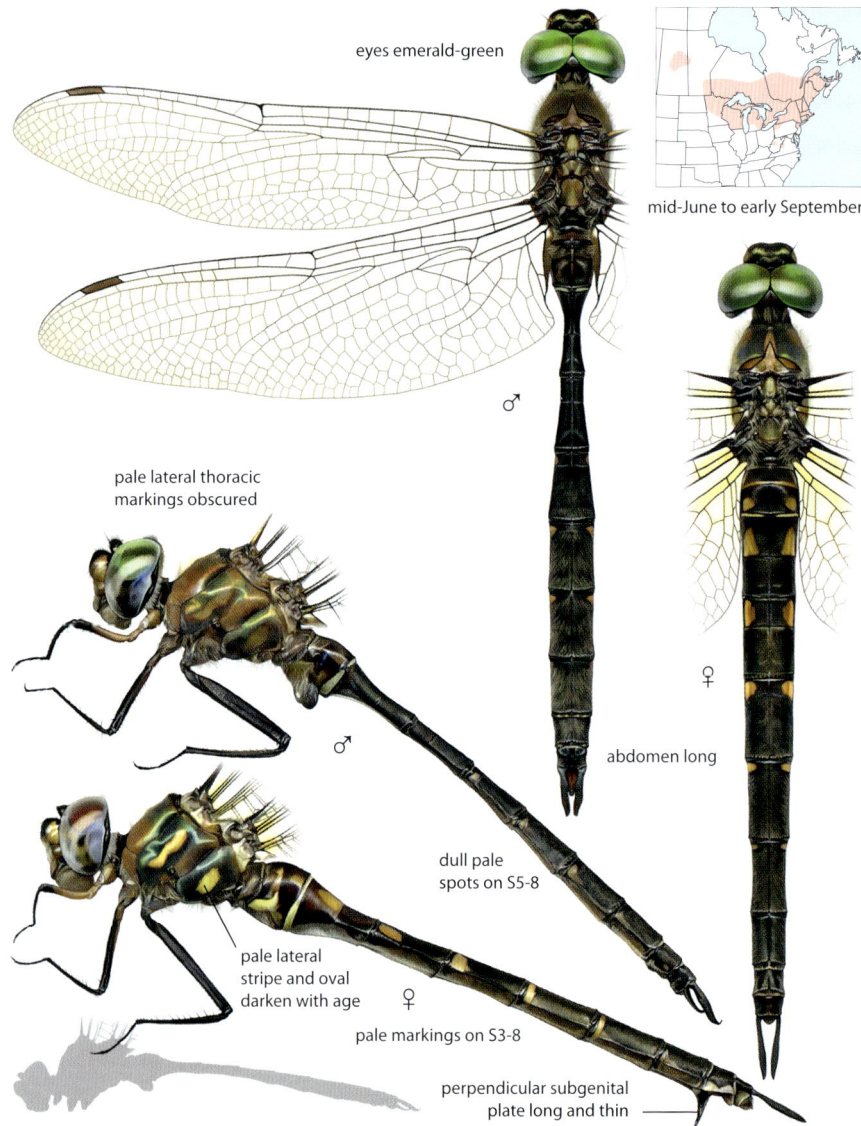

eyes emerald-green

mid-June to early September

♂

pale lateral thoracic markings obscured

♂

abdomen long

dull pale spots on S5-8

♀

pale lateral stripe and oval darken with age ♀

pale markings on S3-8

perpendicular subgenital plate long and thin

Similar species: Kennedy's Emerald thorax usually lacks pale posterior lateral spot, and abdomen lacks pale lateral spots; male abdomen widest at S6, female subgenital plate horizontal. **Clamp-tipped Emerald** female lacks pale abdominal lateral spots, has pale spots on top of thorax between wings. **Ski-tipped Emerald** lateral thoracic stripes do not darken; female lacks pale lateral abdominal spots. **Incurvate Emerald** male has pincer-like cerci; female subgenital plate horizontal.

Ski-tipped Emerald • *Somatochlora elongata*

52–62 mm, 2–2.4 in. Common slender striped emerald found at slow streams, often in bogs and swamps. Eyes emerald-green. Side of thorax with pale yellow anterior stripe and posterior oval spot, which do not darken. Abdomen black, S3 with dorsal pale markings at base. Male cerci long and ski-shaped. Female with triangular subgenital plate.

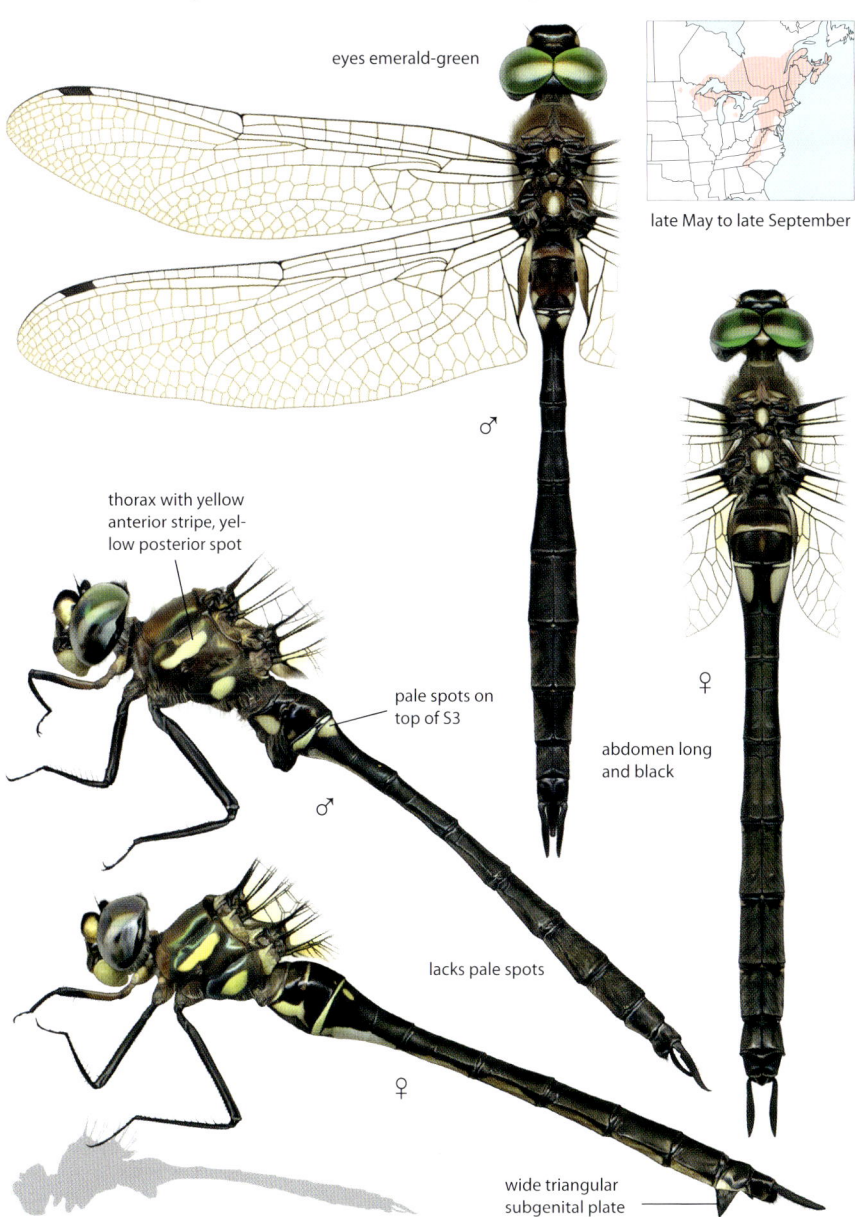

eyes emerald-green

late May to late September

thorax with yellow anterior stripe, yellow posterior spot

♂

pale spots on top of S3

abdomen long and black

♀

♂

lacks pale spots

♀

wide triangular subgenital plate

Similar species: Clamp-tipped Emerald has very different male cerci and female subgenital plate thinner, see pages 424, 429. **Williamson's** and **Incurvate Emeralds** have small pale spots on side of abdomen; lateral thoracic markings often obscured.

Mocha Emerald • *Somatochlora linearis*

58–68 mm, 2.3–2.7 in. Large, dark eastern striped emerald found at small forest streams. Thorax without any pale markings. Abdomen long and slender, middle segments with yellow-orange lateral spots. Male cerci straight with ventral spine near tip. Subgenital plate thorn-shaped and perpendicular to abdomen. Wings brown-tinted with age.

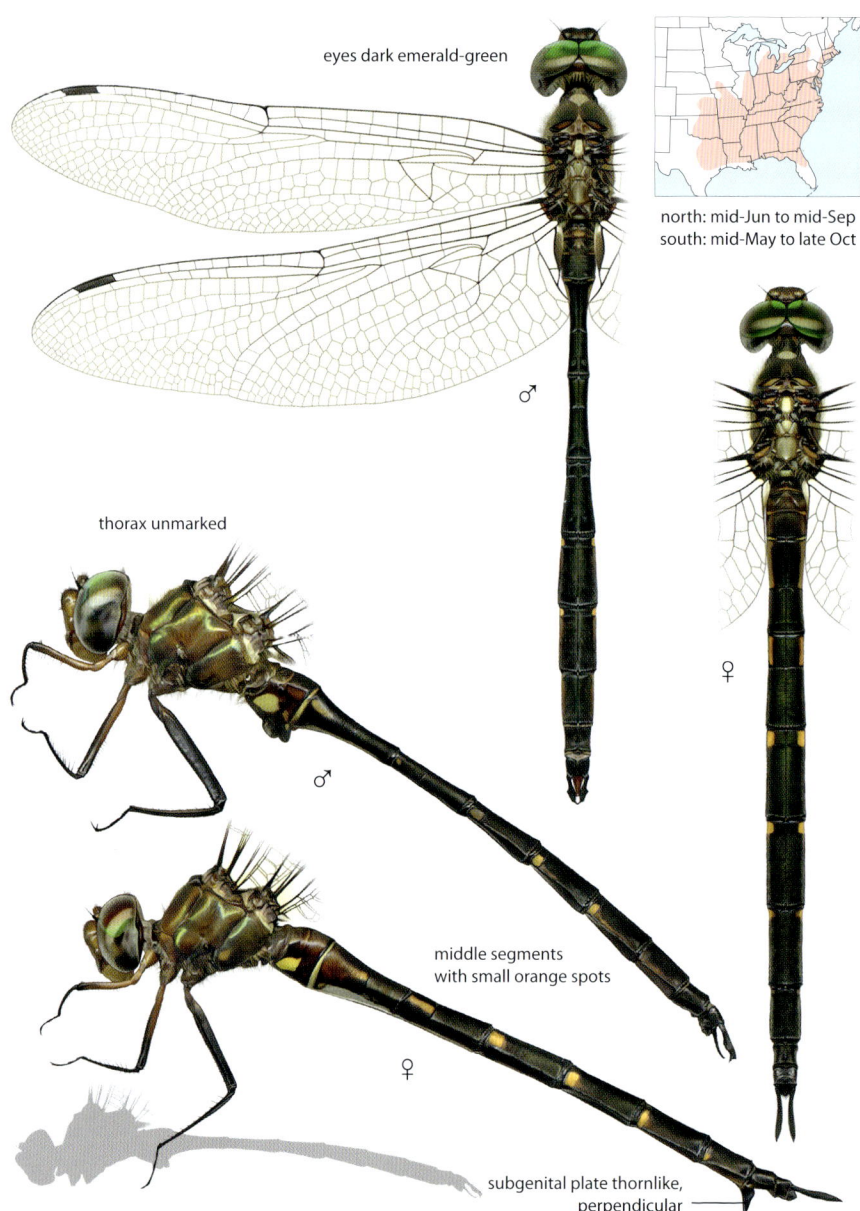

eyes dark emerald-green

north: mid-Jun to mid-Sep
south: mid-May to late Oct

♂

thorax unmarked

♂

♀

middle segments
with small orange spots

♀

subgenital plate thornlike,
perpendicular

Similar species: Williamson's and **Clamp-tipped Emeralds** with pale lateral thoracic markings (though often obscured with age). Male cerci different in shape and lack ventral spine; subgenital plates longer, much longer than S9 (see pages 422, 424, 428–29).

Delicate Emerald • *Somatochlora franklini*

44–54 mm, 1.7–2.1 in. Very slender emerald widespread in the North at mossy bogs and fens with little open water. Thorax with only short pale anterior stripe, often obscured. Brown mark at base of hindwing. Abdomen largely black, male very long. Male cerci pincer-like. Female cerci longer than S9+10; subgenital plate short, horizontal, and scoop-shaped.

eyes emerald-green

late May to mid-August

dark hindwing markings

♂

thorax with short pale anterior stripe, often obscured

male abdomen very long

male cerci pincer-like

♂

♀

female cerci long, longer than S9+10

♀

horizontal subgenital plate short, scoop-like

Similar species: Whitehouse's and **Muskeg Emeralds** also have dark hindwing markings but have shorter abdomens. **Kennedy's Emerald** has shorter abdomen, paler face; lacks dark wing markings.

Forcipate Emerald • *Somatochlora forcipata*

43–51 mm, 1.7–2 in. Slender striped emerald inhabiting small spring-fed streams, primarily in the northeastern US and Canada. Eyes emerald-green. Side of thorax with two pale oval spots. Male usually with small pale lateral spots on S5-8, female with lateral spots on S3-7. Male cerci pincer-like. Female subgenital plate scoop-like, horizontal, reaching end of S9.

eyes emerald-green

early June to early August

thorax with two pale yellow spots

male cerci pincer-like

abdomen long

small pale lateral spots on S5-8

pale lateral spots on S3-7

horizontal, scoop-like subgenital plate

Similar species: Incurvate Emerald has an anterior lateral thoracic stripe instead of an oval spot, thoracic markings usually obscured; female subgenital plate longer, reaching end of S10 (see page 428). **Kennedy's Emerald** usually lacks pale posterior lateral thoracic spot, thoracic markings usually obscured; lacks pale spots on side of abdomen. **Mountain Emerald** has metallic green thorax, lacks pale spots on side of abdomen.

Incurvate Emerald • *Somatochlora incurvata*

49–59 mm, 1.9–2.3 in. Long, slender northeastern striped emerald found at sphagnum bog pools. Eyes emerald-green. Side of thorax with obscured pale anterior stripe and posterior pale oval. Abdomen long, with dull pale lateral spots on S3-8. Male cerci pincer-like. Female subgenital plate scoop-like, horizontal, reaching end of S10.

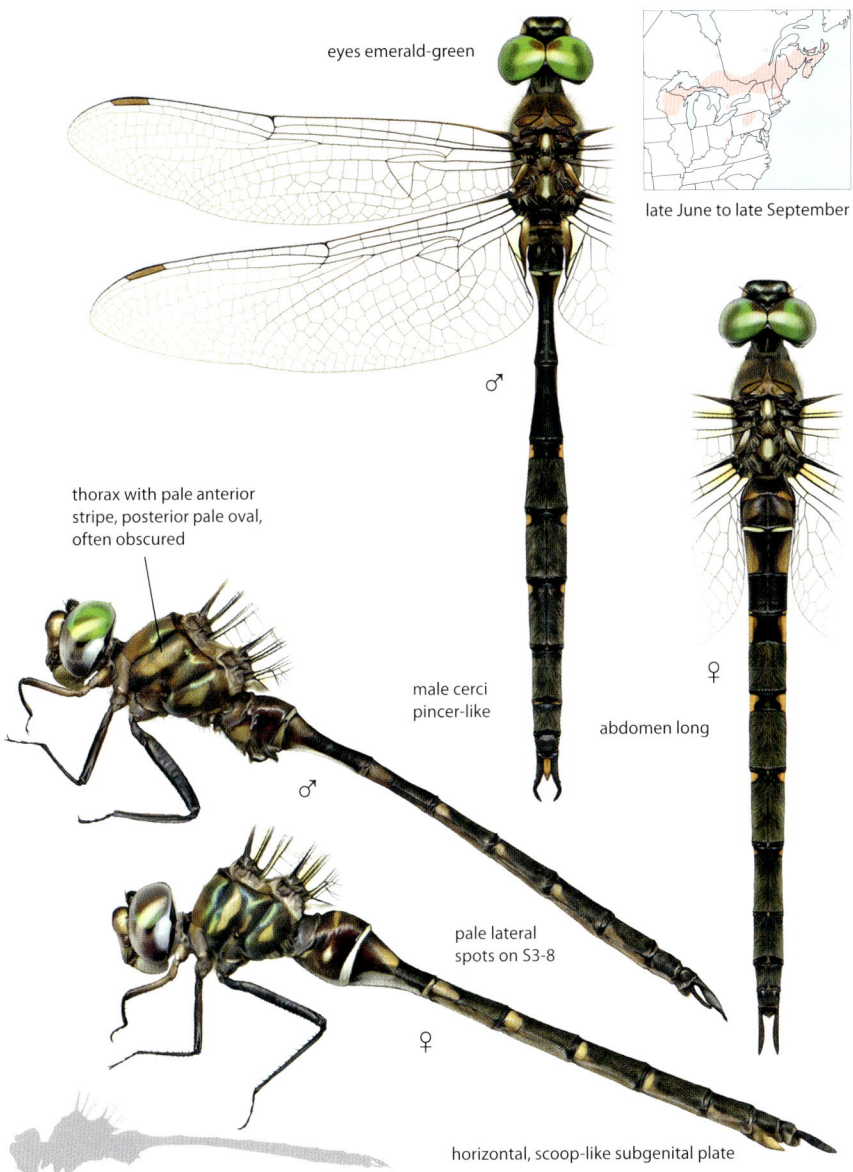

eyes emerald-green

late June to late September

thorax with pale anterior stripe, posterior pale oval, often obscured

♂

male cerci pincer-like

abdomen long

♀

pale lateral spots on S3-8

♂

horizontal, scoop-like subgenital plate

♀

Similar species: Forcipate Emerald thorax has contrasting pale oval lateral spots, female with shorter subgenital plate, reaching end of S9 (see page 428). **Kennedy's Emerald** lacks lateral pale spots on middle segments of abdomen. **Delicate Emerald** has proportionally longer abdomen, dark marks at base of hindwing, lacks pale lateral abdominal spots.

Kennedy's Emerald • *Somatochlora kennedyi*

47–55 mm, 1.9–2.2 in. Dark striped emerald inhabiting open mossy bogs and sedge fens in the northeastern US, Canada, Alaska. Side of thorax usually lacks posterior stripe, has obscure anterior pale stripe. Usually no pale spots on side of abdomen, S2-3 with dull yellow dorsal markings. Male cerci pincer-like. Subgenital plate yellow-orange, short and scoop-like.

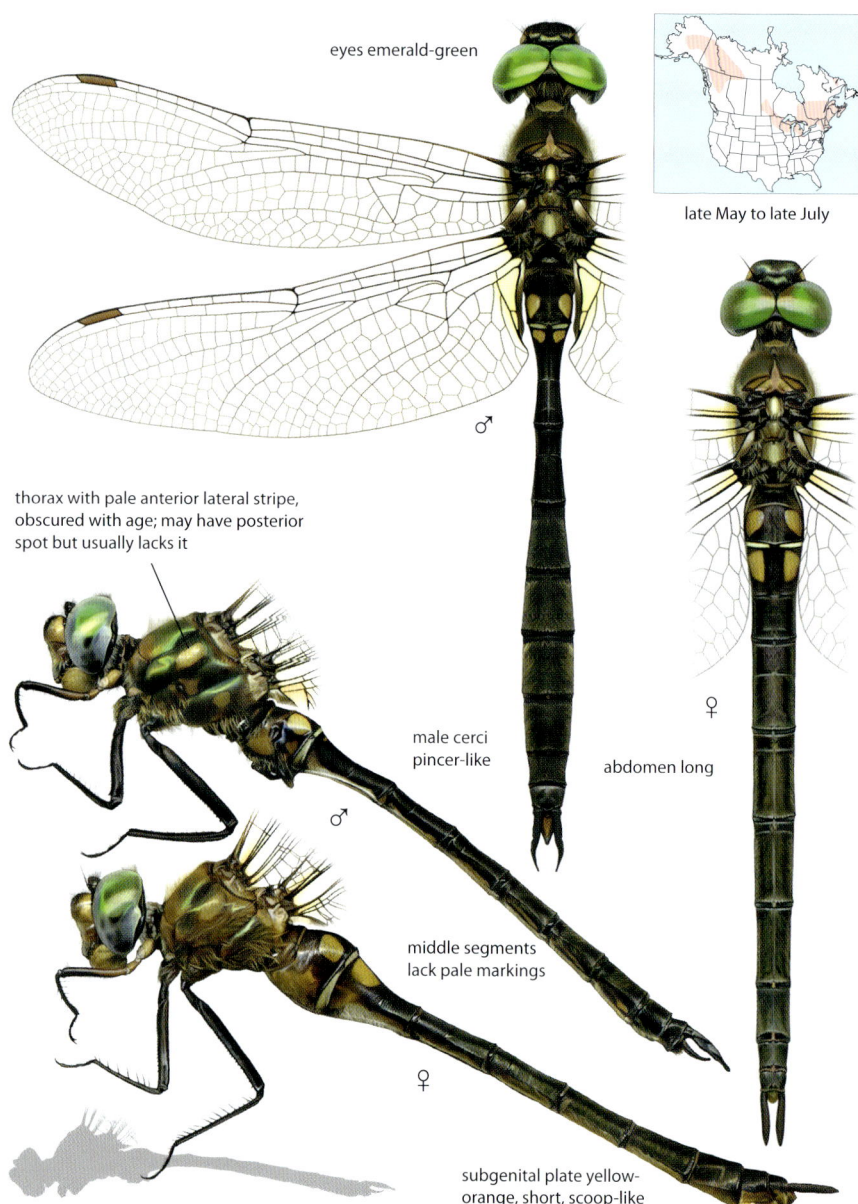

eyes emerald-green

late May to late July

thorax with pale anterior lateral stripe, obscured with age; may have posterior spot but usually lacks it

♂

male cerci pincer-like

abdomen long

♀

middle segments lack pale markings

♀

subgenital plate yellow-orange, short, scoop-like

Similar species: Forcipate, Incurvate, and **Williamson's Emeralds** have pale posterior lateral thoracic spot and pale lateral spots on middle abdominal segments. **Delicate Emerald** more elongated, has dark markings at base of hindwing.

Mountain Emerald • *Somatochlora semicircularis*

47–52 mm, 1.9–2 in. Common western striped emerald found at ponds, bogs, marshes, and wet meadows. Thorax metallic green with two pale oval lateral spots. Abdomen black with only pale lateral markings on S2-3, dorsal yellow spots on S2, female with dorsal yellow spots on S3. Male cerci pincer-shaped. Female subgenital plate short, half length of S9.

eyes emerald-green

late May to early October

thorax metallic green with two pale oval spots

male cerci pincer-like

side of abdomen lacks pale markings

subgenital plate short, half length of S9

♂

♀

Similar species: Brightly marked thorax separates Mountain Emerald from most other striped emeralds within its range. **Forcipate Emerald** has pale spots on sides of middle abdominal segments, face pattern with less contrast. **Ocellated Emerald** has shorter abdomen. Male cerci ski-type; female subgenital plate large and perpendicular.

Clamp-tipped Emerald • *Somatochlora tenebrosa*

48–64 mm, 1.9–2.5 in. Common eastern emerald inhabiting small, shaded forest streams. Thorax with two pale lateral stripes, which fade with age. Abdomen black, S2 with large pale lateral spot, small pale dorsal spots on S3. Male appendages clamp-shaped, showing large circular gap in side view. Subgenital plate long, pointed, slanted slightly rearward.

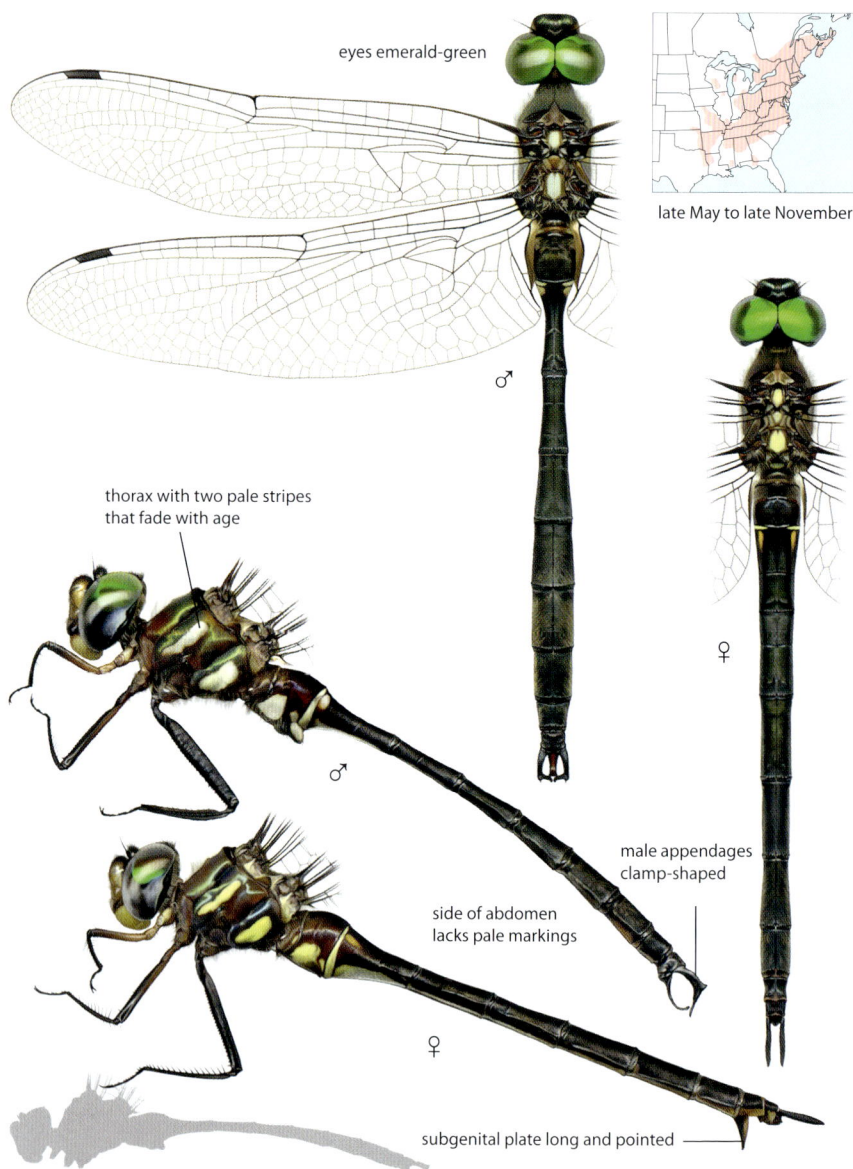

eyes emerald-green

late May to late November

thorax with two pale stripes that fade with age

♂

♂

male appendages clamp-shaped

side of abdomen lacks pale markings

♀

♀

subgenital plate long and pointed

Similar species: Ski-tipped Emerald male cerci straight and ski-shaped, female subgenital plate a wide triangle. **Williamson's Emerald** has pale lateral spots on middle abdominal segments. Male cerci straight; female subgenital plate more perpendicular to abdomen. Top of the thorax between wings darker.

Calvert's Emerald • *Somatochlora calverti*

50–52 mm, 2 in. Local southeastern striped emerald found in sandy pine woods, likely breeding at small, sand-bottomed streams but unknown. Thorax well marked, with two complete pale lateral stripes with pale spots in between, and short pale frontal stripes. Abdomen with narrow pale rings, often with dorsal pale spot on S10.

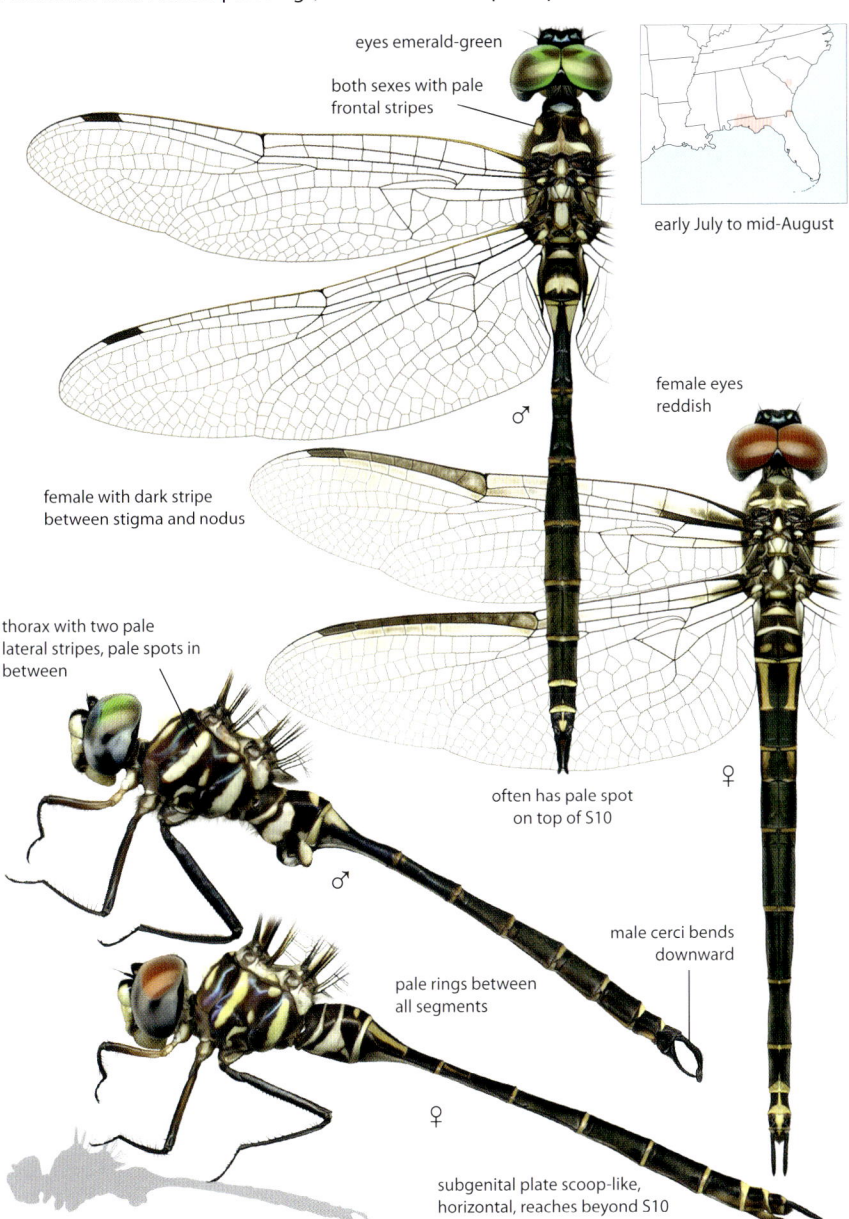

eyes emerald-green

both sexes with pale frontal stripes

early July to mid-August

female eyes reddish

female with dark stripe between stigma and nodus

thorax with two pale lateral stripes, pale spots in between

♂

often has pale spot on top of S10

♀

male cerci bends downward

pale rings between all segments

♀

subgenital plate scoop-like, horizontal, reaches beyond S10

Similar species: Treetop Emerald lacks pale frontal stripes, abdomen with pale rings only on end segments; female wings lack dark stripe between nodus and stigma. **Fine-lined Emerald** thorax has narrower anterior lateral stripe, lacks pale frontal stripes.

Fine-lined Emerald • *Somatochlora filosa*

55–66 mm, 2.2–2.6 in. Long, slender southeastern emerald inhabiting pine woods, likely sandy streams or bog seeps. Thorax with two white lateral stripes, anterior stripe very narrow, sometimes reduced or absent. Abdomen long, often with white rings on S7-9. Male cerci relatively straight. Female subgenital plate thin, skid-like, reaching end of S10.

eyes emerald-green

mid-June to late November

thorax with very narrow pale anterior stripe, sometimes reduced or absent, posteror stripe wider

young females have tinted wing tips

♂

♀

abdomen long and slender

♂

may have white rings on S7-9

female cerci long, longer than S9+10

♀

subgenital plate long, skid-like, reaches to end of S10

Similar species: Lateral thoracic pattern diagnostic when visible. Young females with tinted wing tips recognizable in overhead flight. **Treetop Emerald** has wider pale anterior lateral thoracic stripe, female with longer subgenital plate and shorter cerci, page 429. **Mocha Emerald** lacks pale lateral thoracic markings, has pale spots on side of abdomen.

Treetop Emerald • *Somatochlora provocans*

53–56 mm, 2.1–2.2 in. Well-marked southeastern emerald found at sandy forest seeps and streams. Thorax with two well-defined pale lateral stripes, posterior stripes meeting under thorax. Abdomen black with white rings on S8-9. Female may have dark streaks at wing base. Male cerci bent downward, tips convergent. Subgenital plate long and narrow.

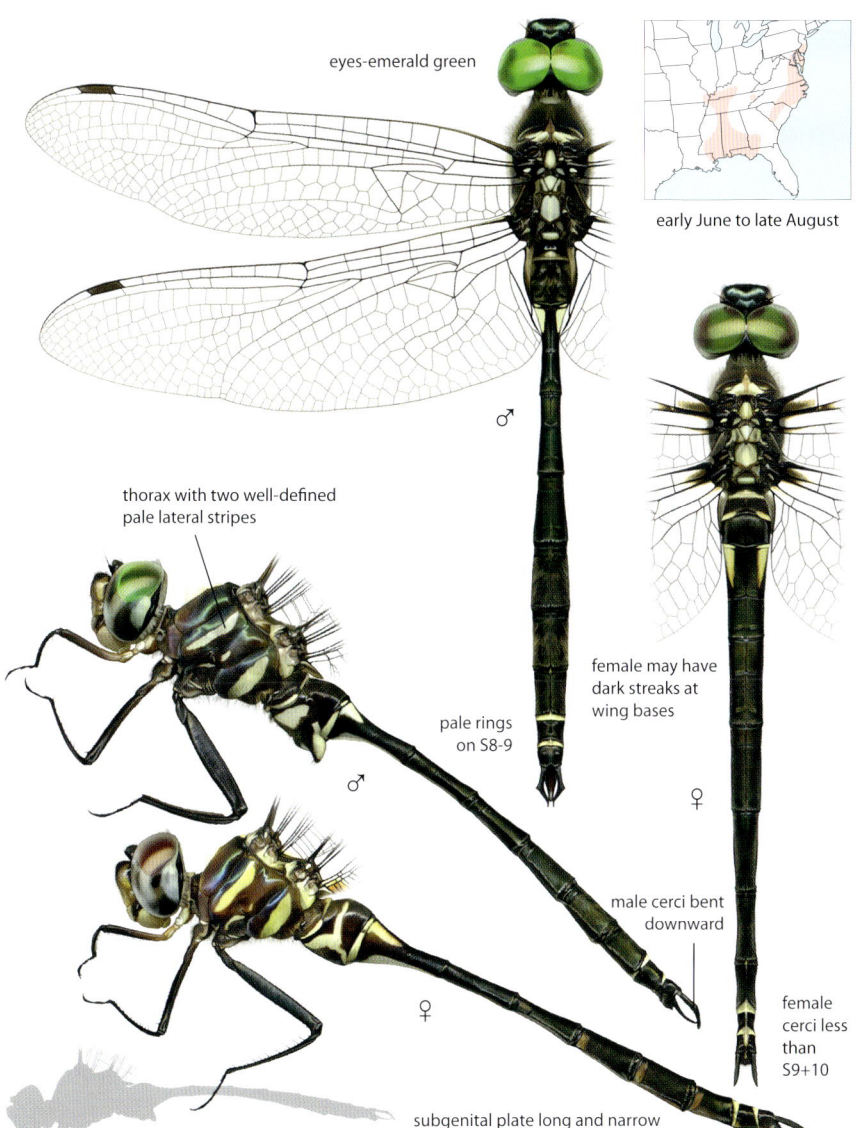

eyes-emerald green

early June to late August

thorax with two well-defined pale lateral stripes

pale rings on S8-9

♂

female may have dark streaks at wing bases

♀

male cerci bent downward

female cerci less than S9+10

♀

subgenital plate long and narrow

Similar species: Calvert's Emerald thorax has pale frontal stripes, two pale spots between lateral stripes; abdomen with pale rings, pale spot on top of S10. Female wings with dark stripe between nodus and stigma. **Ozark** and **Texas Emeralds** very similar, but no known range overlap. Males separable by appendages, page 425; females lack dark wing stripe. **Fine-lined Emerald** has very narrow anterior lateral thoracic stripe; female lacks dark markings at base of wings, has longer cerci.

Texas Emerald • *Somatochlora margarita*

51–59 mm, 2–2.3 in. Well-marked striped emerald found primarily at East Texas pine woods; breeding habitat unknown. Thorax with two strong white lateral stripes, may show pale dash at front. Abdomen with pale dorsal spots on S2, pale dorsal stripes on S3, white rings on S8-9. Male cerci bent downward. Subgenital plate long, narrow, and scoop-shaped.

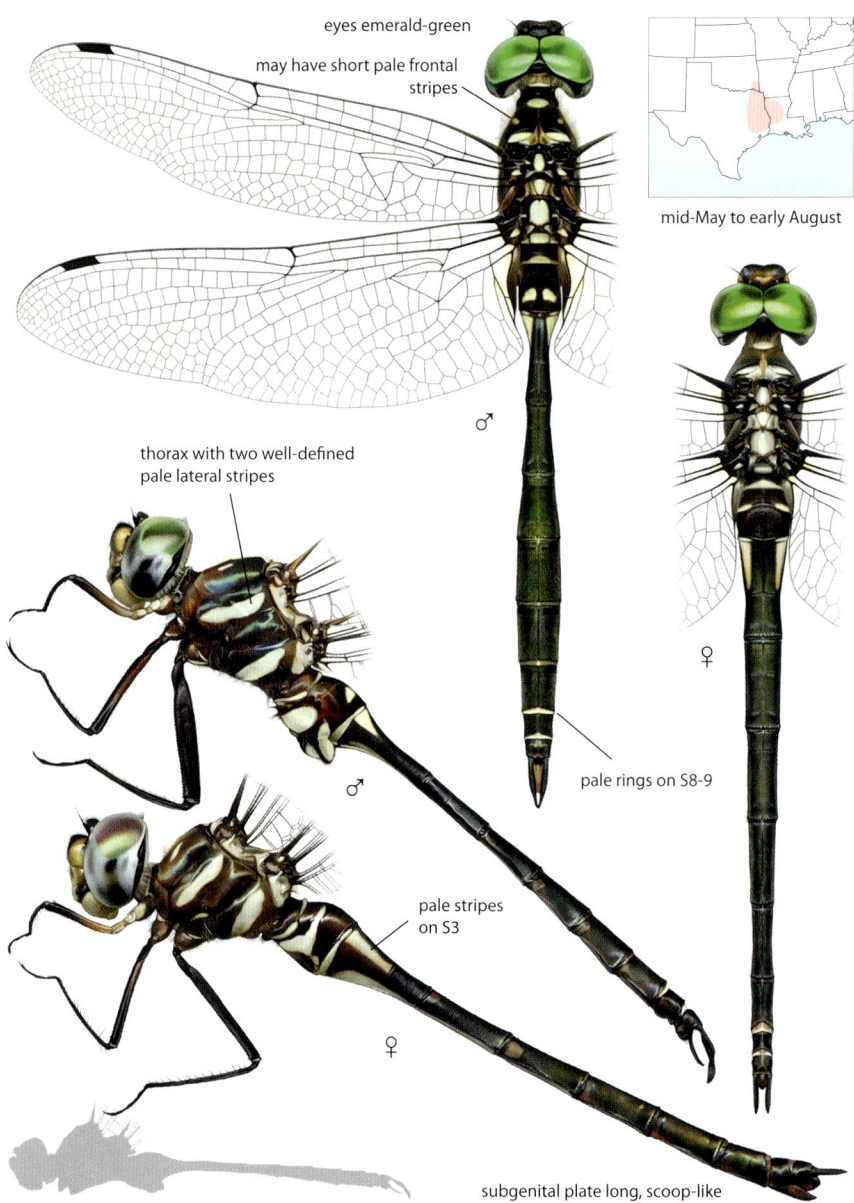

eyes emerald-green

may have short pale frontal stripes

mid-May to early August

thorax with two well-defined pale lateral stripes

♂

♂

♀

pale rings on S8-9

pale stripes on S3

♀

subgenital plate long, scoop-like

Similar species: Strongly patterned thorax unlike any other striped emerald within its range. Identical in pattern to **Treetop** and **Ozark Emeralds,** but no known range overlap. Compare male appendages, female subgenital plates, pages 425, 429–30.

Ozark Emerald • *Somatochlora ozarkensis*

51–58 mm, 2–2.3 in. Well-marked striped emerald found in the Ozark region at forest streams. Thorax with two well-defined white lateral stripes, anterior stripe often narrower than posterior. Abdomen with dorsal pale markings on S2-3, pale rings on S8-9. Male cerci bent downward with widened tip in side view. Subgenital plate long and scoop-like.

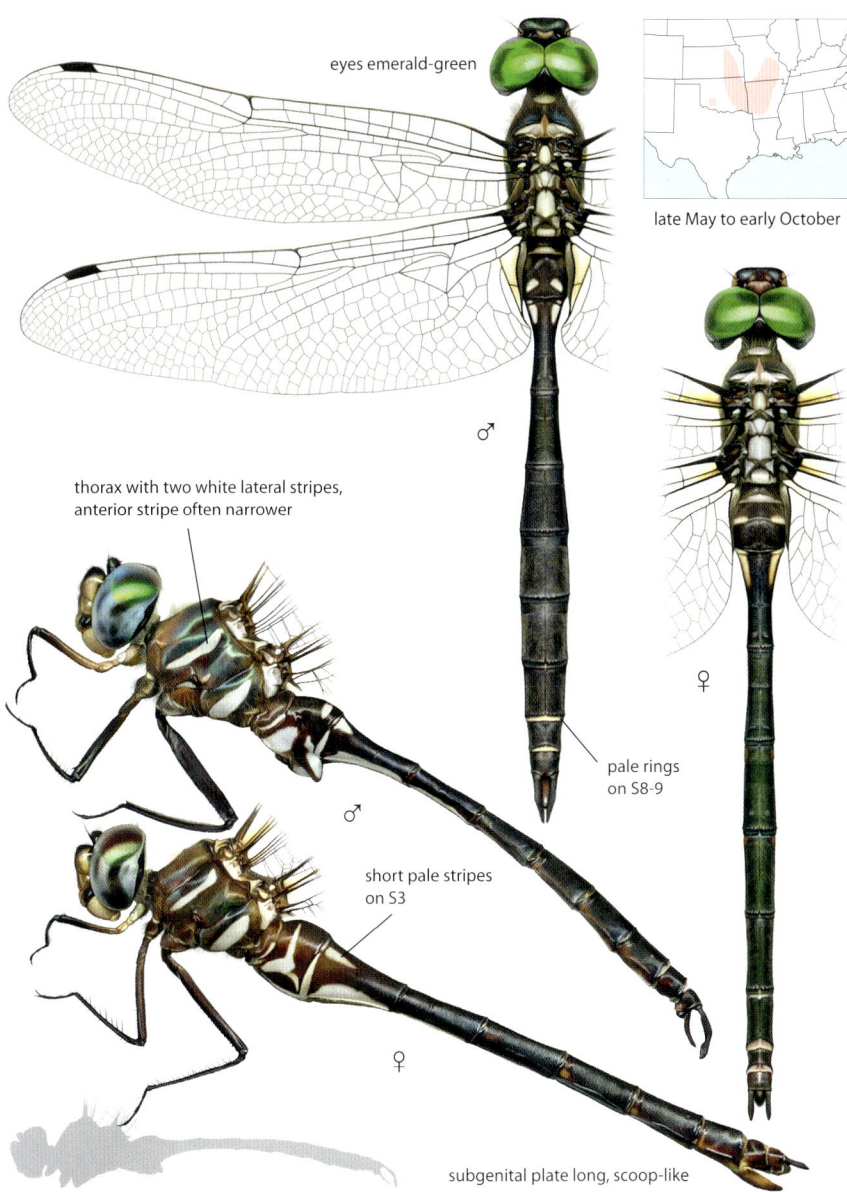

eyes emerald-green

late May to early October

♂

thorax with two white lateral stripes, anterior stripe often narrower

pale rings on S8-9

♀

♂

short pale stripes on S3

♀

subgenital plate long, scoop-like

Similar species: Vividly striped thorax unlike any other striped emerald in its range. **Clamp-tipped Emerald** has different male cerci and female subgenital plate (pages 424, 429). **Texas** and **Treetop Emeralds** nearly identical but no known range overlap. Compare male cerci, female subgenital plates, pages 425, 429.

Hine's Emerald • *Somatochlora hineana*

58–63 mm, 2.3–2.5 in. Large midwestern striped emerald local at sedge and grassy fens with only shallow sheet-water flow. Listed in the US as endangered. Thorax with two yellow lateral stripes, posterior wider. Abdomen black, S3 with dorsal pale markings. Male appendages clamp-like, cerci bent at tip. Subgenital plate oblique, longer than S9.

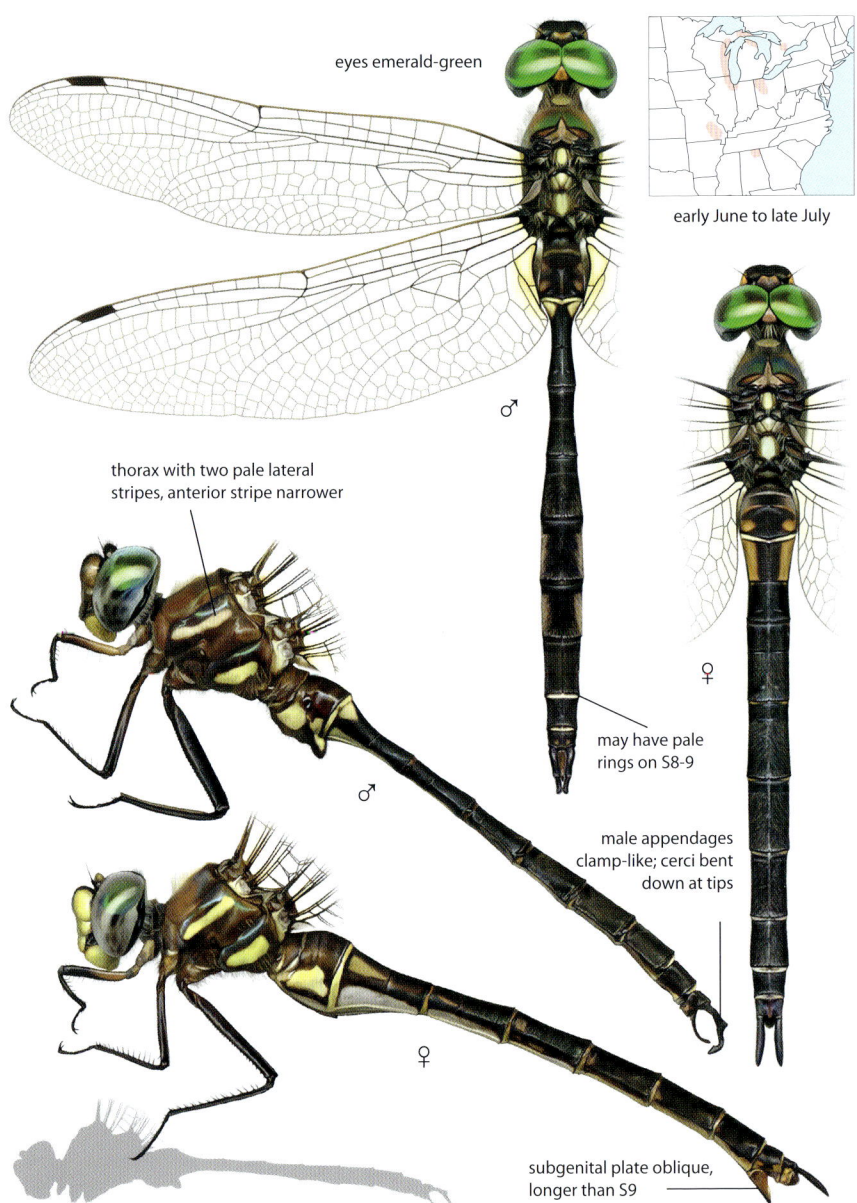

eyes emerald-green

early June to late July

thorax with two pale lateral stripes, anterior stripe narrower

♂

♀

may have pale rings on S8-9

male appendages clamp-like; cerci bent down at tips

♀

subgenital plate oblique, longer than S9

Similar species: Plains Emerald smaller; male with straighter cerci, female with perpendicular subgenital plate. **Clamp-tipped Emerald** male cerci bent downward in middle; gap between cerci and epiproct more circular, subgenital plate more perpendicular.

Plains Emerald • *Somatochlora ensigera*

48–51 mm, 1.9–2 in. Brightly patterned striped emerald found at open and wooded streams, ditches, and rivers in northern Great Plains. Face mostly bright yellow. Thorax with bright yellow anterior lateral stripe and posterior spot. Abdomen black with pale rings on S8-9. Female with full-length pale lateral abdominal stripes that may be obscured with age.

eyes emerald-green

mid-June to late August

♂

side of face bright yellow

side of thorax with bright yellow anterior stripe, posterior spot

♂

pale rings on S8-9

♀

♀

female with pale lateral stripes

female cerci very short

subgenital plate long, perpendicular

Similar species: Hine's Emerald larger; male cerci bent, with oval gap between cerci and epiproct; female subgenital plate obliquely angled, abdomen lacks pale lateral stripes. **Clamp-tipped Emerald** male cerci bent in middle; gap between cerci and epiproct circular. Female lacks pale abdominal stripes and has longer cerci.

Quebec Emerald • *Somatochlora brevicincta*

47–50 mm, 1.9–2 in. Scarce northern striped emerald inhabiting shallow fens and sedge marshes. Eyes emerald-green. Thorax with single short pale anterior lateral stripe. Abdomen with small pale lateral marks, sometimes absent, sometimes forming incomplete rings. Male cerci bent-curl type. Female subgenital plate as long as S9.

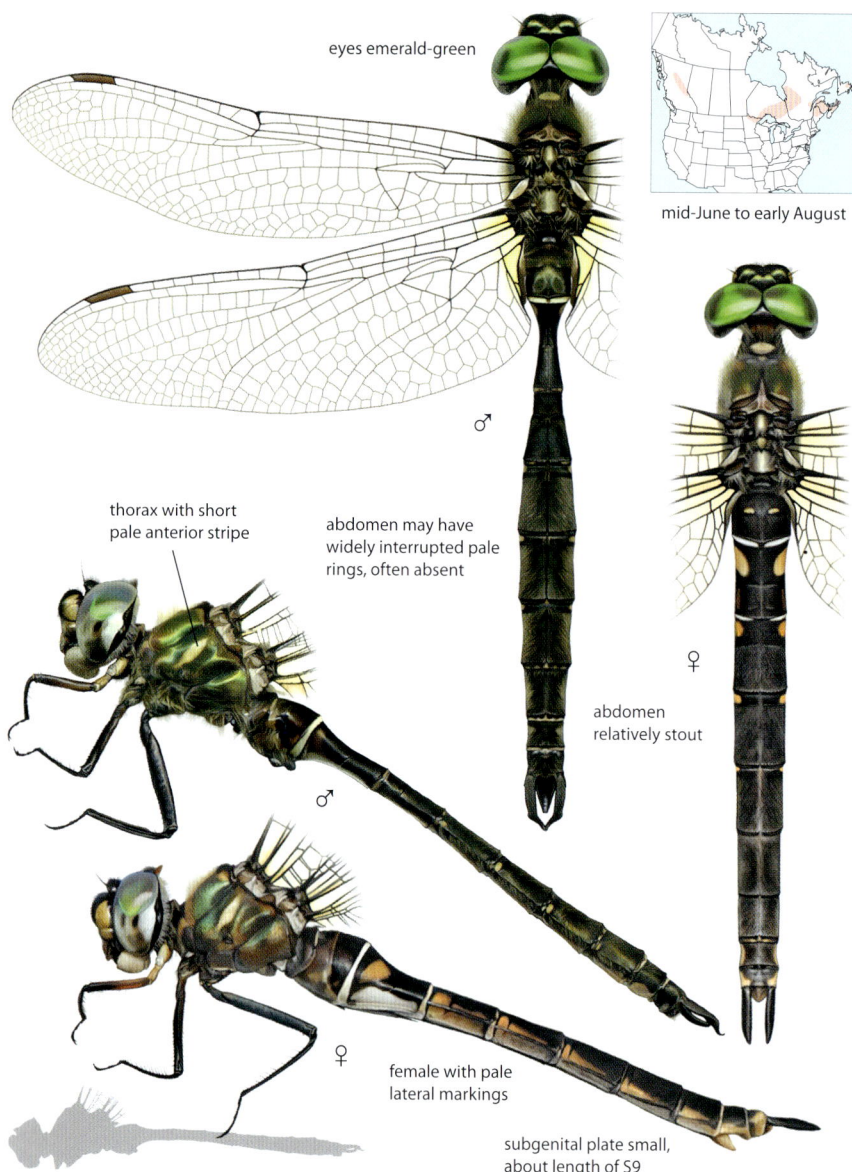

eyes emerald-green

mid-June to early August

♂

thorax with short pale anterior stripe

abdomen may have widely interrupted pale rings, often absent

abdomen relatively stout

♀

♂

♀

female with pale lateral markings

subgenital plate small, about length of S9

Similar species: Hudsonian and **Ringed Emeralds** have more complete pale rings on abdomen. Compare male cerci, page 426, female subgenital plates, pages 430–31. **Lake Emerald** larger; thorax usually without pale markings, abdomen more completely ringed. **Delicate, Muskeg,** and **Whitehouse's Emeralds** have small dark markings at base of hindwing.

Lake Emerald • *Somatochlora cingulata*

55–68 mm, 2.2–2.7 in. Large northern striped emerald inhabiting lakes, occasionally large slow rivers. Thorax lacks pale markings, rarely has obscure brown anterior lateral stripe. Abdomen distinctly ringed, having narrow white rings between each segment. Male cerci bent-curl type. Female largely lacks subgenital plate; cerci long, longer than S9+10.

eyes emerald-green

mid-June to late August

♂

thorax usually unmarked, may have obscure brown anterior stripe

♂

abdomen distinctly ringed, black with narrow white rings

♀

♀

subgenital plate small, does not project

female cerci long, longer than S9+10

Similar species: Male Lake Emeralds often fly over open water, unlike other striped emeralds. **Hudsonian** and **Ringed Emeralds** smaller, have pale anterior lateral stripe on thorax. Dark coloration and emerald-green eyes distinguish Lake Emerald from **mosaic darners.**

Ringed Emerald • *Somatochlora albicincta*

45–52 mm, 1.8–2 in. Widespread northern striped emerald found at sparsely vegetated ponds, lakes, and slow streams. Thorax with single short pale anterior lateral stripe. Abdomen with narrow white rings between all segments. Male cerci bent-curl type. Female subgenital plate small and bilobed.

eyes emerald-green

mid-June to late September

♂

side of thorax with short pale anterior stripe

♂

♀

abdomen black with narrow white rings

♀

brown lateral markings on S2-6, may be reduced or absent

small subgenital plate

Similar species: Hudsonian Emerald identical in pattern; male with large ventral tooth on cercus, female with obliquely angled subgenital plate (see pages 424, 429). **Lake Emerald** larger, usually lacks pale thoracic markings; male epiproct rectangular in ventral view (see page 424). Female subgenital plate not lobed (see page 428).

Hudsonian Emerald • *Somatochlora hudsonica*

50–54 mm, 2–2.1 in. Northern striped emerald inhabiting sedge-margined lakes, ponds, marshes, and muskeg pools in western to central Canada, Rocky Mountains in US. Thorax with short pale anterior lateral stripe. Abdomen with white rings between each segment. Male cerci bent-curl type with large ventral tooth. Subgenital plate short, obliquely angled.

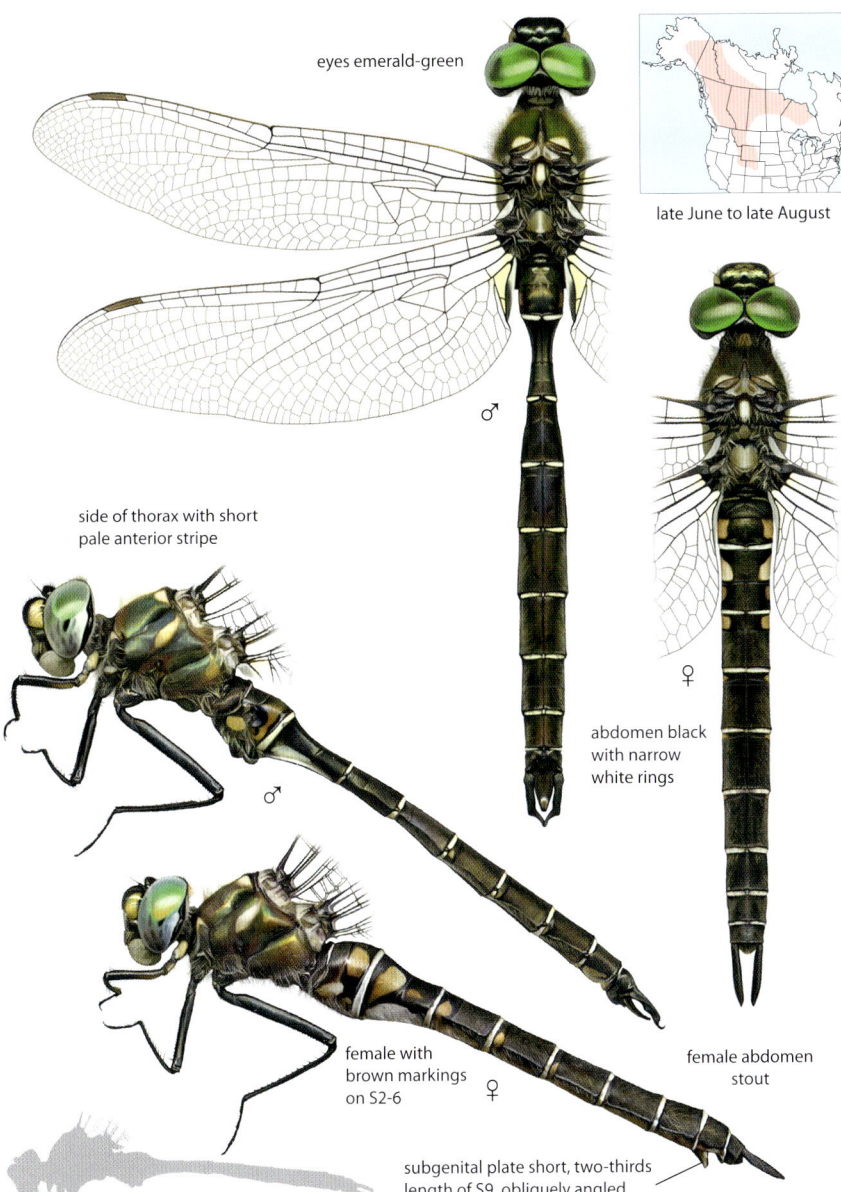

eyes emerald-green

late June to late August

side of thorax with short pale anterior stripe

♂

♀

abdomen black with narrow white rings

female with brown markings on S2-6 ♀

female abdomen stout

subgenital plate short, two-thirds length of S9, obliquely angled

Similar species: Ringed Emerald largely identical; male cerci lacks large ventral tooth, female with small bilobed subgenital plate (see pages 424, 428). **Lake Emerald** larger, usually lacks pale thoracic markings; male cerci lacks ventral tooth, subgenital plate smaller.

Muskeg Emerald • *Somatochlora septentrionalis*

39–48 mm, 1.5–1.9 in. Small northern striped emerald found at open mossy muskeg and fen pools. Thorax with single brown anterior lateral stripe. Hindwing has small dark markings at base. Abdomen relatively short, with indistinct brown lateral spots, especially on female. Male cerci bent-curl type. Female subgenital plate small.

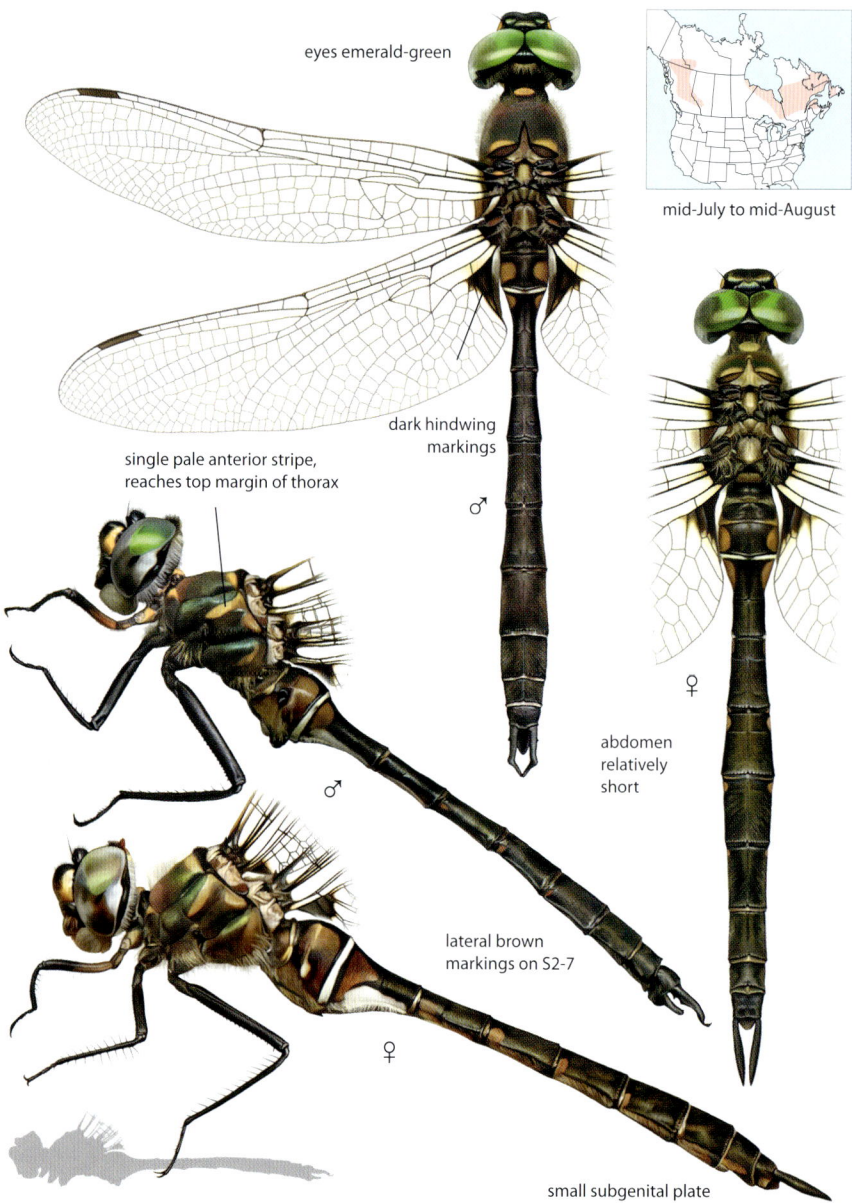

eyes emerald-green

mid-July to mid-August

single pale anterior stripe, reaches top margin of thorax

dark hindwing markings

♂

♀

abdomen relatively short

♂

lateral brown markings on S2-7

♀

small subgenital plate

Similar species: Whitehouse's Emerald nearly identical; female with short, obliquely angled subgenital plate, page 429. Compare male appendages, page 425. **Delicate Emerald** more elongated; male cerci pincer-like, female with longer, horizontal subgenital plate.

Whitehouse's Emerald • *Somatochlora whitehousei*

46–48 mm, 1.8–1.9 in. Small northern striped emerald inhabiting open muskeg ponds and pools. Thorax with single brown anterior lateral stripe. Hindwing with small dark marking at base. Abdomen relatively short, largely lacking pale lateral markings along middle segments. Male cerci bent-curl type. Subgenital plate shorter than S9, obliquely angled.

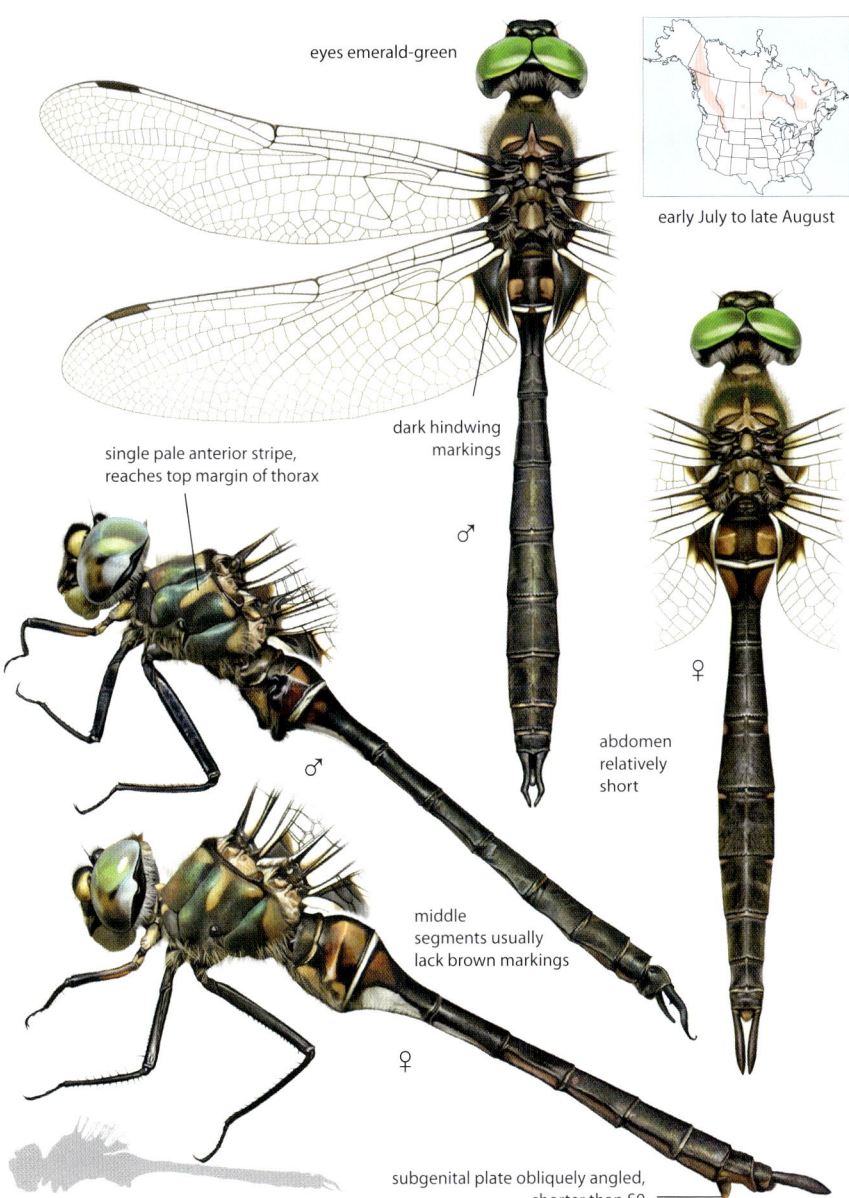

eyes emerald-green

early July to late August

single pale anterior stripe, reaches top margin of thorax

dark hindwing markings

♂

♀

abdomen relatively short

♂

middle segments usually lack brown markings

♀

subgenital plate obliquely angled, shorter than S9

Similar species: Muskeg Emerald nearly identical; female subgenital plate small and deeply notched, page 429. Compare male appendages, page 424. **Delicate Emerald** more elongated; male cerci pincer-type, female with horizontal, scoop-like subgenital plate.

Treeline Emerald • *Somatochlora sahlbergi*

48–50 mm, 1.9–2 in. Dark near-Arctic striped emerald inhabiting cold mossy pools in bogs and fens. Thorax usually without pale markings, may show short brown anterior lateral stripe. Abdomen stout and almost completely black. Male cerci bent-curl type. Female with short bilobed subgenital plate.

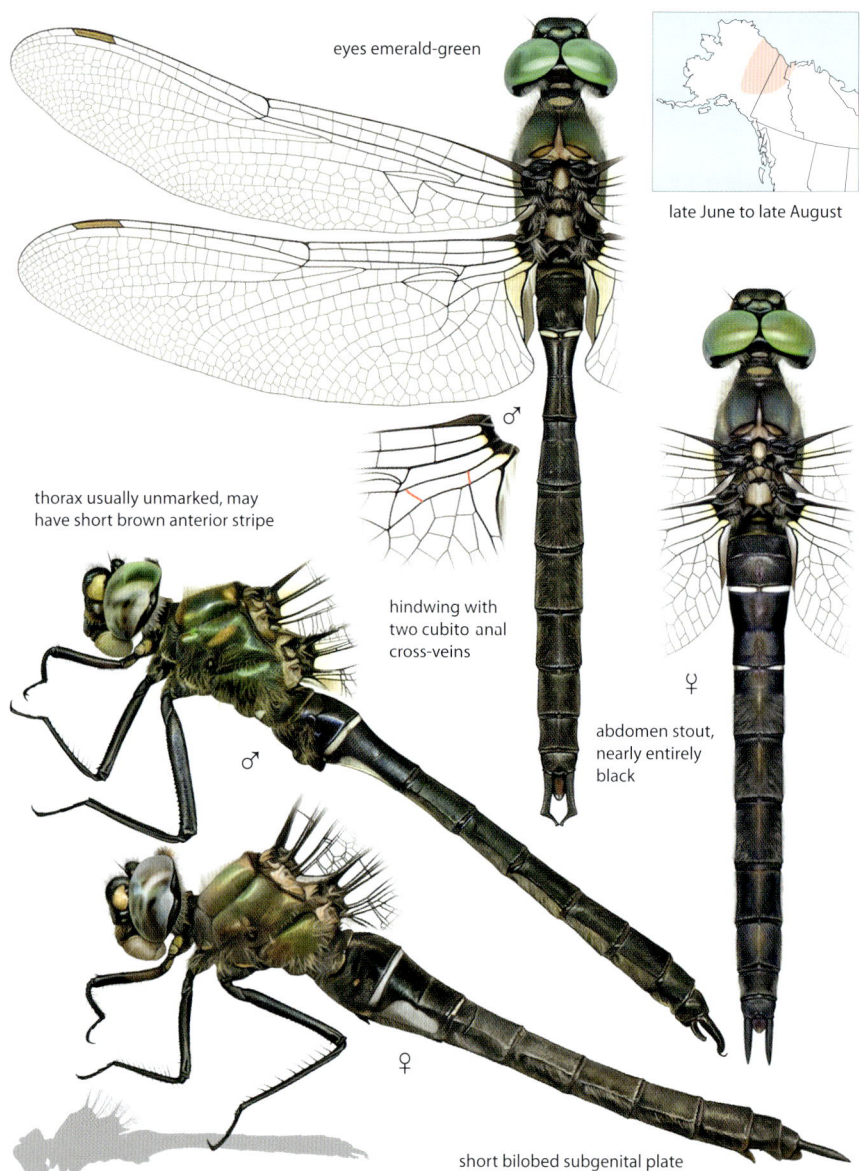

eyes emerald-green

late June to late August

thorax usually unmarked, may have short brown anterior stripe

♂

hindwing with two cubito-anal cross-veins

♂

abdomen stout, nearly entirely black

♀

♀

short bilobed subgenital plate

Similar species: Muskeg and **Whitehouse's Emeralds** have pale anterior lateral thoracic stripe, brown markings on S2-3, dark marks at base of hindwing. **Hudsonian** and **Ringed Emeralds** have pale ringed abdomen. **American Emerald** hindwing has single cubito-anal cross-vein (striped emeralds have two), see page 217.

Coppery Emerald • *Somatochlora georgiana*

48–49 mm, 1.9 in. Small, brown eastern striped emerald inhabiting small, sand-bottomed coastal forest streams. Eyes remain red-brown at maturity. Thorax orange-brown, with two pale lateral stripes. Abdomen mostly brown, with black middorsal markings on S8-10. Male cerci straight. Subgenital plate small and triangular; female cerci short.

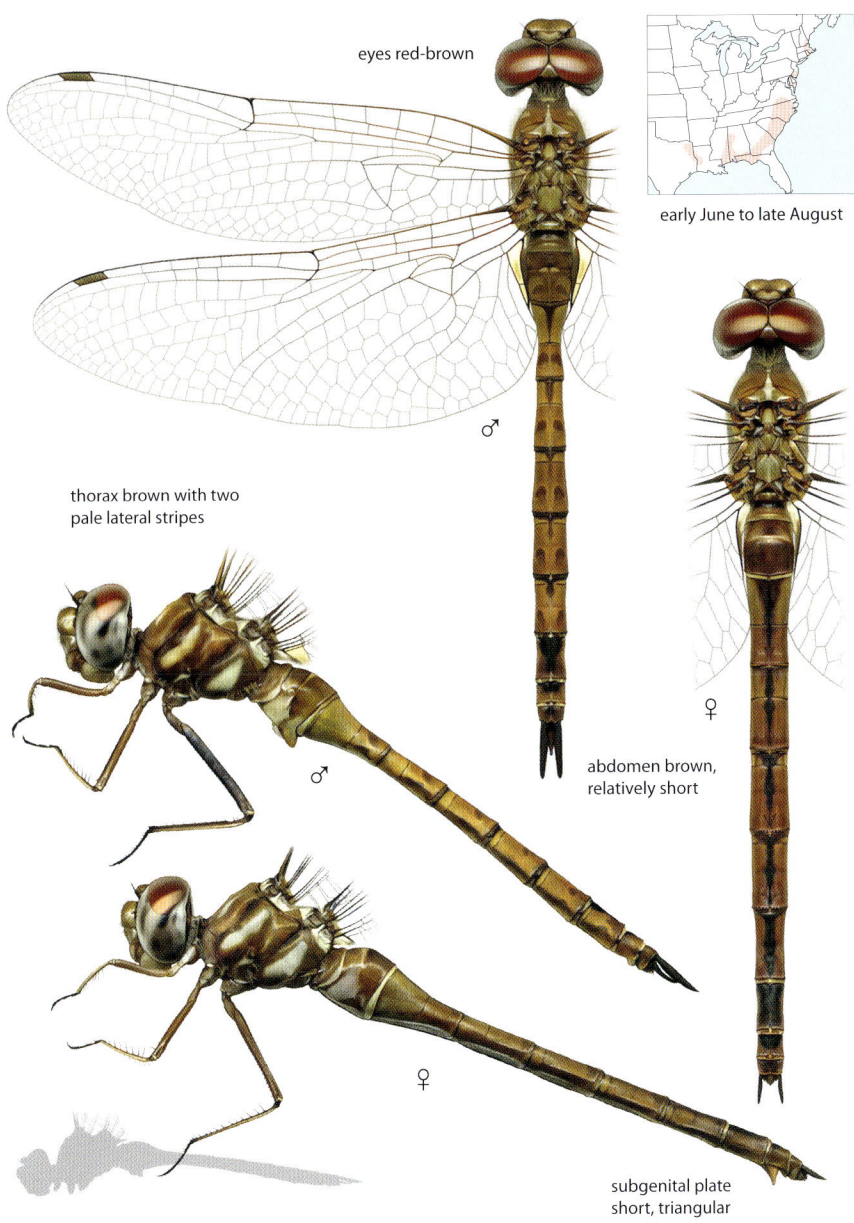

eyes red-brown

early June to late August

thorax brown with two pale lateral stripes

abdomen brown, relatively short

subgenital plate short, triangular

Similar species: Brown coloration distinctive among striped emeralds. **Baskettails** lack pale stripes on side of thorax, have yellow stripes on abdomen.

Uhler's Sundragon • *Helocordulia uhleri*

41–46 mm, 1.6–1.8 in. One of two species of sundragon, found at eastern forest streams and rivers, where males patrol along shore. Thorax unmarked brown. Wings with dark brown and amber streaks at base, small dark dots at cross-veins. Abdomen black; orange ring at base of S3, orange lateral marks on S4-8. Eyes emerald-green when mature.

immature eyes gray, emerald-green when mature

north: mid-Apr to late July
south: mid-Feb to late Aug

dark brown and amber wing markings

thorax unmarked brown

orange ring at base of S3

orange markings on S3-8

Similar species: Selys' Sundragon lacks amber markings at base of wings, has grayer eyes, smaller orange abdominal markings. **Baskettails** have yellow lateral stripes on abdomen. **Shadowdragons** have small yellow spot on side of thorax, abdomen without orange ring on S3. **American** and **petite emeralds** lack dark wing markings. Only a few **striped emeralds** have dark wing markings, largely restricted to base of hindwing when present.

Selys' Sundragon • *Helocordulia selysii*

38–41 mm, 1.5–1.6 in. Southeastern sundragon found at clean woodland streams and small rivers. Thorax unmarked brown. Wings with dark brown marks at bases, dark dots at cross-veins along leading edge. Abdomen black with orange ring at base of S3, small orange lateral marks on S4-8. Eyes pale greenish gray.

eyes green-gray

late February to early May

♂

thorax unmarked brown

orange ring at base of S3

♂

dark brown marks at wing bases; dark dots on leading edge of cross-veins

♀

♀

orange markings on S3-8

Similar species: Uhler's Sundragon has amber markings at wing bases, bright green eyes, larger orange abdominal markings. **Baskettails** have yellow lateral abdominal stripes. **Shadowdragons** have small yellow spot on thorax side; abdomen lacks orange ring on S3.

Prince Baskettail • *Epitheca princeps*

58–78 mm, 2.3–3.1 in. Distinctive large baskettail common throughout the East at slow streams, rivers, also lakes toward the north. Typically seen in flight. Thorax unmarked dark brown. Wings usually with conspicuous dark wing markings at base, nodus, and tip. Abdomen long and dark with pale lateral stripes. Mature male eyes green, female brown.

eyes green when mature

wings with dark markings
at base, nodus, and tip

♂

♀

darker variation

southeastern form *regina*

Wing markings variable even within a single population. Markings often more extensive in the South. Northern populations may have reduced wing markings, but sparsely marked individuals can be found throughout range.

north: late May to mid-Oct
south: late Feb to mid-Nov

pale variation (South Texas) ♂

♂

thorax unmarked brown

typically hangs with abdomen curled upward

♀

Similar species: Many **darner species** similar in size, shape, and flight behavior but lack conspicuous wing markings. **Twelve-spotted Skimmer** and female **Common Whitetail** have similar wing pattern, but abdomen shorter and wider; are perchers, rather than fliers.

Common Baskettail • *Epitheca cynosura*

38–43 mm, 1.5–1.7 in. Most common and widespread eastern baskettail; breeds at ponds, lakes, and slow streams. Thorax brown with two yellow lateral spots. Basal hindwing patch variable. Abdomen with yellow lateral stripes; male moderately constricted at S3, female abdomen broad with short cerci. Eyes brown, bluer or greener when mature.

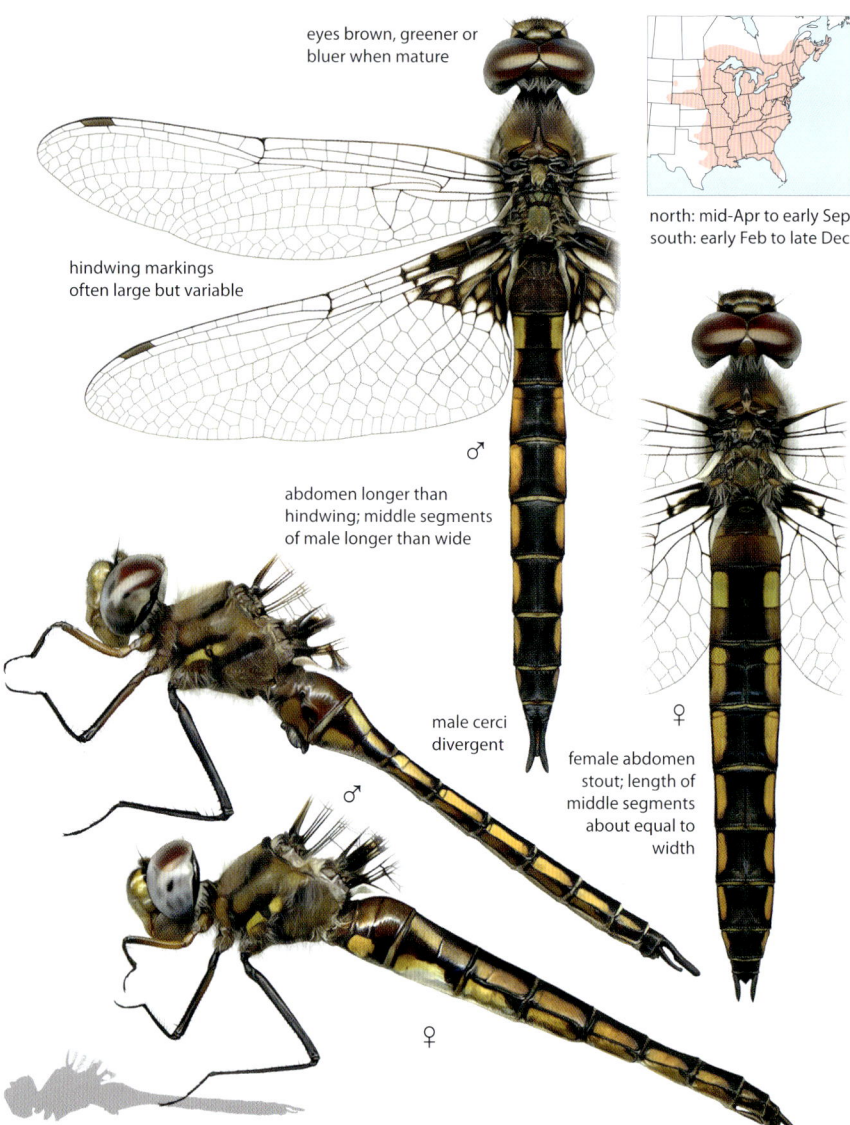

eyes brown, greener or bluer when mature

hindwing markings often large but variable

north: mid-Apr to early Sep
south: early Feb to late Dec

abdomen longer than hindwing; middle segments of male longer than wide

male cerci divergent

female abdomen stout; length of middle segments about equal to width

♂

♀

♂

♀

Similar species: Mantled Baskettail with reduced markings have abdomen shorter than hindwing, middle segments wider than long. **Slender Baskettail** abdomen longer, male more constricted at S3; male cerci longer and straighter. Female abdomen middle segments longer than wide, cerci longer. **Beaverpond Baskettail** has back of head pale. **Spiny Baskettail** male has spindle-shaped abdomen with middle segments wider than long, cerci with small ventral spine; female cerci longer than S9. See **Sepia** and **Florida Baskettails**.

Mantled Baskettail • *Epitheca semiaquea*

32–37 mm, 1.3–1.5 in. Small, stout eastern baskettail found at ponds and lakes, usually with sandy bottoms. Thorax brown with yellow lateral spots. Dark hindwing markings usually large, extending to rear wing margin and to nodus, but some populations have reduced markings. Abdomen short and stubby, particularly in females.

eyes reddish-brown over gray

mid-February to mid-July

♂

middle segments wider than long

abdomen shorter than length of hindwing

hindwing markings extend to nodus and rear edge of wing but variable

♀

♂

♀

Similar species: Very large hindwing markings diagnostic when present. **Common Baskettail** abdomen longer than hindwing, middle segments longer than wide.

Spiny Baskettail • *Epitheca spinigera*

43–47 mm, 1.3–1.9 in. Widespread northern baskettail found at wide variety of ponds, lakes, and slow streams. Thorax brown. Wing markings usually small. Abdomen with yellow lateral stripes, male abdomen spindle-shaped. Male cerci with small ventral spine. Female cerci long, as long as S9+10. Male eyes blue, female greener.

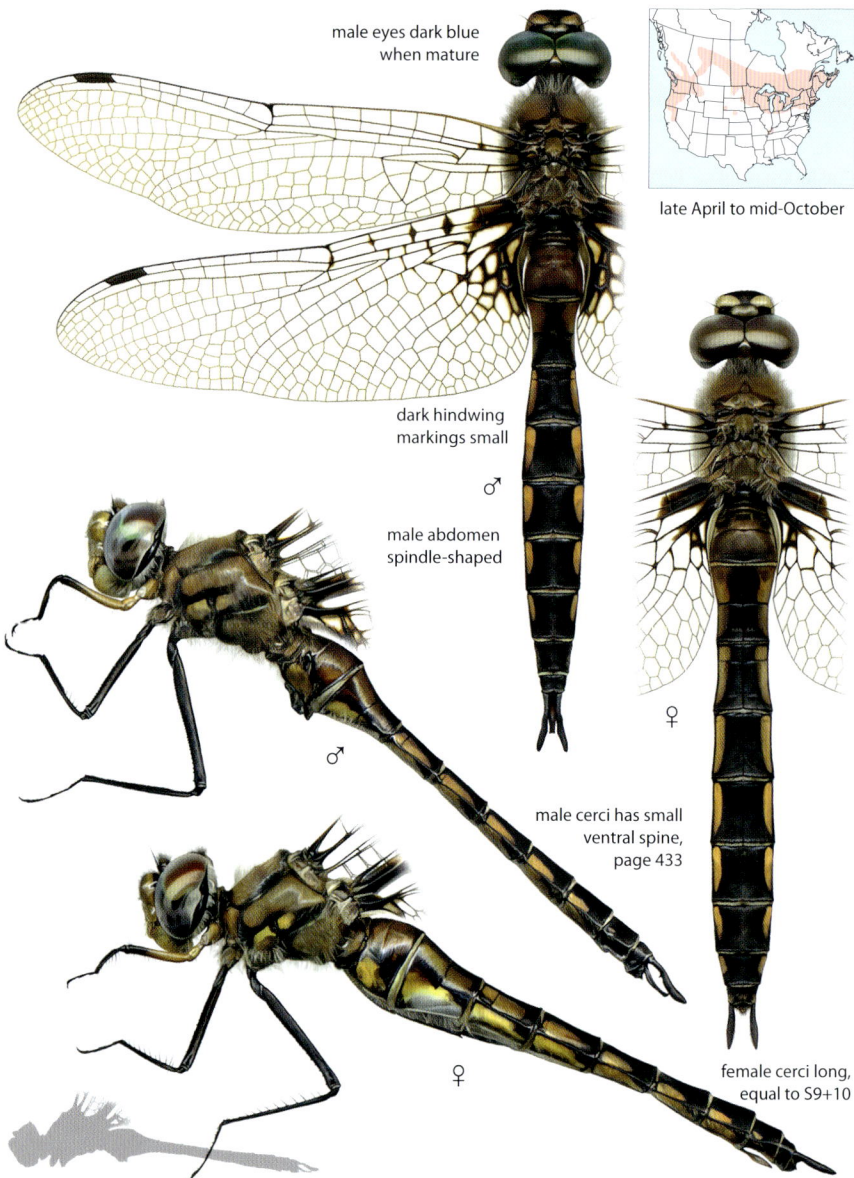

male eyes dark blue when mature

late April to mid-October

dark hindwing markings small

♂

male abdomen spindle-shaped

♂

male cerci has small ventral spine, page 433

♀

♀

female cerci long, equal to S9+10

Similar species: Beaverpond Baskettail has back of head pale; frons lacks T-spot. Male cerci downturned at tip. **Common Baskettail** with small wing markings not easily separated; male abdomen less spindle-shaped, with middle segments longer than wide. Female cerci shorter; male cerci lacks ventral spine.

Beaverpond Baskettail • *Epitheca canis*

43–48 mm, 1.7–1.9 in. Northern baskettail common at bog and marsh ponds, lakes, and slow streams. Back of head pale; face lacks T-spot. Dark hindwing markings small. Mature female has heavily tinted wings. Abdomen with yellow lateral stripes, male abdomen spindle-shaped. Male cerci downturned at tip. Eyes green to blue when mature.

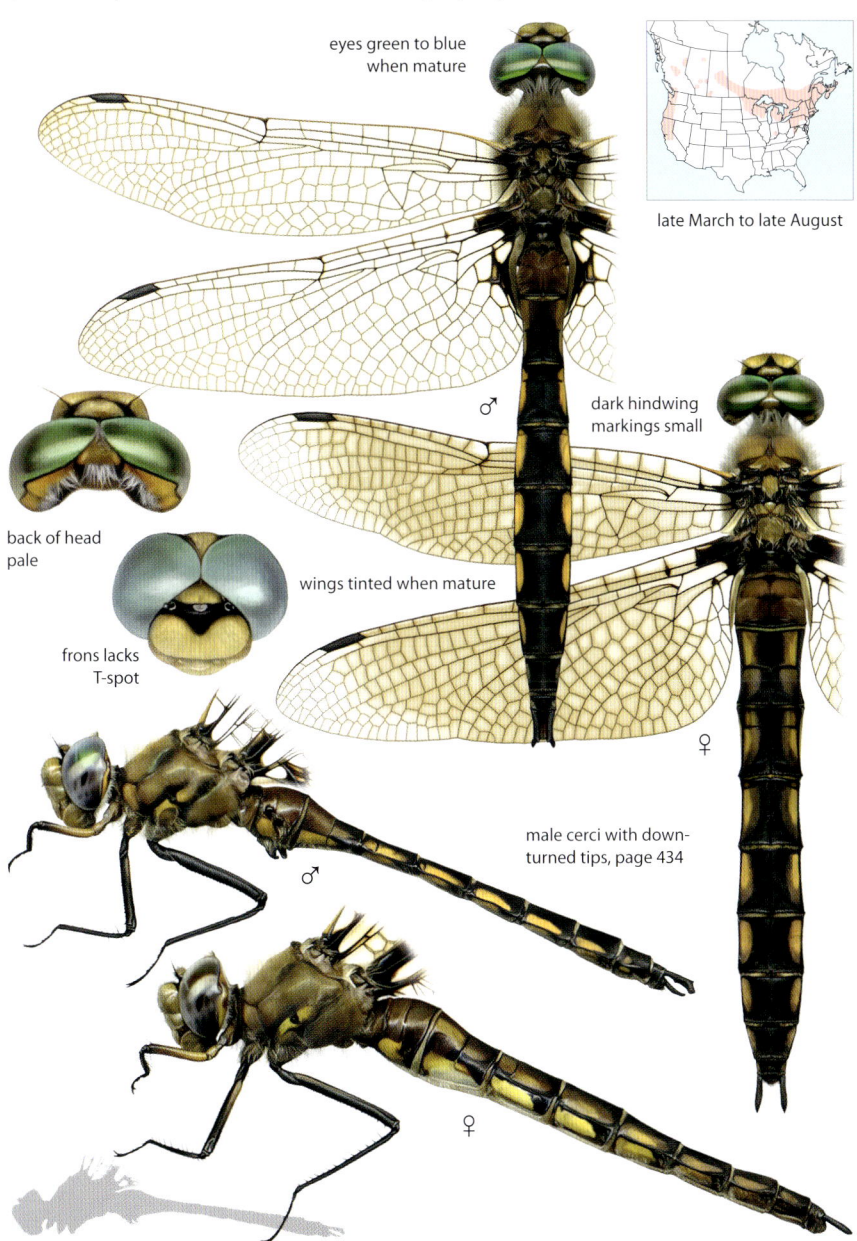

eyes green to blue when mature

late March to late August

back of head pale

frons lacks T-spot

dark hindwing markings small

wings tinted when mature

♂

♀

male cerci with down-turned tips, page 434

♂

♀

Similar species: Other baskettails have back of head black. **Spiny Baskettail** has frons with T-spot; male cerci relatively straight with pointed tip.

Slender Baskettail • *Epitheca costalis*

38–48 mm, 1.5–1.9 in. Narrow-bodied southeastern baskettail found at sand-bottomed lakes, ponds, and streams. Thorax brown. Dark hindwing markings usually small, females rarely with dark stripes along leading edge. Abdomen long and slender, male abdomen constricted at S3. Male cerci long, often more parallel than other species. Female cerci long.

eyes blue-gray

dark hindwing markings small

late January to early July

♂

abdomen slender; male constricted at S3

females may have dark stripes along leading edge of wings, rare in the Southeast

male cerci long, relatively straight

♀

female middle segments (S4-5) longer than wide,

♂

♀

female cerci long

Similar species: Common Baskettail has wider abdomen, male only slightly constricted at S3; cerci shorter with tips divergent. Female with length of middle segments about equal to width; cerci shorter. **Florida Baskettail** nearly identical, separable by shape of male cerci (see page 433); female cerci shorter. See **Dot-winged Baskettail**.

Dot-winged Baskettail • *Epitheca petechialis*

41–43 mm, 1.6–1.7 in. Slender baskettail found in the south-central US at ponds, lakes, and slow streams. Thorax brown. Wings with short dark streaks at base, dark dots at cross-veins along leading edge and at nodus. Abdomen long and slender, male abdomen constricted at S3. Eyes brown to blue-gray.

eyes blue-gray

early February to mid-August

♂

abdomen slender; male constricted at S3

wings with dark markings at base, dark dots along leading edge and at nodus

♀

proportionally similar to Slender Baskettail

♂

♀

Similar species: Dark wing dots diagnostic when present. Individuals without dots largely identical to **Slender Baskettail**; male separable by shape of cerci, page 432. **Common Baskettail** has wider abdomen, male abdomen only slightly constricted S3; female cerci shorter.

Florida Baskettail • *Epitheca stella*

40–48 mm, 1.6–1.9 in. Our slenderest baskettail; inhabits ponds, lakes, coastal and saw-grass marshes, primarily in the Florida peninsula, including the Everglades. Thorax brown. Dark hindwing markings very limited. Abdomen long and slender. Eyes red-brown to greenish. Male cerci straight; female cerci short.

eyes red-brown, greenish when mature

late January to late December

dark hindwing markings minimal

♂

abdomen very slender

♂

♀

♀

female cerci short

Similar species: Slender Baskettail nearly identical, but dark wing markings slightly larger; separable by shape of male cerci (see page 432); female cerci longer. **Common Baskettail** has wider abdomen, dark wing markings usually larger. **Sepia Baskettail** has wider, shorter abdomen.

Sepia Baskettail • *Epitheca sepia*

35–45 mm, 1.4–1.8 in. Small, stocky southeastern baskettail found at sand-bottomed lakes and slow streams. Thorax red-brown. Hindwing nearly unmarked. Abdomen short, tapered, with yellow lateral markings forming nearly continuous stripes along abdomen length. Female cerci short. Eyes of both sexes red.

eyes red

early Mar to early December

thorax warm brown

dark hindwing markings very small

♂

♂

♀

abdomen short

female cerci short

♀

abdomen with full-length yellow stripes, form nearly continuous stripe

Similar species: Common Baskettail has dark wing markings. Male abdomen longer, slightly constricted at S3; female abdomen more parallel-sided, with yellow lateral abdominal markings forming dashes, rather than continuous stripe. **Slender** and **Florida Baskettails** have longer, narrower abdomens.

Robust Baskettail • *Epitheca spinosa*

42–47 mm, 1.7–1.9 in. Large, stout southeastern baskettail found at swamps and boggy ponds and lakes. Thorax covered with long whitish hairs. Hindwing with two small dark markings at base. Abdomen notably wide and relatively short. Male cerci bent downward at tip. Mature male eyes blue-green.

eyes blue-green

late February to mid-May

dark hindwing markings small

thorax pale with long whitish hairs

♂

abdomen wide, relatively short

♀

male cerci has tips downturned, page 433

♂

♀

Similar species: Often recognizable in flight by size; heavily built with very wide abdomen. White-haired thorax often appears paler than other baskettails. **Common Baskettail** has longer, slenderer abdomen.

Broad-tailed Shadowdragon • *Neurocordulia michaeli*

40–43 mm, 1.6–1.7 in. Wide-bodied shadowdragon found at clean northeastern rivers. Crepuscular. Wings with amber dots along leading edge from base to nodus. Thorax brown with yellow lateral spot; top of thorax between wings pale yellow. Dark brown abdomen distinctly wide, with lateral yellow markings. Male eyes yellow-green. Legs dark.

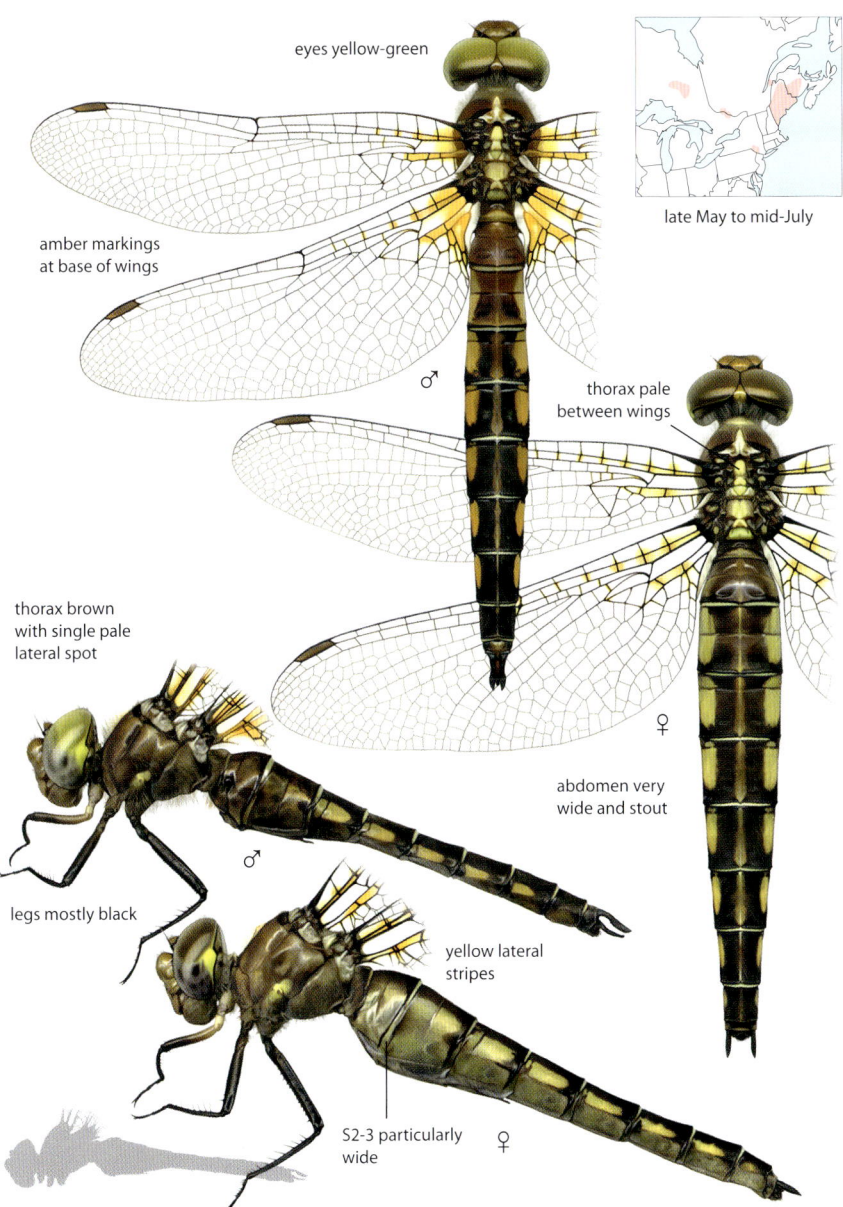

eyes yellow-green

amber markings at base of wings

late May to mid-July

thorax pale between wings

♂

thorax brown with single pale lateral spot

♀

legs mostly black

abdomen very wide and stout

♂

yellow lateral stripes

S2-3 particularly wide

♀

Similar species: Stygian and **Umber Shadowdragons** have dark patch at base of wing; abdomen slenderer, legs paler. **Baskettails** lack amber wing markings, top of thorax between wings uniformly brown.

Stygian Shadowdragon • *Neurocordulia yamaskanensis*

45–55 mm, 1.8–2.2 in. Large eastern shadowdragon inhabiting large rivers and clean lakes. Crepuscular. Wings with amber markings at base; large amber patch with dark brown veins at base of hindwing. Thorax brown with small yellow lateral spot, front of thorax darkened. Abdomen brown with yellow stripes on sides of S4-8. Eyes brownish-green.

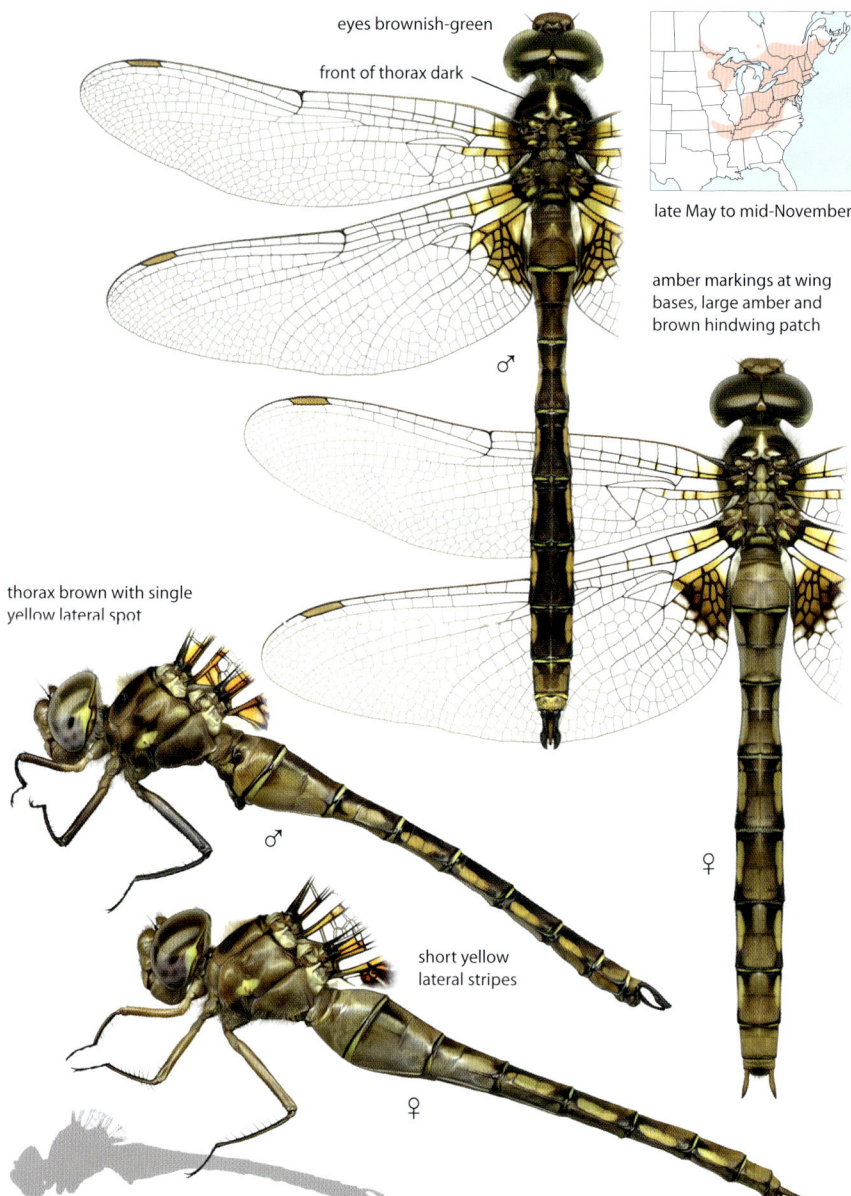

eyes brownish-green

front of thorax dark

late May to mid-November

amber markings at wing bases, large amber and brown hindwing patch

thorax brown with single yellow lateral spot

short yellow lateral stripes

Similar species: Umber Shadowdragon lacks amber wing markings, has dark patch at base of forewing. **Baskettail species** lack amber wing markings, have thorax with two yellow lateral spots, blacker legs.

Umber Shadowdragon • *Neurocordulia obsoleta*

43–48 mm, 1.7–1.9 in. Well-marked eastern shadowdragon found at clean, often rocky streams and rivers. Crepuscular. All wings with brown patch at base, brown spots along leading edge from base to nodus. Thorax brown with yellow lateral spot, front of thorax darkened. Abdomen brown with vague dorsal dark stripes, pale lateral stripes.

eyes gray-brown

brown dots between wing base and nodus

late March to late July

dark patches at base of each wing

♂

♀

thorax brown with single yellow lateral spot; front of thorax dark

♂

vague pale lateral stripes

legs pale brown

♀

Similar species: Dark patches at base of each wing distinctive. **Stygian Shadowdragon** lacks dark patch at base of forewing, has amber wing makings, more conspicuous yellow stripes on abdomen. Small **baskettails** lack dark forewing markings.

Smoky Shadowdragon • *Neurocordulia molesta*

45–53 mm, 1.8–2.1 in. Eastern shadowdragon found at large rivers, sometimes large streams. Crepuscular. Wings with brown spots from base to nodus. Thorax dull brown with elongated lateral yellow spot bordered by black, front of thorax darkened. Abdomen gray-brown with vague dark stripes, inconspicuous yellow rings between segments.

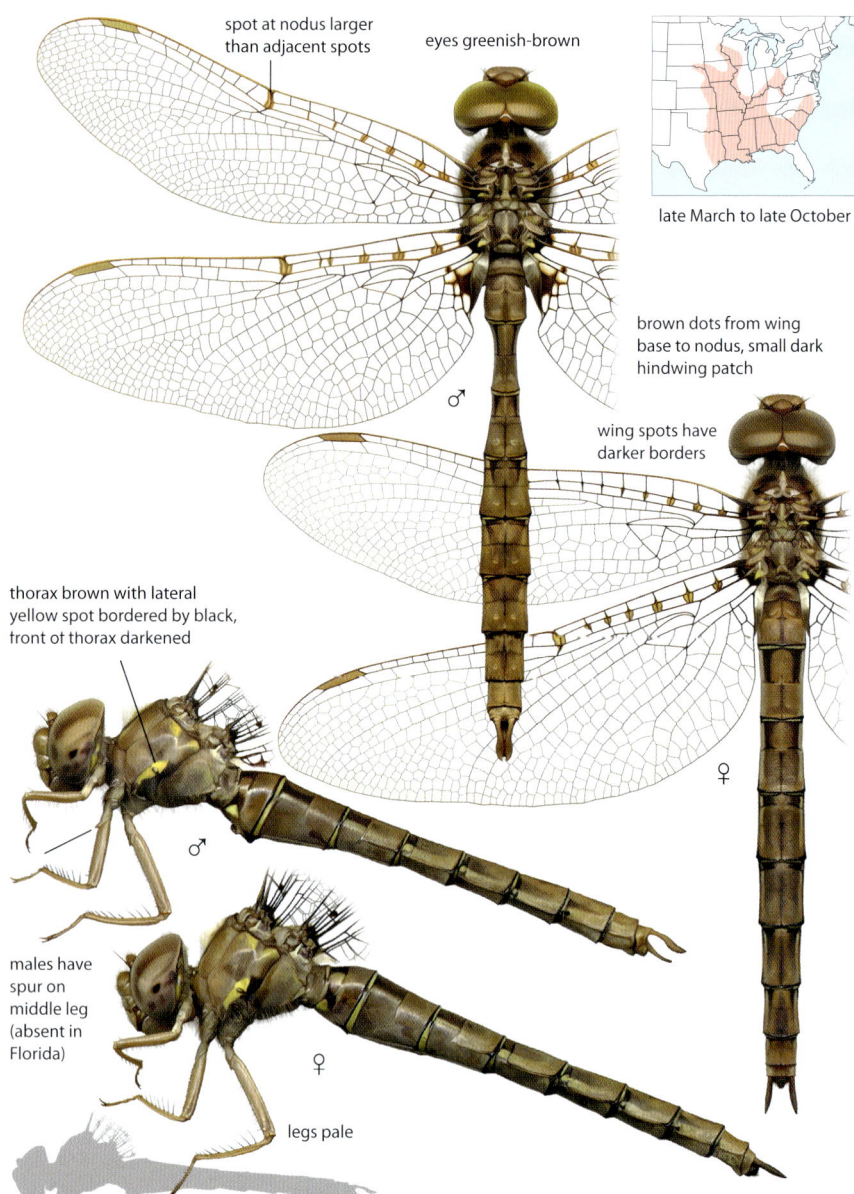

spot at nodus larger than adjacent spots

eyes greenish-brown

late March to late October

brown dots from wing base to nodus, small dark hindwing patch

wing spots have darker borders

thorax brown with lateral yellow spot bordered by black, front of thorax darkened

males have spur on middle leg (absent in Florida)

legs pale

♂

♀

Similar species: Cinnamon Shadowdragon has amber markings only at base of wings; yellow lateral thoracic spot lacks black border. **Alabama Shadowdragon** more orange, lacks yellow lateral thoracic spot, wings with amber wing spots from base to stigma.

Orange Shadowdragon • *Neurocordulia xanthosoma*

48–50 mm, 1.9–2 in. Shadowdragon with heavily marked wings found at small streams to large rivers in the southern Great Plains. Wings with dark markings at base, orange spots from base to stigma along leading edge. Male wings at least one-fourth orange at base, often full wing. Male body orange-brown, female brown; thorax with diffused lateral yellow spot.

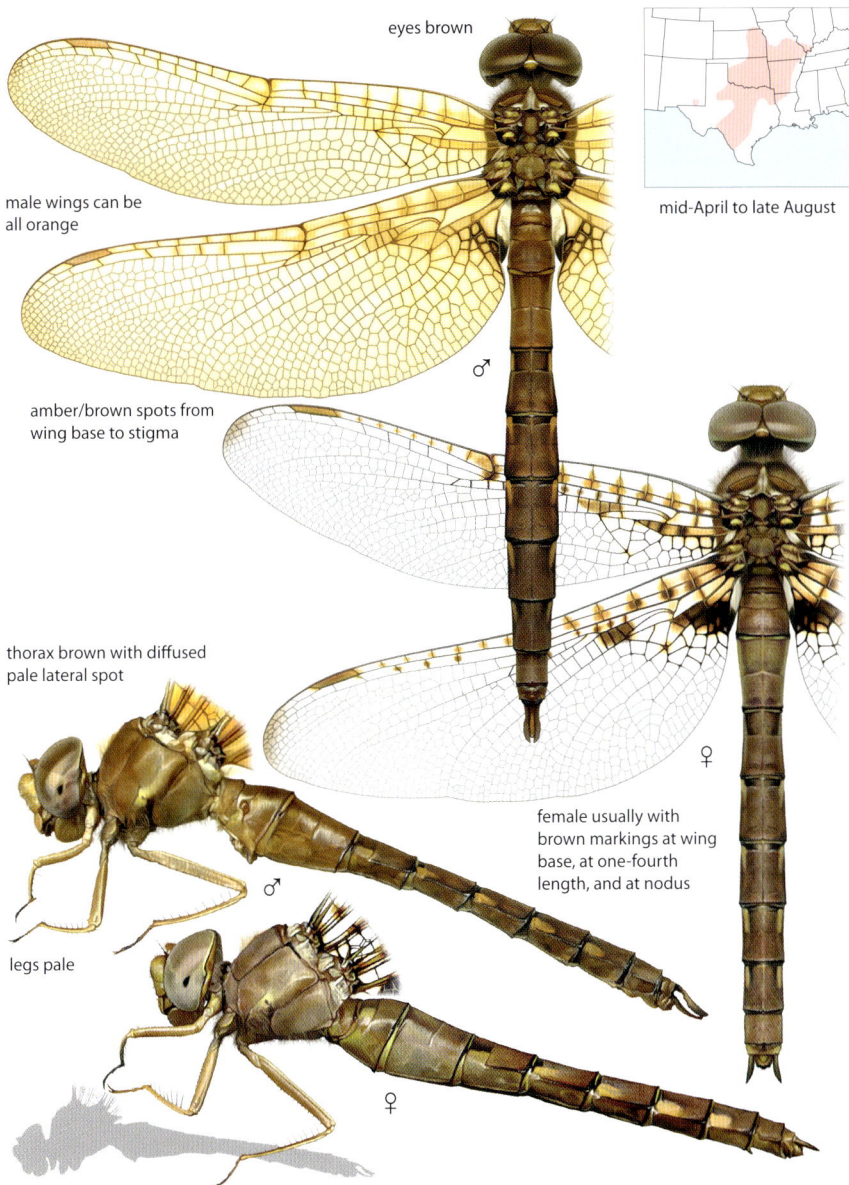

eyes brown

male wings can be all orange

amber/brown spots from wing base to stigma

thorax brown with diffused pale lateral spot

legs pale

♂

♀

female usually with brown markings at wing base, at one-fourth length, and at nodus

♂

♀

mid-April to late August

Crepuscular. Roosts in low foliage, so often more easily found than other shadowdragons.

Similar species: Extensively orange-marked wings distinctive. **Umber Shadowdragon** lacks orange wing markings; no range overlap.

Alabama Shadowdragon • *Neurocordulia alabamensis*

42–46 mm, 1.7–1.8 in. Pale southeastern shadowdragon inhabiting clear, slow woodland streams, usually with sand bottoms. Crepuscular. Wings with row of amber dots running full length along leading edge; mature wings tinted brown. Thorax orange-brown with vague pale yellowish lateral stripes. Abdomen uniformly orange-brown. Eyes pale brown.

amber spots along leading edge between wing base and stigma, may be subtle

early May to early July

eyes pale brown

thorax warm brown with vague pale lateral stripe

♂

♀

♂

legs pale

abdomen uniformly orange-brown

♀

Similar species: Cinnamon and **Smoky Shadowdragons** darker brown, side of thorax with yellow spot; lack amber spots along outer wings between nodus and stigma.

Cinnamon Shadowdragon • *Neurocordulia virginiensis*

42–49 mm, 1.7–1.9 in. Common southeastern shadowdragon found at clean, flowing rivers. Crepuscular. Wings least marked among shadowdragons, with small amber markings restricted to wing bases. Thorax brown with single yellow lateral spot. Abdomen brown with vague yellow-brown lateral stripes. Eyes warm brown.

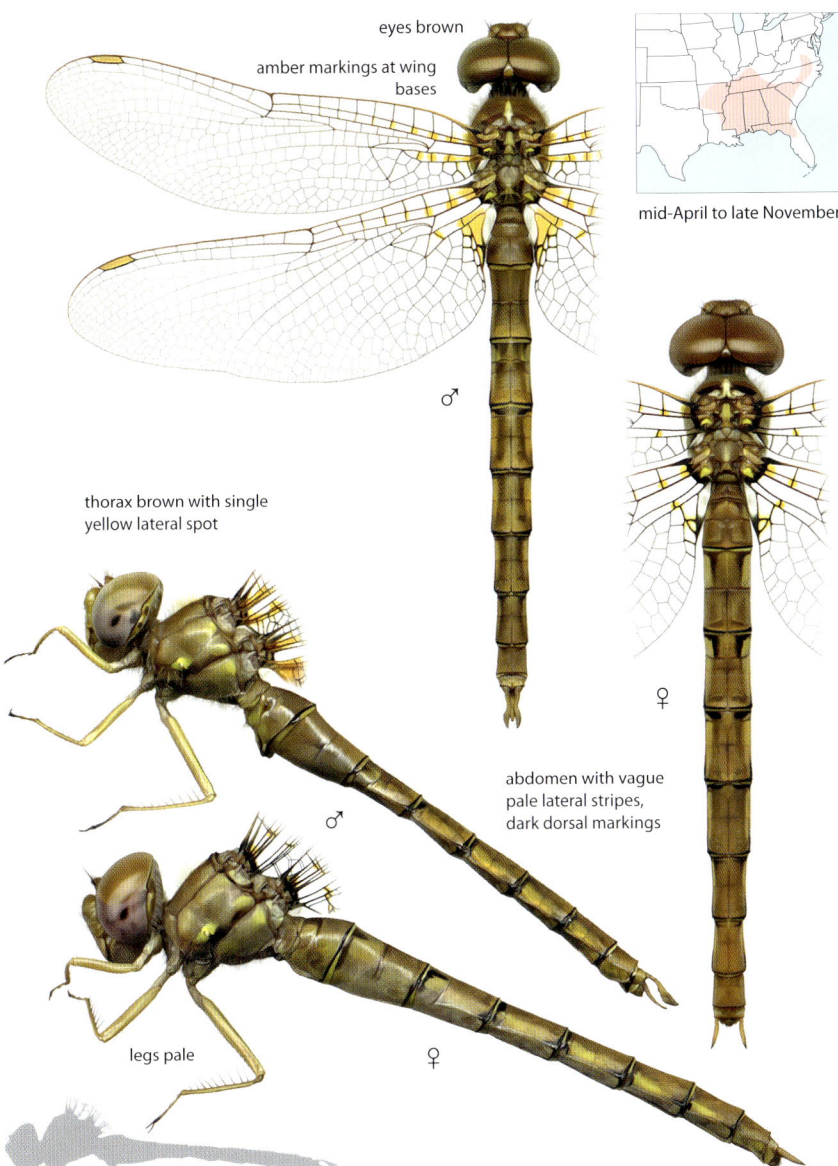

eyes brown

amber markings at wing bases

mid-April to late November

thorax brown with single yellow lateral spot

♂

♀

abdomen with vague pale lateral stripes, dark dorsal markings

♂

legs pale

♀

Similar species: Alabama Shadowdragon more orange in color, lacks yellow lateral spot on thorax, wings with row of amber dots running full length of wing. **Smoky Shadowdragon** has yellow spot on side of thorax bordered by black, dark wing spots between base of wings to nodus.

Ebony Boghaunter • *Williamsonia fletcheri*

29–35 mm, 1.1–1.4 in. Very small, dark northeastern emerald inhabiting acid bog pools and fens. Percher, unlike other emeralds. Mature male eyes blue-green, female green-gray. Thorax dark brown with narrow white markings between wings. Abdomen black with narrow white rings at base of S3 and S4, sometimes S5. Face black.

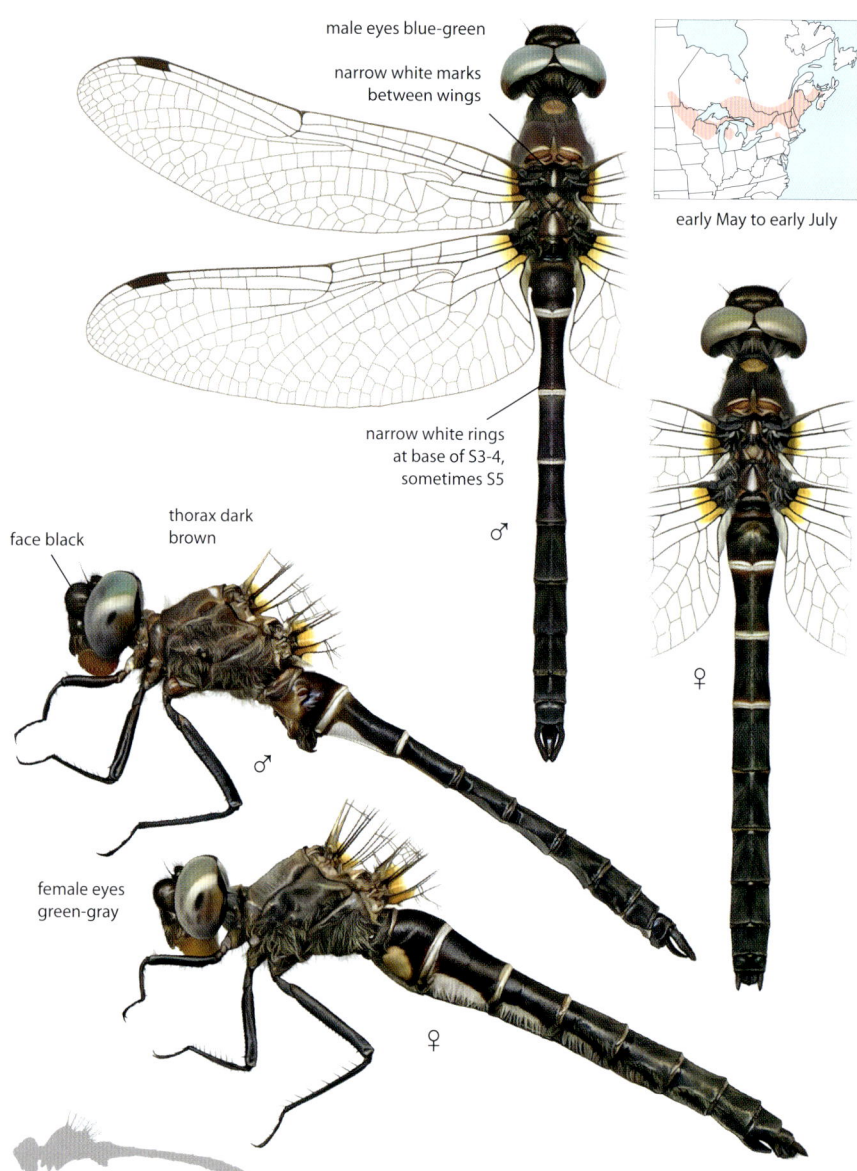

male eyes blue-green

narrow white marks between wings

early May to early July

narrow white rings at base of S3-4, sometimes S5

♂

thorax dark brown

face black

♂

♀

female eyes green-gray

♀

Similar species: Ringed Boghaunter has dull orange rings on abdomen, pale face. **Black Meadowhawk** lacks white abdominal rings; immature and females with pale markings on side of thorax and on top of most, if not all, abdominal segments. **Whitefaces** have white faces; most have pale lateral thoracic markings and dorsal abdominal markings.

Ringed Boghaunter • *Williamsonia lintneri*

31–34 mm, 1.2–1.3 in. Very small, dark early-season emerald found at acid bogs and fens with shallow pools in the Northeast. Eyes brown to gray. Thorax dark brown. Abdomen black with dull orange rings on S2-9. Face pale yellow-brown. Unlike other emeralds, boghaunters are perchers, rather than fliers, often perching low or directly on the ground.

eyes brown-gray

mid-April to mid-June

♂

♀

thorax brown

♂

abdomen with orange-brown bands on S2-9

face yellow-brown

♀

Similar species: Ebony Boghaunter has narrow white rings on S3-5, with rest of abdomen black; dark face. **Whitefaces** have white face; most have middorsal pale spots on abdomen.

Skimmers • Family Libellulidae

The skimmers are the largest family of dragonflies in North America, with well over a hundred species. They include many of our most common species and are the dragonflies most likely encountered by beginners. The vast majority breed at still waters, while only a few require running water. Most are perchers, sometimes on the ground but most often on plant stems, where they alight obliquely or horizontally at the stem's tip. Some skimmers—the saddlebags, gliders, clubskimmers, and sylphs—are fliers, and these often hang vertically at rest. Of over 1,000 species worldwide, 113 are North American.

Skimmers vary greatly in size and appearance. The most conspicuous are large and showy, many exhibiting bright colors and strong patterns. Many skimmers have boldly marked wings, a feature rare in other families. Others are more muted in appearance, making them easy to overlook, and the family includes some of our smallest dragonfly species. Unlike other families, many skimmers are marked in red and can have red, orange, green, or white stigmas. Many species are white or bluish in appearance, often due to the development of a waxy bloom called pruinosity. As a dragonfly matures, pruinosity may develop on the abdomen, thorax, and/or even parts of the wings.

All skimmers have eyes that meet at a seam at the top of the head, and their hindwings typically have a foot-shaped anal loop. Males and females have similarly shaped hindwings, with males lacking auricles on the side of S2 that other dragonfly families have. The emeralds are considered the most closely related family, but their eyes have a posterior bump, making the rear margin of the eyes more sinuous, and the hindwing anal loop is more club-like than foot-like.

Skimmer head dorsal view
eyes meet at seam

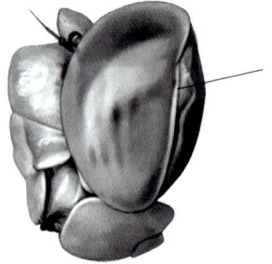

Skimmer head lateral view
rear margin relatively straight

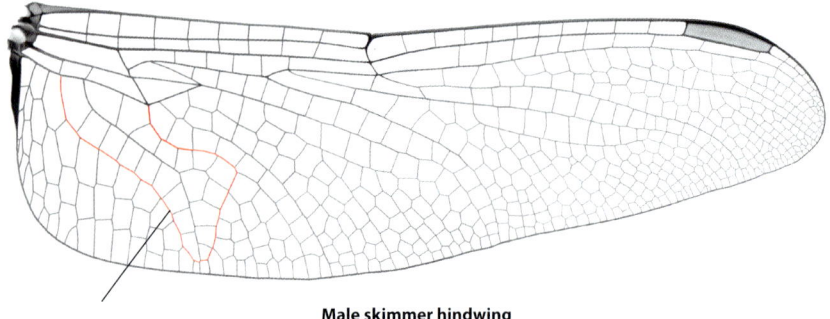

Male skimmer hindwing
anal loop foot-like in shape

Amberwings, genus *Perithemis*, page 275

A genus of a dozen Neotropical species, the amberwings are easily identified by the coloration of their wings and their very small size. Male wings are entirely yellow-orange, while females are patterned in amber and brown. The wings are short and rounded, and the hindwing anal loop is complete but relatively straight, its shape likened to the foot of a ballet dancer *en pointe*. The abdomen is short and spindle-shaped. Of the 12 species, three are North American.

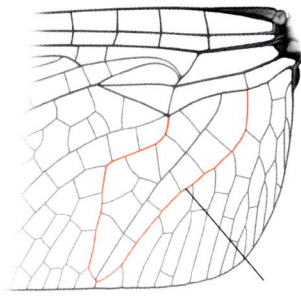

Amberwing anal loop

Whitetails, genus *Plathemis*, page 278

Both whitetail species occur in North America. They are closely related to king skimmers, but males have a large forked structure underneath S1 that king skimmers lack. Both whitetail species are stocky, with a wide abdomen. They are similarly patterned. The eyes and body are mostly brown, the abdomen with pale dashes in immatures and females, becoming conspicuously white in adult males. The wings are heavily marked, but males and females have very different patterns. Both species habitually perch on the ground, logs, and other low perches, often in the open.

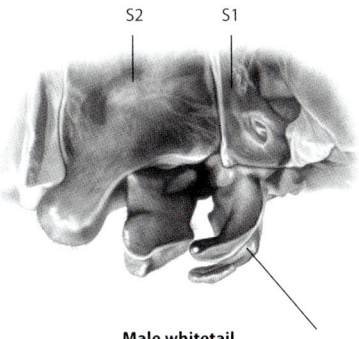

Male whitetail
forked structure under S1

Corporals, genus *Ladona*, page 280

Small genus consisting of three North American species. Their name comes from having pale stripes at the front of the thorax that suggest the insignia worn to denote military rank. They are dark, stout skimmers, with small dark markings at the base of the wings. They fly early in the year and are typically among the first dragonflies on the wing. They are closely related to king skimmers, with which they had previously been combined, but corporals typically perch flat on the ground, which king skimmers rarely do.

King Skimmers, genus *Libellula*, page 283

These large dragonflies are among our most common and familiar. Conspicuous at freshwater ponds throughout the continent, they are showy species, many having distinctive and colorful wing patterns. While they often perch, usually obliquely on sticks and stems, they are strong fliers, with males aggressively chasing each other as they defend their breeding territories. Their head is fairly large, the thorax is robust, and the wings are long and densely veined. The abdomen is stout, somewhat flattened, and shorter than the wings. Immatures of both sexes are similarly patterned. As they mature, males and some females dramatically change in appearance. Many males change color, some to bright red, but the majority of our species develop pruinosity on the abdomen and thorax, so white, pale blue, and gray are common colors. Some male skimmers develop white wing markings; they are our only dragonflies other than whitetails to do so. Many of our species are easily identified by their color and pattern, but a number of similar species pairs and groups require greater scrutiny. Nineteen of the 28 species of king skimmer have been found in North America.

Tropical King Skimmers, genus *Orthemis,* page 302
Separated by details of wing venation, tropical king skimmers are similar to king skimmers in size and structure while largely replacing them in the New World tropics. Only three of the 28 species range north of Mexico into our area. Mature males of our species are easily recognizable to genus by their large size and striking red or purple coloration. Immatures and females are brown and rather similar, but the species are usually separable by their lateral thoracic patterns.

Scarlet Skimmers, genus *Crocothemis*, page 305
An Old World genus consisting of 10 species, one of which has been introduced into North America. Males of most but not all species are predominately red or brown. Ours is a bright red skimmer with an established population on the Florida peninsula.

Narrow-winged Skimmers, genus *Cannaphila*, page 306
Small genus consisting of three Neotropical species, with one, the Gray-waisted Skimmer, ranging to our area. They are most related to king and tropical king skimmers, but the base of the hindwing is narrower, resulting in a blunter, less foot-shaped anal loop.

Pondhawks, genus *Erythemis*, page 307
Medium to large-sized skimmers, the pond-hawks are named for their rapacious preda-tory habits, often taking prey as large as them-selves, including other dragonflies. To aid in capturing such big prey, pondhawks have ex-tra long spines on the femurs of the middle and hind legs. Pondhawks inhabit a wide variety of still-water habitats, most flying and perching low, often on the ground or on low or floating vegetation. Seven of the 10 New World species occur in North America. In shape and size, they most resemble some king skimmers, but their ground-perching behavior often sets them

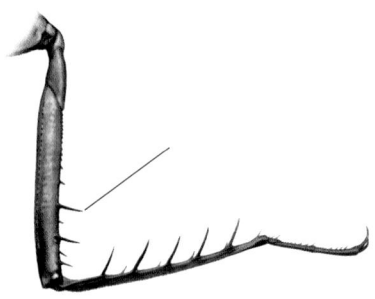

Pondhawk leg
long spines on middle and hind leg femurs

apart. Their patterns are relatively simple; the thorax is usually a single solid color, varying among our species from bright green, pale blue, and deep red to brown and black.

Scarlet-tails, genus *Planiplax*, page 314
The scarlet-tails are five medium-sized Neotropical species, of which one has been found in South Texas. They are stout-bodied skimmers with very long legs inhabiting open ponds and lakes with emergent vegetation. The adult males have striking patterns. Their face and eyes are dark, the thorax is black with varying amounts of bluish pruinosity, and the abdomen is partially to completely bright red. Immatures and females are mostly brown.

Tropical Pennants, genus *Brachymesia*, page 315
All three species in this genus are found in the southern US. They share an unpatterned thorax, and the base of the abdomen is highly expanded in side view. Females and im-matures are primarily brown in color, but adult males are strikingly different: one species remains brown, one becomes all black, while the third develops a bright red abdomen.

Metallic Pennants, genus *Idiataphe*, page 318
Small New World genus consisting of four species, with only one reaching the southern US. They are small, slender skimmers that typically perch horizontally at the tips of stems

and branches. They are dark black and/or brown overall, including the eyes, with the thorax showing some metallic reflections at close range.

Coastal Pennants, genus *Macrodiplax*, page 319
This genus consists of two species that live primarily at coastal habitats, one in the Old World, the other in the New. Like other pennants, they habitually perch at the tip of stems, but coastal pennants have relatively spare wing venation. Ours is a dark species, with a distinctive round dark patch at the base of the hindwing.

Small Pennants, genus *Celithemis*, page 320
The small pennants are a group of eight North American species. They are named for their habit of perching at the tip of stems like a flag on a flagpole, often with their forewings held higher than the hindwings. They include some of our most ornate and colorful dragonflies, the majority of species having heavily marked wings. Their flight is fluttery, and their wings are long and wide, compared to a short abdomen and small thorax. They are found primarily in the eastern and central US, breeding at marshy ponds and lakes.

Whitefaces, genus *Leucorrhinia*, page 329
Whitefaces are small northern skimmers with a conspicuous, clear white face. They inhabit primarily vegetated ponds and lakes, where they fly and perch low. The white face contrasts with their overall dark appearance: dark eyes, thorax and abdomen mostly black with red or yellow markings. The legs are black, and all whitefaces have a small dark triangular marking at the base of the hindwing. Meadowhawks are similarly small and marked with red and yellow but lack the dark hindwing marking and have significantly less black on the thorax and abdomen. Species identification of whitefaces usually starts with abdomen pattern, but changes in appearance due to age can be significant, as pale markings of some species become obscured. Others develop a large amount of white pruinosity, particularly on the

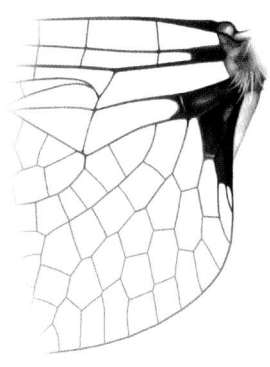

Whiteface hindwing
dark triangular mark at base

abdomen. To identify the most similar species, close examination of wing venation, the male's hamules, and the female's subgenital plate may be necessary. Of 14 species worldwide, seven are North American.

Dragonlets, genus *Erythrodiplax*, page 336
A large Neotropical genus consisting of 61 described species, with seven found north of Mexico. They are small to medium-sized skimmers, the majority inhabiting marshy ponds and lakes. They are most similar to the meadowhawks but have differences in venation and abdominal structure. Like meadowhawks, some dragonlets are marked in red, but males of other species are all black or have a dark thorax coupled with a pale blue abdomen. Similar species can be difficult to differentiate, particularly females and immatures.

Meadowhawks, genus *Sympetrum*, page 344
The meadowhawks are a large genus of small, slim skimmers distributed throughout the Northern Hemisphere. Of the 56 described species, 13 are found in North America. Except for a single black species, red is the dominant color of the males (and some females) of our species. The combination of small size and red coloration makes meadowhawks fairly easy to recognize. The similarly small whitefaces may also have red markings, but they are

darker in overall appearance, having considerably more black on the thorax and abdomen, along with a dark triangular patch at the base of the hindwing. With several species similar in appearance, identifying meadowhawks to species is often challenging. Some species do have distinctive patterns, but others have to be scrutinized closely. Important clues include the color of the face, legs, and wing veins, but some meadowhawks can be separated only by examining genital structures under magnification.

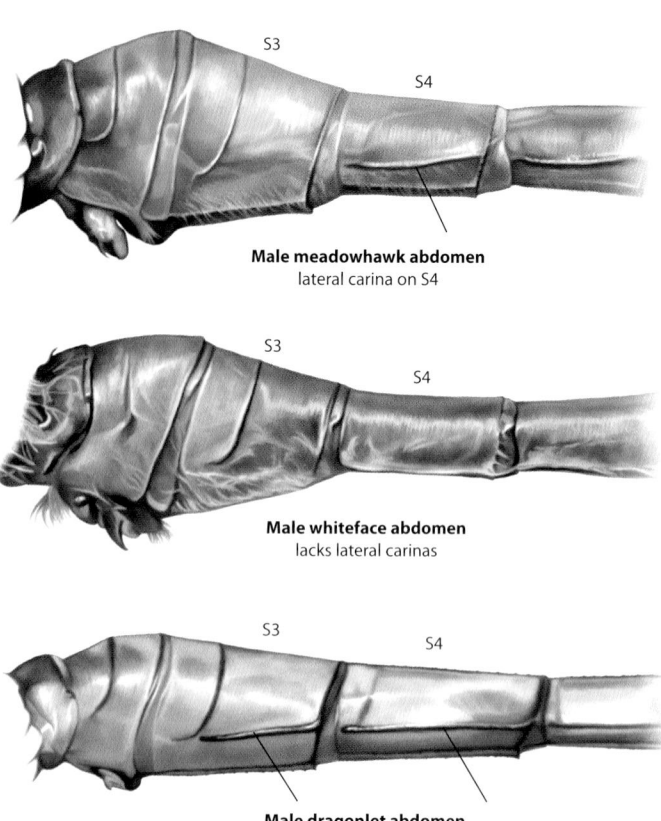

Male meadowhawk abdomen
lateral carina on S4

Male whiteface abdomen
lacks lateral carinas

Male dragonlet abdomen
lateral carinas on S3 and S4

Elfin Skimmer, genus ***Nannothemis***, page 360
At less than an inch long, the Elfin Skimmer is the smallest North American dragonfly and the only member of its genus. An inhabitant of eastern bogs, its tiny size alone is distinctive, but the hindwing has an incomplete anal loop, a feature unique among our skimmers.

Blue Dasher, genus ***Pachydiplax***, page 361
This genus consists of a single species, the Blue Dasher, one of the most common dragonflies in North America. It is found at most still-water habitats within its wide range. It is a small skimmer with a white face and a

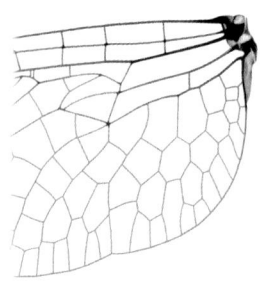

Elfin Skimmer
anal loop incomplete

striped thorax, mature males with green eyes. Males and some females develop pale blue pruinosity on the abdomen but retain a black abdomen tip. The Blue Dasher's wing venation differs from all other dragonflies by having only a single cross-vein below the stigma bordering an exceptionally long empty cell.

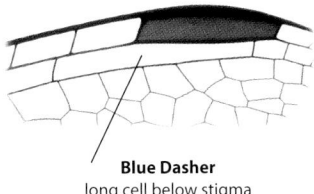

Blue Dasher
long cell below stigma

Tropical Dashers, genus *Micrathyria*, page 362
The tropical dashers are a genus of small Neotropical skimmers. Of 48 described species, only four have been found in the southern US. They are separated from other small skimmers by having wings with two bridge cross-veins rather than one. Tropical Dashers have bright green to blue eyes and a white face. The thorax is well striped, and almost all species have conspicuous pale markings on the top of S7. Males have a slightly clubbed abdomen with a rather short S10.

Setwings, genus *Dythemis*, page 366
The setwings are a group of seven Neotropical skimmers, four of which range into the southern US. They are named for perching in a "get-set" position like a sprinter before a race: head inclined downward with the abdomen raised, wings held down and forward (while the pose is characteristic, setwings do not always perch in this manner). Our species are quite varied in appearance, most with slender abdomens, some males slightly clubbed. Some have intricately patterned bodies; others are plain but boldly colored. All show wing markings, some conspicuously, others only barely.

Rock Skimmers, genus *Paltothemis*, page 370
A small Neotropical genus consisting of three stream-dwelling species, one occurring north of Mexico. The hindwing is very broad at the base. The trait is usually an indication of gliding flight, but rock skimmers are perchers that typically rest flat on rocks. Our species is intricately patterned and quite unlike any other North American dragonfly.

Filigree Skimmer, genus *Pseudoleon*, page 371
The only species in its genus, the Filigree Skimmer is a dark southwestern skimmer of rocky streams and rivers. Its pattern is unique. The dragonfly appears dark overall, but its entire body, including the eyes, is intricately patterned with fine pale stripes, and large dark lace-like markings cover the wings.

Evening Skimmers, genus *Tholymis*, page 372
This genus consists of two tropical species, one Old World, the other New. They are similar to shadowdragons in their plain appearance and habits, hanging in shade by day, and flying for short periods primarily at dusk and dawn. Their wing venation is unusual, with the anal loop reaching the hind margin of the hindwing, and the female's subgenital plate is unique among all dragonflies, being shaped like an inverted trough in which the eggs are held by stiff hairs.

Clubskimmers, genus *Brechmorhoga*, page 373
Genus of fairly large Neotropical skimmers. Of 16 species, three have been recorded in North America in the Southwest, where they inhabit clear streams. They are seen mostly in flight, feeding in open areas or males patrolling low over the water. At rest, they typically hang vertically. The species of clubskimmers are similar in color and pattern. Eyes are pale blue in mature males and some females. The thorax is dark brown with pale green-tinted

stripes. The abdomen is slightly clubbed, less so in the female, and mostly black, with large pale dorsal spots on S7 that can be seen in flight.

Sylphs, genus *Macrothemis*, page 376
The sylphs are a Neotropical genus of small, slender, stream-dwelling skimmers. They are fliers, feeding in clearings with males patrolling up and down streams. They typically hang vertically at rest; some species also perch on top of leaves. Of the 42 New World species, four reach North America. Among these four, three have conspicuous pale spots on top of S7. Males of these spotted species also have clubbed abdomens, noticeable in overhead flight.

Rainpool Gliders, genus *Pantala*, page 380
The two species of rainpool glider both occur in North America. One is wide-ranging in the New World; the other has a near global range, having been recorded on every continent except Antarctica. With wide hindwings adapted for sustained gliding, they are capable of remaining aloft for hours and even days while traveling thousands of miles during migratory flight. They breed primarily at temporary pools and ponds, where their larvae develop rapidly. Unsurprisingly, rainpool gliders are most often seen in flight, feeding in open areas. While intricately patterned, they appear plain brown or yellow on the wing, with reddish eyes and largely clear wings. When perched, usually on tree branches, they hang vertically, unlike most other skimmers.

Pasture Gliders, genus *Tauriphila*, page 382
The pasture gliders are a Neotropical genus of five species, three of which occur in North America. They are associated with still waters with floating vegetation, in particular water hyacinth and water lettuce. Like their close relatives, the saddlebags and hyacinth gliders, pasture gliders are fliers with long wings, the hindwing very broad at the base. Their venation is considerably denser than the hyacinth gliders, and compared to the saddlebags, pasture gliders are smaller with shorter cerci. At rest, pasture gliders hang vertically. The abdomen is constricted, narrowest between S3 and 4, and spindle-shaped, particularly in the male. Our species have a dark patch at the base of each hindwing.

Hyacinth Gliders, genus *Miathyria*, page 385
A Neotropical genus consisting of two species, with one found north of Mexico. They are fliers with broad hindwings, similar in habits and closely related to saddlebags and pasture gliders, but are separated due to their greatly reduced wing venation. They are closely associated with habitats with floating vegetation, primarily water hyacinth and water lettuce.

Saddlebags, genus *Tramea*, page 386
The saddlebags are 21 species of highly aerial skimmers distributed throughout the world, primarily in the tropics. Seven species have been recorded in North America. They are typically seen in constant flight over water or feeding in open areas. At rest, they often perch at the top of stems, similar to pennants, but they may also hang vertically from branches. The genus is named for having dark hindwing patches that suggest of the shape of saddlebags on a horse. Our species can be divided into two groups: one with broad hindwing patches, the other with narrow. As with other gliders, the wings are long, with the hindwing very wide at the base. Wing venation and longer cerci help separate saddlebags from the similar pasture and hyacinth gliders.

Mexican Amberwing • *Perithemis intensa*

23–26 mm, 0.9–1 in. Very small, brightly colored southwestern skimmer inhabiting ponds, lakes, ditches, and slow streams. Side of thorax unmarked yellow. Abdomen orange with small dark dorsal dots. Male wings amber, stigma orange. Female wings with brown dots and amber bands, stigma brown.

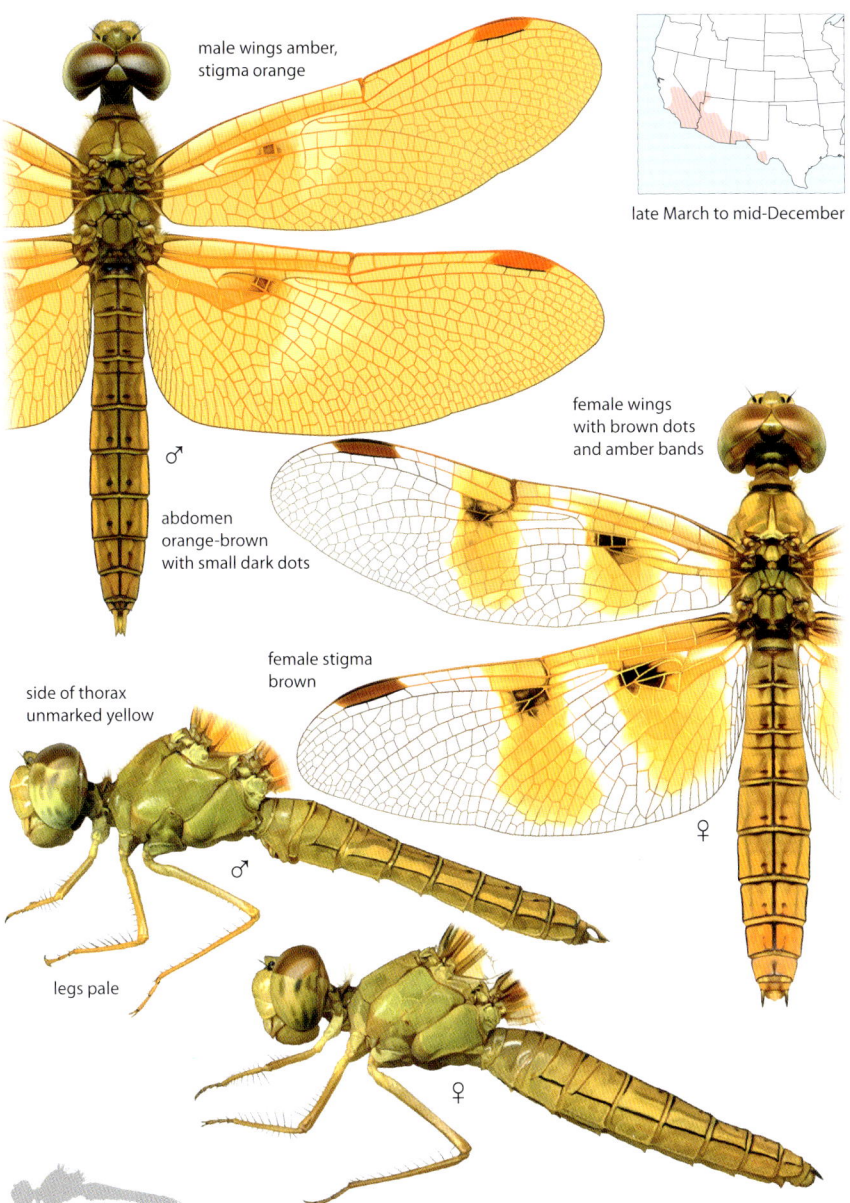

male wings amber, stigma orange

late March to mid-December

female wings with brown dots and amber bands

abdomen orange-brown with small dark dots

female stigma brown

side of thorax unmarked yellow

legs pale

Similar species: Eastern Amberwing has two pale stripes on side of thorax, triangular pale markings on top of abdomen. **Slough Amberwing** has three pale stripes on side of thorax, abdomen with pairs of narrow dorsal stripes; female may have dark wing tips.

Eastern Amberwing • *Perithemis tenera*

20–25 mm, 0.8–1 in. Common amberwing found at a wide variety of still or slow-flowing habitats throughout the East. Tiny and wasplike. Thorax brown with two pale lateral stripes. Abdomen brown with pale dorsal triangular markings. Stigmas red-orange. Male wings amber. Female wings with brown and amber bands but variable.

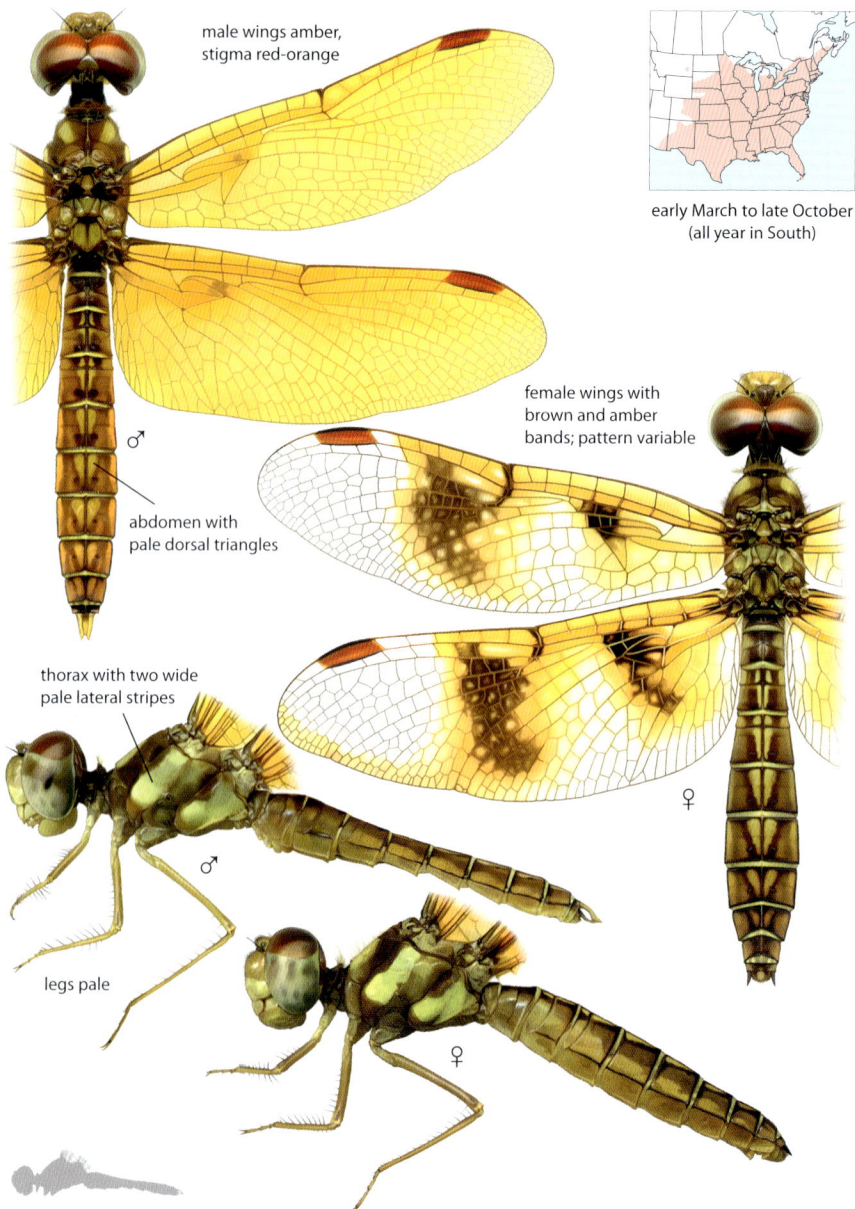

male wings amber, stigma red-orange

early March to late October (all year in South)

female wings with brown and amber bands; pattern variable

abdomen with pale dorsal triangles

thorax with two wide pale lateral stripes

legs pale

♂

♀

♀

Similar species: Mexican Amberwing has side of thorax unmarked, abdomen with small dark dots. **Slough Amberwing** has three pale lateral stripes on thorax, abdomen with pairs of parallel pale stripes.

Slough Amberwing • *Perithemis domitia*

22–25 mm, 0.9–1 in. Tiny southwestern amberwing found at woodland stream pools, sloughs, and shaded ponds. Thorax brown with three pale lateral stripes. Abdomen with pairs of pale narrow dorsal stripes. Male wings amber, stigma red. Female wings variable: usually basal half amber with brown margin, wing tip dark, stigma dark brown.

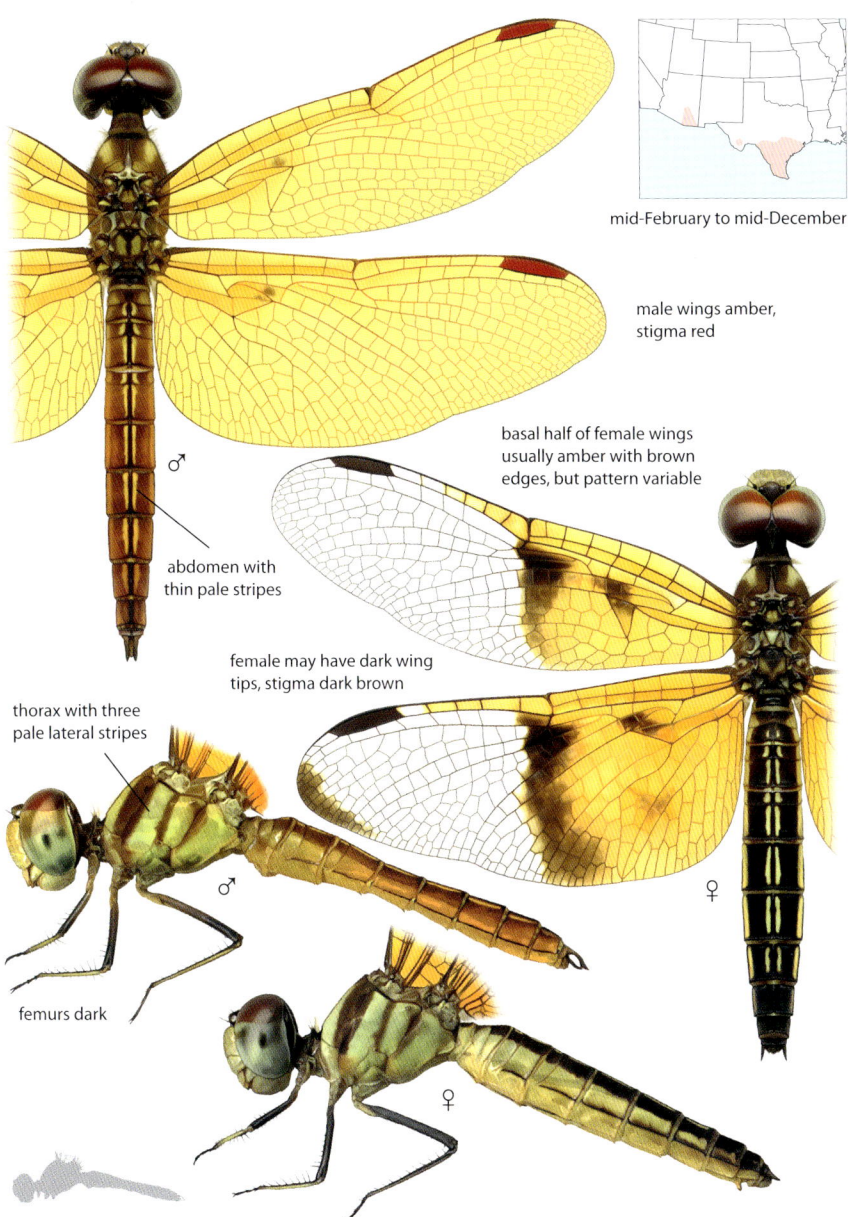

mid-February to mid-December

male wings amber, stigma red

basal half of female wings usually amber with brown edges, but pattern variable

♂

abdomen with thin pale stripes

female may have dark wing tips, stigma dark brown

thorax with three pale lateral stripes

♂

♀

femurs dark

♀

Similar species: Eastern Amberwing has two pale lateral thoracic stripes, pale triangular markings on top of abdomen. **Mexican Amberwing** has side of thorax unmarked yellow; top of abdomen marked with small dark dots.

Common Whitetail • *Plathemis lydia*

42–48 mm, 1.7–1.9 in. Abundant, widespread whitetail found at most still and slow-flowing waters, including degraded habitats. Typically perches on ground. Mature male with wide brown wing bands; abdomen pruinose white. Immature and female abdomen brown with pale dashes. Female wings with dark markings at base, nodus, and tip.

male with wide dark band between nodus and stigma

north: late Apr to mid-Nov
south: early Feb to late Dec

female wings with dark stripe at base, dark patch at nodus, and dark wing tips

mature male abdomen pruinose white

♂

♀

female and immatures with short pale dashes

♂

abdomen stout

♀

thorax with yellow and white lateral stripes

Similar species: Desert Whitetail male has inner half of wings mostly white, dark wing band paler in middle; female lacks dark wing tips. **Twelve-spotted Skimmer** female has yellow abdominal markings, forming continuous stripes.

Desert Whitetail • *Plathemis subornata*

40–51 mm, 1.6–2 in. Western whitetail inhabiting desert pools, ponds, slow streams, and hot springs. Immatures and female brown with contrasting yellow abdominal spots, wings with dark mark at base, dark zigzag bands at nodus and stigma. Area between bands darkens to form wide band in mature male; male inner half of wing and abdomen white.

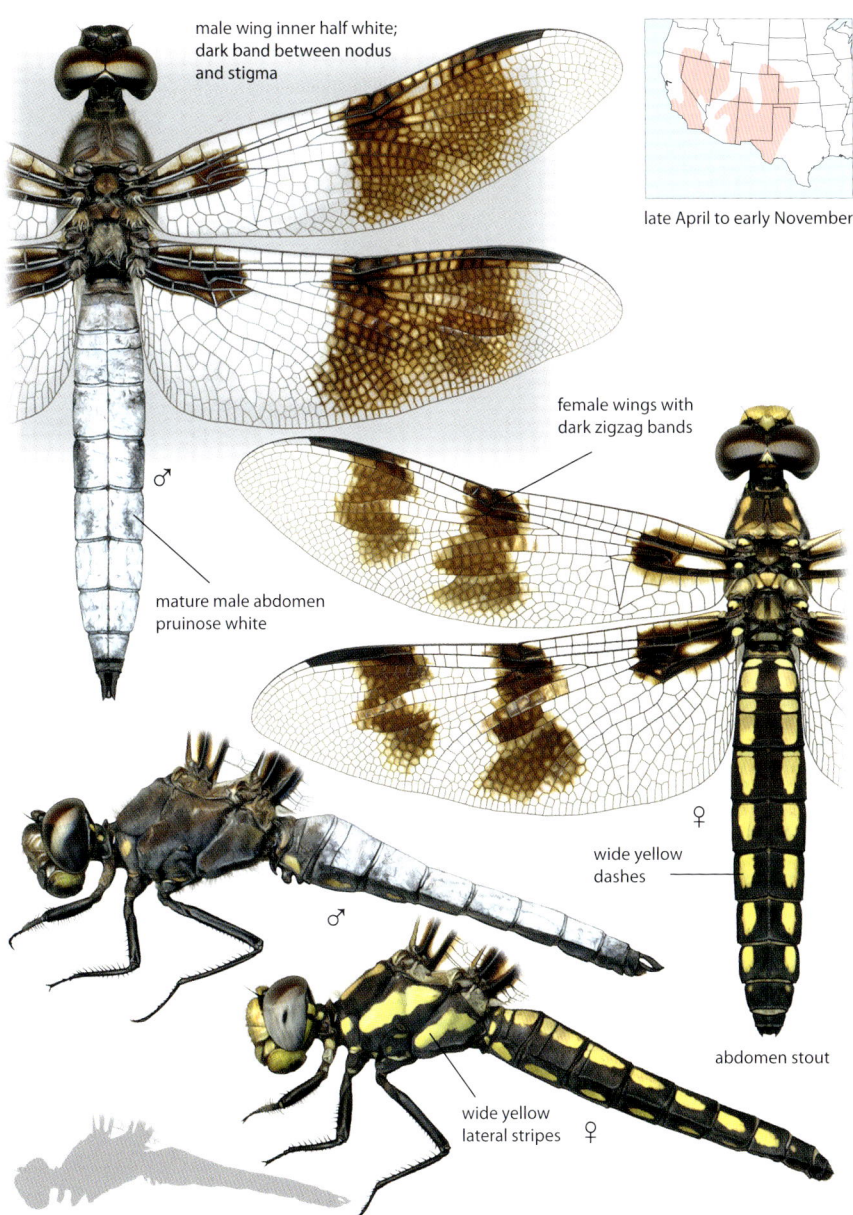

male wing inner half white; dark band between nodus and stigma

late April to early November

female wings with dark zigzag bands

♂

mature male abdomen pruinose white

♀

wide yellow dashes

♂

abdomen stout

wide yellow lateral stripes ♀

Similar species: Common Whitetail male has only small white patch on hindwing, no white on forewing. Both sexes of **Eight-spotted Skimmer** have white patch below stigma; dark wing band narrower and does not reach stigma. Female with single dark wing band.

Blue Corporal • *Ladona deplanata*

34–38 mm, 1.3–1.5 in. Small, dark corporal found at eastern ponds, lakes, and slow streams. Mature male front of thorax and abdomen gray-blue. Female and immatures brown; thorax with pale shoulder stripe bordering black shoulder stripe, abdomen with black triangular dorsal markings. Hindwing with two dark marks at base.

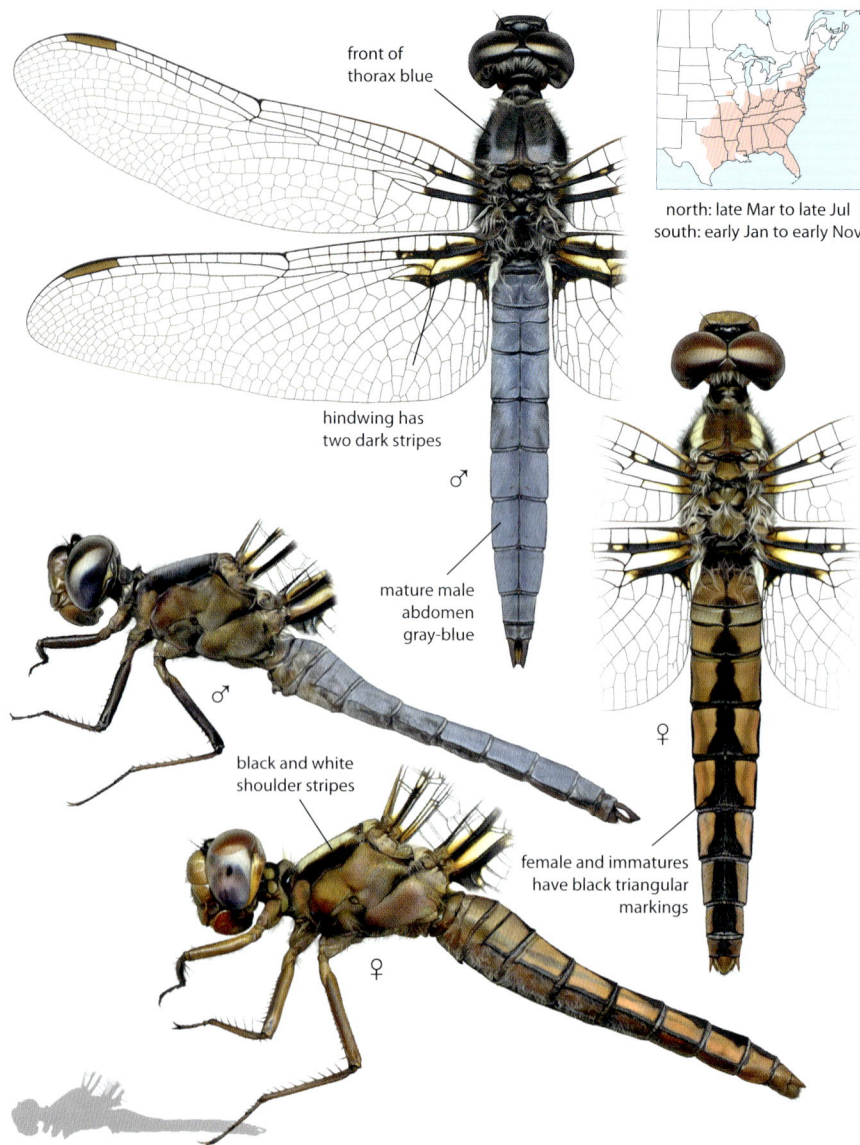

front of thorax blue

north: late Mar to late Jul
south: early Jan to early Nov

hindwing has two dark stripes

♂

mature male abdomen gray-blue

♂

black and white shoulder stripes

♀

female and immatures have black triangular markings

♀

Flies in early spring; in many areas the first dragonfly on the wing.
Similar species: White Corporal has narrow black middorsal stripe on abdomen, single large triangular dark marking at base of hindwing; mature male with white abdomen.
Chalk-fronted Corporal has triangular dark mark at base of hindwing; dark abdominal markings less triangular in shape and form wide dorsal stripe.

Chalk-fronted Corporal • *Ladona julia*

41–45 mm, 1.6–1.8 in. Common northern corporal found at woodland and open ponds, lakes, slow streams, and often acidic bog ponds. Mature male front of thorax and base of abdomen pruinose white. Female and immature brown; thorax with pale shoulder stripes, abdomen with wide black middorsal stripe. Dark triangular mark at base of hindwing.

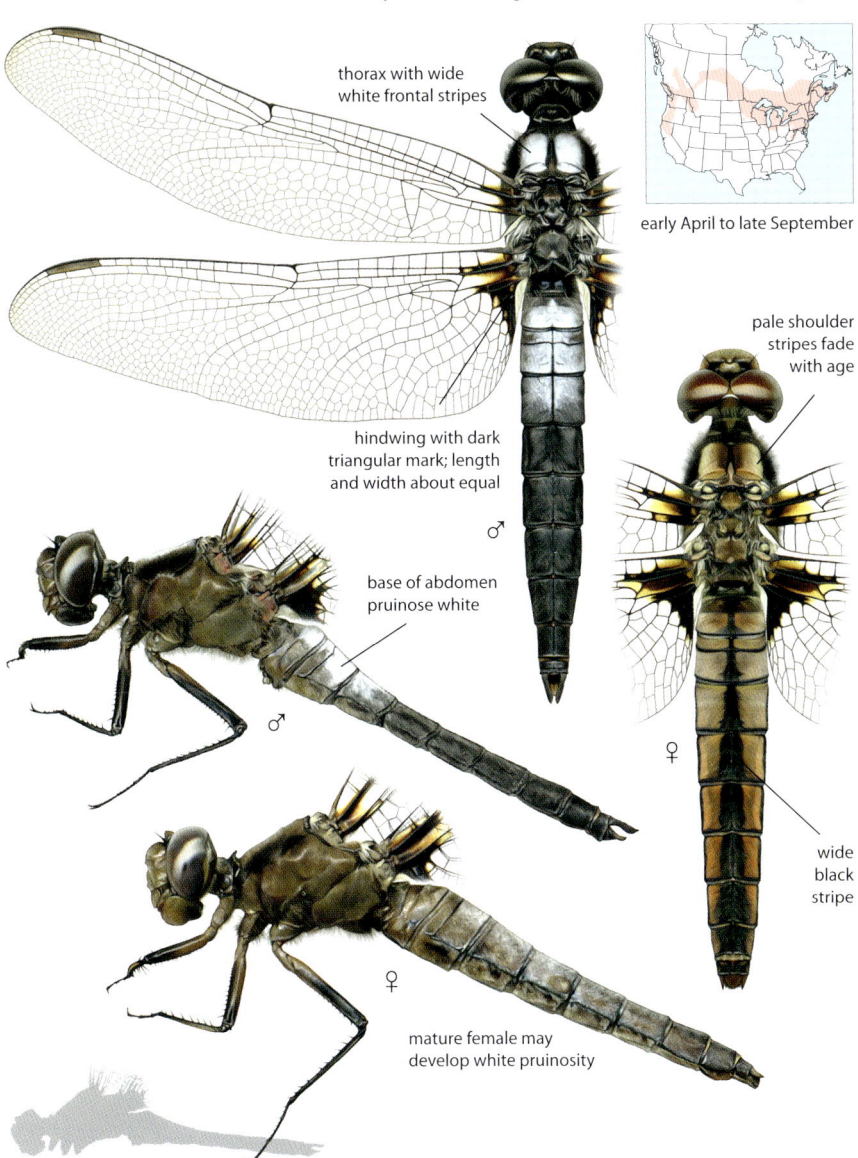

thorax with wide white frontal stripes

early April to late September

hindwing with dark triangular mark; length and width about equal

♂

base of abdomen pruinose white

♂

pale shoulder stripes fade with age

♀

wide black stripe

♀

mature female may develop white pruinosity

Similar species: Blue Corporal has triangular black markings on top of abdomen, and base of hindwing has two dark streaks; mature male with blue-gray abdomen. **White Corporal** has narrow black middorsal stripe on abdomen. Dark hindwing mark extends twice as long as wide. Adult male with abdomen almost completely white, lacks wide white markings at thorax front. **Belted** and **Frosted Whitefaces** smaller and slimmer, with white faces.

White Corporal • *Ladona exusta*

37–43 mm, 1.5–1.7 in. Northeastern coastal corporal inhabiting sandy ponds, lakes, and bogs. Mature male abdomen white; thorax with pale shoulder stripes. Female and immatures brown; thorax with pale shoulder stripe bordering black shoulder stripe, abdomen with narrow black middorsal stripe. Hindwing marking singular dark triangle.

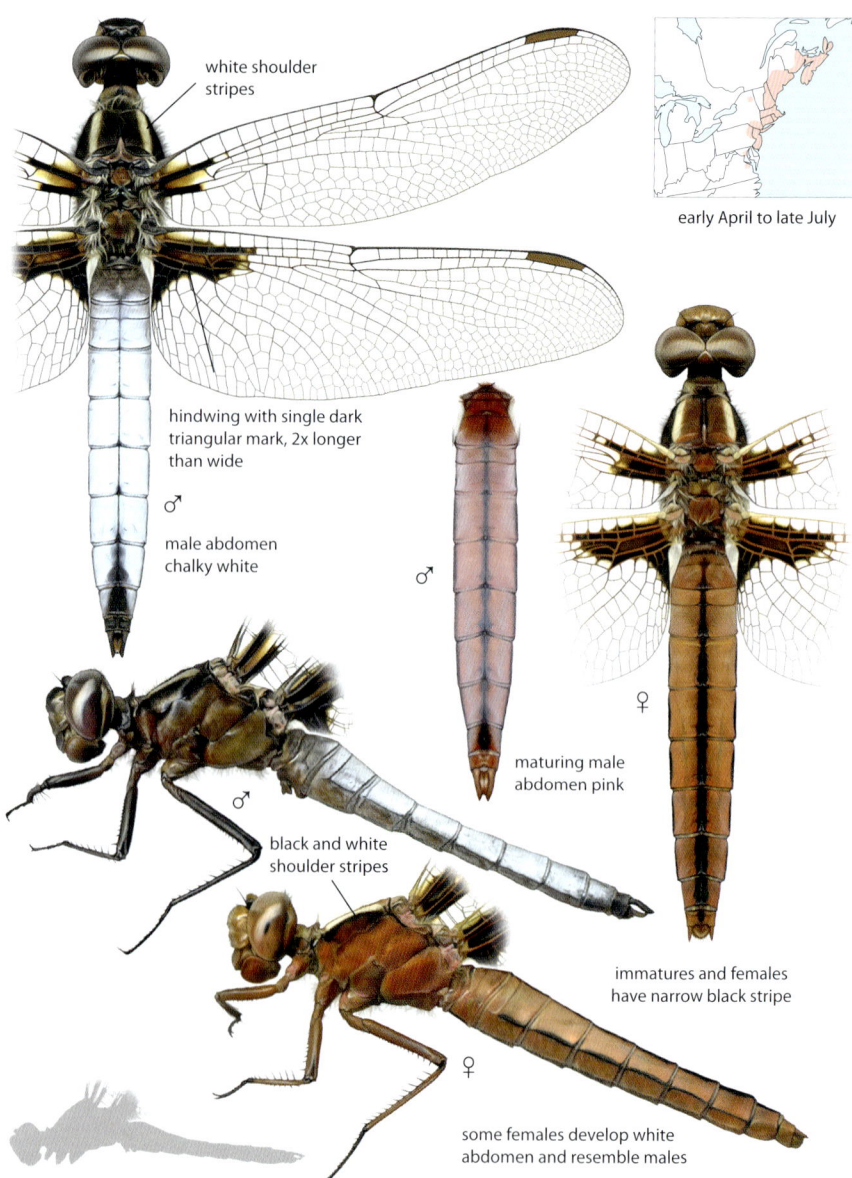

white shoulder stripes

early April to late July

hindwing with single dark triangular mark, 2x longer than wide

♂

male abdomen chalky white

♂

maturing male abdomen pink

♀

♂

black and white shoulder stripes

immatures and females have narrow black stripe

♀

some females develop white abdomen and resemble males

Similar species: Blue Corporal has wide black triangular markings on top of abdomen, and hindwing has two dark streaks at base; mature male abdomen blue-gray. **Chalk-fronted Corporal** has wide black middorsal stripe on abdomen, and length of dark hindwing triangular marking is about equal to width; mature male with front of thorax white.

Widow Skimmer • *Libellula luctuosa*

42–50 mm, 1.7–2 in. Common, widespread skimmer inhabiting a wide variety of ponds, lakes, and slow streams. Both sexes with distinctive wide, dark bands at base of each wing; male bands wider, extending from wing base to nodus. Mature male develops white wing bands, white pruinosity on abdomen and front of thorax.

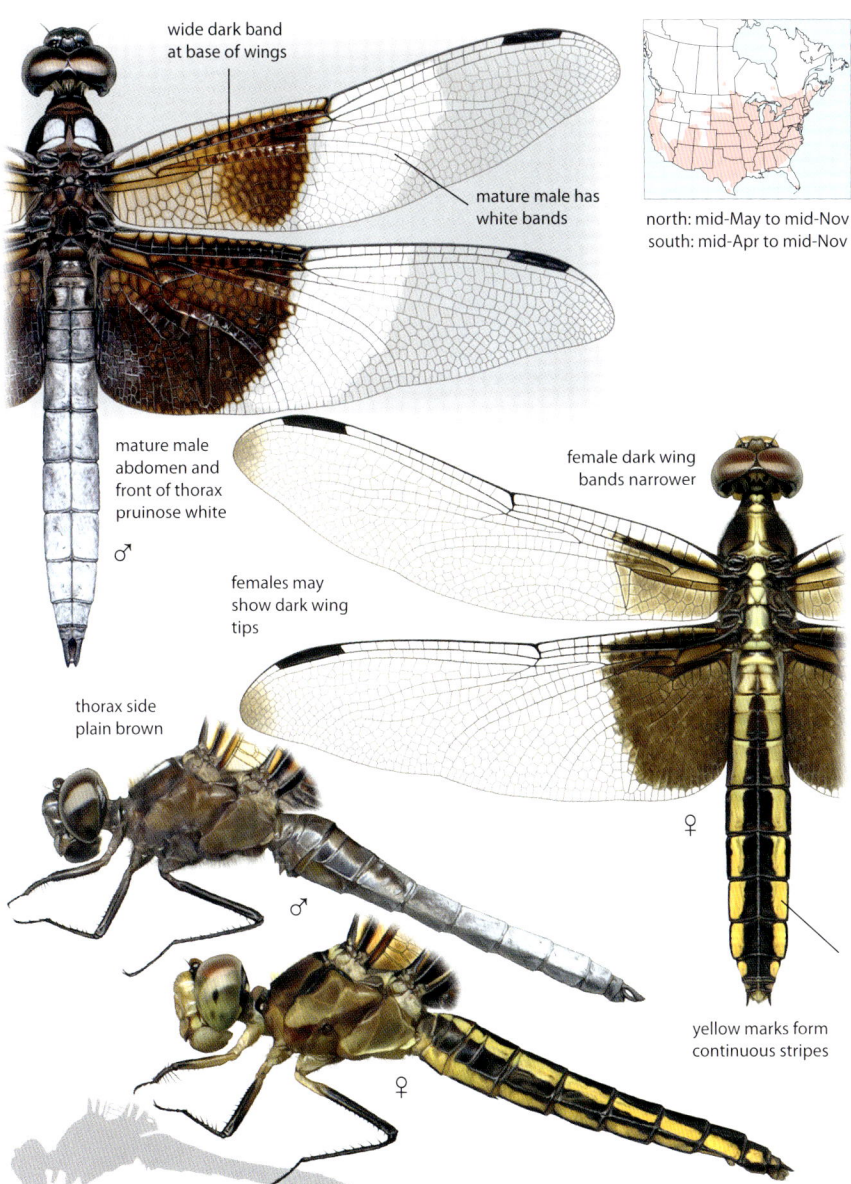

wide dark band at base of wings

mature male has white bands

north: mid-May to mid-Nov
south: mid-Apr to mid-Nov

mature male abdomen and front of thorax pruinose white

♂

female dark wing bands narrower

females may show dark wing tips

thorax side plain brown

♂

♀

♀

yellow marks form continuous stripes

Similar species: Wing pattern unique among king skimmers. **Black Saddlebags** hind-wing patch smaller, with more jagged margin, and forewing lacks dark band; abdomen slender with yellow dorsal spots. **Black-winged Dragonlet** wing patches larger, extending well beyond nodus. Abdomen longer and thinner; appendages pale.

Twelve-spotted Skimmer • *Libellula pulchella*

52–57 mm, 2–2.2 in. Common, widespread skimmer found at vegetated ponds, lakes, slow streams. Wings with large dark marks at base, nodus, and tip. Mature male with white spots between dark markings and at base of hindwing; abdomen pruinose white. Immature and female abdomen dark brown with yellow markings, forming continuous stripes.

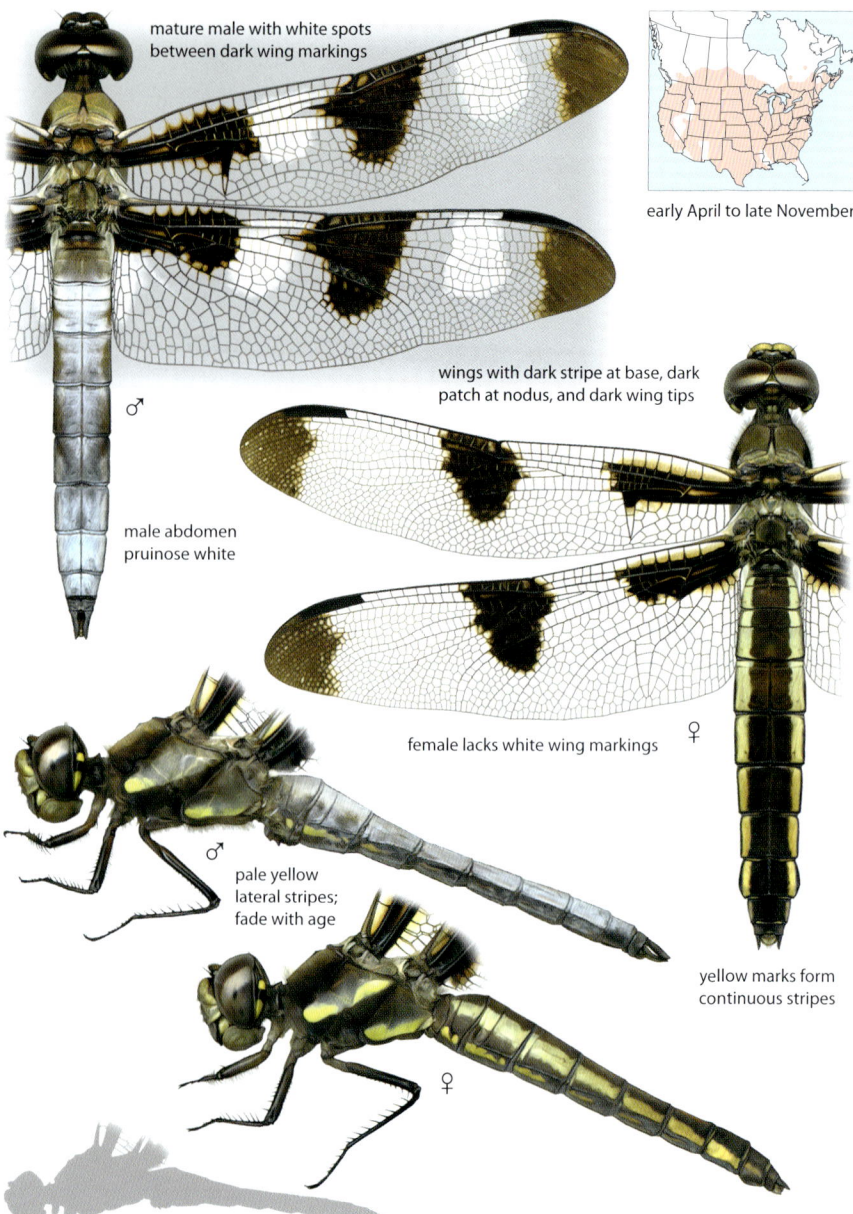

mature male with white spots between dark wing markings

early April to late November

male abdomen pruinose white

wings with dark stripe at base, dark patch at nodus, and dark wing tips

female lacks white wing markings

pale yellow lateral stripes; fade with age

yellow marks form continuous stripes

Similar species: Eight-spotted and **Hoary Skimmers** lack dark wing tip markings. Female **Common Whitetail** has more isolated short pale dashes on abdomen.

Eight-spotted Skimmer • *Libellula forensis*

49–51 mm, 1.9–2 in. Common western skimmer inhabiting a large variety of ponds, lakes, and slow streams. Wings of both sexes with wide dark brown stripe at base bordered by white, brown band at nodus, white patch below stigma. Mature male thorax front and abdomen pruinose white. Immature and female brown; abdomen with short yellow dashes.

dark stripe at wing base bordered by white

dark band at nodus

white patch below stigma

mid-April to mid-October

wing tips clear

♂

mature male top of thorax and abdomen pruinose white

female wing pattern similar to male, but some females may lack white markings

♀

♂

♀

abdomen with short yellow dashes

Similar species: Twelve-spotted Skimmer has dark wing tips; female has longer yellow abdominal stripes, forming continuous stripes, and lacks white wing markings. Female **Common Whitetail** has dark wing tips, lacks white wing markings. **Hoary Skimmer** has small dark spot at nodus, lacks white patch below stigma; female lacks white wing markings.

Hoary Skimmer • *Libellula nodisticta*

46–52 mm, 1.8–2 in. Western skimmer inhabiting marshy slow streams, ponds, springs, hot springs in the north. All wings with wide dark stripe at base and small dark spot at nodus. Mature male thorax and abdomen pruinose white; white at wing bases. Immature and female thorax brown with four pale yellow lateral spots, abdomen with yellow dashes.

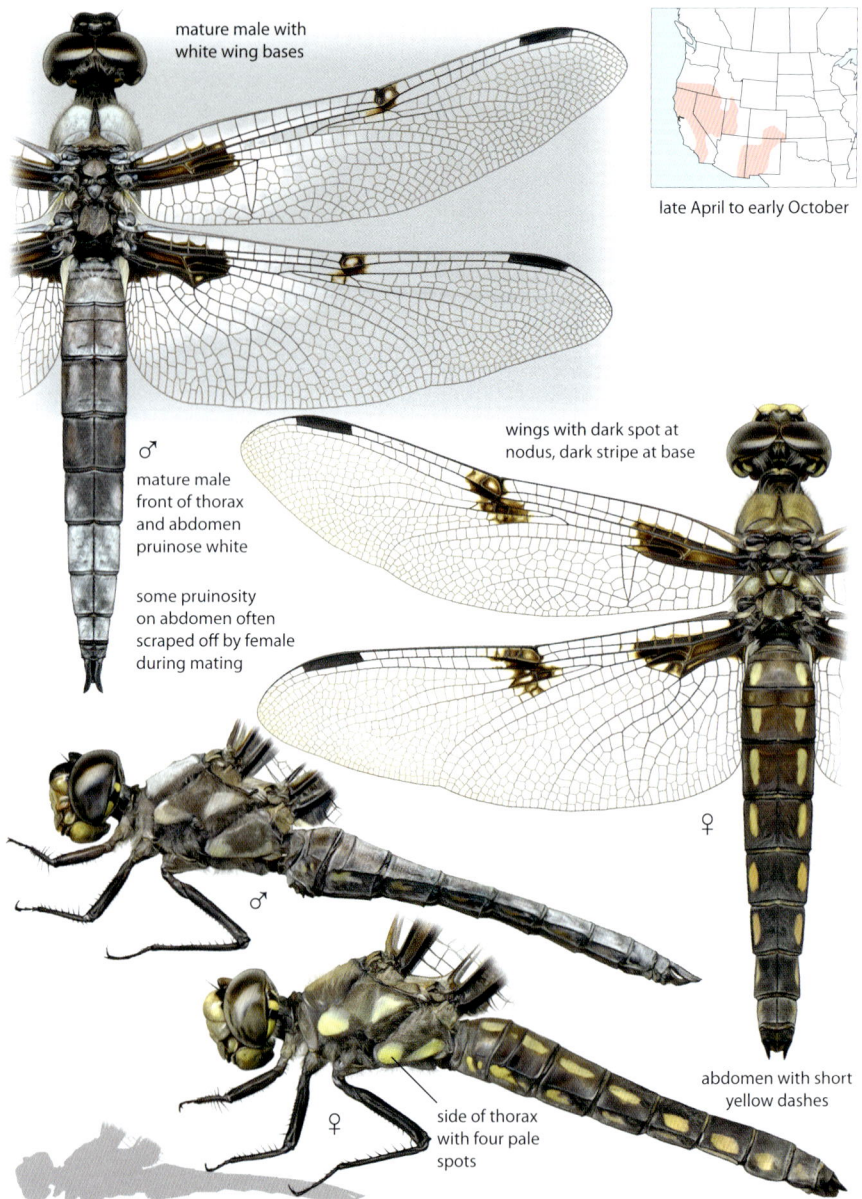

mature male with white wing bases

late April to early October

♂

mature male front of thorax and abdomen pruinose white

some pruinosity on abdomen often scraped off by female during mating

wings with dark spot at nodus, dark stripe at base

♀

♂

♀

side of thorax with four pale spots

abdomen with short yellow dashes

Similar species: Eight-spotted Skimmer has brown band at nodus, and both sexes have white patches below stigmas. **Four-spotted Skimmer** yellow-brown, lacks dark stripe at base of forewing. **Twelve-spotted Skimmer** has dark wing tips.

Bleached Skimmer • *Libellula composita*

42–49 mm, 1.7–1.9 in. Pale western skimmer inhabiting desert springs, alkaline lakes and marshes. Wings with brown/amber marking at base, may have brown spot at nodus, costal vein white. Face white, eyes gray. Mature male thorax and abdomen pruinose blue. Immature and female thorax pale, with brown frontal stripes and black lateral stripes.

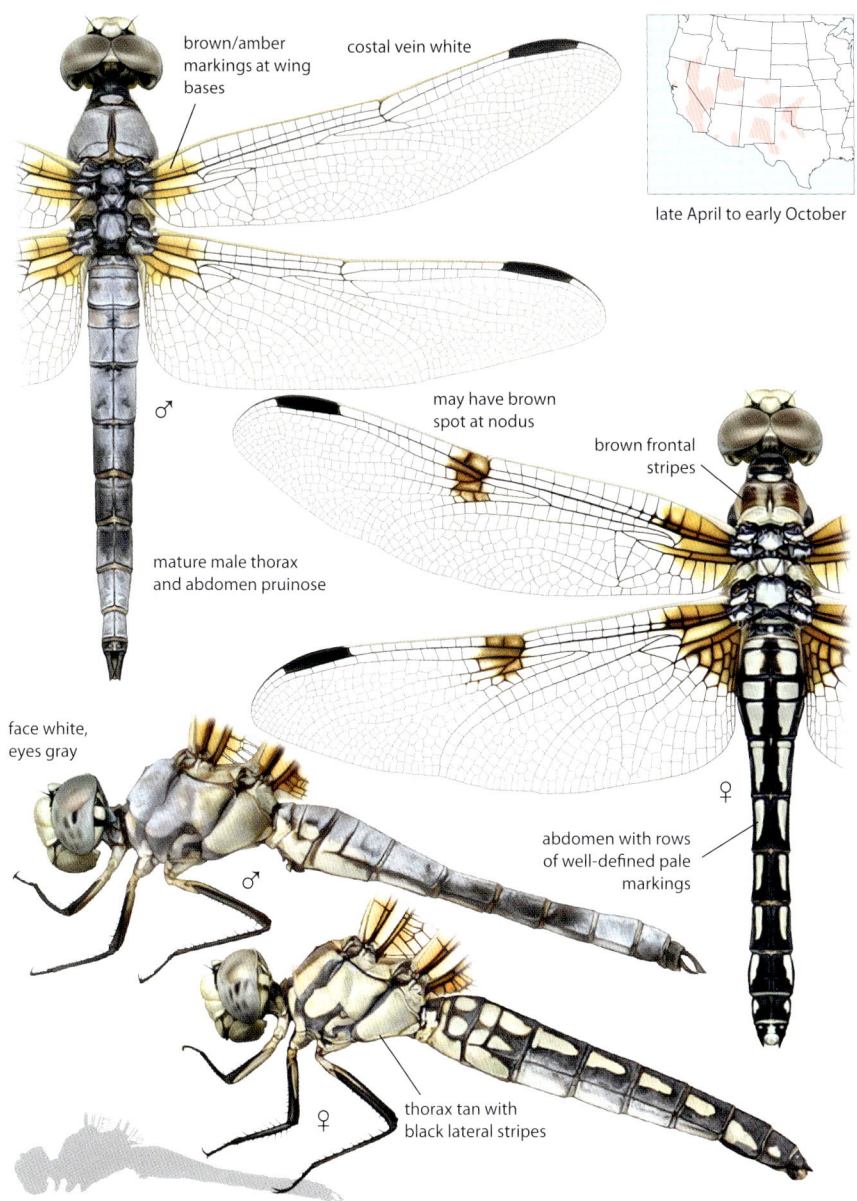

brown/amber markings at wing bases

costal vein white

late April to early October

♂

may have brown spot at nodus

brown frontal stripes

mature male thorax and abdomen pruinose

face white, eyes gray

♀

abdomen with rows of well-defined pale markings

♂

♀

thorax tan with black lateral stripes

Similar species: Comanche Skimmer has black and white stigmas, dark costal vein, lacks dark markings at base of wings; female with single dark lateral thoracic stripe. **Hoary Skimmer** has dark stripe at base of each wing and dark costal vein; side of thorax mostly dark.

Spangled Skimmer • *Libellula cyanea*

41–46 mm, 1.6–1.8 in. Small eastern skimmer found at vegetated ponds, lakes, and slow streams. Both sexes with white and black stigmas and dark streaks at wing bases. Mature male blue, with dark green eyes and black face. Female dark brown, with two large contrasting pale patches on side of thorax; face tan, wing tips dark.

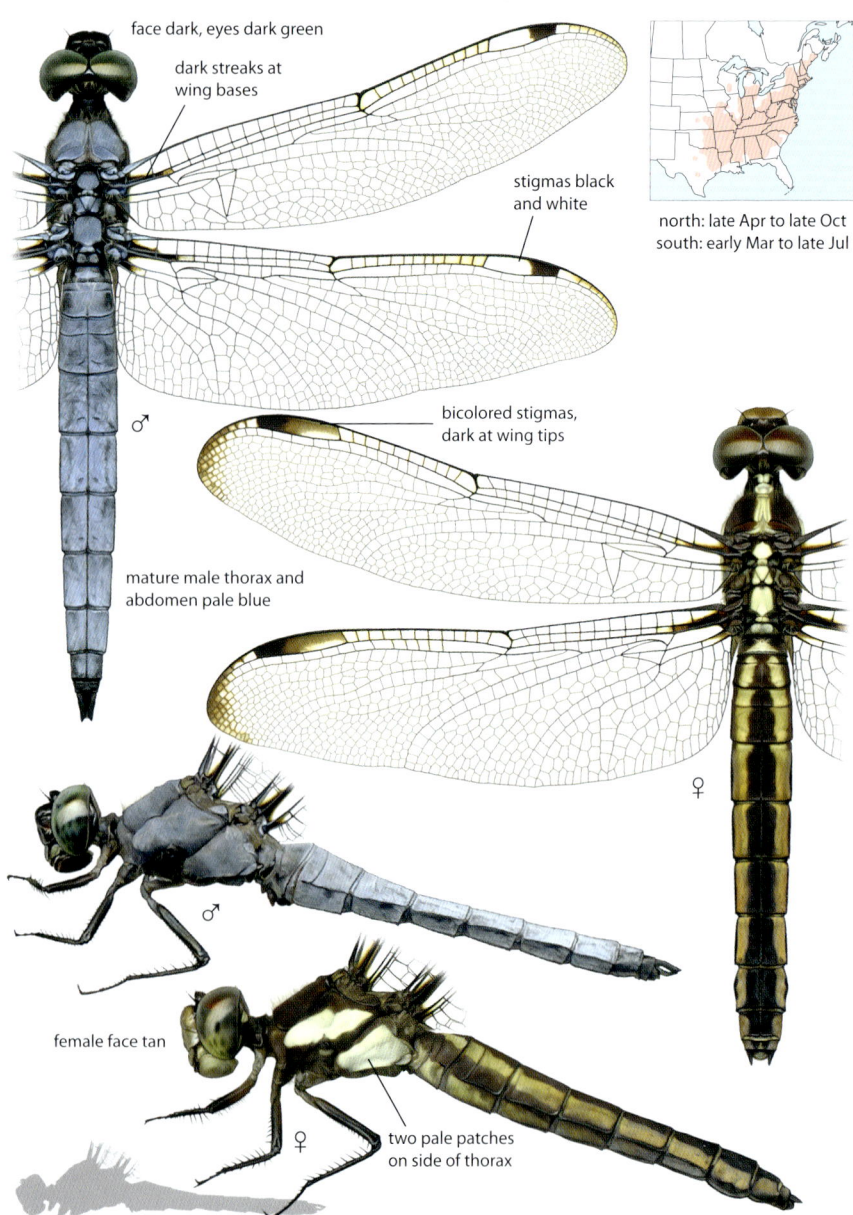

face dark, eyes dark green

dark streaks at wing bases

stigmas black and white

north: late Apr to late Oct
south: early Mar to late Jul

bicolored stigmas, dark at wing tips

♂

mature male thorax and abdomen pale blue

♀

♂

female face tan

♀

two pale patches on side of thorax

Similar species: Yellow-sided Skimmer has brown stigmas, mostly pale femurs. **Comanche Skimmer** has white face, paler blue eyes; both pale lateral patches extend to top margin of thorax. Lacks dark streaks at wing bases. Male **Eastern Pondhawk** has green face.

Comanche Skimmer • *Libellula comanche*

47–55 mm, 1.9–2.2 in. Pale southwestern skimmer inhabiting springs, seeps, and slow portions of streams. Both sexes with white and black stigmas, white face, gray-blue eyes. Mature male pale blue. Immature and female thorax brown with mostly pale sides; abdomen with yellow stripes.

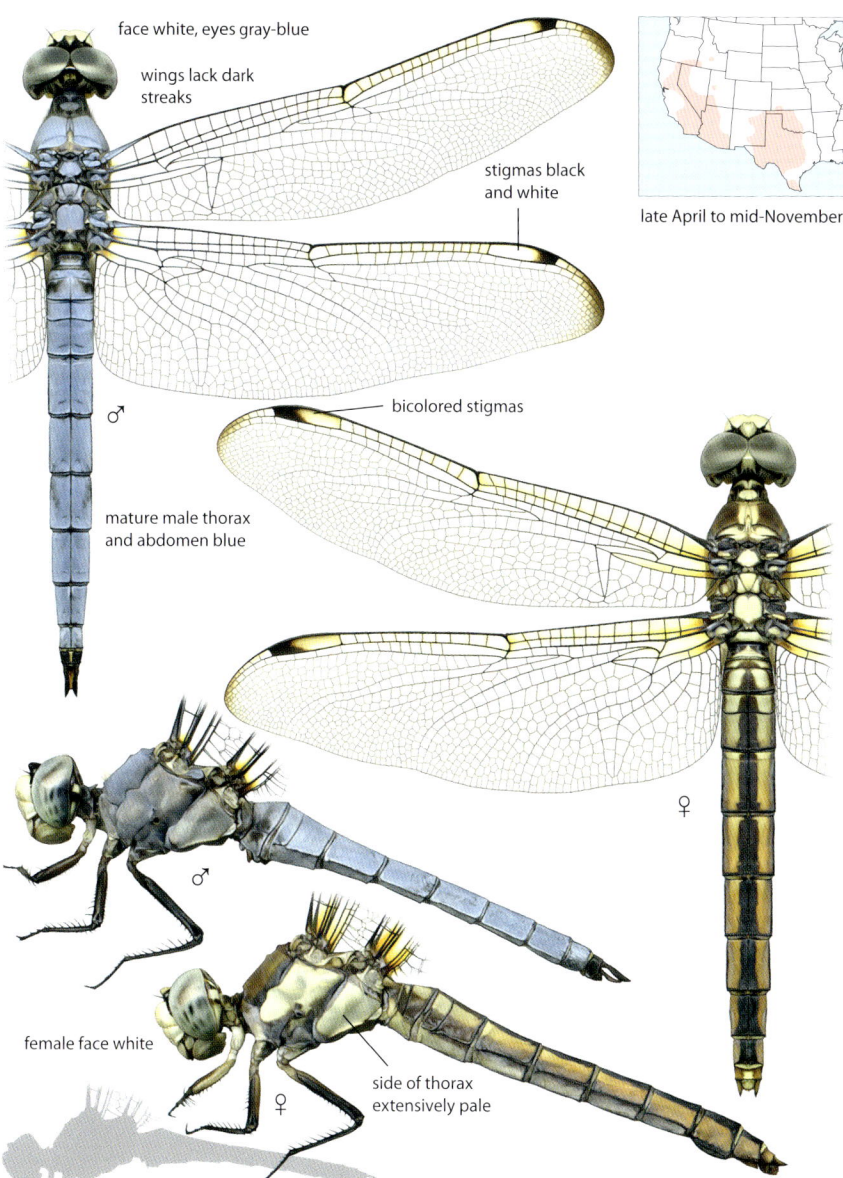

face white, eyes gray-blue

wings lack dark streaks

stigmas black and white

late April to mid-November

♂

bicolored stigmas

mature male thorax and abdomen blue

♀

♂

female face white

♀

side of thorax extensively pale

Similar species: Spangled Skimmer has dark streaks at base of wing; pale patch on side of thorax smaller. Face black in male, tan in female. **Yellow-sided Skimmer** has brown stigmas; male with dark face, female with large dark wing tips. **Bleached Skimmer** has brown/amber markings at base of wings, black stigmas.

Yellow-sided Skimmer • *Libellula flavida*

48–51 mm, 1.9–2 in. Pale southeastern skimmer inhabiting spring seeps, boggy ponds, slow streams. Immature and female side of thorax yellow with single dark stripe; stigma yellow, leading edge of wings amber, femurs mostly pale. Mature male pale blue; face and eyes dark, wings with dark streaks at base, stigma brown. Female wing tips dark.

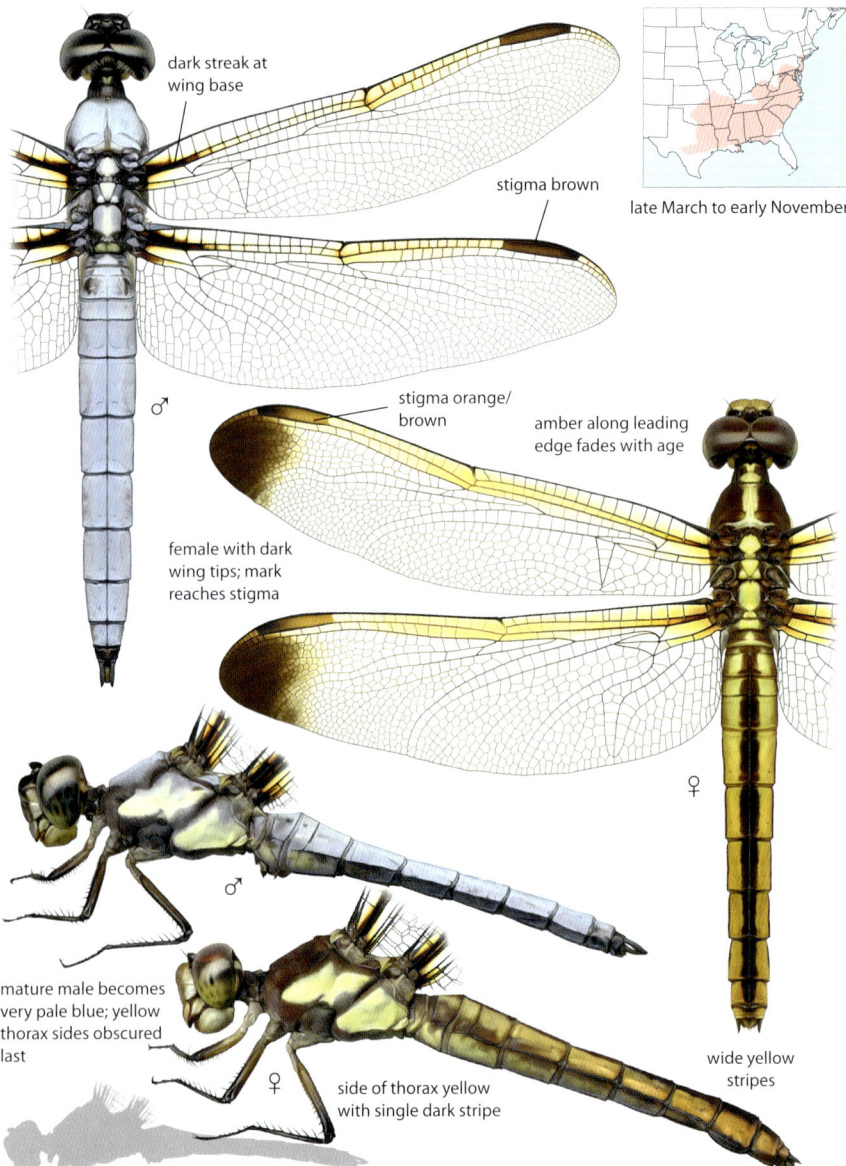

dark streak at wing base

stigma brown

late March to early November

stigma orange/ brown

amber along leading edge fades with age

female with dark wing tips; mark reaches stigma

♂

♀

mature male becomes very pale blue; yellow thorax sides obscured last

♂

♀

side of thorax yellow with single dark stripe

wide yellow stripes

Similar species: Spangled and **Comanche Skimmers** have bicolored stigmas; femurs dark. **Great Blue Skimmer** has white face, dark spot at nodus, black stigmas. Side of thorax more extensively pale. **Bar-winged** and **Slaty Skimmers** have black stigmas. Pale area on side of thorax extends forward of the first lateral suture; femurs black. Adult males much darker.

Great Blue Skimmer • *Libellula vibrans*

56–63 mm, 2.2–2.5 in. Large eastern skimmer inhabiting wooded swamps, slow forest streams, and temporary ponds. Face white. Wings with dark streak at base, small spot at nodus, dark tips. Stigma black. Mature male eyes, thorax and abdomen blue. Immature and female thorax brown, with extensive white sides; femurs mostly pale.

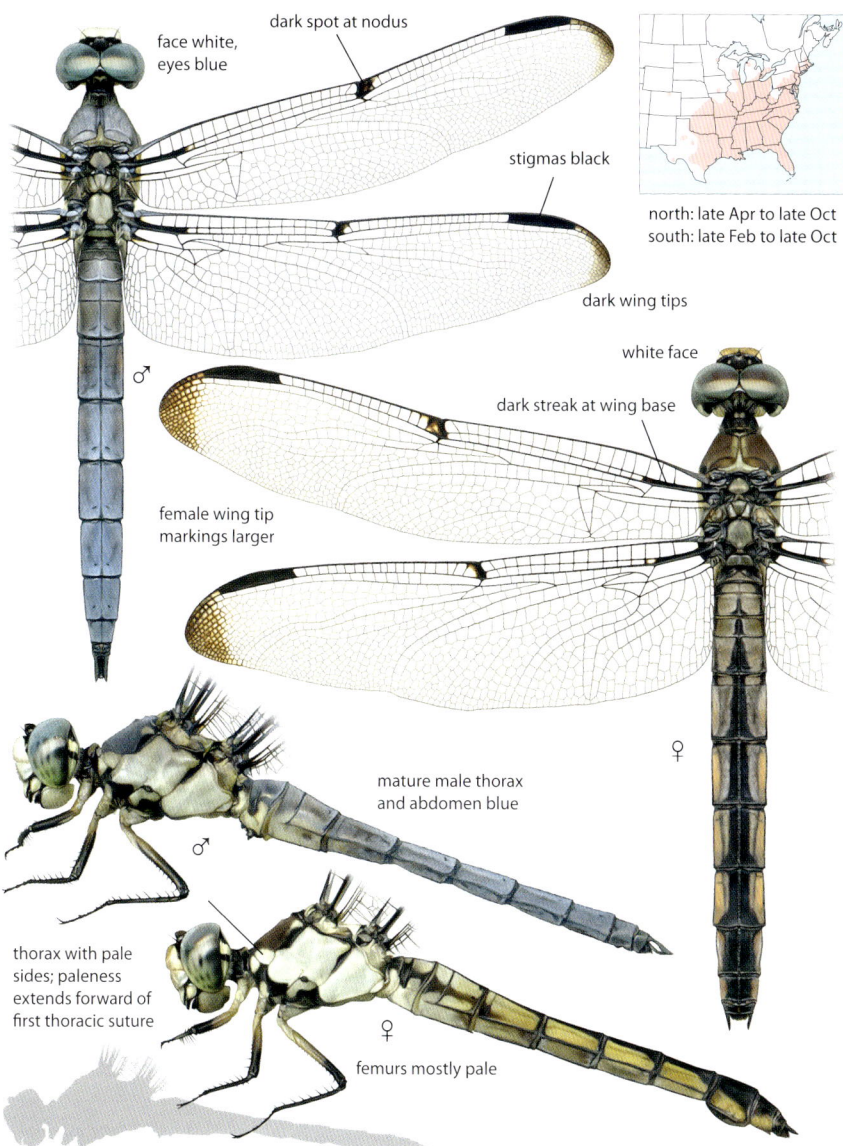

face white, eyes blue

dark spot at nodus

stigmas black

north: late Apr to late Oct
south: late Feb to late Oct

dark wing tips

white face

dark streak at wing base

female wing tip markings larger

mature male thorax and abdomen blue

♂

thorax with pale sides; paleness extends forward of first thoracic suture

♀

femurs mostly pale

Similar species: Bar-winged Skimmer side of thorax has dark triangle below forewing; femurs black. Adult male darker overall with dark face, white at base of hindwing. **Slaty Skimmer** has black femurs, darker face and eyes. **Yellow-sided Skimmer** stigmas brown. Pale area on side of the thorax does not extend forward of first thoracic suture; lacks white face and dark spot at nodus.

Slaty Skimmer • *Libellula incesta*

50–52 mm, 2 in. Common dark eastern skimmer found at wide variety of ponds, lakes, and slow streams. Mature male dark gray; face and eyes dark, wings unmarked, stigma black. Immature and female thorax dark brown with pale sides; some females darken like male. Female wing pattern variable, often with dark streaks at base and dark tips.

male wings largely unmarked

north: late May to early Oct
south: early Apr to early Nov

female wings usually with dark streak at base and dark tip

♂

mature male thorax and abdomen dark gray

female wing pattern variable, ranging from unmarked to having continuous dark stripe with wide dark tip

♀

♂

side of thorax pale; paleness extends forward of first thoracic suture

legs black   ♀

mature female brown, sometimes blackish like male

Similar species: Bar-winged Skimmer has similar wolf's-head pattern on side of thorax, but dark mark below hindwing larger and widens below; males have dark wing markings, with white at base of hindwing. **Great Blue Skimmer** has white face; femurs mostly pale, and side of thorax lacks dark triangle below forewing.

Bar-winged Skimmer • *Libellula axilena*

60–62 mm, 2.4 in. Large dark eastern skimmer found at forest pools, ditches, and slow woodland streams. Both sexes with thin dark streak at wing base, spot at nodus, dark streak or spot between nodus and stigma, dark wing tips. Mature male has black face; abdomen and top of thorax pruinose gray.

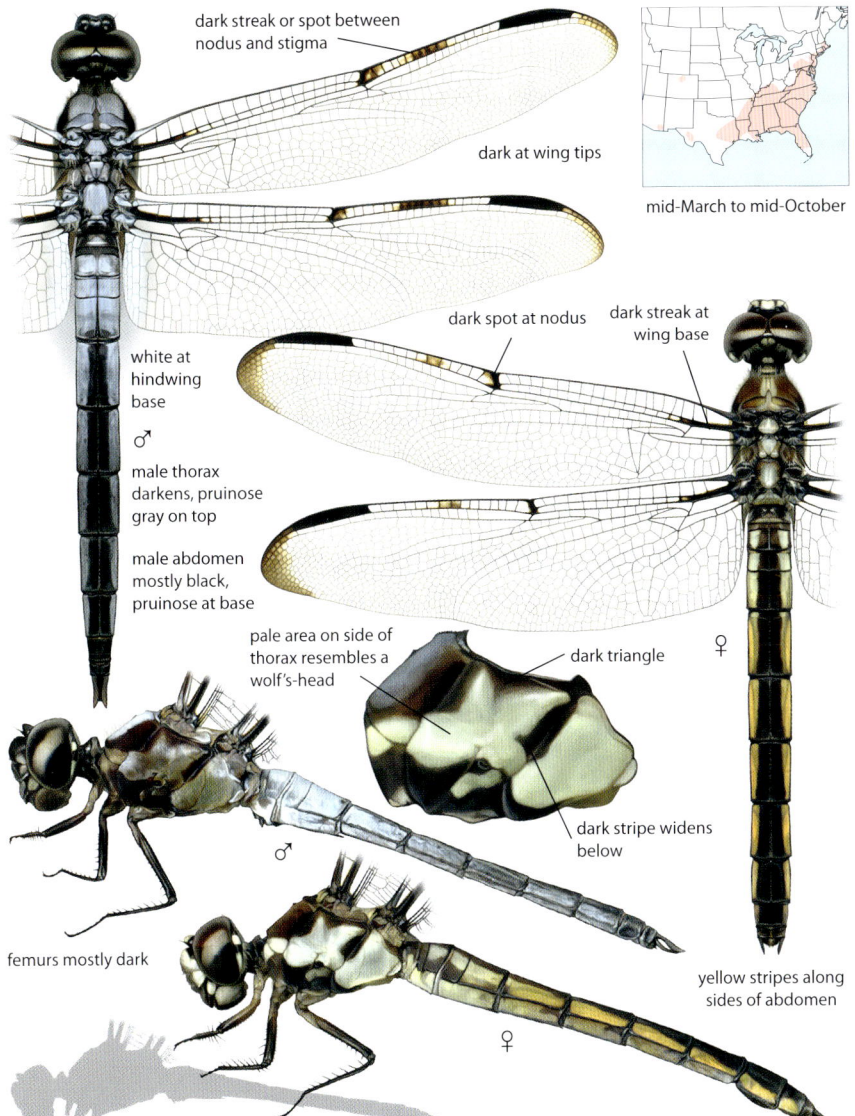

dark streak or spot between nodus and stigma

dark at wing tips

mid-March to mid-October

white at hindwing base

♂

male thorax darkens, pruinose gray on top

male abdomen mostly black, pruinose at base

dark spot at nodus

dark streak at wing base

♀

pale area on side of thorax resembles a wolf's-head

dark triangle

dark stripe widens below

♂

femurs mostly dark

yellow stripes along sides of abdomen

♀

Similar species: Slaty Skimmer male uniformly dark gray, lacks dark streaks at wing bases and white at base of hindwing. Female has less contrasting face pattern, narrow dark mark on side of thorax under hindwing, and often has larger dark wing tips. **Great Blue Skimmer** has white face, lacks dark marking between nodus and stigma. Mature male with blue eyes; abdomen and thorax uniformly blue. Immature and female side of thorax pale, lacking dark triangle below forewing; femurs mostly pale.

Purple Skimmer • *Libellula jesseana*

50–57 mm, 2–2.2 in. Uniquely colored king skimmer local to some clear, white-sand-bottomed lakes in Florida. No other species has blue body with orange wings. Immature and most females resemble Golden-winged Skimmer, with brown thorax and yellow-striped abdomen. Some females become blue, like mature male.

stigma orange

wings orange-tinted

early May to mid-September

♂

mature male thorax
and abdomen blue

♀

female wings vary from orange-tinted like
male to clear (but with some orange veins)

♂

femurs brown

♀

some females blue like male,
but usually brown/yellow

Similar species: Mature male unmistakeable. Blue females with clear wings similar to female **Slaty Skimmer** but have orange stigma and at least some orange veins. Immatures and brown/yellow females identical to **Golden-winged Skimmer** but may show darker-edged stigma.

Painted Skimmer • *Libellula semifasciata*

45–48 mm, 1.8–1.9 in. Brightly patterned eastern skimmer found at vegetated woodland ponds and slow streams. Thorax pale brown with white lateral stripe. Wing pattern distinctive: amber patch with dark streaks at base, brown spot at nodus, brown stripe below brown to orange stigma. Mature male wing veins red, making wings appear orange.

amber wing bases

brown spot at nodus

stigma dark orange

brown stripe below stigma

north: mid-Apr to late Sep
south: late Feb to late Oct

hindwing base with two dark streaks

♂

female stigma brown

♀

♂

thorax brown with white anterior stripe, yellow posterior stripe

abdomen brown with yellow-orange lateral stripes

♀

Similar species: Four-spotted Skimmer has large dark triangular marking at base of hindwing, less amber at base of wings. Usually lacks brown stripe below stigma; stigma black. Lacks white stripe on side of thorax. **Calico Pennant** smaller, with pale, heart-shaped markings on top of slenderer abdomen. **Halloween Pennant** wings completely amber.

Golden-winged Skimmer • *Libellula auripennis*

51–58 mm, 2–2.3 in. Brightly colored eastern skimmer inhabiting grassy ponds and lakes. Immature and female thorax brown with vague pale stripes, abdomen yellow with dark middorsal stripe, stigma yellow. Costal vein uniformly pale from wing base to stigma. Hind leg tibia black. Mature male abdomen orange, most wing veins and stigma orange.

costal vein single pale color

stigma orange

wings orange-tinted

early March to mid-October

♂

male has most veins orange

mature male abdomen orange

stigma orange

female wings lightly tinted

♀

♂

side of thorax pale; paleness does not extend forward of first thoracic suture

♀

hind leg tibia black

Similar species: Needham's Skimmer has dark costal vein between wing base and nodus (often more noticeable on female). Pale side of thorax unstriped and extends forward of first thoracic suture; hind leg tibia brown. Mature male redder on front of thorax and abdomen; wing veins orange only along front of wing. Yellow-form female **Purple Skimmer** usually not separable, but border of stigma may be darker.

Needham's Skimmer • *Libellula needhami*

53–56 mm, 2.1–2.2 in. Brightly colored, mainly coastal skimmer inhabiting lakes, marshes, canals, tidal rivers, often brackish. Wings orange-tinted. Paleness on side of thorax extends forward of first thoracic suture. Costal vein darker between wing base and nodus, pale beyond nodus. Hind leg tibia brown. Mature male abdomen and thorax front orange-red.

costal vein changes color at nodus, darker toward wing base (often obscured on mature male)

stigmas orange

wings orange-tinted

early March to mid-November

♂

mature male front of thorax and abdomen red-orange

orange veins only along leading edge; most veins dark

♀

♂

pale side of thorax extends forward of first suture

♀

female may be yellow or orange

femurs pale, hind leg tibia brown

Similar species: Golden-winged Skimmer costal vein uniformly pale from base of wing to stigma, and pale area on side of thorax does not extend forward of first thoracic suture; hind leg tibia black. Mature male with mostly orange wing veins.

Four-spotted Skimmer • *Libellula quadrimaculata*

42–46 mm, 1.7–1.8 in. Common, widespread northern skimmer found at mud-bottomed ponds and lakes, particularly acidic bog ponds. Dull yellow-brown overall; tapering abdomen with pale yellow lateral dashes and black tip. Wings with small dark spot at nodus, amber at bases, large dark triangular patch at hindwing base. Stigma black.

small dark mark at each nodus

mid-April to early October

stigmas black

♂

tapered abdomen brown with yellow lateral dashes and black tip

form *praenubilia* has dark band below stigma; rare

base of hindwing with amber and dark triangular marking

♀

♂

thorax yellow-brown

♀

Similar species: Painted Skimmer more orange overall, and has different wing pattern with more amber at base, two dark streaks at hindwing base, larger brown markings at nodus, brown stripe under stigma. **Hoary Skimmer** has wide dark streak at base of all wings. Female **Bleached Skimmer** has side of thorax pale, pale stripes on top of all abdominal segments.

Red-mantled Skimmer • *Libellula gaigei*

51–53 mm, 2–2.1 in. Brightly colored skimmer straying into South Texas. Mature male face, top of thorax, and abdomen bright red. Both sexes with large, well-defined amber patches at base of wings, reaching nodus only along leading edge. Immature and female brown, female with brown wing tips; some females become bright red, like the male.

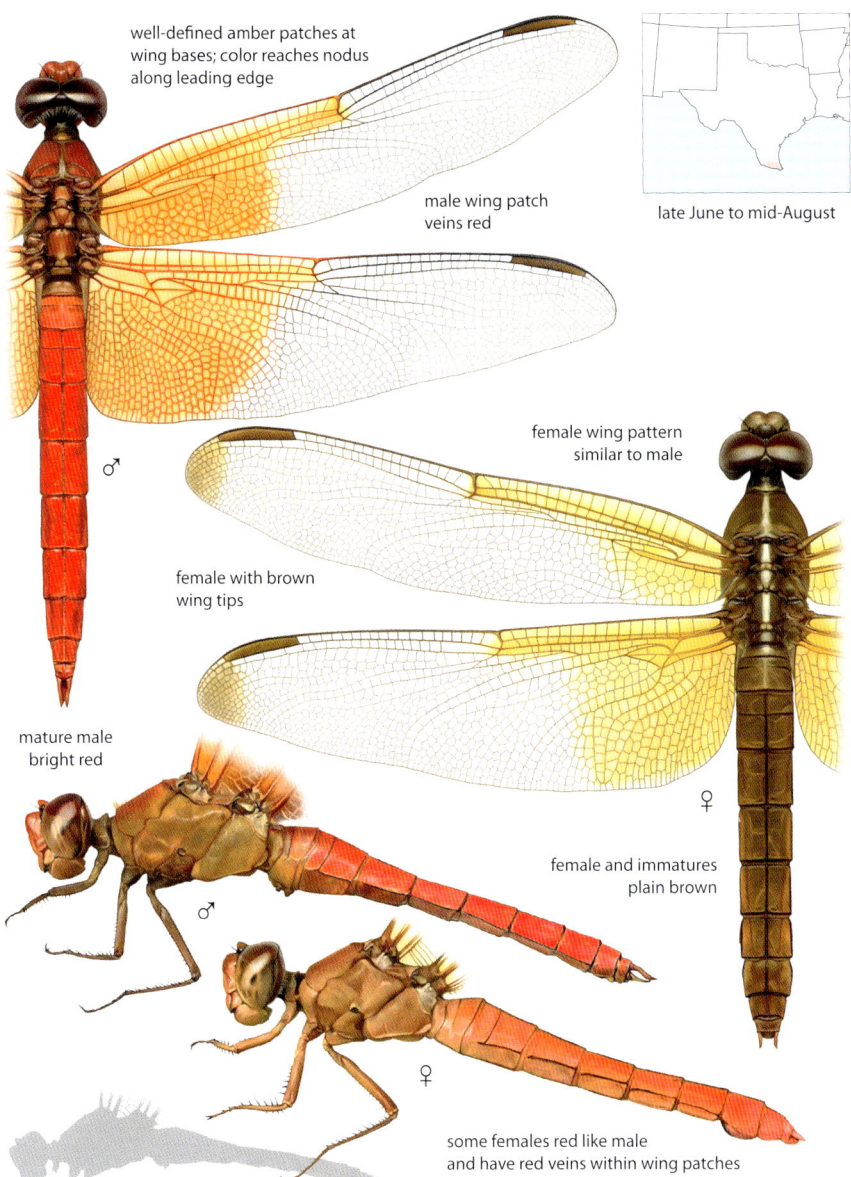

well-defined amber patches at wing bases; color reaches nodus along leading edge

male wing patch veins red

late June to mid-August

female wing pattern similar to male

female with brown wing tips

mature male bright red

female and immatures plain brown

some females red like male and have red veins within wing patches

Similar species: Flame Skimmer has wider abdomen. Male wing patches larger, with dark streaks at base; female with less defined wing markings and amber coloration extending along leading edge to stigma, lacks dark tips. **Neon Skimmer** has much wider abdomen. Male wing patch smaller and less defined; female lacks wing coloration.

Flame Skimmer • *Libellula saturata*

52–61 mm, 2–2.4 in. Common showy western skimmer found at ponds, lakes, clean streams, ditches, and hot springs. Mature male with red face, red-brown thorax, orange-red abdomen; wings orange from base to nodus. Female and immature unmarked brown. Female wings can be similar to male but are typically less amber.

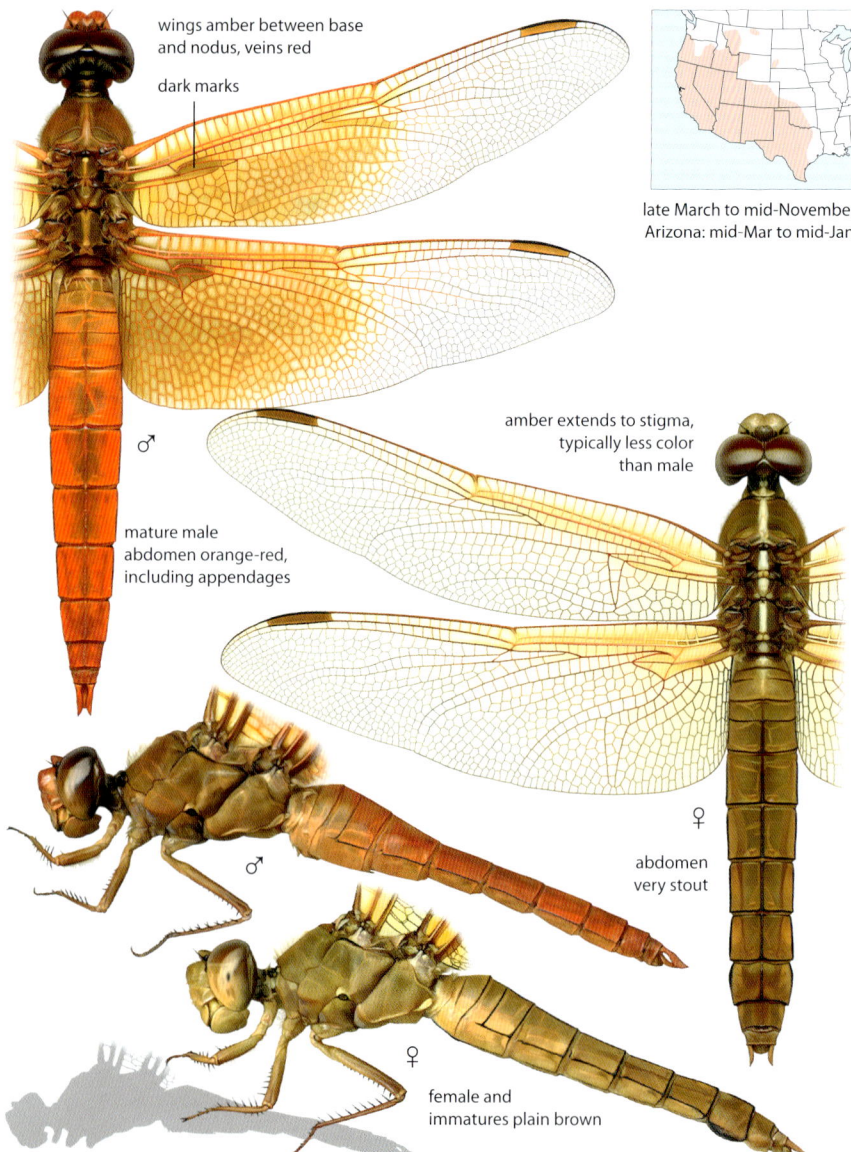

wings amber between base and nodus, veins red

dark marks

♂

mature male abdomen orange-red, including appendages

late March to mid-November
Arizona: mid-Mar to mid-Jan

amber extends to stigma, typically less color than male

♀

abdomen very stout

♂

♀

female and immatures plain brown

Similar species: Neon Skimmer male has less amber in wings (extending to nodus only along leading edge), lacks dark streaks at wing bases; front of thorax bright red. Female has clear wings; flange on side of S8 much larger. **Red-mantled Skimmer** has narrower abdomen. Male wing patches reach nodus only along leading edge of wing; lacks dark streaks at wing bases. Female wing pattern similar but has dark wing tips.

Neon Skimmer • *Libellula croceipennis*

55–57 mm, 2.2 in. Brightly colored southwestern skimmer inhabiting small, clean streams and ditches. Mature male face, front of thorax, and abdomen bright red; wings amber at base, with amber extending to nodus only along leading edge. Immature and female plain brown. Female notably heavy-bodied with wings clear.

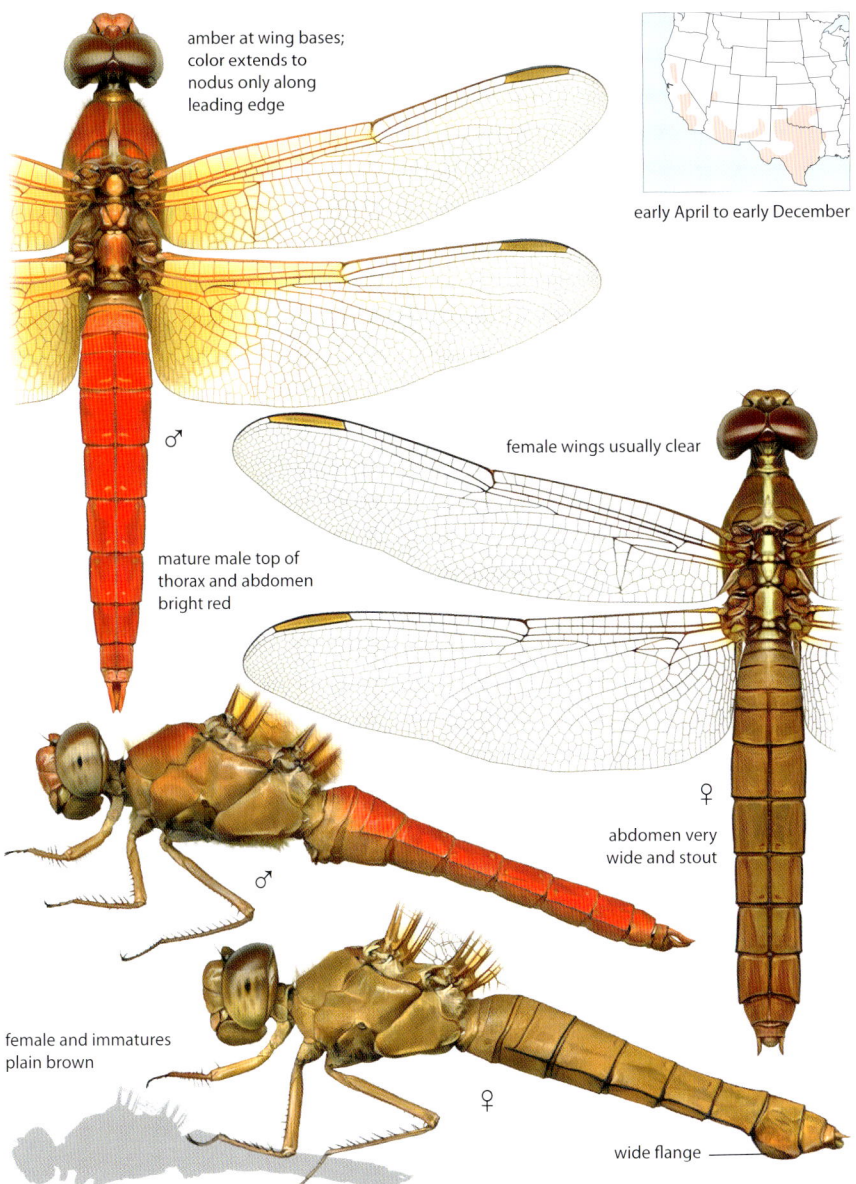

amber at wing bases; color extends to nodus only along leading edge

early April to early December

female wings usually clear

mature male top of thorax and abdomen bright red

abdomen very wide and stout

♂

♀

female and immatures plain brown

♂

♀

wide flange

Similar species: Flame Skimmer male has entire inner half of wing orange with dark streaks at base; front of thorax brown, abdomen orange-red. Female with some amber coloration in wings; lateral flange on S8 smaller. **Red-mantled Skimmer** has well-defined wing patches; abdomen narrower.

Roseate Skimmer • *Orthemis ferruginea*

52–55 mm, 2–2.2 in. Large, pink southern skimmer common at a wide variety of still-water habitats, including marginal ones. Mature male top of face metallic purple, eyes dark purple, thorax purple, abdomen pink, including appendages. Immature and female brown with white lateral thoracic markings.

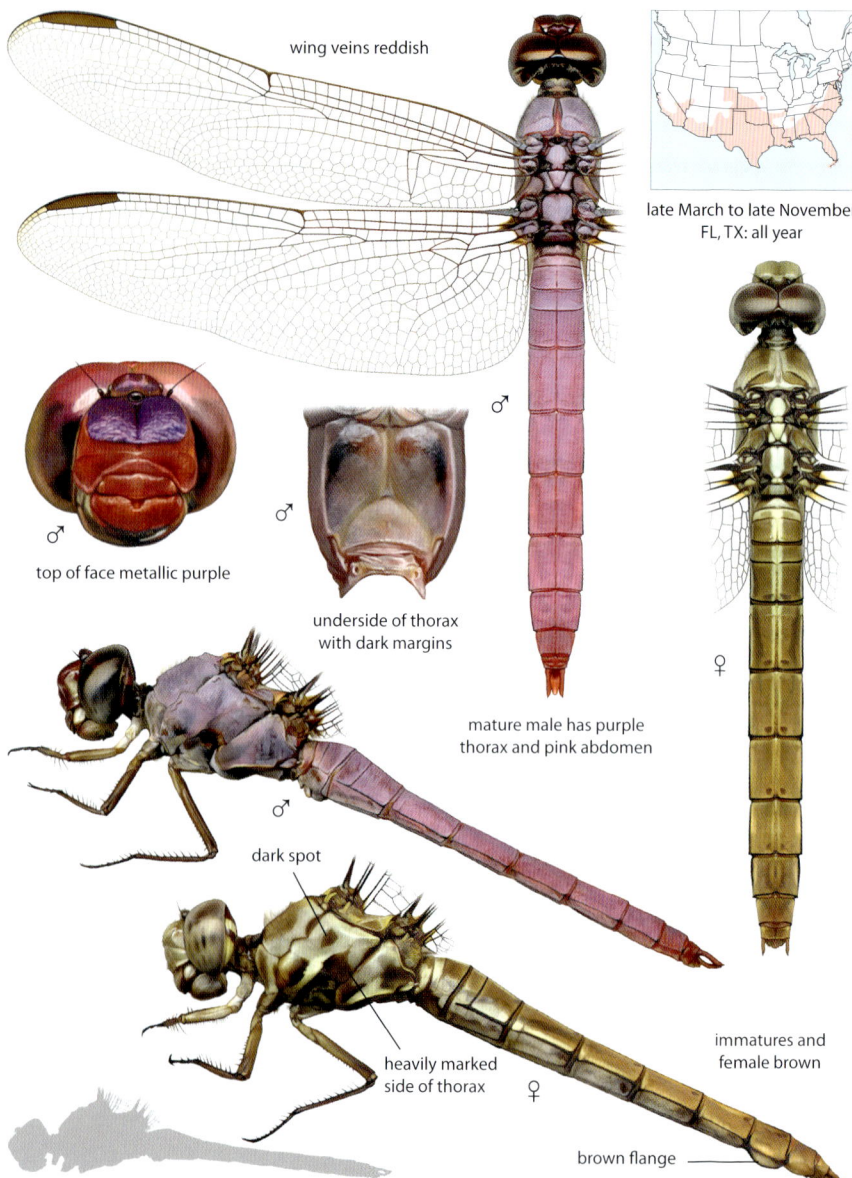

wing veins reddish

late March to late November
FL, TX: all year

♂

♂

top of face metallic purple

♂

underside of thorax
with dark margins

♀

mature male has purple
thorax and pink abdomen

♂

dark spot

immatures and
female brown

heavily marked
side of thorax ♀

brown flange

Similar species: Carmine Skimmer male has uniformly red face; eyes, thorax, and abdomen redder. Female thorax with subtle pale parallel lateral stripes; flange on S8 dark.
Antillean Skimmer male has uniformly red face, red abdomen; female has narrow pale lateral thoracic stripes.

Carmine Skimmer • *Orthemis discolor*

38–53 mm, 1.5–2.1 in. Large, reddish-pink skimmer found at muddy woodland ponds and slow streams; southwestern, primarily southern Texas. Mature male face uniformly red, eyes red-purple, thorax red-purple, abdomen rose red. Immature and female brown; side of thorax lightly marked with pale parallel stripes.

wing veins black

early March to early January

face uniformly red

underside of thorax uniformly brown

mature male has red-purple thorax, rose-colored abdomen; typically redder than Roseate Skimmer

side of thorax with subtle parallel pale stripes

female and immatures brown

S8 flange dark

Similar species: Roseate Skimmer male has purple and red face, dark purple eyes, abdomen more pale pink; female with more contrasting lateral thoracic pattern, flange on side of S8 brown.

Antillean Skimmer • *Orthemis species*

48–54 mm, 1.9–2.1 in. Large, red skimmer, local to South Florida ponds and lakes. Mature male with red face, purple thorax, and red abdomen. Female brown with narrow pale lateral thoracic stripes. Once considered a form of Roseate Skimmer, it is nearly identical to the South American species *Orthemis schmidti,* but its taxonomy remains unresolved.

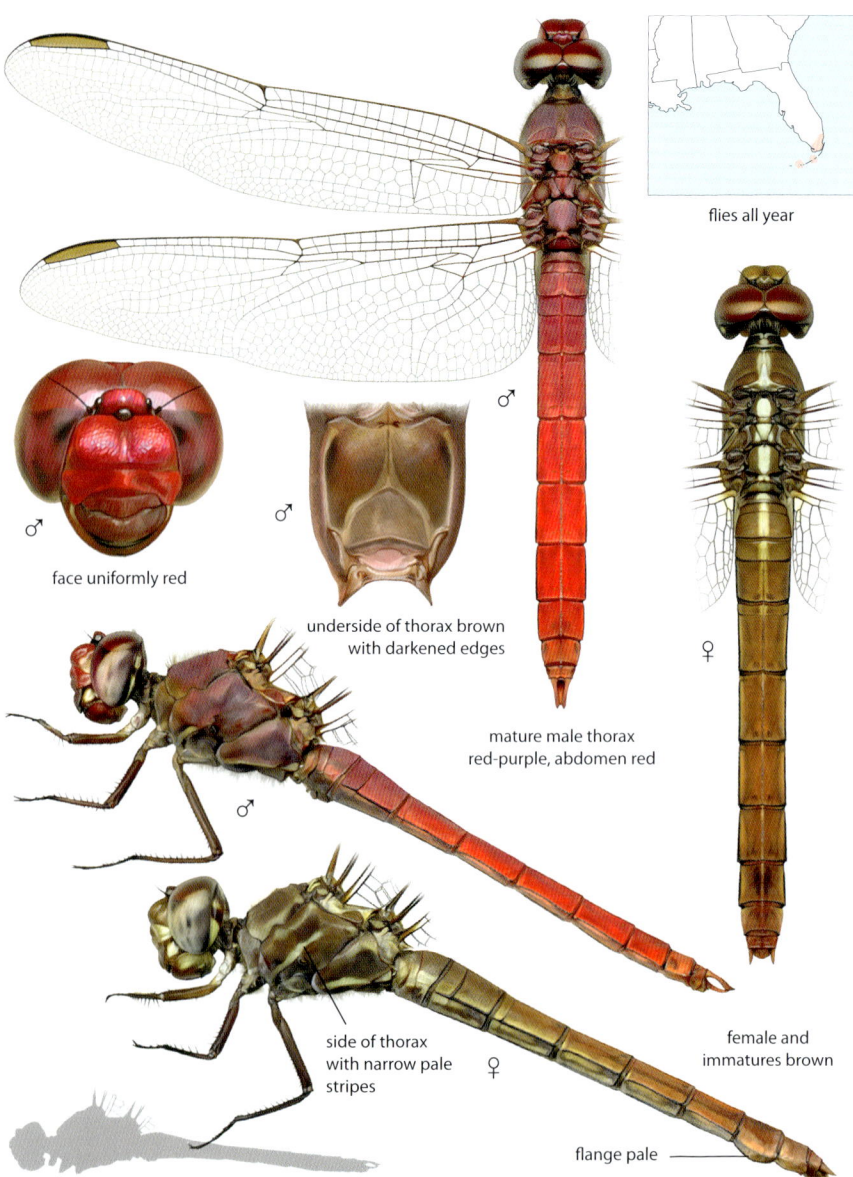

flies all year

♂

face uniformly red

♂

underside of thorax brown with darkened edges

♂

mature male thorax red-purple, abdomen red

♀

side of thorax with narrow pale stripes

♀

female and immatures brown

flange pale

Similar species: Roseate Skimmer male has metallic purple and red face, dark purple eyes, abdomen pink. Female with wider, more contrasting pale lateral thoracic stripes. **Scarlet Skimmer** smaller, has dark stripe on top of abdomen, amber spot at base of hindwing, paler legs. Female with side of thorax yellow, lacks flange on side of S8.

Scarlet Skimmer • *Crocothemis servilia*

37–43 mm, 1.5–1.7 in. Medium-sized, bright red skimmer introduced from Asia, common at southern Florida lakes and canals. Abdomen with narrow black dorsal stripe. Hindwing with amber spot at base. Legs pale. Mature male all red. Immature and female yellow with pale shoulder stripes.

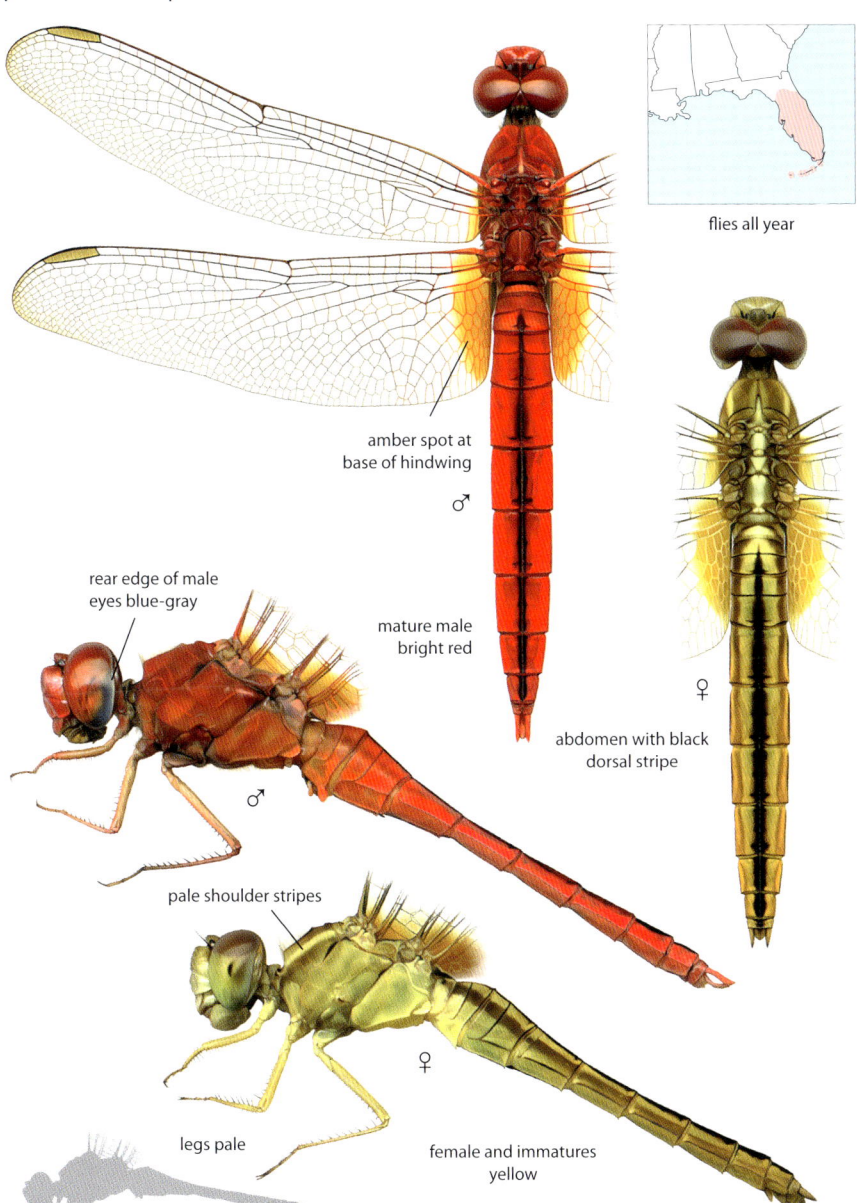

flies all year

amber spot at
base of hindwing

♂

mature male
bright red

rear edge of male
eyes blue-gray

♂

abdomen with black
dorsal stripe

♀

pale shoulder stripes

♀

legs pale

female and immatures
yellow

Similar species: Red-tailed Pennant has brown thorax, no pale shoulder stripes; abdomen lacks black dorsal stripe. **Antillean Skimmer** larger, lacks dorsal black stripe on abdomen and amber hindwing patches.

Gray-waisted Skimmer • *Cannaphila insularis*

37–44 mm, 1.5–1.7 in. Narrow-winged skimmer found at marshes, swamps, and slow shaded streams, primarily in southern Texas. Hindwing narrows at base. Thorax dark with metallic reflections and contrasting pale yellow stripes. Abdomen tapering. Immatures brown, black when mature; male develops gray pruinosity at base. Male eyes blue-green.

top of face dark, male eyes blue-green

often with dark wing tips

early March to early October

base of hindwing narrow

♂

tapered black abdomen, pruinose at base

thorax dark with metallic reflections, contrasting pale yellow stripes

♀

♂

female abdomen may be black-striped or mostly brown

♀

♀

large flange side of S8

Similar species: Black Setwing has thinner abdomen; mature male face all dark, lacks pale thoracic stripes. **Blue Dasher** smaller, with white face; dark lateral thoracic stripes more separated. Immature and female with pale dashes on S2-7; female lacks abdominal flange.

Great Pondhawk • *Erythemis vesiculosa*

56–59 mm, 2.2–2.3 in. Our largest pondhawk, found at a wide variety of still-water habitats in the South, particularly marshy ponds. Both sexes with bright green face and solid green thorax. Abdomen long and slender, widely inflated at the base in side view, green with black bands. Femurs and appendages pale.

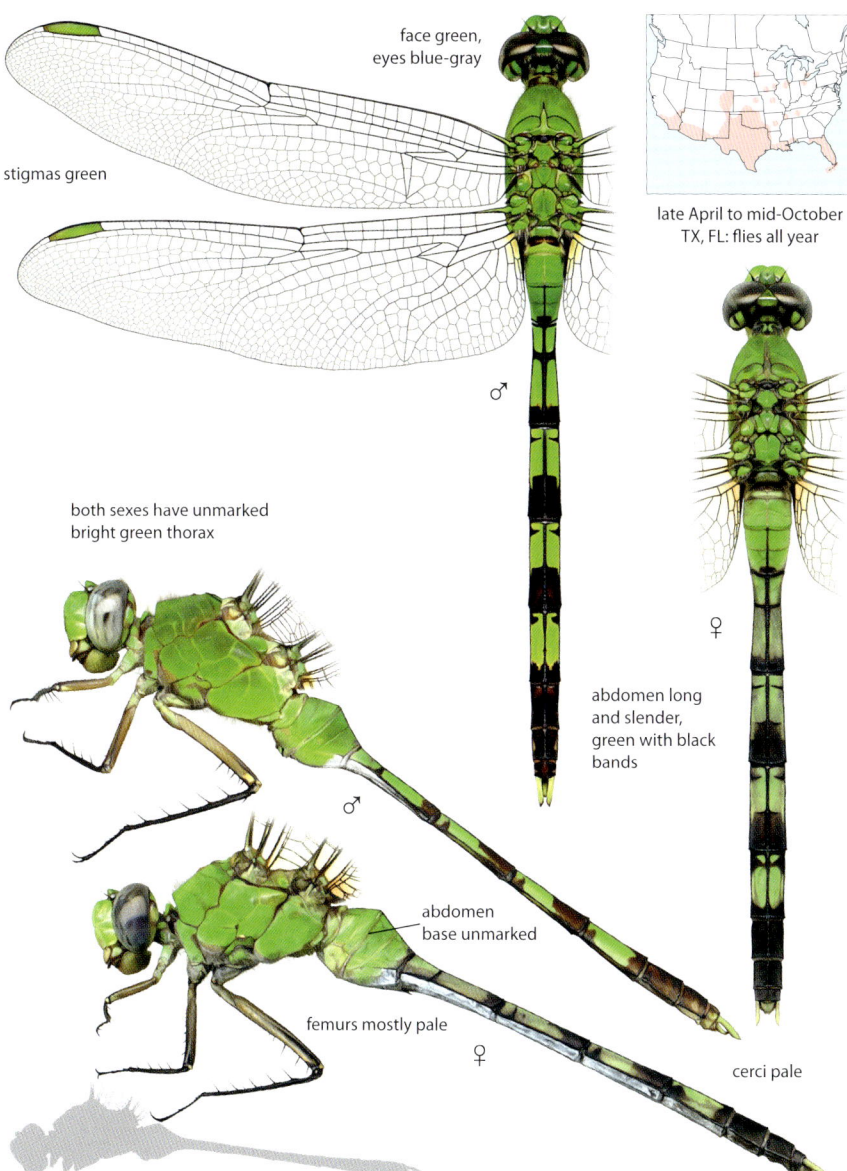

face green,
eyes blue-gray

stigmas green

late April to mid-October
TX, FL: flies all year

♂

both sexes have unmarked
bright green thorax

♀

abdomen long
and slender,
green with black
bands

♂

abdomen
base unmarked

femurs mostly pale

♀

cerci pale

Similar species: Eastern and **Western Pondhawks** smaller with shorter, wider abdomens. Mature males blue, females with fine dark lines at base of abdomen and spout-like subgenital plates. **Darners with green thoraxes** lack wide green bands on the abdomen, are fliers instead of perchers, and hang vertically at rest.

Eastern Pondhawk • *Erythemis simplicicollis*

38–44 mm, 1.5–1.7 in. Common to abundant eastern pondhawk at most still-water habitats with emergent vegetation. Face and eyes green. Immature and female green with black abdominal bands; abdomen pattern regionally variable. Mature male thorax and abdomen pruinose blue, combination of green and blue during transition. Cerci pale.

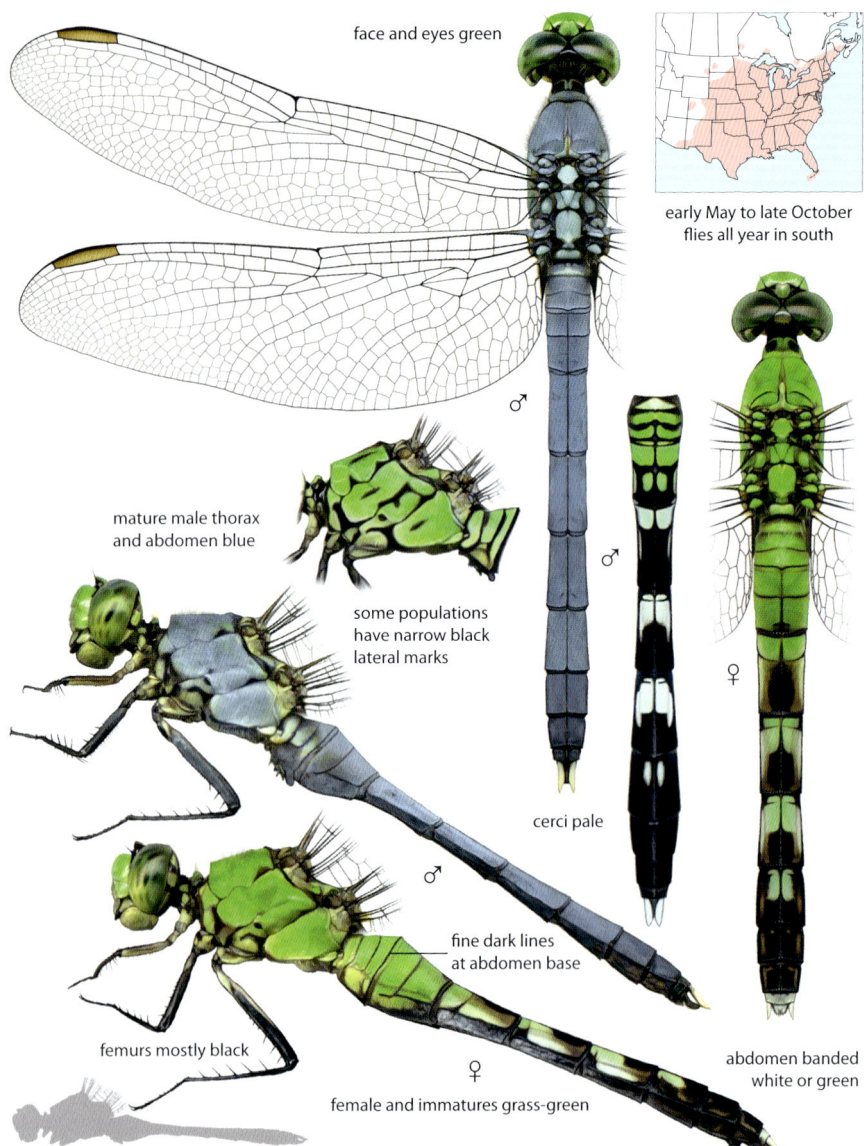

face and eyes green

early May to late October
flies all year in south

mature male thorax
and abdomen blue

some populations
have narrow black
lateral marks

cerci pale

fine dark lines
at abdomen base

femurs mostly black

abdomen banded
white or green

female and immatures grass-green

Similar species: Western Pondhawk with slightly wider abdomen; mature male with dark cerci, female with dark middorsal stripe on abdomen. Intermediates in pattern and abdomen shape suggest that Eastern and Western Pondhawks hybridize or may be the same species. **Great Pondhawk** larger, with longer, narrow abdomen. Stigmas green; female lacks projecting subgenital plate. **Blue Dasher** has white face, striped thorax, dark cerci.

Western Pondhawk • *Erythemis collocata*

40–42 mm, 1.6–1.7 in. Common western pondhawk found at a wide variety of still and slow-running water, including ponds, marshes, slow streams, canals, and hot spring streams. Similar to Eastern Pondhawk, but abdomen broader; male with blue eyes, dark cerci. Female abdomen green and tan with black middorsal stripe, femurs pale.

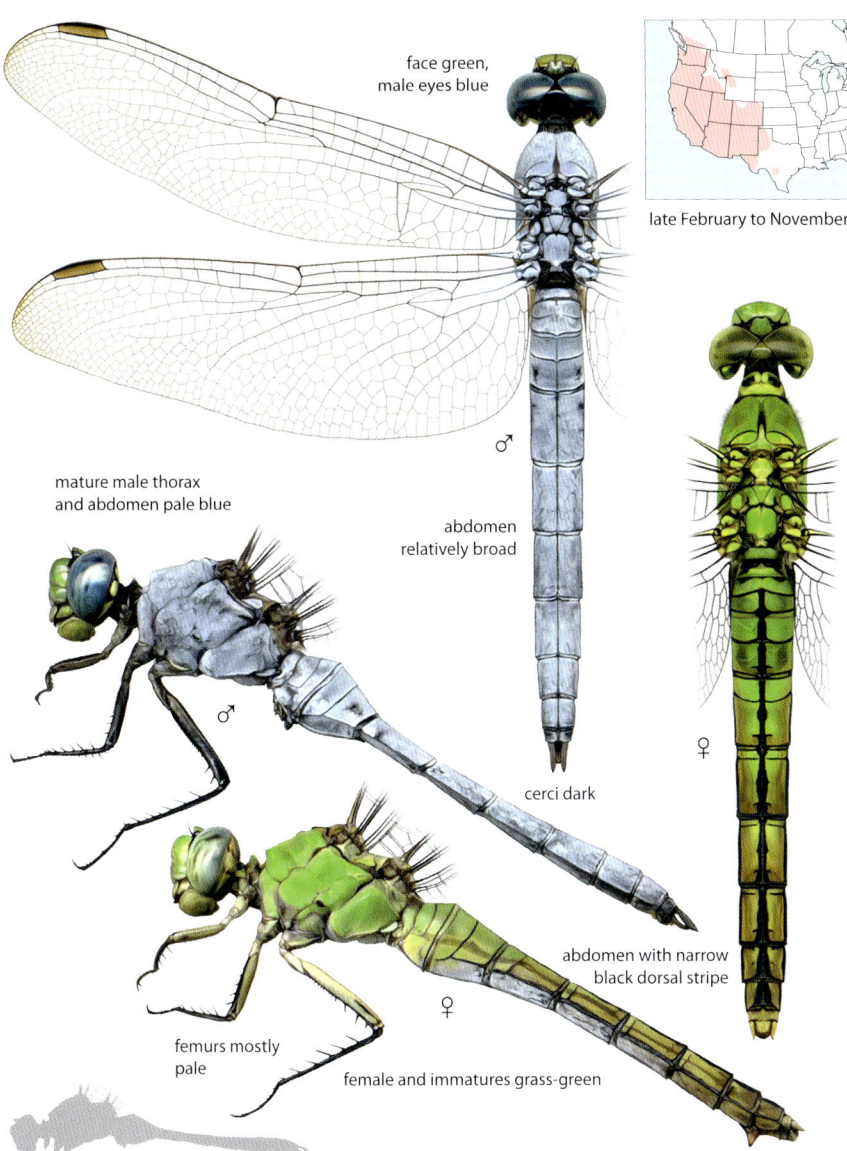

face green,
male eyes blue

late February to November

mature male thorax
and abdomen pale blue

abdomen
relatively broad

♂

abdomen
relatively broad

cerci dark

♀

abdomen with narrow
black dorsal stripe

femurs mostly
pale

female and immatures grass-green

♂

♀

Similar species: Eastern Pondhawk abdomen narrower. Mature male with green eyes, pale cerci; female with black bands on abdomen, middle and hind leg femurs dark. **Great Pondhawk** larger, abdomen longer and narrower, stigma green; female lacks projecting subgenital plate. **Blue Dasher** has white face, black back of head, amber markings at base of wings. **Comanche Skimmer** has white face, black-and-white stigmas.

Black Pondhawk • *Erythemis attala*

42–44 mm, 1.7 in. Dark-bodied southern pondhawk inhabiting heavily vegetated ponds, swamps, and slow streams. Mature male all black, including face and eyes. Female brown; abdomen with large pale squarish spots that darken with age. Hindwing with dark rounded patch, may be smaller, amber, or absent on female. Appendages tan.

face and eyes dark

late April to early November

large dark
hindwing patch

♂

mature male all black

female hindwing
markings variable

♀

♂

♀

female thorax
plain brown

♀

squarish pale markings
darken with age

Similar species: Pin-tailed Pondhawk has much narrower abdomen. **Band-winged Dragonlet** male slenderer; wings with dark bands between nodus and stigma. **Four-spotted Pennant** male has longer, slenderer abdomen, white stigmas, dark wing spots between stigma and nodus. **Marl Pennant** has slimmer abdomen; perches at tip of stems, instead of flat on the ground like the pondhawk.

Pin-tailed Pondhawk • *Erythemis plebeja*

42–47 mm, 1.7–1.9 in. Southern pondhawk found at open vegetated ponds, marshes, canals. Similar to Black Pondhawk, but both sexes with very slender abdomen. Mature male all black. Female and immature brown; thorax with dark shoulder stripes, abdomen with black bands. Some females darken with age. Hindwing with dark patch at base. Cerci tan.

flies all year

dark hind-wing patch

♂

mature male all black

♂

unmarked side of thorax

♀

mature dark female

♀

both sexes have narrow abdomen, pale cerci

♀

wide pale dorsal abdominal markings

Similar species: Black Pondhawk has wider abdomen. **Black Setwing** lacks dark hind-wing patch; immature and female have lateral thoracic stripes. **Band-winged Dragonlet** has wider abdomen, mature male with dark wing bands.

Claret Pondhawk • *Erythemis mithroides*

36–46 mm, 1.4–1.8 in. Brightly colored southern pondhawk inhabiting vegetated swamps. Mature male face red, eyes and thorax deep red-brown, abdomen bright red, including appendages; hindwing with dark brown band at base. Immature and female unmarked brown, with reddish abdomen, amber patch at base of hindwing. Legs brown.

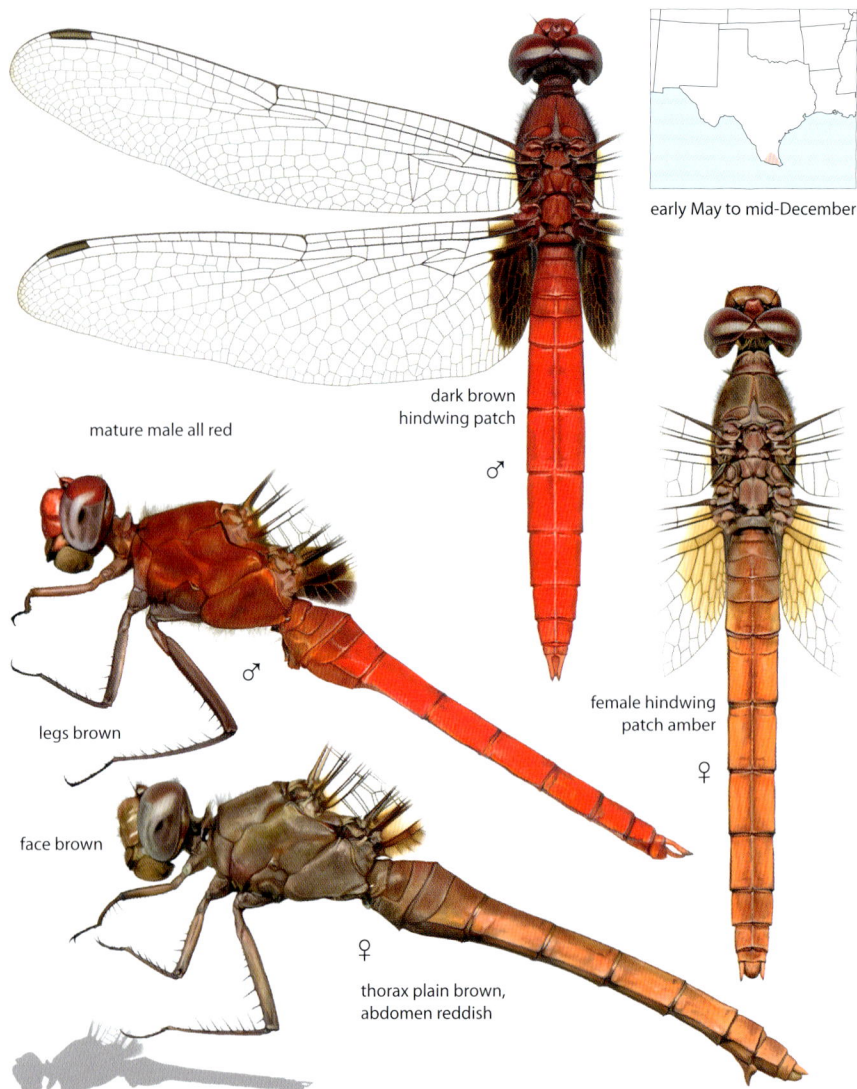

early May to mid-December

mature male all red

dark brown hindwing patch

♂

♂

legs brown

face brown

♀

female hindwing patch amber

♀

thorax plain brown, abdomen reddish

Similar species: Red-tailed Pennant has dark legs, smaller amber hindwing patch; male thorax brown. **Mexican Scarlet-tail** female has long black legs, large brown patch at base of hindwing; abdomen shorter and wider. **Flame** and **Neon Skimmers** lack dark hindwing patch; females have large lateral flange on S8. **Carmine Skimmer** has base of hindwing clear; mature male more rose-purple, female with large lateral flange on S8. **Arch-tipped Glider** is a flier, with narrower abdomen constricted at S3-4. **Saddlebags** species have black markings on S8-9, long appendages and legs, and are fliers.

Flame-tailed Pondhawk • *Erythemis peruviana*

36–43 mm, 1.4–1.7 in. Brightly marked pondhawk inhabiting heavily vegetated ponds and marshes; single Texas record. Mature male face and eyes dark, thorax and abdomen base deep pruinose blue, S4-10 and appendages bright red. Immature and female brown; thorax with dark shoulder and pale middorsal stripe, abdomen tan with brown dashes.

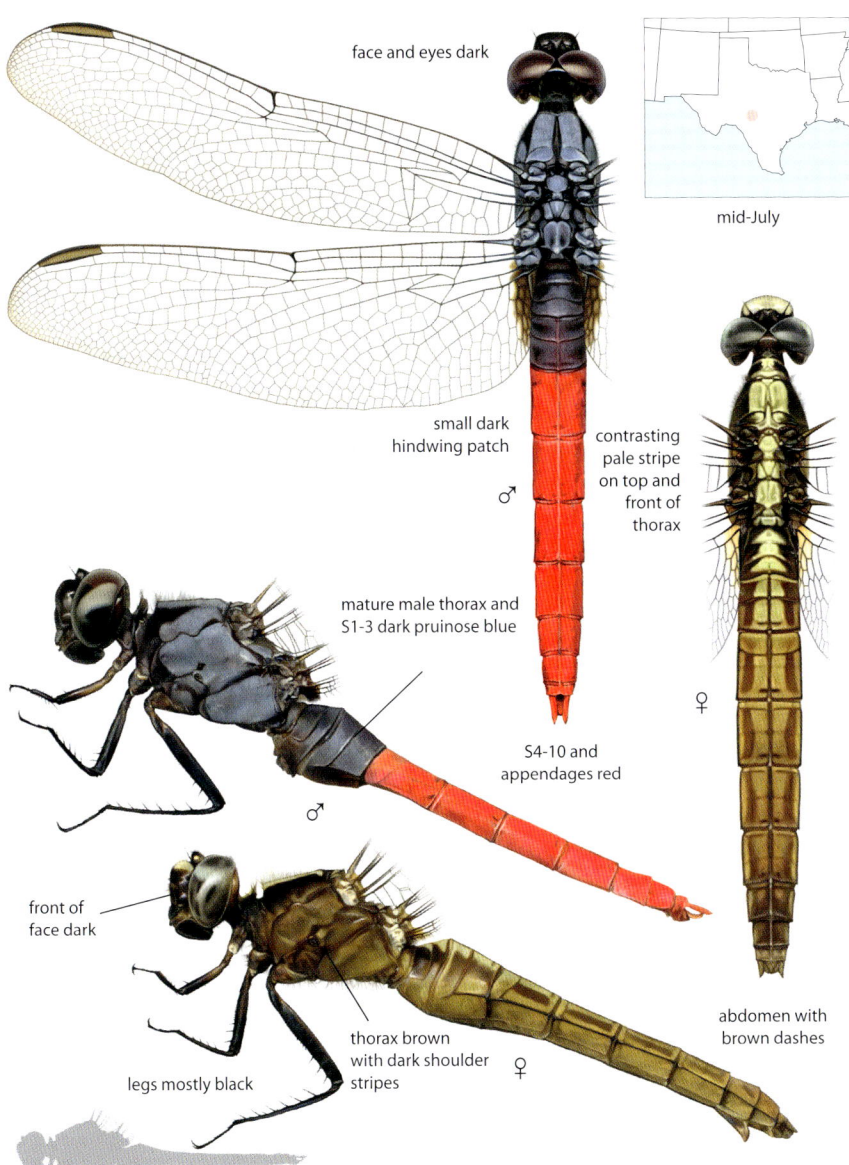

face and eyes dark

mid-July

small dark
hindwing patch

contrasting
pale stripe
on top and
front of
thorax

♂

mature male thorax and
S1-3 dark pruinose blue

S4-10 and
appendages red

♀

front of
face dark

abdomen with
brown dashes

thorax brown
with dark shoulder
stripes   ♀

legs mostly black

Similar species: Mexican Scarlet-tail male has S3 red, dark cerci; dark hindwing patch larger. Female has large dark hindwing patch, thorax uniformly dark brown, abdomen plain red-brown. **Claret Pondhawk** female has all-brown thorax, abdomen unmarked orange-brown.

Mexican Scarlet-tail • *Planiplax sanguiniventris*

37–42 mm, 1.5–1.7 in. Stout-bodied skimmer recorded in South Texas inhabiting vegetated ponds, lakes, and sloughs. Male face and eyes dark, thorax blue-black, abdomen bright red, with S1-2 and cerci dark. Female thorax unmarked brown, abdomen orange-brown. Both sexes with large dark hindwing patch and long black legs.

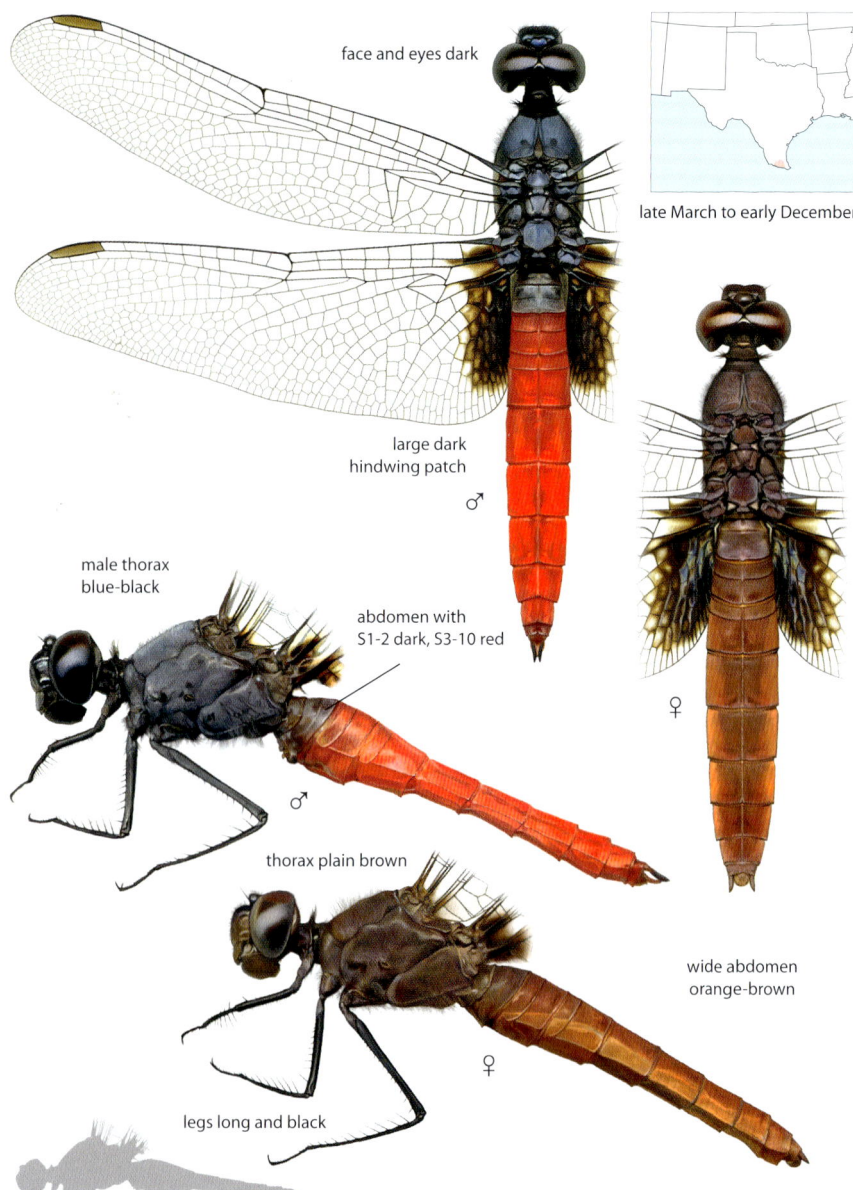

face and eyes dark

late March to early December

large dark hindwing patch

♂

male thorax blue-black

abdomen with S1-2 dark, S3-10 red

♂

♀

thorax plain brown

wide abdomen orange-brown

♀

legs long and black

Similar species: Flame-tailed Pondhawk male has S3 dark, red cerci, smaller hindwing patch, shorter legs, femurs with extra long spines. **Claret Pondhawk** female has top of face and legs pale, slenderer abdomen with spout-like subgenital plate.

Red-tailed Pennant • *Brachymesia furcata*

41–46 mm, 1.6–1.8 in. Small, common tropical pennant found at ponds and lakes, including brackish water; southern. Thorax unmarked brown. Mature male with all-red abdomen and red face. Immatures and female plain brown; female can show red on abdomen. Wings with small amber spot at base. Eyes red over blue-gray. Face pale. Legs black.

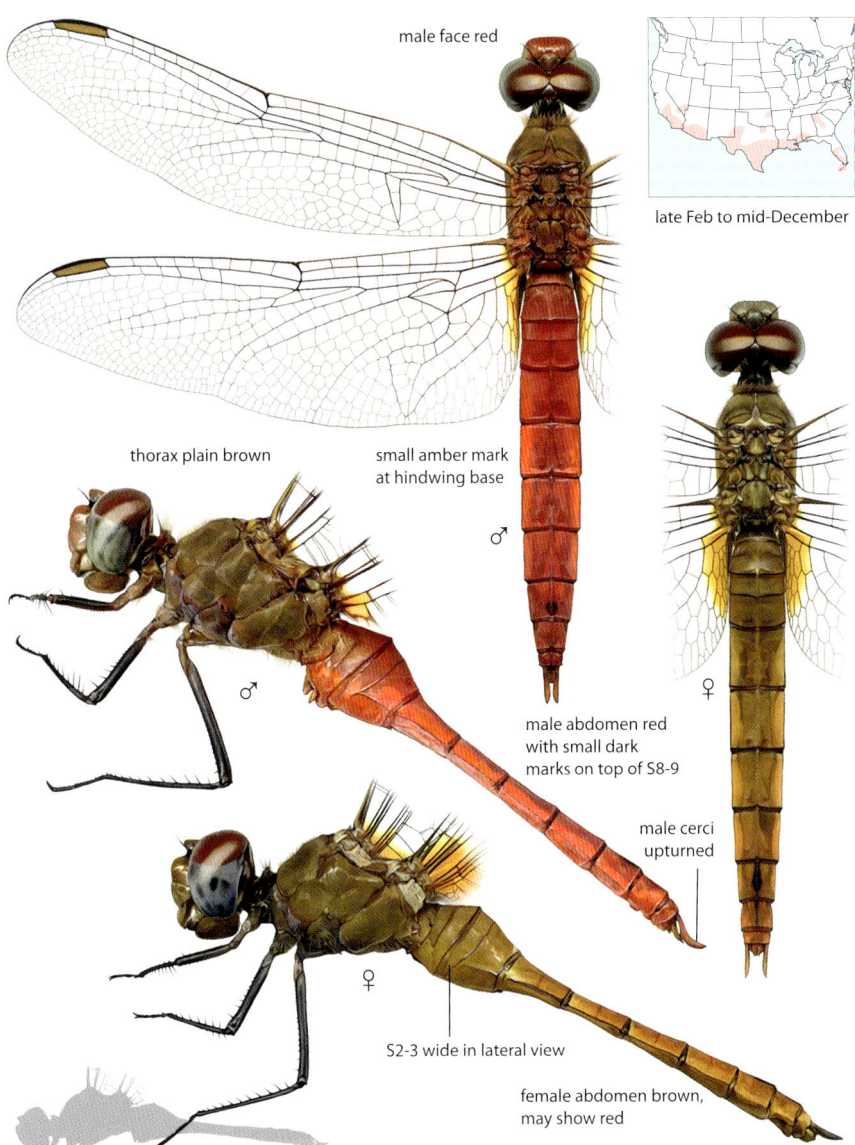

male face red

late Feb to mid-December

thorax plain brown

small amber mark
at hindwing base

♂

male abdomen red
with small dark
marks on top of S8-9

male cerci
upturned

♂

S2-3 wide in lateral view

♀

female abdomen brown,
may show red

Similar species: Scarlet Skimmer has black middorsal stripe on abdomen; male thorax red. **Tropical king skimmers** larger. Mature males with purple or red thorax; immatures and females with pale lateral and middorsal thoracic stripes, female abdomen with lateral flanges on S8. **Meadowhawks** have black lateral markings on abdomen and/or have lateral thoracic markings.

Four-spotted Pennant • *Brachymesia gravida*

50–54 mm, 2–2.1 in. Dark common tropical pennant inhabiting ponds and lakes, including alkaline and brackish waters, in the coastal southern US. All-white stigmas unique. Immatures brown with clear wings. Both sexes darken to black when mature, female retaining pale facial markings. Both sexes develop wing markings, darker and more defined in male.

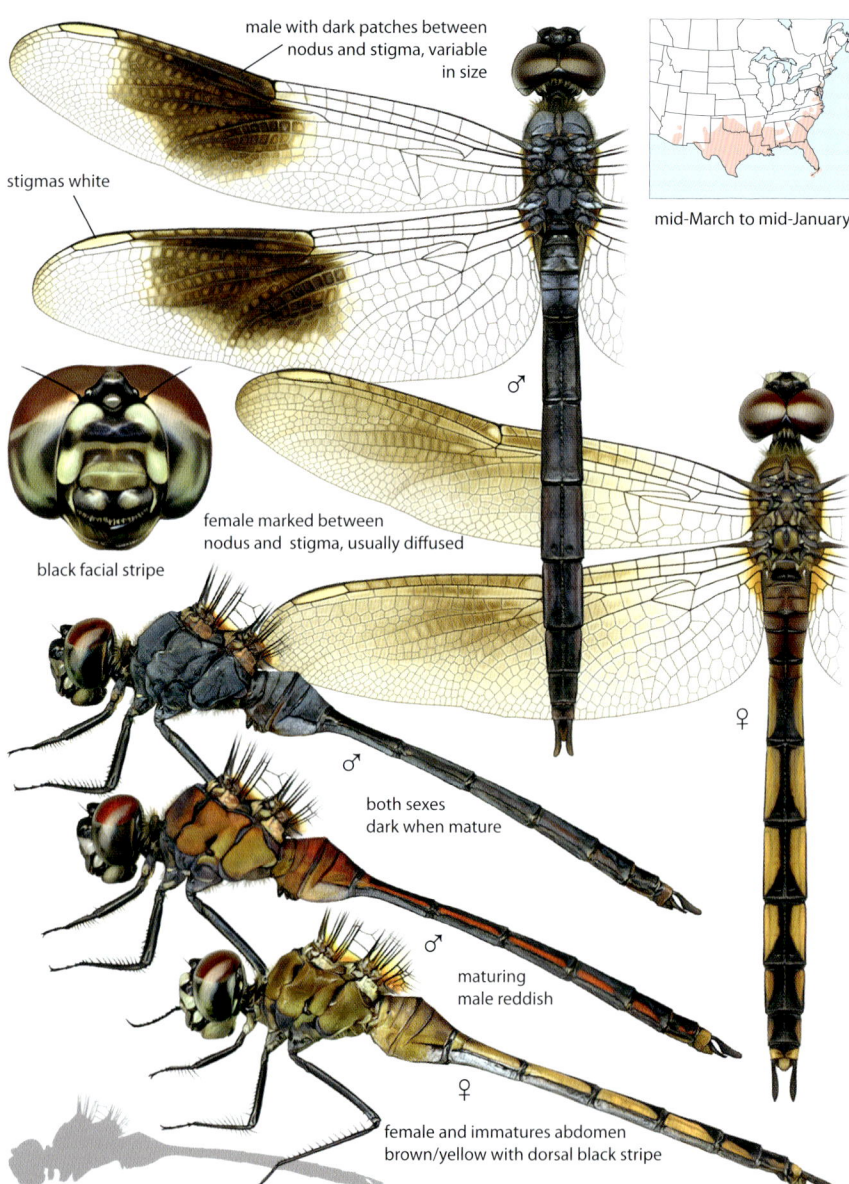

male with dark patches between nodus and stigma, variable in size

stigmas white

mid-March to mid-January

black facial stripe

female marked between nodus and stigma, usually diffused

♂

♀

both sexes dark when mature

♂

maturing male reddish

♀

female and immatures abdomen brown/yellow with dorsal black stripe

Similar species: Tawny Pennant has tan face, tan stigmas; wing markings more diffused. **Band-winged Dragonlet** has dark stigmas, pale appendages. Male with dark bands running completely across wings; female and immatures with larger dark lateral markings on abdomen, top of thorax between wings pale.

Tawny Pennant • *Brachymesia herbida*

43–46 mm, 1.7–1.8 in. Plain brown tropical pennant found at marshes, ponds, and lakes, including brackish waters. Both sexes with unmarked brown thorax. Abdomen yellow-brown with black dorsal stripe. Stigmas tan. Both sexes with diffused wing markings. Face mostly pale. Eyes red over gray.

markings diffused from wing base to tip

stigmas tan

late April to mid-January (recorded in all months)

front of face pale

thorax unmarked brown

♂

♀

abdomen yellow-brown with black dorsal stripe

♂

♀

Similar species: Four-spotted Pennant blacker when mature; stigmas white, wing markings more concentrated between nodus and stigma, face with black stripe.

Metallic Pennant • *Idiataphe cubensis*

36–41 mm, 1.4–1.6 in. Small, dark, slender species inhabiting open ponds and lakes, sometimes brackish, in southern Florida and Texas. Eyes dull brown over gray, top of male face black with metallic reflections. Thorax metallic black and brown, abdomen black with brown sides. Hindwing with small amber spot at base.

late May to early November
Florida: all year

small amber spot at
base of hindwing

♂

thorax black and brown with
metallic green reflections

♂

♀

♀

abdomen black
with brown sides

Similar species: Black Setwing abdomen proportionally longer and slenderer, thorax lacks metallic reflections, no amber at hindwing base. **Seaside Dragonlet** male all black, lacks metallic reflections on thorax and brown coloration on side of abdomen.

Marl Pennant • *Macrodiplax balteata*

37–42 mm, 1.5–1.7 in. Medium-sized coastal pennant found at open brackish and fresh-water ponds, and lakes, including marl ponds. Both sexes with large, rounded dark hind-wing patch. Mature male all black. Immature and female yellow-brown with black lateral thoracic stripes; abdomen with black rings, black dorsal and lateral stripes, S8-10 black.

male face and eyes dark

early April to mid-November
Florida: all year

large, rounded
hindwing patch

♂

mature male
blackish

female face pale

♀

abdomen yellow-
brown with black
bands; tip black

♂

thorax yellow-
brown with black
lateral stripes

♀

black dorsal stripe

♀

Similar species: Double-ringed Pennant smaller, with smaller hindwing markings; female abdomen mostly black. **Martha's Pennant** hindwing patch wider, covering hindwing triangle; female abdomen with yellow dorsal spots. **Hyacinth Glider** has dark hindwing band, not rounded spot. **Saddlebags species** have larger, irregularly shaped patches or straight-margined hindwing bands.

Double-ringed Pennant • *Celithemis verna*

32–35 mm, 1.3–1.4 in. Dark southeastern pennant found at open ponds, lakes, and borrow pits with emergent vegetation along edges. Hindwing with very small patch at base. Immatures and female thorax yellow with black lateral stripes; abdomen with yellow rings on S3 and 4, remainder of abdomen black. Mature male all black.

late March to mid-August

small dark
hindwing patch

♂

mature male all black

♀

female abdomen
black with yellow
rings on S3-4

♂

black striping on
side of thorax

♀

Similar species: Martha's Pennant has larger hindwing patch; thorax lacks black lateral stripes; abdomen with yellow spots on top of S4-7. **Seaside Dragonlet, Metallic Pennant,** and **Black Setwings** lack dark hindwing patch. **Four-spotted Pennant** larger, with large dark wing spots between nodus and stigma.

Martha's Pennant • *Celithemis martha*

25–33 mm, 1–1.3 in. Small, dark pennant inhabiting ponds and lakes with emergent shore-line vegetation, often sand-bottomed, along the northeastern Atlantic coast. Large hind-wing patch with three dark bands. Side of thorax unmarked. Female and immature marked with yellow. Maturing male reddish, then becomes all black with darkened hindwing patch.

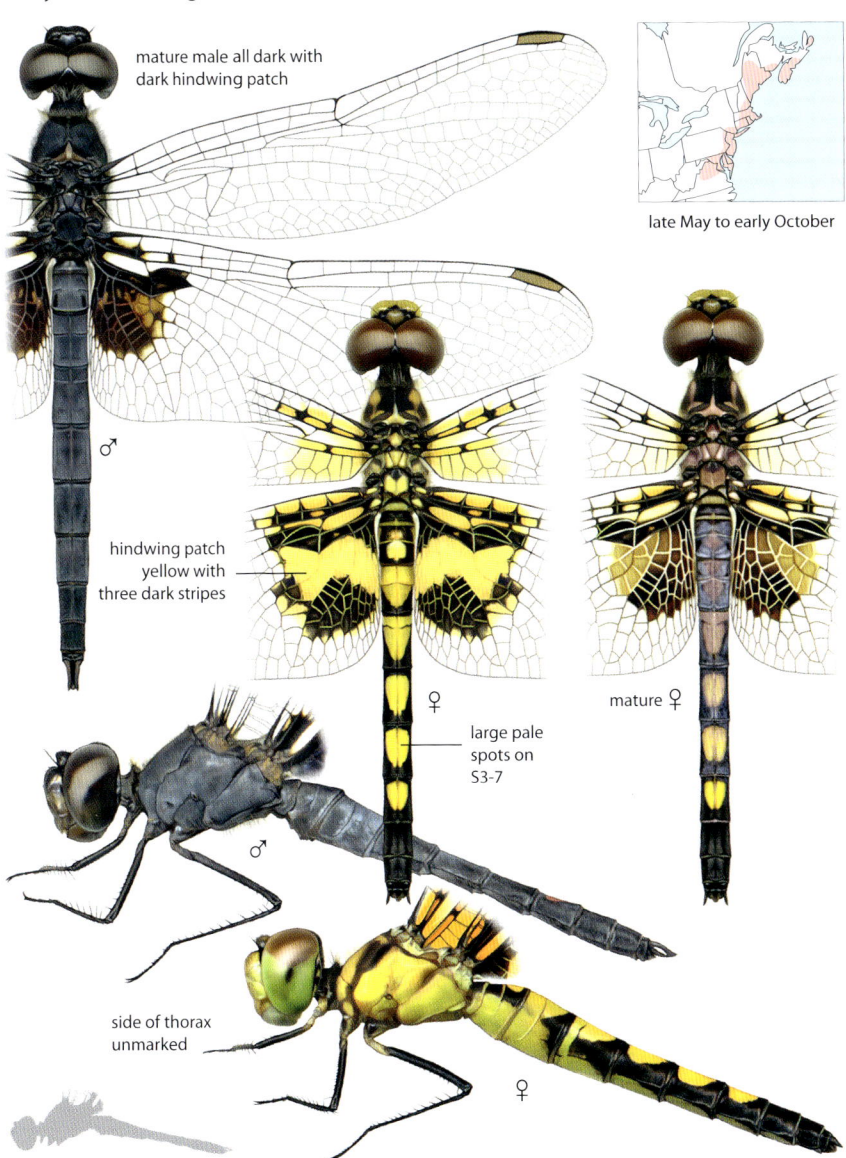

mature male all dark with dark hindwing patch

late May to early October

hindwing patch yellow with three dark stripes

large pale spots on S3-7

mature ♀

side of thorax unmarked

Similar species: Double-ringed Pennant hindwing patch smaller, has black lateral thoracic stripes, lacks yellow dorsal markings on S5-7. **Ornate Pennant** has black lateral thoracic stripes. **Seaside Dragonlet** lacks dark hindwing patch. **Marl Pennant** hindwing patch narrower, does not cover hindwing triangle. **Four-spotted Pennant** larger, has large dark wing spots between nodus and stigma.

Banded Pennant • *Celithemis fasciata*

30–38 mm, 1.2–1.5 in. Dark, well-patterned pennant inhabiting ponds and lakes, typically permanent and sand-bottomed. Wings marked with dark brown; pattern variable but always has large marking extending from base to nodus, patch between nodus and stigma, and dark wing tips. Immature and female body marked with yellow; mature male black.

wing pattern variable but always well defined

late March to late October

dark markings extend from wing base to nodus

dark wing tips

dark patch between stigma and nodus

♂

mature male black

southern populations have yellow at base of wings

♀

♂

side of thorax heavily striped ♀

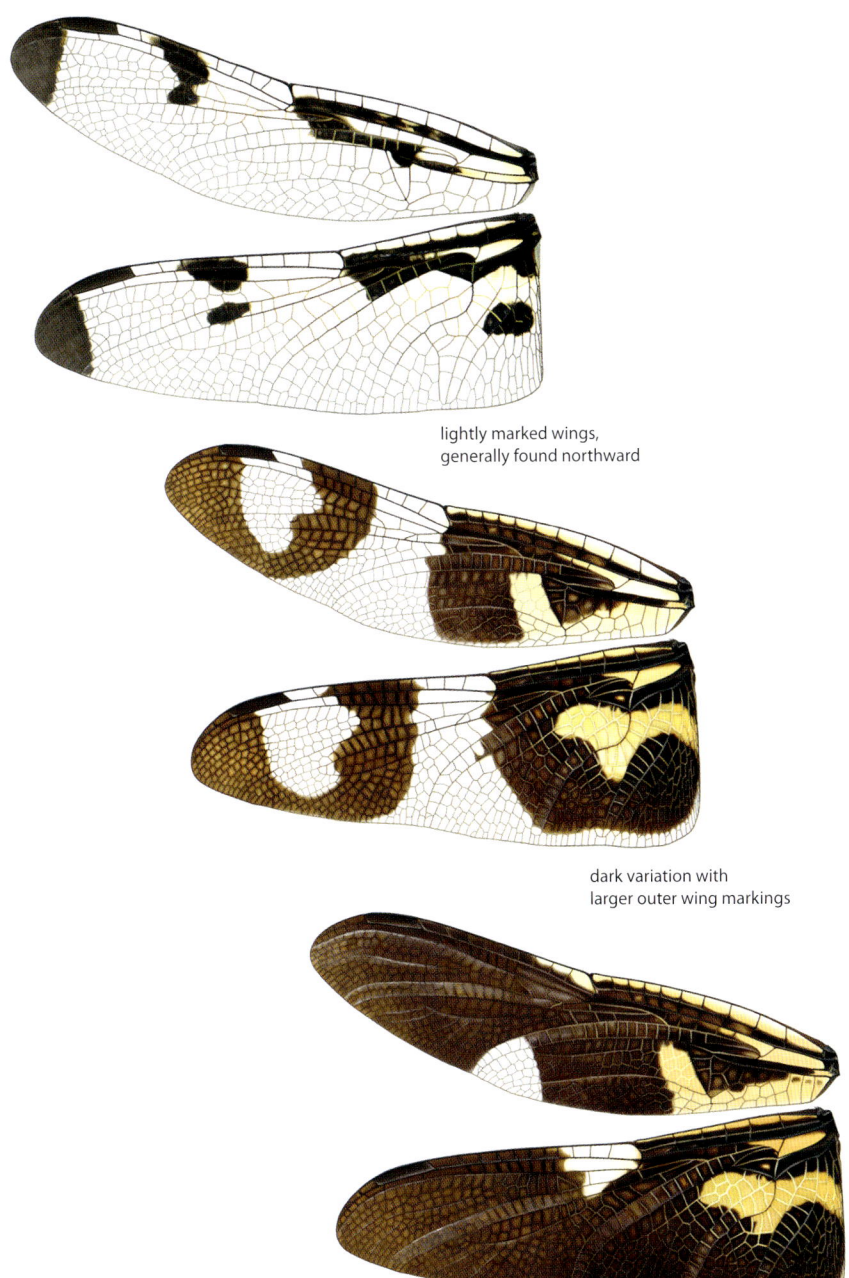

lightly marked wings,
generally found northward

dark variation with
larger outer wing markings

dark extreme

Similar species: Calico Pennant dark markings at base of wings do not extend to nodus; stigmas pale, pale markings on top of abdomen wider and heart-shaped; mature male red. **Halloween Pennant** has entire wing tinted yellow, stigmas pale; male with red wing veins.

Ornate Pennant • *Celithemis ornata*

33–35 mm, 1.3–1.4 in. Dark southeastern pennant found at ponds and lakes with emergent grass or sedges. Hindwing with large patch at base; patch amber with three dark brown bands. Thorax with black lateral stripes. Juveniles and female with yellow markings. Mature male thorax brown, abdomen with deep red spots; red veins in hindwing patch.

eyes dark red

mid-January to late October

♂

mature male very dark; pale markings subtle, dark red

♂

hindwing patch amber with three dark stripes

♀

black lateral thoracic stripes joined at top

♀

dorsal pale marks relatively narrow

lateral pale stripes on S3-4

Similar species: Martha's Pennant lacks black stripes on side of thorax; male lacks red coloration. **Amanda's Pennant** has larger hindwing patch with two dark bands, lacks black lateral stripes on thorax. **Red-veined Pennant** usually has very small hindwing patch, dark lateral thoracic stripes unjoined at top, pale stripes on sides of S4-5.

Amanda's Pennant • *Celithemis amanda*

27–31 mm, 1.1–1.2 in. Brightly patterned southeastern pennant inhabiting open ponds and lakes with some emergent vegetation. Very large patch at hindwing base, yellow/amber with two dark bands. Thorax largely unmarked. Abdomen with wide, pale dorsal spots. Immature and female with yellow markings. Mature male markings red.

mid-April to late October

large hindwing patch; color nearly reaches hindwing margin

♂

mature male with red face, red thoracic and abdominal markings, red veins in hindwing patch

large hindwing patch amber with two dark stripes

♀

♂

side of thorax largely unmarked

♀

pale dorsal markings wide

Similar species: Ornate Pennant has smaller hindwing patch with three dark bands, thorax with black lateral stripes; dorsal abdominal marking narrower. **Martha's Pennant** hindwing patch usually smaller but variable; patch has three dark bands. Thorax with dark shoulder stripe. Male lacks red coloration. **Calico Pennant** has dark wing spots between nodus and stigma and at wing tip.

Red-veined Pennant • *Celithemis bertha*

28–36 mm, 1.1–1.4 in. Brightly colored southeastern pennant found at sand-bottomed ponds and lakes with some emergent grass or sedges. Hindwing patch usually very small, may be larger in South Florida. Thorax with black lateral stripes. Abdomen with narrow pale dorsal dashes. Juveniles and females marked with yellow, adult male red.

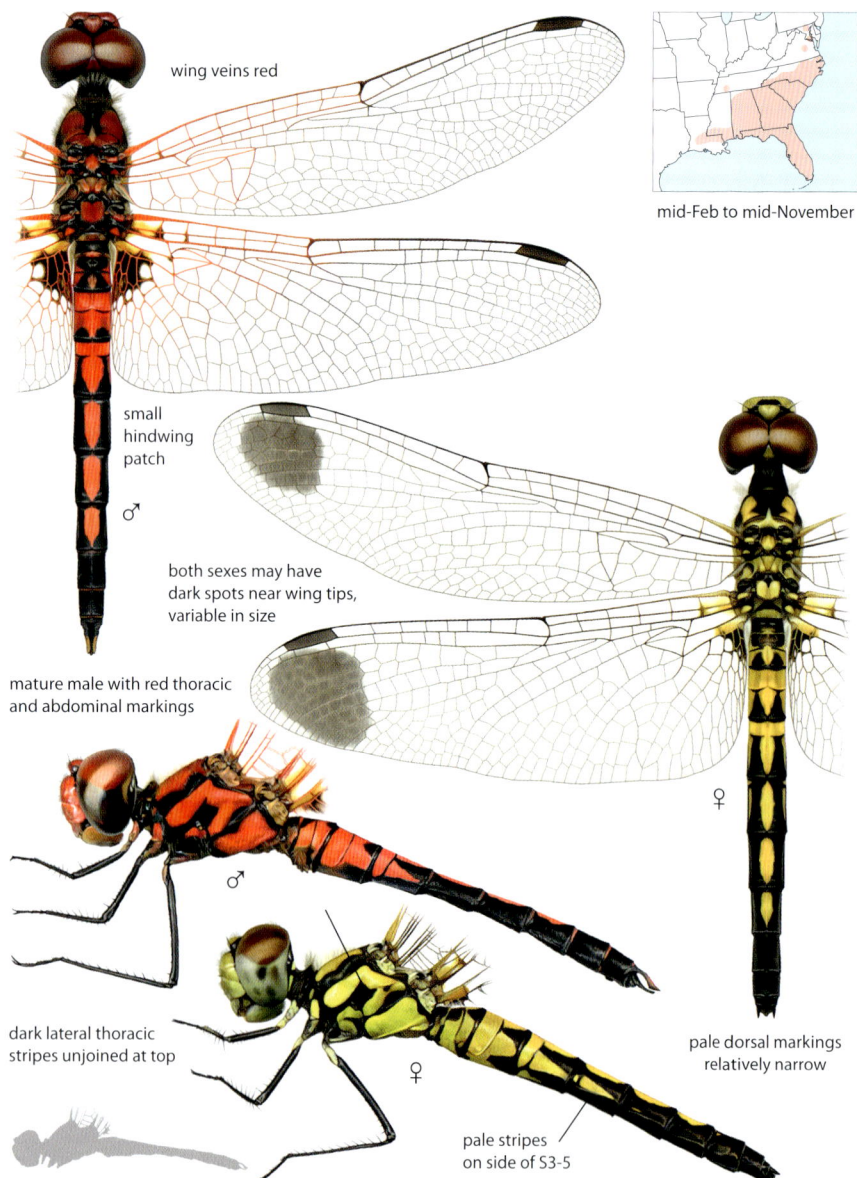

wing veins red

mid-Feb to mid-November

small
hindwing
patch

♂

both sexes may have
dark spots near wing tips,
variable in size

mature male with red thoracic
and abdominal markings

♀

♂

dark lateral thoracic
stripes unjoined at top

♀

pale dorsal markings
relatively narrow

pale stripes
on side of S3-5

Similar species: Ornate Pennant has larger hindwing patch, black lateral thoracic stripes joined at top, lacks pale lateral stripe on S5; male dark red and brown. **Double-ringed Pennant** lacks pale dorsal markings on S5-7 and any red coloration. **Martha's** and **Amanda's Pennants** have larger hindwing patch, lack black lateral thoracic stripes.

Calico Pennant • *Celithemis elisa*

29–34 mm, 1.1–1.3 in. Common eastern pennant inhabiting ponds and lakes with emergent and shoreline vegetation. Large yellow and dark brown patch at base of hindwing. Usually has dark brown spots between nodus and stigma and at wing tips. Abdomen with heart-shaped dorsal markings. Immature and female markings yellow, adult male red.

typically has dark spots at wing tips and between nodus and stigma, can be light

male stigmas red

north: late May to late Sept
south: late Mar to mid-Nov

large dark hind-wing patch with red veins

♂

mature male with red thoracic and abdominal markings

immature and female markings yellow

♀

♂

cerci pale

♀

pale dorsal markings heart-shaped

Similar species: Banded Pennant has dark wing markings extending from base to nodus, lacks or has narrower pale abdominal markings, lacks red coloration. **Amanda's** and **Ornate Pennants** lack dark markings beyond base of wing.

Halloween Pennant • *Celithemis eponina*

36–42 mm, 1.4–1.7 in. Common brightly colored and patterned pennant found at ponds, open lakes, and marshes. Wings entirely yellow with conspicuous dark brown bands and large spots; male wings redder, with red veins and stigma. Mature male body brown and black, immature and female with brighter pale markings.

mature male has red stigmas and red wing veins

north: late May to late Nov
south: early Apr to late Nov
Florida: all year

wings entirely yellow

southern populations often have more contrasting thorax pattern

dark bands below stigma and nodus, dark spots at wing bases

♂

♀

♂

♀

Similar species: Only small pennant with wings entirely colored yellow. **Calico Pennant** has yellow at wing bases, but rest of wing clear. **Banded Pennant** may have yellow on inner half of wing, but outer half clear.

Canada Whiteface • *Leucorrhinia patricia*

24–29 mm, 0.9–1.1 in. Very small far-northern whiteface found at bog ponds and fens, typically with floating mats of moss. The smallest whiteface. Immatures yellow, mature male red. Male similar to Belted and Crimson-ringed Whiteface but slenderer and delicate. Female with short, narrow pale streaks on top of S3-6, sometimes absent; S7 black.

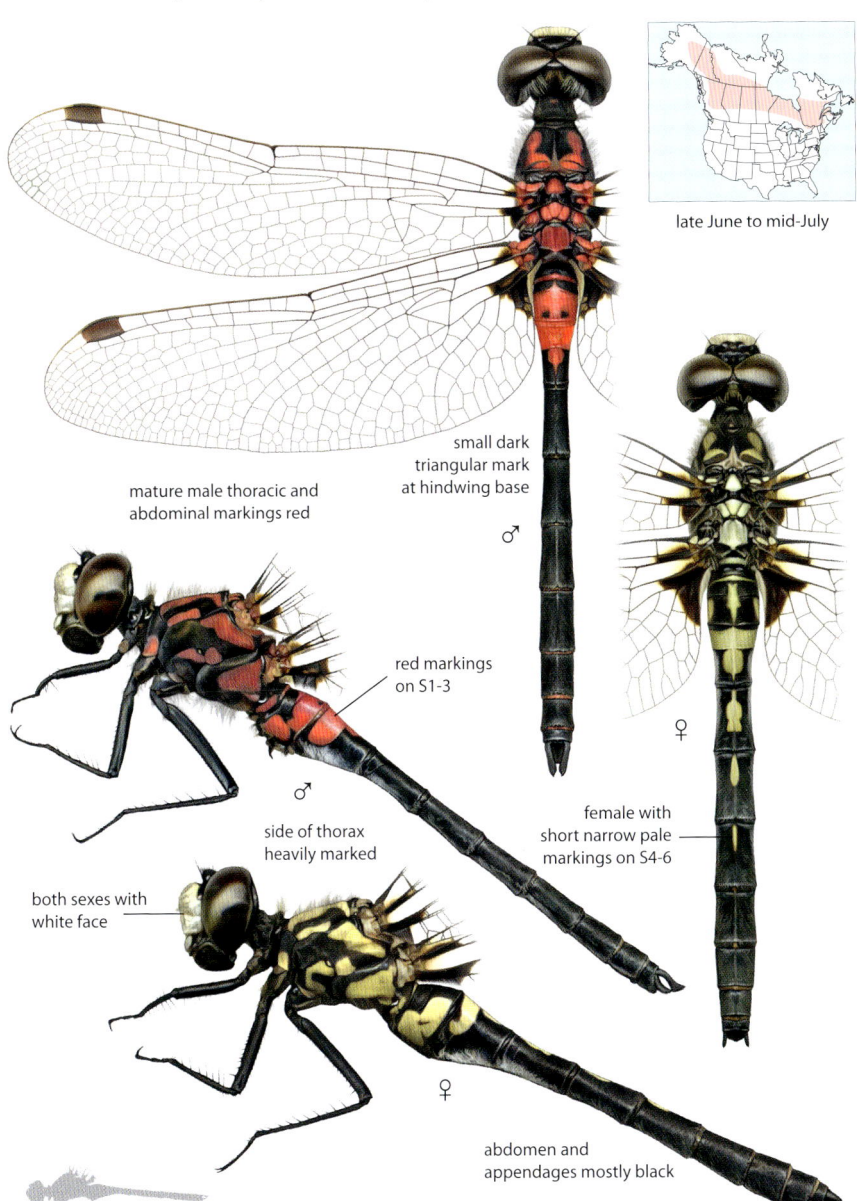

late June to mid-July

small dark triangular mark at hindwing base

♂

mature male thoracic and abdominal markings red

red markings on S1-3

♂

side of thorax heavily marked

female with short narrow pale markings on S4-6

♀

both sexes with white face

abdomen and appendages mostly black

♀

Similar species: Belted and **Crimson-ringed Whitefaces** larger, abdomen thicker, have three cells adjacent to forewing triangle (two in Canada); females with pale spot on S7. Female **Frosted Whiteface** has thicker, stubby abdomen and side of thorax pale.

Crimson-ringed Whiteface • *Leucorrhinia glacialis*

34–35 mm, 1.3–1.4 in. Common, widespread whiteface inhabiting boggy ponds and lakes. Face white. Dark triangular mark at hindwing base. Immatures yellow, mature male and some females red. Mature male abdomen black with red only on S2-3. Female with short, narrow pale dashes on top of S4-6, small pale triangular mark on S7. Labium all black.

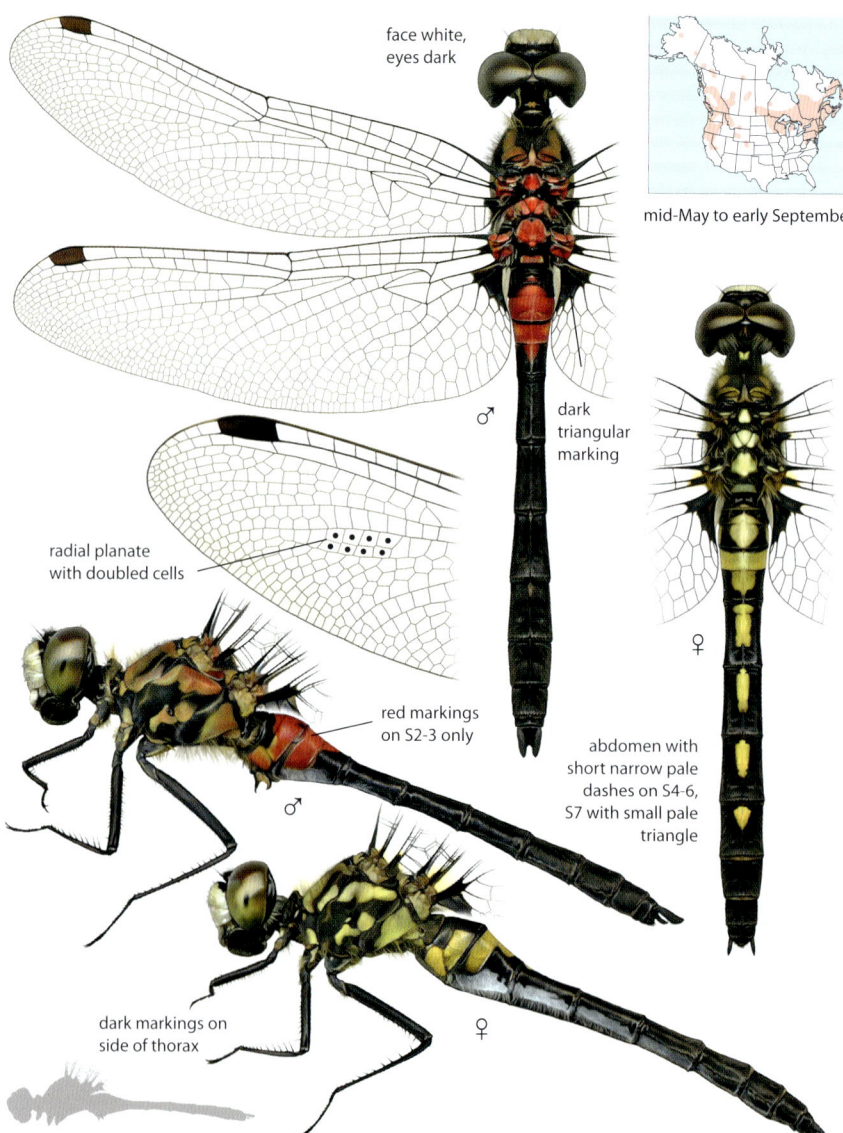

face white, eyes dark

mid-May to early September

dark triangular marking

radial planate with doubled cells

red markings on S2-3 only

♂

♂

♀

abdomen with short narrow pale dashes on S4-6, S7 with small pale triangle

dark markings on side of thorax

♀

Similar species: Belted Whiteface has single row of cells in radial planate, female with white spot on labium. Also compare male hamules, page 434. **Canada Whiteface** smaller, has radial planate with single row of cells; female with short narrow streaks on abdomen, S7 black. **Hudsonian Whiteface** has wider pale abdominal markings, usually has pale veins within dark hindwing patch. **Dot-tailed Whiteface** has paler side of thorax; S7 pale spot squarish and wider than S6 marking.

Belted Whiteface • *Leucorrhinia proxima*

33–36 mm, 1.3–1.4 in. Common, widespread whiteface found at boggy and marshy ponds and lakes. Mature eastern male has base of abdomen pruinose, thorax with red or brown between wings (may be subtle). Western male and female pattern nearly identical to Crimson-ringed Whiteface. Female and some males have white spot on labium.

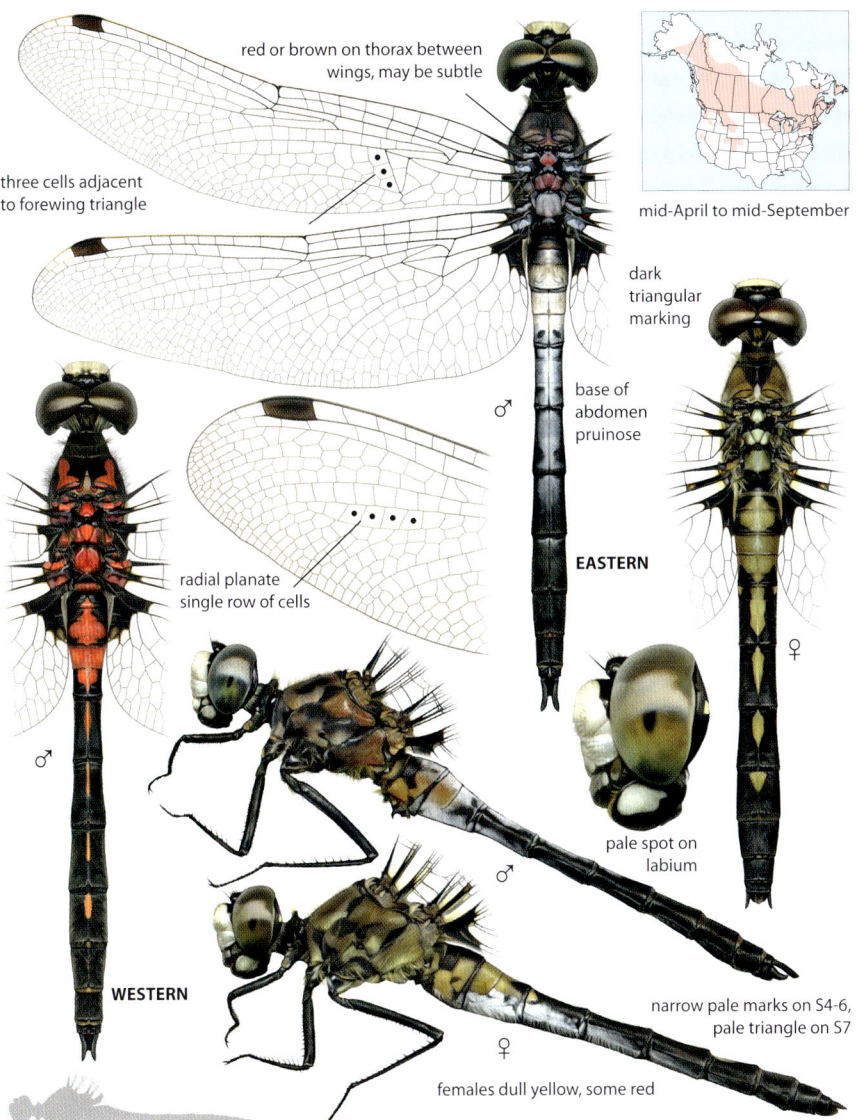

red or brown on thorax between wings, may be subtle

three cells adjacent to forewing triangle

mid-April to mid-September

dark triangular marking

♂

base of abdomen pruinose

EASTERN

radial planate single row of cells

♂

♀

pale spot on labium

♂

WESTERN

narrow pale marks on S4-6, pale triangle on S7

♀

females dull yellow, some red

Similar species: Crimson-ringed Whiteface has two rows of cells in radial planate, labium black. Compare male hamules, page 434. **Frosted Whiteface** abdomen shorter, has two cells adjacent to forewing triangle; male lacks red or brown coloration between wings. Also compare hamules, page 434. **Canada Whiteface** smaller, has two rows of cells adjacent to forewing triangle; female with narrow streaks on abdomen, S7 black, labium black. Also compare male hamules, page 434.

Frosted Whiteface • *Leucorrhinia frigida*

28–32 mm, 1.1–1.3 in. Common dark northeastern whiteface inhabiting vegetated ponds and lakes, often bog pools. Small with short, stubby abdomen. Mature male black with white pruinosity on S2-4. Female and immature side of thorax mostly pale, abdomen with short yellow streaks on top of S4-6. Female develops pruinosity on abdomen like male.

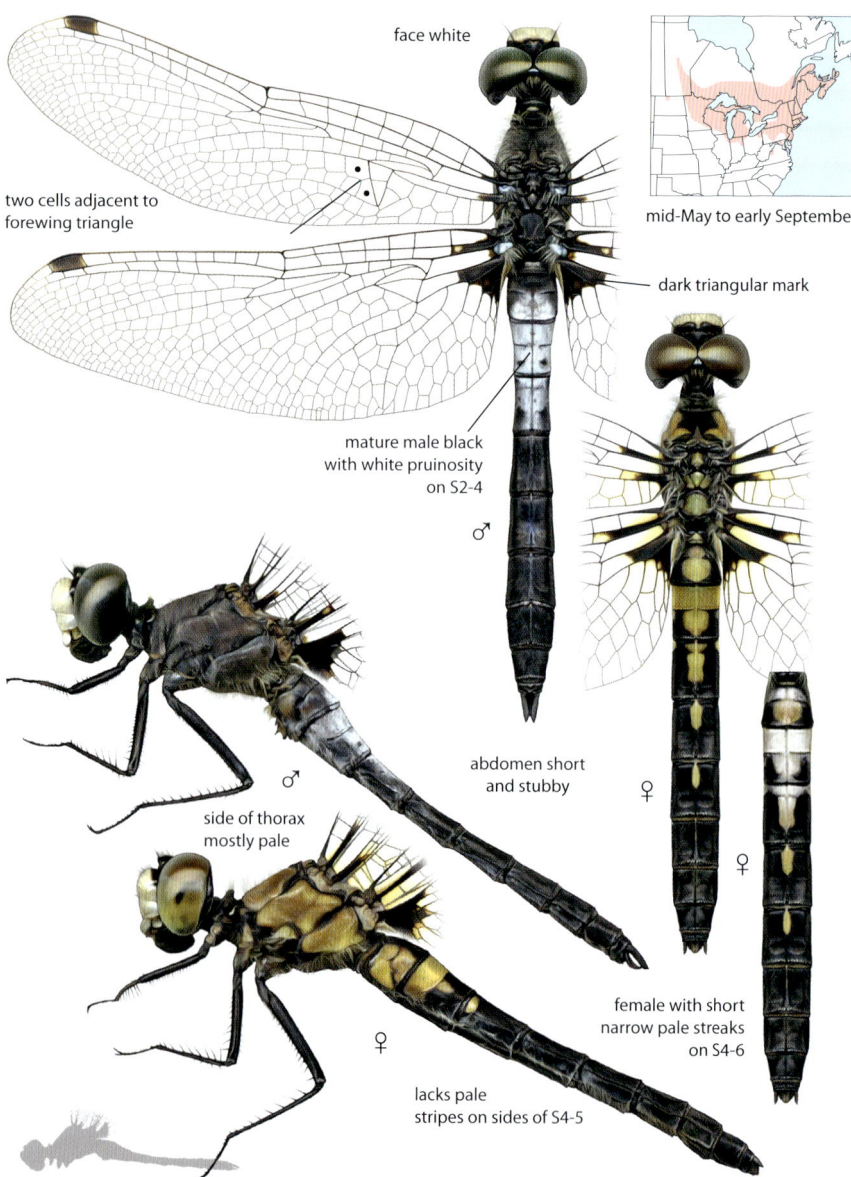

face white

two cells adjacent to forewing triangle

mid-May to early September

dark triangular mark

mature male black with white pruinosity on S2-4

♂

abdomen short and stubby

♀

♂

side of thorax mostly pale

♀

female with short narrow pale streaks on S4-6

♀

lacks pale stripes on sides of S4-5

Similar species: Dot-tailed Whiteface has yellow spot on S7. **Belted Whiteface** abdomen longer, has three cells adjacent to forewing triangle (two in Frosted). Eastern male has some red or brownish coloration on top of thorax between wings, often subtle. Compare hamules, page 434. **Other similar whitefaces** have more black markings on side of thorax.

Dot-tailed Whiteface • *Leucorrhinia intacta*

29–33 mm, 1.1–1.3 in. Small dark whiteface found at vegetated ponds and lakes, common and widespread. Immature thorax lightly marked in black, abdomen with full-length yellow markings on top of S3-6; squarish S7 marking wider than S6 mark. Both sexes darken when mature, abdomen retaining only S7 marking.

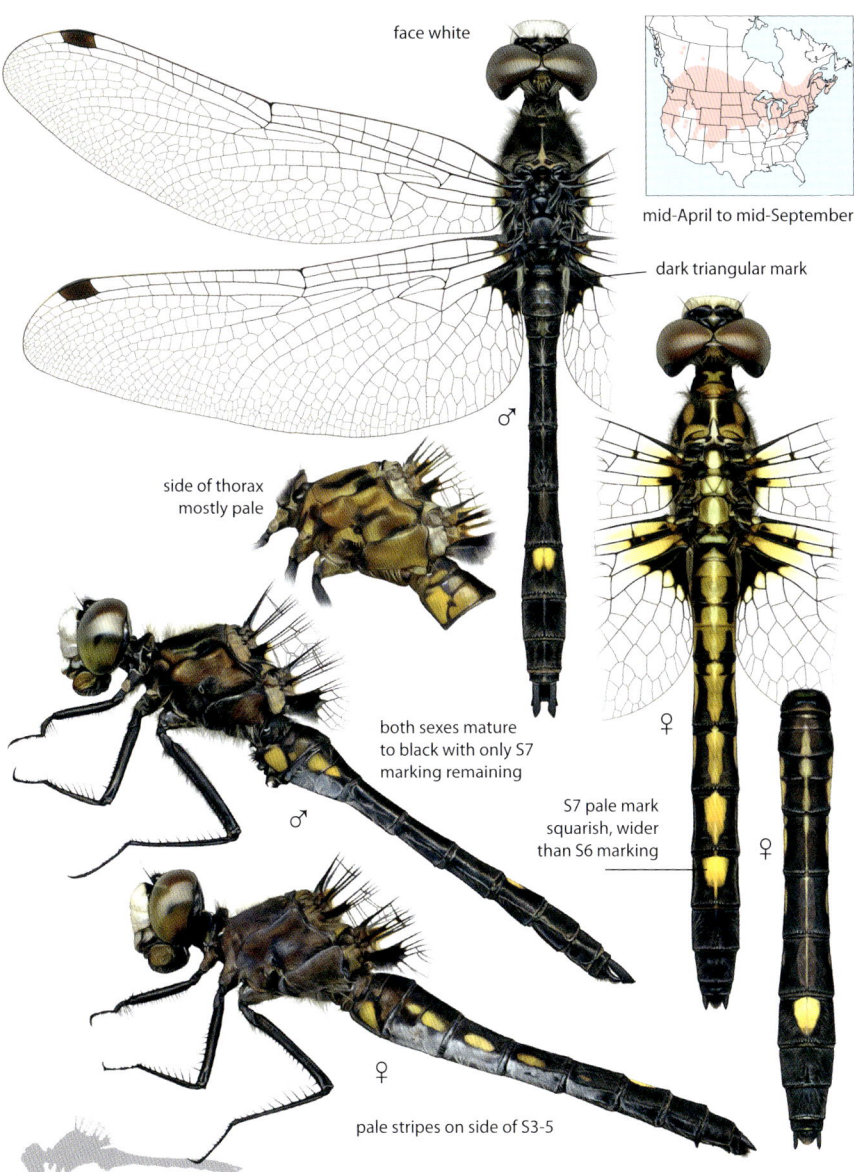

face white

mid-April to mid-September

dark triangular mark

side of thorax
mostly pale

♂

both sexes mature
to black with only S7
marking remaining

♂

S7 pale mark
squarish, wider
than S6 marking

♀

♀

♀

pale stripes on side of S3-5

Similar species: Frosted Whiteface lacks yellow spot on S7. **Hudsonian Whiteface** has more black on side of thorax, pale spot on S8 equal in width with S7 spot, usually has pale veins within dark patch at base of hindwing. **Crimson-ringed** and **Belted Whitefaces** have more black on side of thorax, lack pale stripes on side of S4-5; S8 spot small and triangular.

Hudsonian Whiteface • *Leucorrhinia hudsonica*

27–32 mm, 1.1–1.3 in. Common, widespread whiteface inhabiting marshes, bog ponds, and vegetated lakes. Immatures yellow, mature male and some females red. Both sexes typically with dorsal pale markings on S2-7; S4-7 markings equal in width. Often has pale veins within the dark triangular hindwing patch.

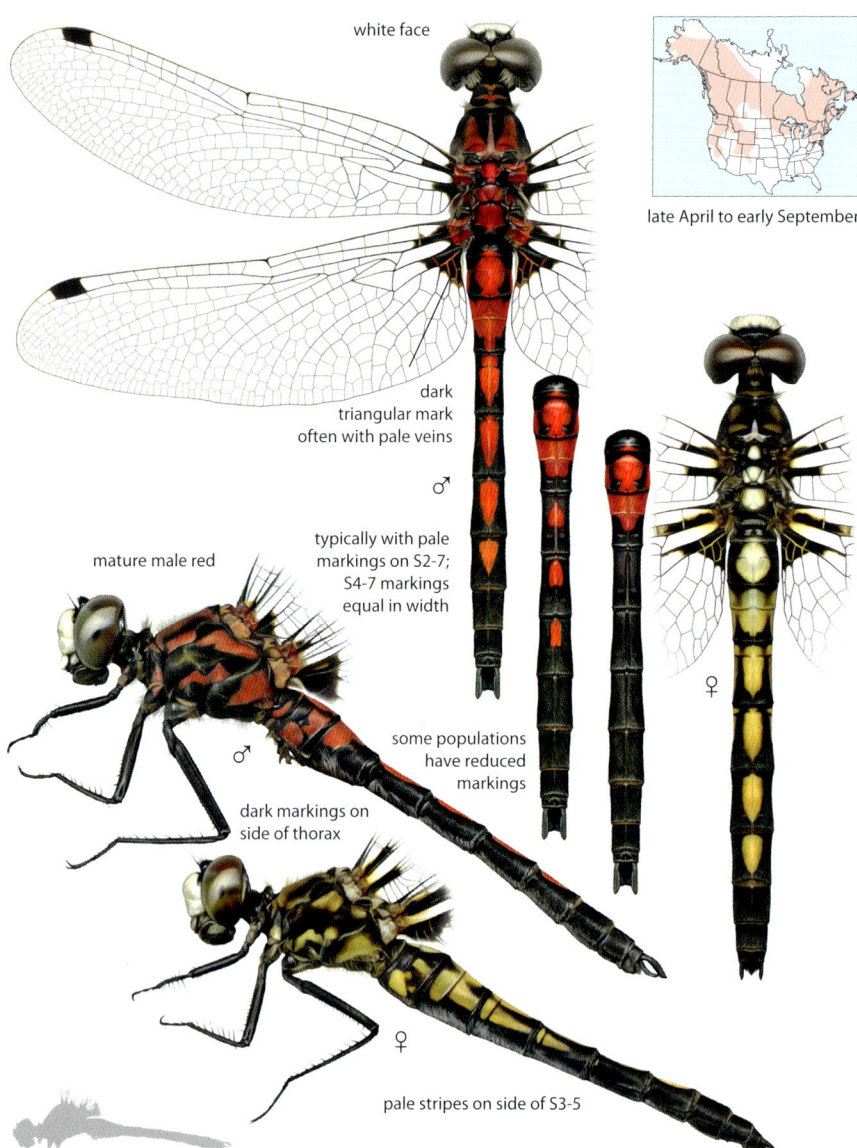

white face

late April to early September

dark triangular mark often with pale veins

♂

typically with pale markings on S2-7; S4-7 markings equal in width

♀

mature male red

♂

some populations have reduced markings

dark markings on side of thorax

♀

pale stripes on side of S3-5

Similar species: Boreal Whiteface has wider, full-length dorsal markings on S3-7, lacks pale veins in dark hindwing patch. Male with pale spot on S8; female lacks lateral pale stripes on S3-5. **Dot-faced Whiteface** has paler side of thorax, dorsal spot on S7 wider than S6 spot; lacks pale veins in hindwing patch. **Other yellow-marked whitefaces** have narrower dorsal abdominal markings, lack pale veins in hindwing patch.

Boreal Whiteface • *Leucorrhinia borealis*

44–46 mm, 1.7–1.8 in. Extensively red-marked whiteface found at sedge marshes, lakes, and bog and prairie ponds in the Northwest and Rocky Mountains. The largest whiteface. Abdomen with wide, full-length pale markings on top of S3-7, suggesting continuous stripe. Male often with red spot on S8. Immatures yellow, mature female often red.

face white

late May to early August

dark triangular mark

mature male red,
mature female often red

side of thorax
heavily marked

male often with
spot on S8

♂

abdomen with wide, full-
length pale markings on S3-7

♀

♂

♀

sides of S3-5 lack pale stripes

Similar species: Hudsonian Whiteface has smaller, narrower dorsal abdominal markings, pale veins within dark hindwing marking. Male lacks pale spot on S8; female with lateral pale stripes on S3-5.

Band-winged Dragonlet • *Erythrodiplax umbrata*

38–45 mm, 1.5–1.8 in. Common large dragonlet found at shallow vegetated ponds and marshes, often temporary. Mature male all black; cerci pale, wings with dark brown band between stigma and nodus. Immature and female green-gray; abdomen with dark middorsal stripe and lateral markings. Some mature females darken, may develop dark wing bands.

male with wide dark bands between nodus and stigma

early Feb to late December
Florida: all year

mature male black, cerci pale

single row of cells in medial planate

♂

♀

female and immatures greenish; abdomen with dark middorsal stripe

side of thorax pale ♀

dark dashes on side of abdomen

Similar species: Four-spotted Pennant has dark wing spots, white stigmas. **Black-winged Dragonlet** mature male has mostly black wings; some females also have black wings or dark spot at hindwing base. Clearer-winged females nearly identical, have two rows of cells in medial planate, dark markings on side of abdomen more rectangular in shape.

Black-winged Dragonlet • *Erythrodiplax funerea*

38–42 mm, 1.5–1.7 in. Large, dark dragonlet inhabiting temporary open ponds and marshes; rare southwestern vagrant. Mature male body all black, cerci pale, wings mostly black. Immature and female yellow-brown, abdomen with dark markings. Some mature females black with dark wings like male or have dark spot at hindwing base.

forewing with wide dark band at nodus

late July to early September

hindwing mostly black

♂

female wings may be mostly clear, mostly black like male, or have dark patch at hindwing base

two rows of cells in medial planate

mature male all black, cerci pale

♀

♂

♀

♀

abdomen with squarish dark lateral markings

Similar species: Band-winged Dragonlet mature male has dark bands between stigma and nodus; female nearly identical, has single row of cells in medial planate, dark markings on side of abdomen more pointed in shape. **Filigree Skimmer** has more complex wing pattern, pale-striped eyes and body, dark appendages. **Widow Skimmer** dark wing patches do not extend past nodus; abdomen stouter.

Little Blue Dragonlet • *Erythrodiplax minuscula*

25–27 mm, 1–1.1 in. Very small southeastern dragonlet inhabiting marshy ponds and lakes. Mature male eyes and thorax usually blue, face dark, very small dark patch at hindwing base, abdomen powder blue with black tip, cerci usually pale. Female and immatures yellow, minimal amber at hindwing base, abdomen with dark dorsal stripe, S8-9 black.

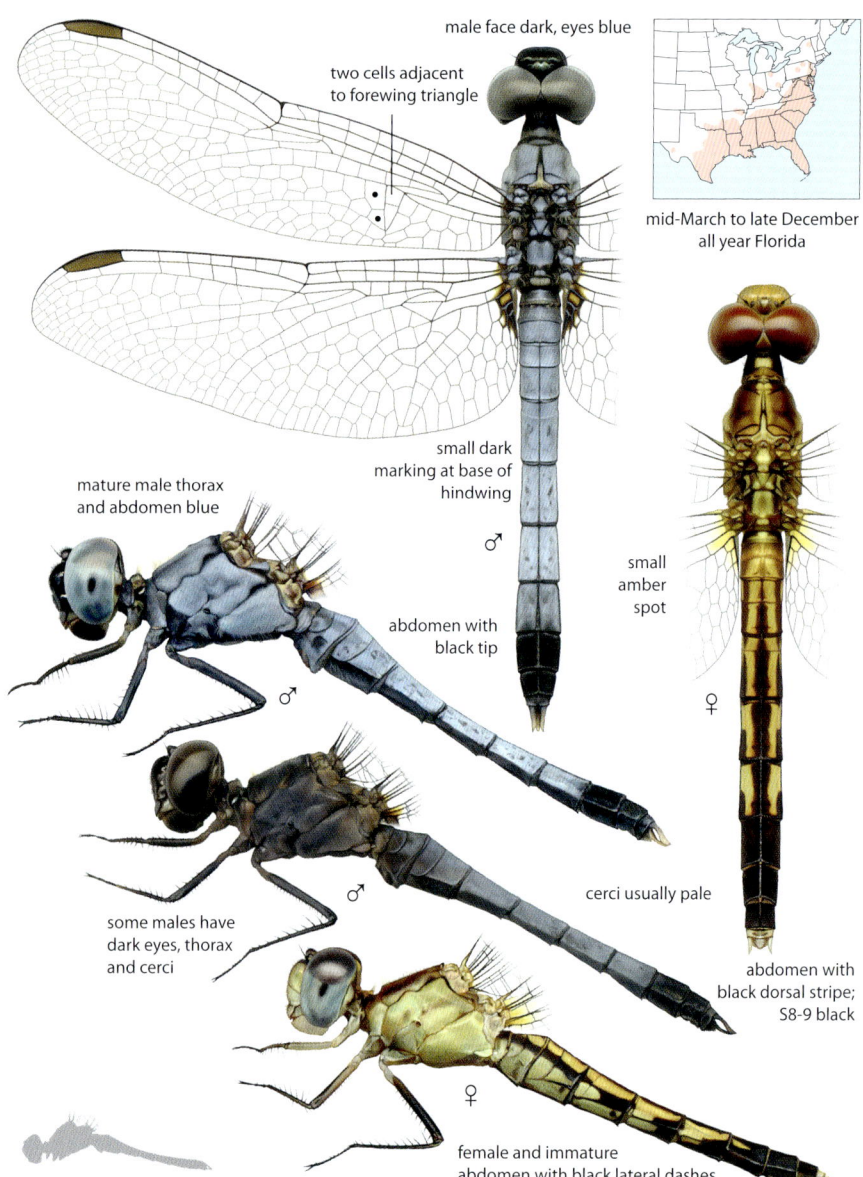

male face dark, eyes blue

two cells adjacent to forewing triangle

mid-March to late December all year Florida

small dark marking at base of hindwing

♂

mature male thorax and abdomen blue

small amber spot

abdomen with black tip

♂

♀

some males have dark eyes, thorax and cerci

cerci usually pale

♂

abdomen with black dorsal stripe; S8-9 black

♀

female and immature abdomen with black lateral dashes

Similar species: Blue Dasher has striped black and pale green thorax, green eyes, and white face. **Eastern Pondhawk** larger, with green face. **Red-faced** and **Plateau Dragonlets** have larger amber hindwing patch, female with pale stripes on S8-9. **Elfin Skimmer** smaller, has white face. **Meadowhawks** lack dark middorsal abdominal stripe.

Plateau Dragonlet • *Erythrodiplax basifusca*

26–30 mm, 1–1.2 in. Small southwestern dragonlet found at shallow marshes, marshy ponds, lakes, sometimes stream pools. Mature male face and eyes dark, thorax black, abdomen blue with black tip, cerci dark. Immature and female yellow, similar to Little Blue Dragonlet but with reduced dark abdominal markings and often more amber at base of hindwing.

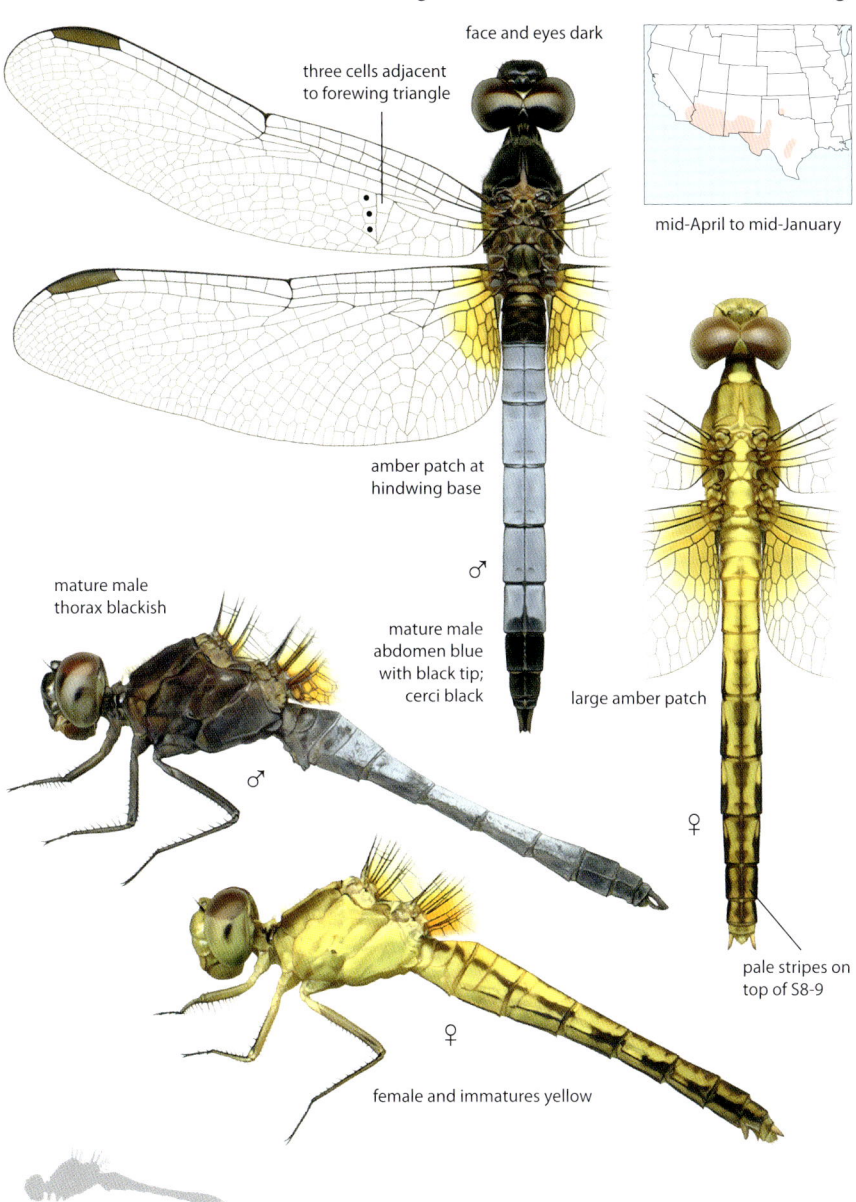

three cells adjacent to forewing triangle

face and eyes dark

mid-April to mid-January

amber patch at hindwing base

mature male thorax blackish

♂

mature male abdomen blue with black tip; cerci black

large amber patch

♀

pale stripes on top of S8-9

♀

female and immatures yellow

Similar species: Red-faced Dragonlet mature male has red face and brown thorax; female identical but may have reddish abdomen. **Meadowhawks** lack dark middorsal stripe on abdomen. **Blue Dasher** has white face, striped thorax; male with green eyes.

Seaside Dragonlet • *Erythrodiplax berenice*

31–35 mm, 1.2–1.4 in. Only North American dragonfly that breeds in salt water; found at coastal marshes, mangrove swamps, alkaline lakes in the West. Mature male all black with clear wings. Immatures and female heavily striped yellow and black, maturing females highly variable in pattern. Some females with brown wing markings.

mature male
all black

stigmas brown

north: late May to late Sept
south: mid-Feb to early Nov
Florida: all year

♂

♀

♀

some females have
brown wing markings

Similar species: Metallic Pennant has brown patch at hindwing base, thorax with metallic green reflections, side of abdomen base brown. **Black Setwing** has longer, narrower abdomen, darker stigmas. **Pin-tailed Pondhawk** abdomen narrower, male cerci pale; has dark hindwing patch. **Double-ringed** and **Martha's Pennants** have dark hindwing patch.

♂ mature male all black,
can be slightly pruinose

♀ immature female tiger-striped,
yellow-orange and black

♀

♀

♀ mature females
can be all black

Red-faced Dragonlet • *Erythrodiplax fusca*

24–35 mm, 0.9–1.4 in. Small dragonlet found locally in South Texas at shallow, marshy ponds and lakes. Mature male face and eyes red, thorax red-brown, abdomen red-brown at base, middle segments pale blue, tip dark, cerci pale. Immature and female yellowish; abdomen with prominent dark middorsal stripe. Both sexes with large amber hindwing marking.

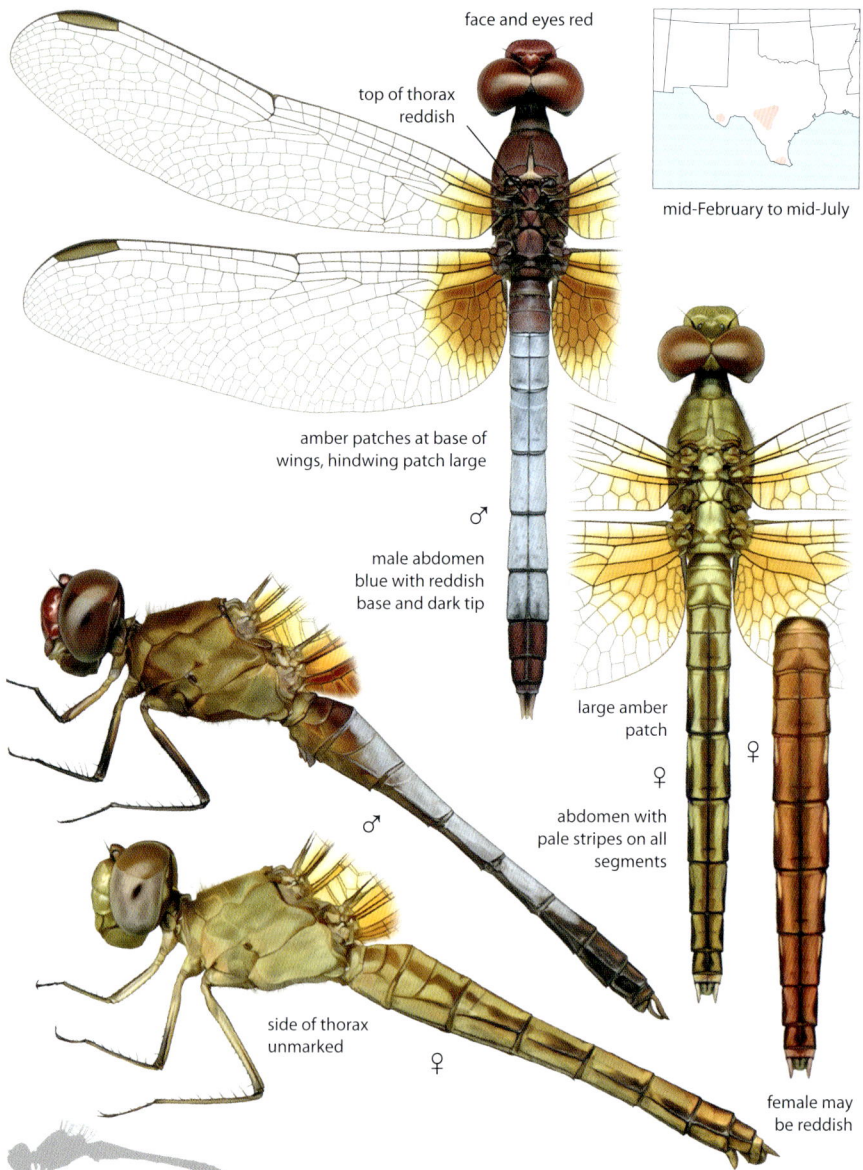

face and eyes red

top of thorax reddish

mid-February to mid-July

amber patches at base of wings, hindwing patch large

♂

male abdomen blue with reddish base and dark tip

large amber patch

♀

abdomen with pale stripes on all segments

♀

side of thorax unmarked

♂

♀

female may be reddish

Similar species: Plateau Dragonlet mature male has black face and thorax; female identical to yellow female Red-faced Dragonlet and not separable. **Blue Dasher** has white face, striped thorax, male with green eyes. **Meadowhawks** lack dark middorsal abdominal stripe.

Red-mantled Dragonlet • *Erythrodiplax fervida*

32–36 mm, 1.3–1.4 in. Well-marked dragonlet inhabiting shallow vegetated ponds, lakes, and slow-moving waters; single South Texas record. Male face, eyes, and thorax dark red, abdomen red and black, appendages pale red. Immature and female pale yellow, abdomen black with pale dashes. Both sexes with amber patch at base of both fore- and hindwings.

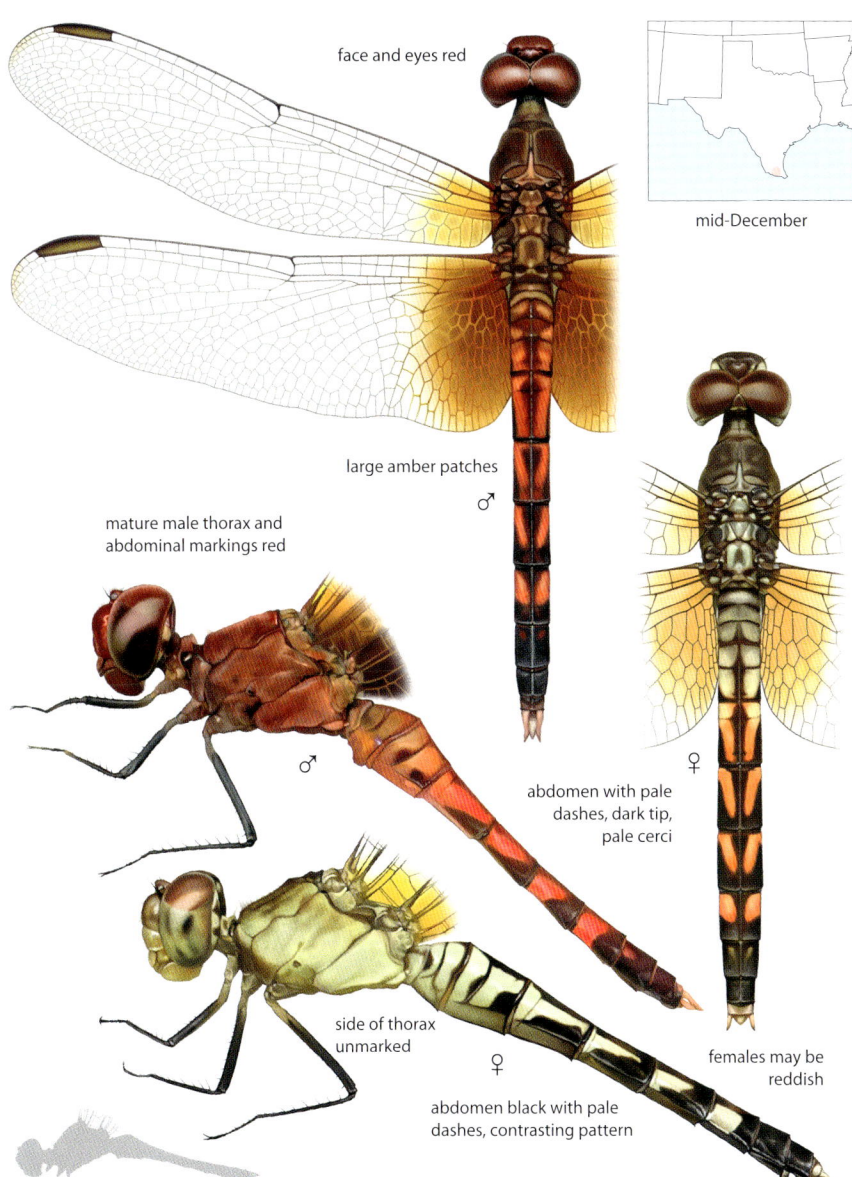

face and eyes red

mid-December

large amber patches

♂

mature male thorax and abdominal markings red

♂

abdomen with pale dashes, dark tip, pale cerci

♀

side of thorax unmarked

♀

females may be reddish

abdomen black with pale dashes, contrasting pattern

Similar species: Red-faced Dragonlet female abdomen with less contrasting pattern; dorsal pale markings parallel, forming continuous stripes. **Red Rock Skimmer** has more intricate abdominal pattern, including pale markings on top of S8-9; thorax with dark lateral stripes.

Black Meadowhawk • *Sympetrum danae*

30–33 mm, 1.2–1.3 in. Small, dark meadowhawk inhabiting marshes, bogs, fens, and vegetated lake borders. Our only meadowhawk without red markings. Immature and female thorax black with two lateral yellow stripes flanking two or three small yellow spots; front of thorax brown, abdomen black with yellow markings. Mature male becomes all black.

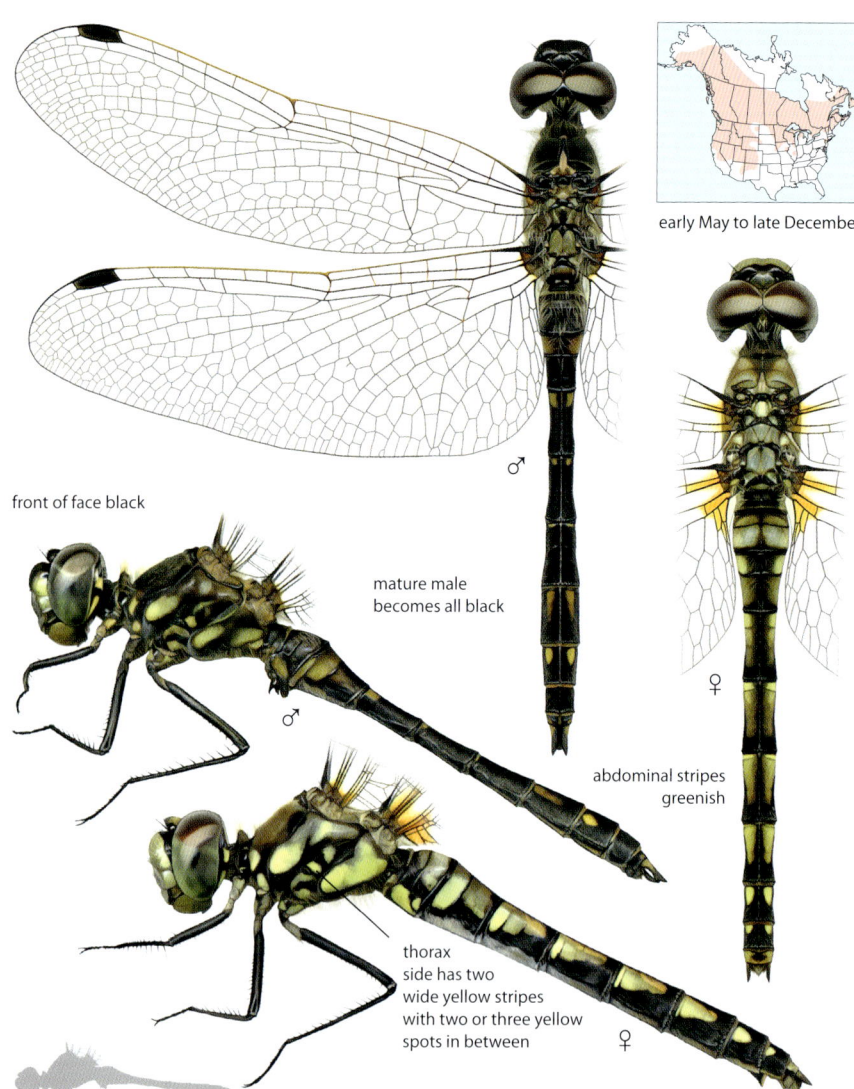

early May to late December

front of face black

mature male
becomes all black

♂

♀

abdominal stripes
greenish

thorax
side has two
wide yellow stripes
with two or three yellow
spots in between

♀

Similar species: Seaside Dragonlet found in very different coastal marsh habitat; wings with denser venation and brown stigmas, immature and females with multiple yellow lateral thoracic stripes but not spots. **Double-ringed Pennant** has dark patch at base of hindwing. **Dot-tailed** and **Frosted Whitefaces** have white faces, small dark triangular mark at base of hindwing. **Elfin Skimmer** smaller, eye with vertical brown stripe, abdomen with yellow markings along top. **Ebony Boghaunter** has blue-green-gray eyes, narrow white rings at abdomen base.

Blue-faced Meadowhawk • *Sympetrum ambiguum*

36–38 mm, 1.4–1.5 in. Uniquely colored eastern meadowhawk found at marshes and woodland ponds, including temporary ones. Thorax unmarked gray. Legs tan. Abdomen with diffused black markings forming dark rings. Mature male with distinctive green-blue face. Mature male and some females develop red coloration on abdomen and top of thorax.

male face green-blue,
eyes blue-gray

north: late May to mid-Nov
south: early Apr to late Dec

♂

thorax
unmarked gray

♂

abdomen with
diffused black rings

♀

female may be
red or brown

legs pale

♀

Similar species: Autumn Meadowhawk has brown or yellow/brown thorax; abdomen lacks black rings.

White-faced Meadowhawk • *Sympetrum obtrusum*

31–39 mm, 1.2–1.5 in. Widespread meadowhawk common at marshes, bogs, forest ponds, and lakes, including temporary ones. Face clear white when mature. Mature male and some females red. Thorax with vague yellow lateral stripes that fade with age. Abdomen with black lateral triangles. Legs and wing veins black.

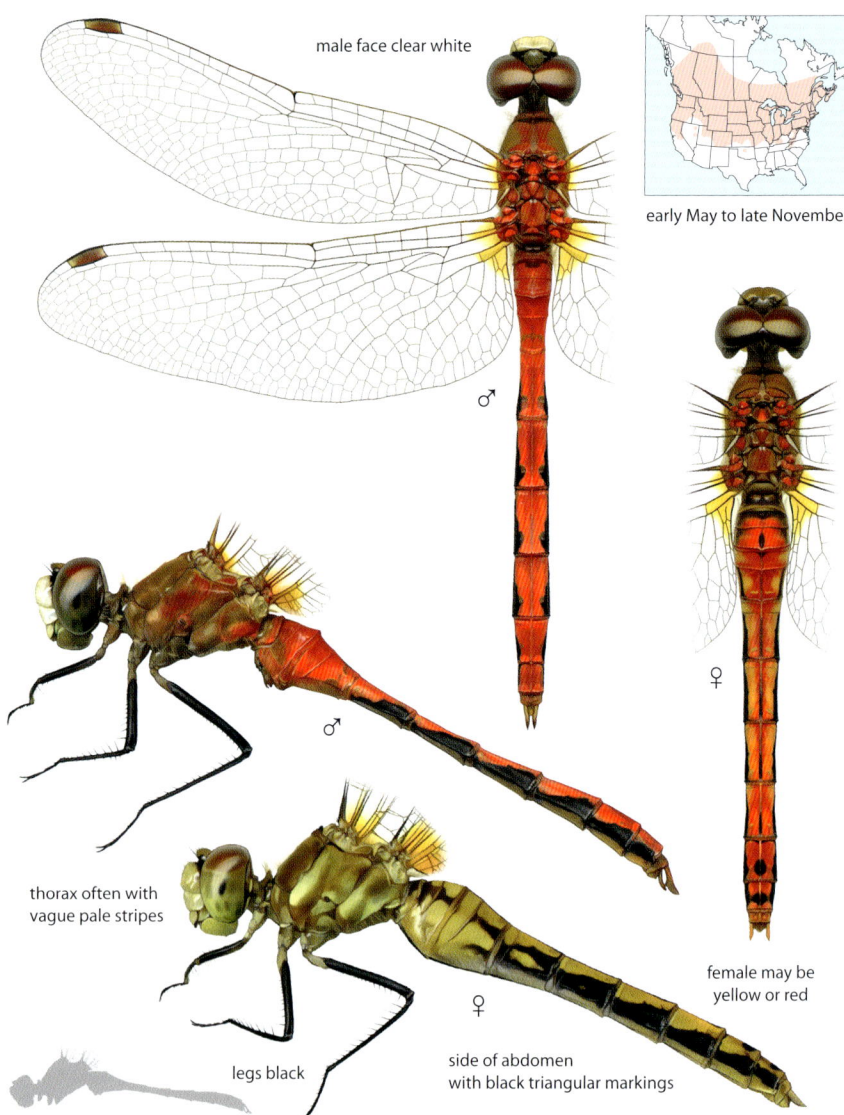

male face clear white

early May to late November

♂

♀

♂

thorax often with vague pale stripes

legs black

♀

female may be yellow or red

side of abdomen with black triangular markings

Similar species: Western form of **Cherry-faced Meadowhawk** male has red face, anterior wing veins orange; eastern-form male has dirty white face. Immature and female only separable by structure of male hamules and female subgenital plate, page 435. **Ruby Meadowhawk** has tan face; immature and female separable only by male hamules and female subgenital plate, page 435. **Striped Meadowhawk** has better defined pale lateral thoracic stripes, dirty white face; black abdominal markings often reduced but variable.

Ruby Meadowhawk • *Sympetrum rubicundulum*

33–34 mm, 1.3 in. Common meadowhawk inhabiting open marshes, marshy lakes, ponds, including temporary ones, from the East to the northern Great Plains. Abdomen with black lateral triangles. Legs black. Face yellow to brown. Mature male and some females red. Western form (Minnesota, Iowa) with extensive amber wing bases; eastern form clear-winged.

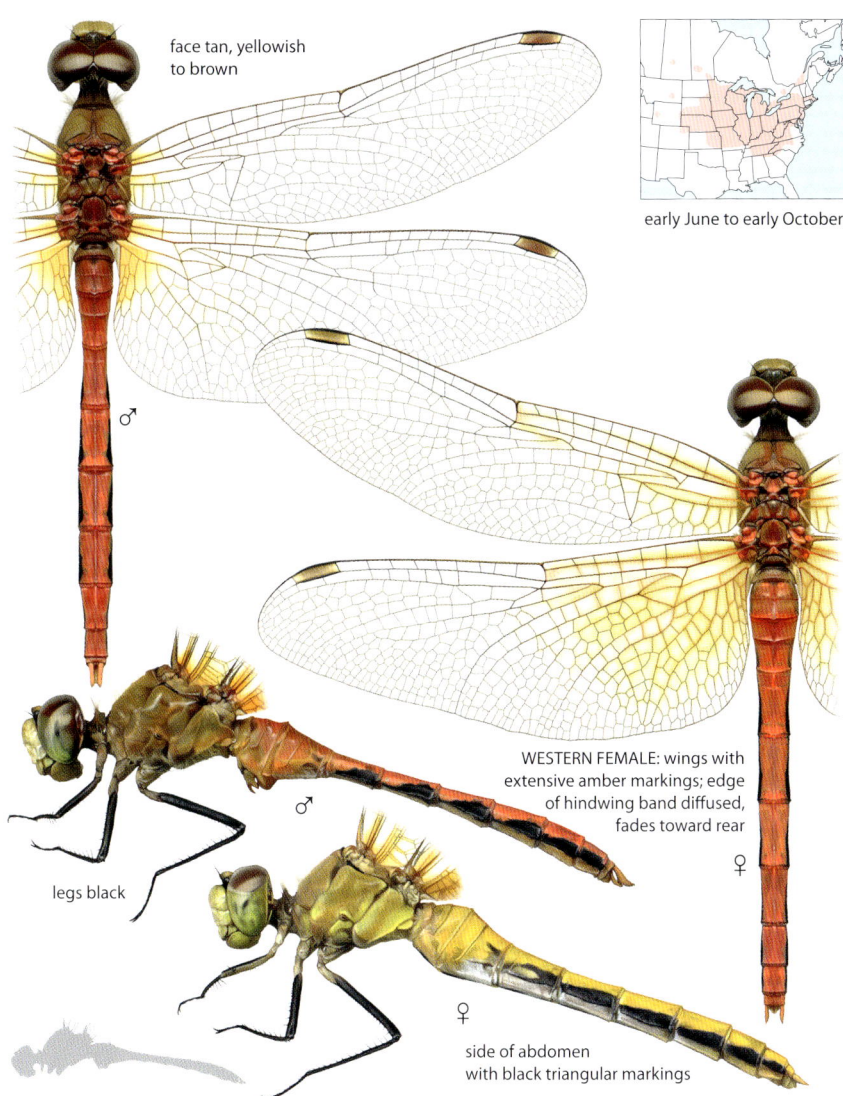

face tan, yellowish to brown

♂

early June to early October

WESTERN FEMALE: wings with extensive amber markings; edge of hindwing band diffused, fades toward rear

♀

♂

legs black

♀

side of abdomen with black triangular markings

Similar species: Band-winged Meadowhawk similar to western Ruby, but edge of wing band more defined with brown cross-band. Western form of **Cherry-faced Meadowhawk** male has red face, anterior wing veins orange; females with amber wing bases similar, should be separated by subgenital plate, page 435. Eastern **Cherry-faced Meadowhawk** identical, separable only by male hamules and female subgenital plate, page 435. **White-faced Meadowhawk** has clean white face when mature; immatures separable only by male hamules and female subgenital plate, page 435.

Cherry-faced Meadowhawk • *Sympetrum internum*

21–36 mm, 0.8–1.4 in. Common, widespread meadowhawk inhabiting shallow ponds, lakes, and marshes. Abdomen with black lateral triangular markings. Legs black. **Western form** (Ohio west) has anterior wing veins orange; some females with basal half of wing amber, mature male face red.

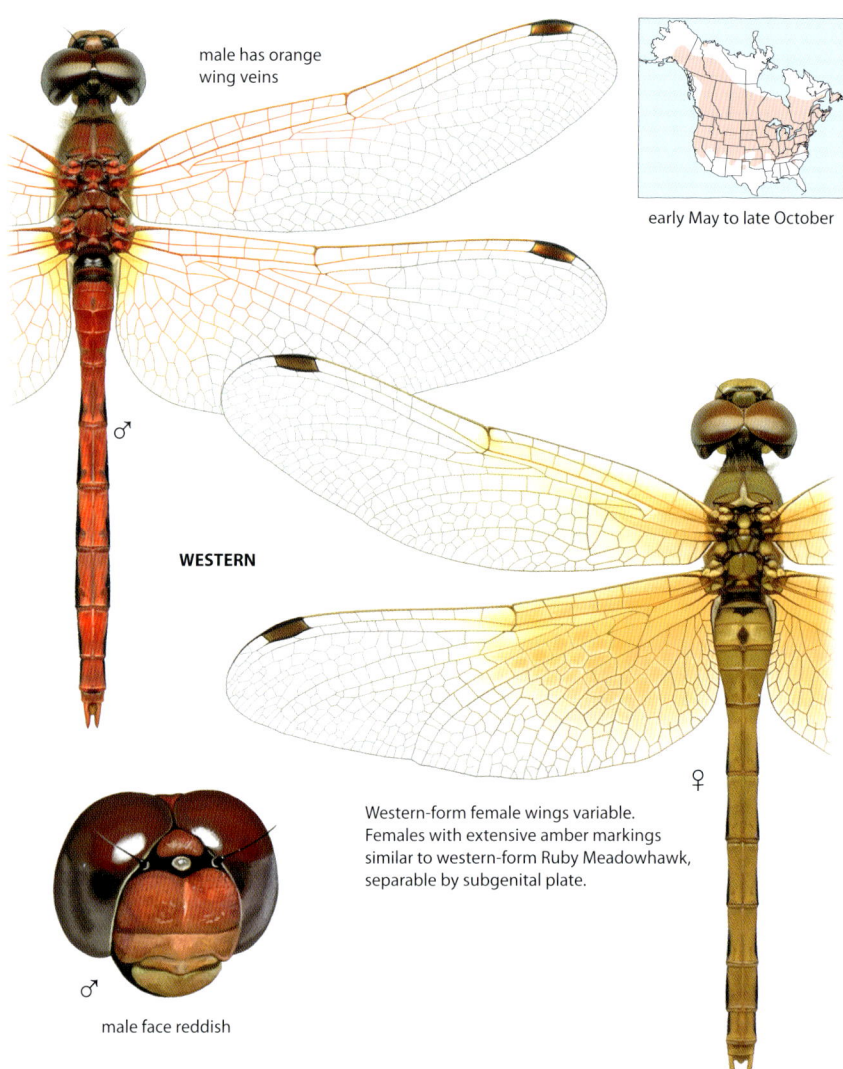

male has orange wing veins

early May to late October

WESTERN

Western-form female wings variable. Females with extensive amber markings similar to western-form Ruby Meadowhawk, separable by subgenital plate.

male face reddish

Similar species: Reddish face separates male Cherry-faced from similarly patterned **White-faced** and **Ruby Meadowhawks**. **White-faced Meadowhawk** male has clean white face and dark wing veins; immatures and females separable only by male hamules and female subgenital plates, page 435. **Ruby Meadowhawk** male lacks red face and orange wing veins; female largely identical, separable by subgenital plate, page 435. **Saffron-winged, Autumn,** and **Red-veined Meadowhawks** have less black on side of abdomen. **Band-winged Meadowhawk** has black stripes on side of abdomen; edge of wing bands more defined.

Eastern form has black wing veins; face usually dirty yellow-white to brown, but males of some northern populations have reddish face.

wing veins black

♂ ♀

face dirty white;
some eastern males have reddish face

EASTERN

♂

black triangles
on side of abdomen

♀

legs black

female may be brown or red

Similar species: Eastern form with dirty white face nearly identical to **White-faced** and **Ruby Meadowhawks**. Males should be separated by the structure of their hamules, females by their subgenital plates, page 435.

Band-winged Meadowhawk • *Sympetrum semicinctum*

24–40 mm, 0.9–1.4 in. Widespread meadowhawk with several distinct forms, inhabits shallow open ponds and marshes, more common in the West. Basal half of wings amber; hindwing band complete from leading edge to rear margin, edge of band straight. Abdomen with black lateral stripe. Legs black. **Eastern form** has unmarked thorax.

mature male
face reddish

north: early Jun to late Dec
south: early May to early Nov

wings with amber at bases; hindwing band extends from base to nodus, edge of band straight

♂

♀

♂

side of thorax
unmarked

♀

black stripes on side of abdomen

legs black

Western forms have black markings on side of thorax.
Northwestern form, **occidentale,** has evenly colored wing bands. Great Plains and Rocky Mountains form, **fasciatum,** has darker bands on other edge of wing patch. Southwestern form, **californicum,** has wing markings reduced, sometimes absent.

subspecies **occidentale**

♂

subspecies **fasciatum**

♂

♂

side of thorax marked with black

♀

Similar species: Western forms of **Ruby** and **Cherry-faced Meadowhawks** can have similar but more diffused wing markings, lack black lateral thoracic markings; black lateral abdominal markings triangular. **Black Meadowhawk** immatures and female have darker lateral thoracic pattern.

Saffron-winged Meadowhawk • *Sympetrum costiferum*

31–37 mm, 1.2–1.5 in. Common, widespread meadowhawk found at marshes, vegetated ponds, and lakes, including both alkaline and acidic waters. Wings with amber stripe along leading edge that fades with age, retained by female longest. Abdomen with black lateral stripes (may be reduced or absent). Legs variably striped.

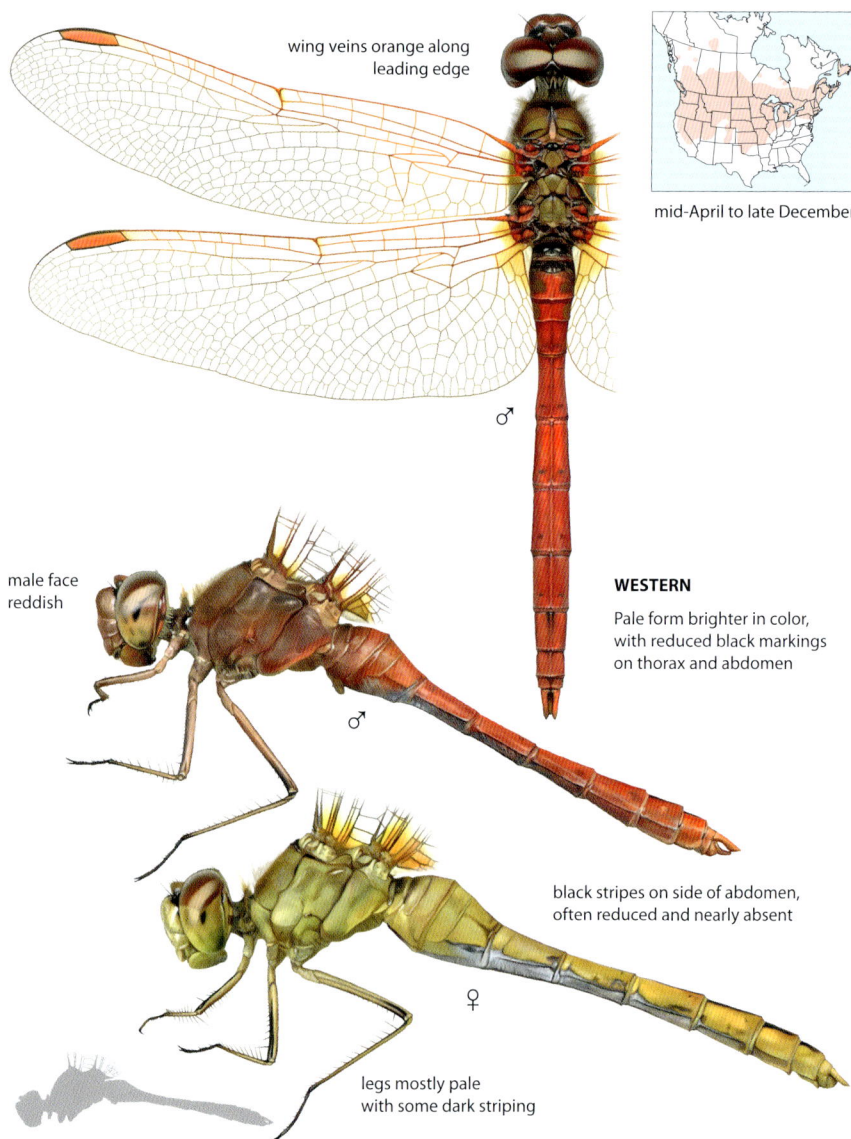

wing veins orange along leading edge

mid-April to late December

male face reddish

WESTERN

Pale form brighter in color, with reduced black markings on thorax and abdomen

black stripes on side of abdomen, often reduced and nearly absent

♂

♀

legs mostly pale with some dark striping

Similar species: Autumn Meadowhawk has all pale legs (Saffron-winged usually with some black striping) and amber only at wing bases. **Red-veined Meadowhawk** has black legs, thorax with white lateral stripes when young, male with red face. **Cherry-faced, White-faced,** and **Ruby Meadowhawks** have black legs, black triangular markings on side of abdomen.

Eastern populations considerably darker, with dark markings on thorax and black stripe on side of abdomen. Mature male dark red; face brown, wing veins darker, legs black. Female legs usually with some pale markings.

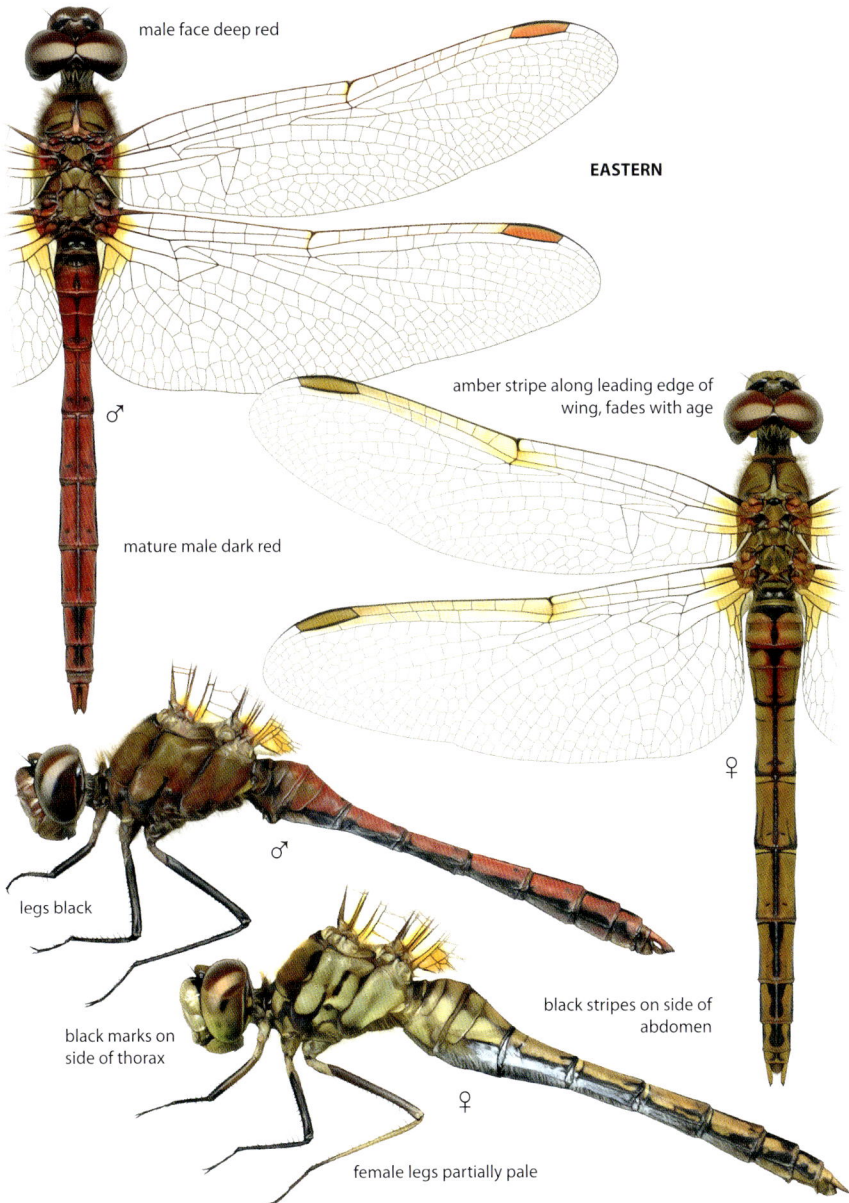

male face deep red

EASTERN

amber stripe along leading edge of wing, fades with age

♂

mature male dark red

♀

legs black

♂

black stripes on side of abdomen

black marks on side of thorax

♀

female legs partially pale

Similar species: Dark red male Saffron-winged Meadowhawk distinctive in the East. **Cherry-faced, White-faced,** and **Ruby Meadowhawks** have black legs, black triangular markings on side of abdomen; males brighter red.

Red-veined Meadowhawk • *Sympetrum madidum*

42–45 mm, 1.7–1.8 in. Large northwestern meadowhawk found at vegetated temporary ponds, including brackish waters. Immature and female thorax brown with two white lateral stripes. Abdomen with white lateral stripes bordered by black stripes, wings with amber stripe along leading edge; stripes fade with age. Wings tinted brown when mature.

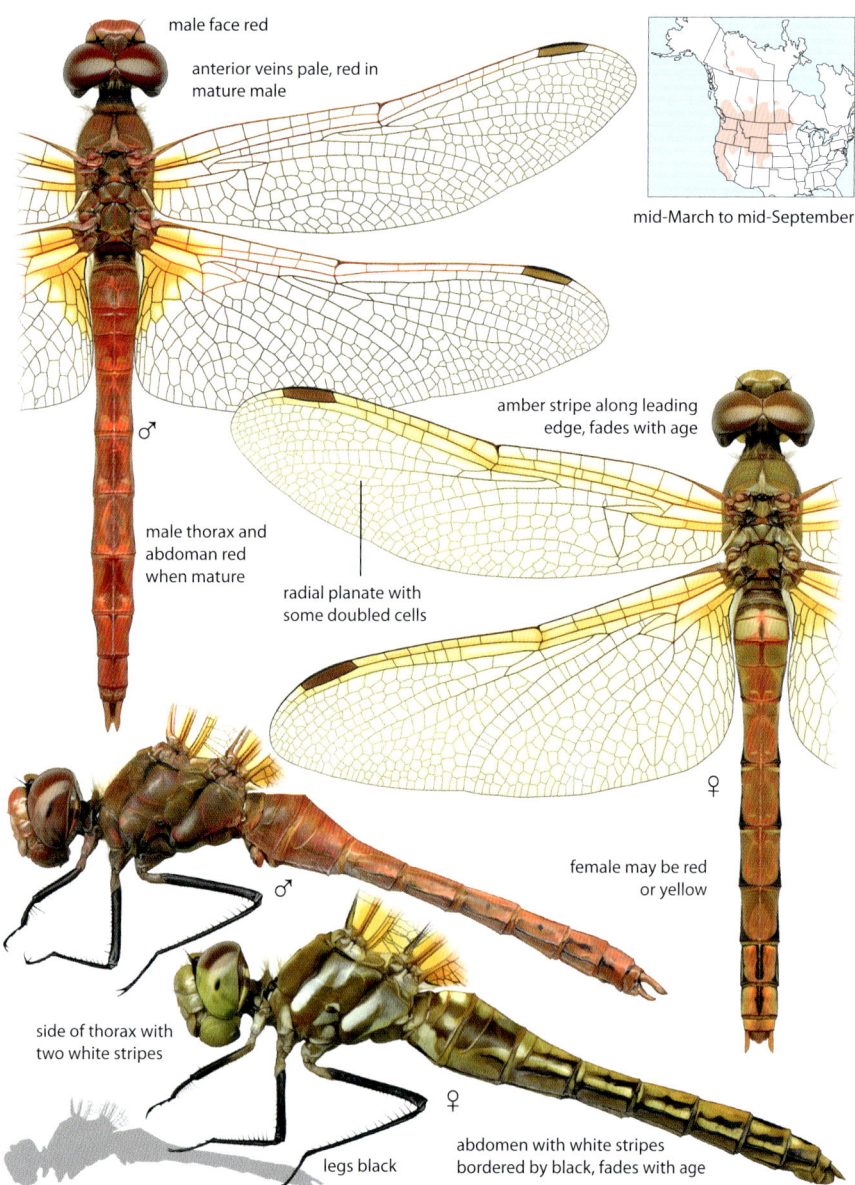

male face red

anterior veins pale, red in mature male

mid-March to mid-September

amber stripe along leading edge, fades with age

♂

male thorax and abdoman red when mature

radial planate with some doubled cells

♀

female may be red or yellow

♂

side of thorax with two white stripes

♀

legs black

abdomen with white stripes bordered by black, fades with age

Similar species: Striped Meadowhawk lacks white lateral markings on abdomen, usually has single row of cells in radial planate; mature male with tan face. **Variegated Meadowhawk** has yellow dots on side of thorax, legs striped black and brown. **Cardinal Meadowhawk** has dark markings at base of wings.

Striped Meadowhawk • *Sympetrum pallipes*

34–38 mm, 1.3–1.5 in. Common western meadowhawk inhabiting ponds and shallow marshes, including those formed by seasonal flooding. Thorax brown with whitish frontal stripes and two white lateral stripes. Abdomen usually without black dorsal markings, black lateral markings variable; drier interior populations paler.

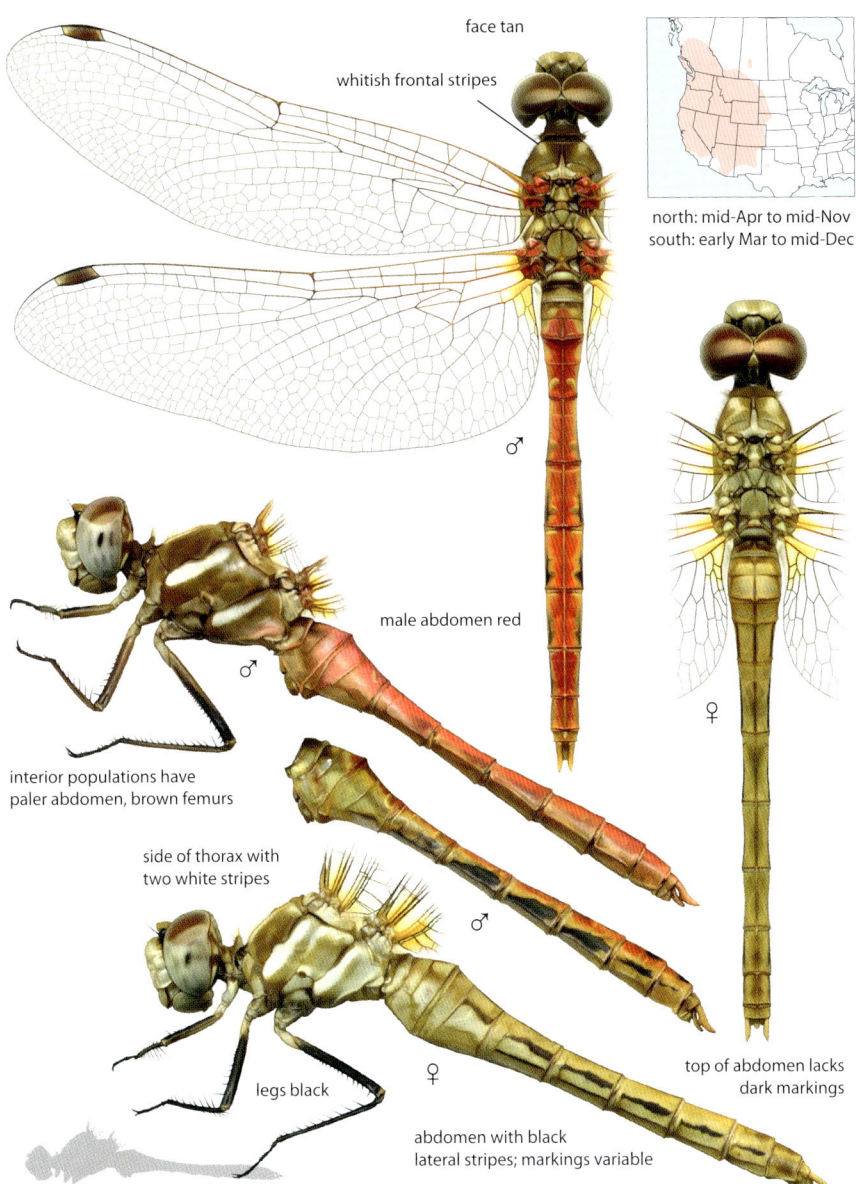

face tan

whitish frontal stripes

north: mid-Apr to mid-Nov
south: early Mar to mid-Dec

♂

male abdomen red

interior populations have
paler abdomen, brown femurs

side of thorax with
two white stripes

♂

♀

legs black

♀

abdomen with black
lateral stripes; markings variable

top of abdomen lacks
dark markings

Similar species: Red-veined Meadowhawk has two rows of cells in radial planate; mature male with red face, female with white lateral stripes on abdomen, immatures with amber stripe along leading edge of wings. **Variegated Meadowhawk** has yellow dots on side of thorax, abdomen with white lateral markings.

Variegated Meadowhawk • *Sympetrum corruptum*

39–42 mm, 1.5–1.7 in. Complexly patterned meadowhawk found at wide variety of ponds, lakes, and stream pools; common in the West, highly migratory. Thorax gray-brown with two white lateral stripes with yellow spots at stripe bases. Abdomen brown with white lateral marks, black stripes on top of S8-9; male with red dorsal stripe and cross-bands.

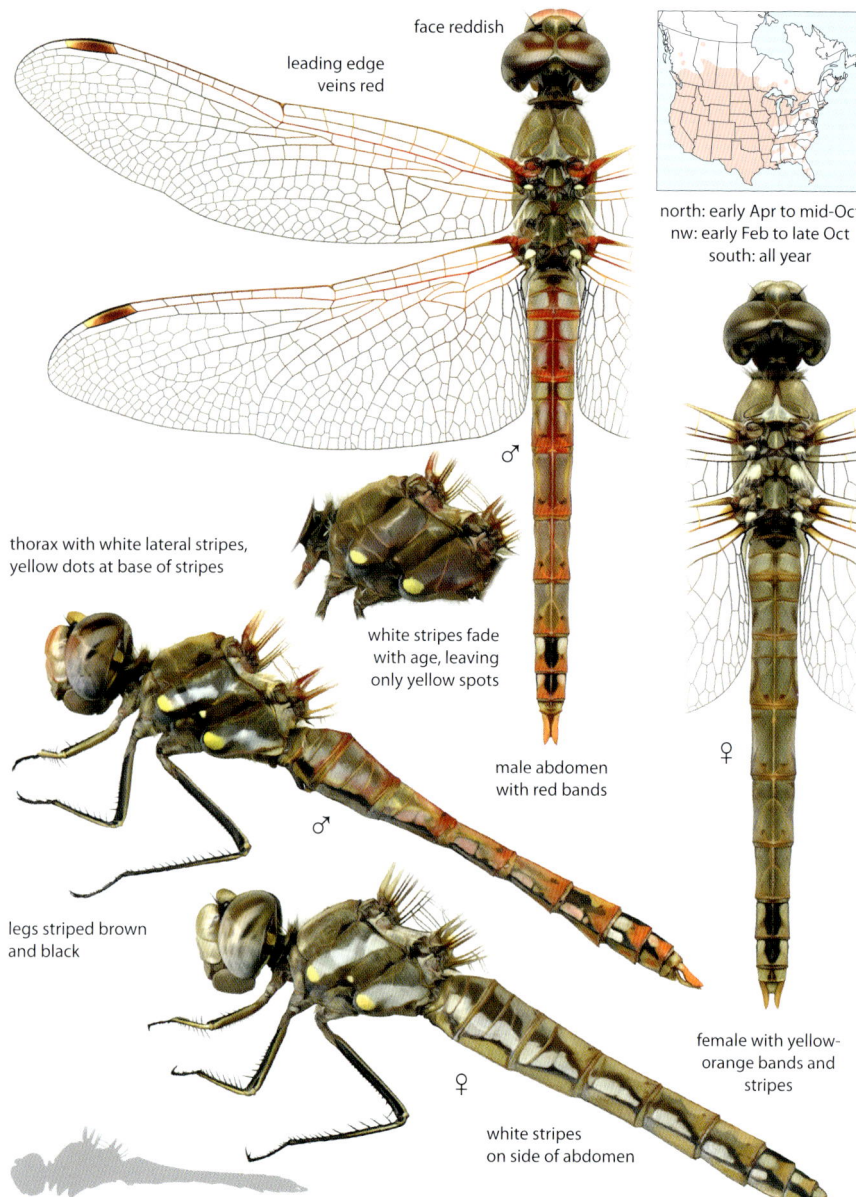

face reddish

leading edge veins red

north: early Apr to mid-Oct
nw: early Feb to late Oct
south: all year

thorax with white lateral stripes, yellow dots at base of stripes

white stripes fade with age, leaving only yellow spots

male abdomen with red bands

legs striped brown and black

♂

♀

white stripes on side of abdomen

female with yellow-orange bands and stripes

Similar species: Red-veined Meadowhawk lacks yellow dots on side of thorax, has all-black legs, wings tinted amber or brown. **Striped Meadowhawk** lacks yellow dots on side of thorax; abdomen often lacks white laterally, black dorsally.

Cardinal Meadowhawk • *Sympetrum illotum*

38–40 mm, 1.5–1.6 in. Stout western meadowhawk common at ponds, lakes, marshes, and stream pools. Thorax with two lateral white spots. Wings with amber bases and dark streaks, amber extending to nodus along leading edge. Abdomen lacks dark markings. Immatures brown, male and some females red upon maturity.

early April to mid-December
(recorded in all months)

both sexes with dark streaks and
amber patches at base of wings

♂

♀

thorax with two
white lateral spots

♀

Similar species: White-spotted thorax distinctive. No other meadowhawk has dark streaks at base of wings.

Autumn Meadowhawk • *Sympetrum vicinum*

31–35 mm, 1.2–1.4 in. Common, widespread species at marshes, vegetated ponds, and lakes. Often the last dragonfly seen each year. Immature thorax brown with yellow sides. Abdomen with limited dark markings. Legs pale brown. Mature male and some females develop red abdomen. Mature male face red. Female with spout-like subgenital plate.

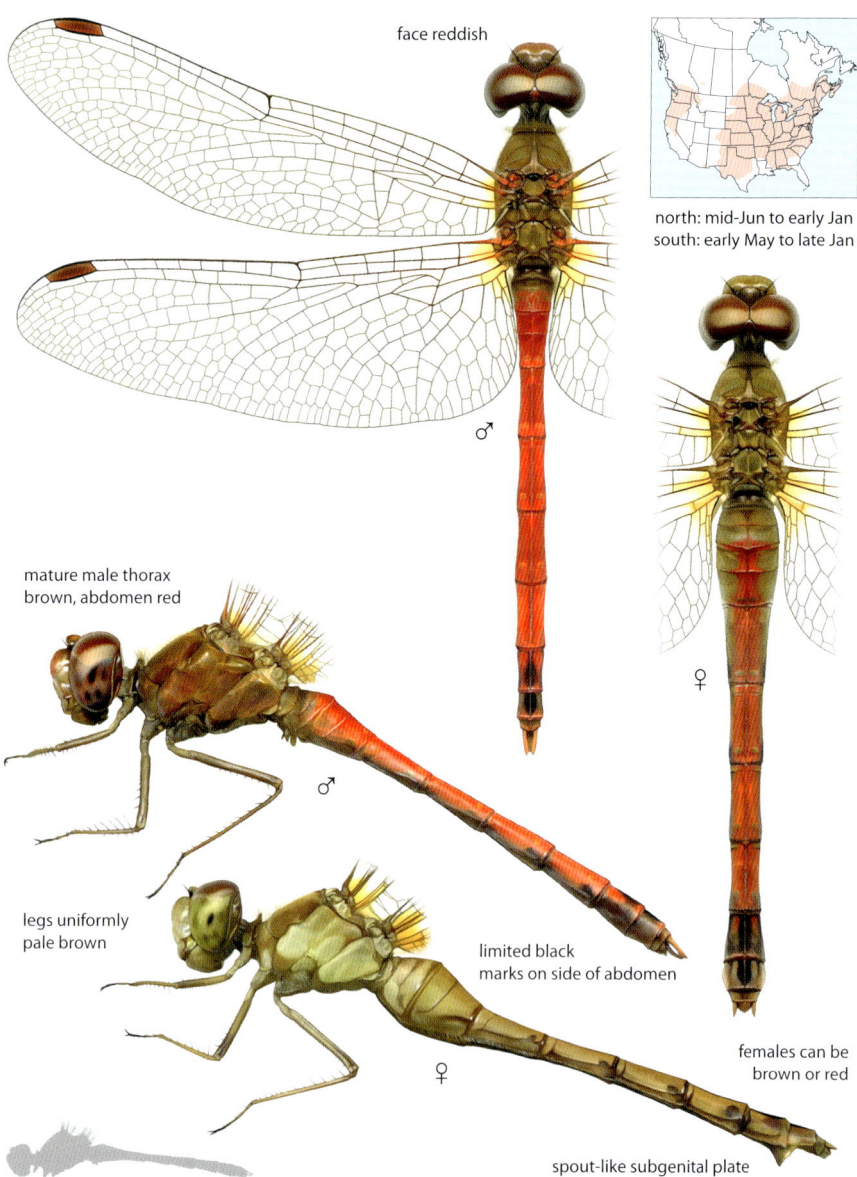

face reddish

north: mid-Jun to early Jan
south: early May to late Jan

♂

♀

mature male thorax
brown, abdomen red

♂

legs uniformly
pale brown

limited black
marks on side of abdomen

females can be
brown or red

♀

spout-like subgenital plate

Similar species: All-pale legs separate Autumn Meadowhawk from most other meadowhawks. **Saffron-winged Meadowhawk** usually has some dark striping on legs, immatures and females with amber stripe along leading edge of wing; female lacks spout-like subgenital plate. **Blue-faced Meadowhawk** has gray thorax, abdomen with dark bands.

Spot-winged Meadowhawk • *Sympetrum signiferum*

33–42 mm, 1.3–1.7 in. Southwestern meadowhawk inhabiting vegetated stream pools, ponds, marshes. Wings with amber bases, hindwing with small dark patch, stigmas dark. Thorax unmarked, abdomen with small black lateral spots. Legs pale brown. Mature male and some females with red abdomen. Mature male face red, eyes red over blue-gray.

face red

stigmas dark

early Aug to late November

wings amber at bases, hindwing with small dark patch

♂

mature male red

legs pale

♀

♂

♀

short black marks on side of abdomen

Similar species: Cardinal Meadowhawk has dark streaks at the base of each wing, amber patches more extensive, thorax with two white lateral spots, abdomen without dark markings. **Autumn Meadowhawk** range not known to overlap; lacks dark hindwing patch.

Elfin Skimmer • *Nannothemis bella*

18–20 mm, 0.7–0.8 in. Smallest North American dragonfly; local to eastern sphagnum bogs, lake margins, and marl fens. Tiny. Face white, eye with vertical brown stripe. Mature male pruinose bluish-white, abdomen slightly clubbed. Female striped yellow and black, wings often with amber at base.

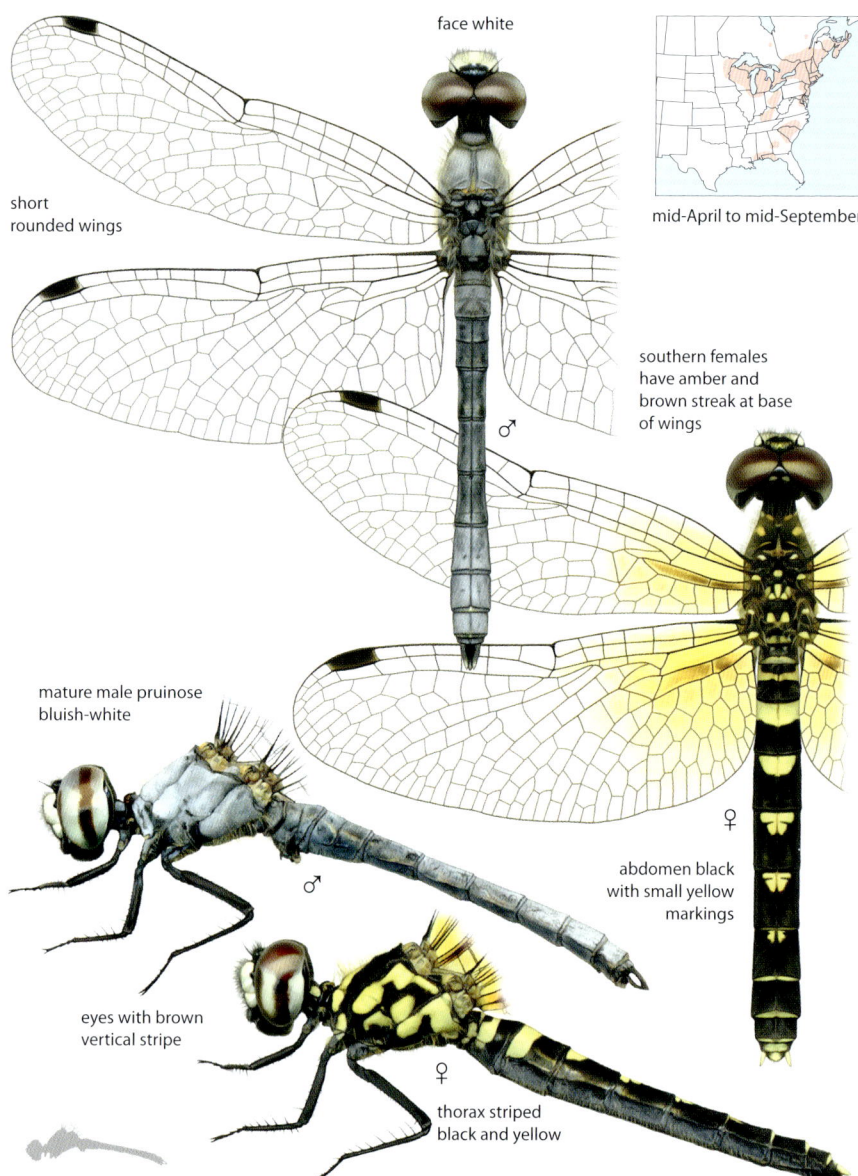

face white

short
rounded wings

mid-April to mid-September

southern females
have amber and
brown streak at base
of wings

mature male pruinose
bluish-white

♂

♀

abdomen black
with small yellow
markings

♂

eyes with brown
vertical stripe

♀

thorax striped
black and yellow

Similar species: Little Blue Dragonlet male larger; face black, appendages usually pale, eyes lack brown vertical stripe. **Whitefaces** larger, have dark triangular patch at hindwing base, dorsal abdominal markings mostly longer than narrow. Female **Eastern Amberwing** with brown abdomen, pale legs, brown wing markings, red stigma.

Blue Dasher • *Pachydiplax longipennis*

28–45 mm, 1.1–1.8 in. Abundant small skimmer widespread at almost all still-water habitats. Face white, eyes green, thorax yellow-green with three dark lateral stripes. Immature and female abdomen black with pale yellow dashes. Abdomen of mature male and some females pruinose blue with black tip. Thorax of mature western male pruinose blue.

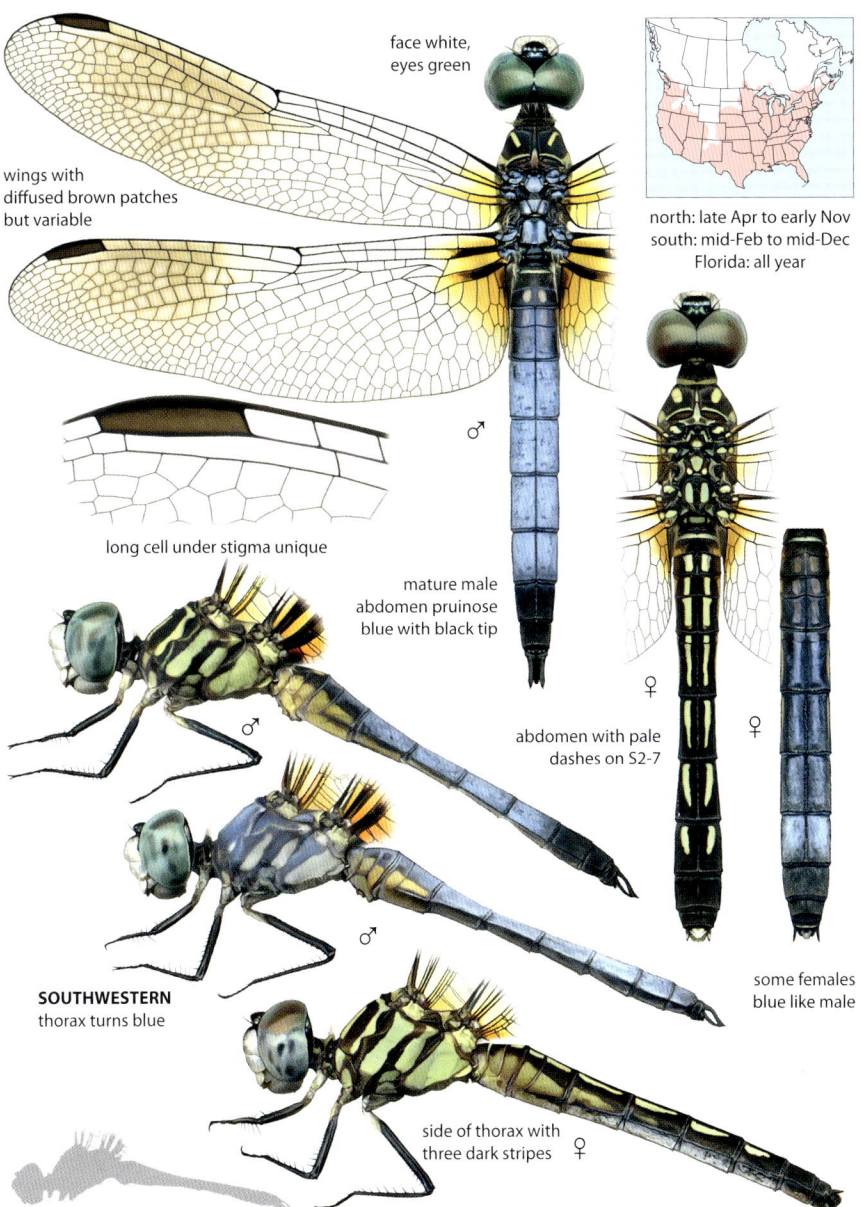

face white,
eyes green

wings with
diffused brown patches
but variable

north: late Apr to early Nov
south: mid-Feb to mid-Dec
Florida: all year

long cell under stigma unique

♂

mature male
abdomen pruinose
blue with black tip

♀

♀

abdomen with pale
dashes on S2-7

♂

SOUTHWESTERN
thorax turns blue

♂

some females
blue like male

side of thorax with
three dark stripes ♀

Similar species: Eastern and **Western Pondhawks** have green face, lack wing coloration, often perch on ground. **Little Blue** and **Plateau Dragonlets** lack dark lateral thoracic stripes. **Tropical dashers** have conspicuous pale markings on S7.

Thornbush Dasher • *Micrathyria hagenii*

33–35 mm, 1.3–1.4 in. Small, fairly common tropical dasher inhabiting marshes, ponds, lakes, and canals, primarily in Texas. Face white, eyes blue-green. Thorax pale green with dark lateral stripes resembling the letters *IYI*. Abdomen black with rows of pale dashes, large square-shaped pale spots on S7; male abdomen slightly clubbed.

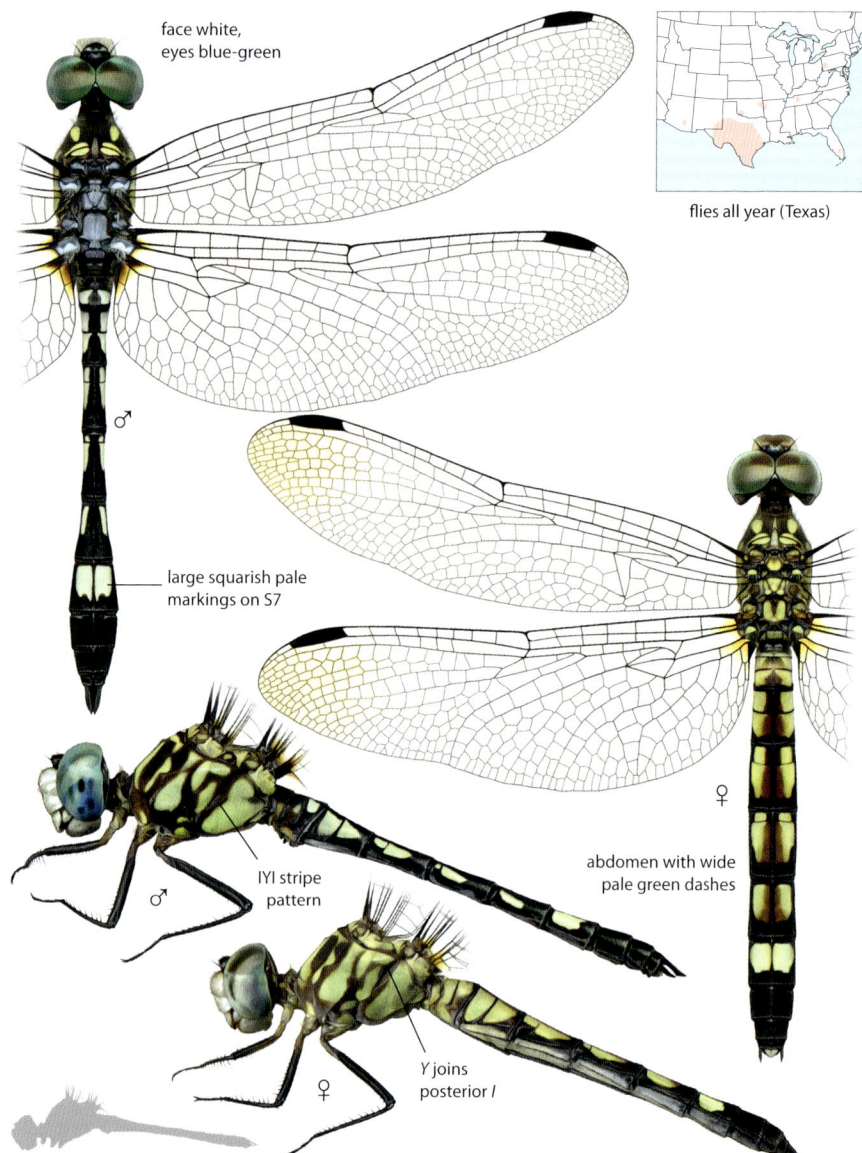

face white,
eyes blue-green

flies all year (Texas)

large squarish pale
markings on S7

♂

IYI stripe
pattern

♂

Y joins
posterior *I*

♀

♀

abdomen with wide
pale green dashes

Similar species: Caribbean Dasher lateral thoracic pattern IYI, but *Y* does not join posterior *I*; abdomen longer and slenderer, S7 pale markings larger and trapezoidal, male S6 black, female pale abdominal stripes narrower. **Spot-tailed Dasher** with IIW lateral thoracic pattern, S7 spots smaller and triangular. **Three-striped Dasher** with III lateral thoracic pattern. **Blue Dasher** has III lateral thoracic pattern and lacks large pale markings on S7.

Caribbean Dasher • *Micrathyria dissocians*

37–40 mm, 1.5–1.6 in. Medium-sized tropical dasher recorded in South Texas. Face white, eyes green to blue. Thorax pale green with dark lateral stripes resembling the letters *IYI*. Abdomen slender, black with narrow pale stripes, large trapezoidal pale marks on S7, male abdomen slightly clubbed.

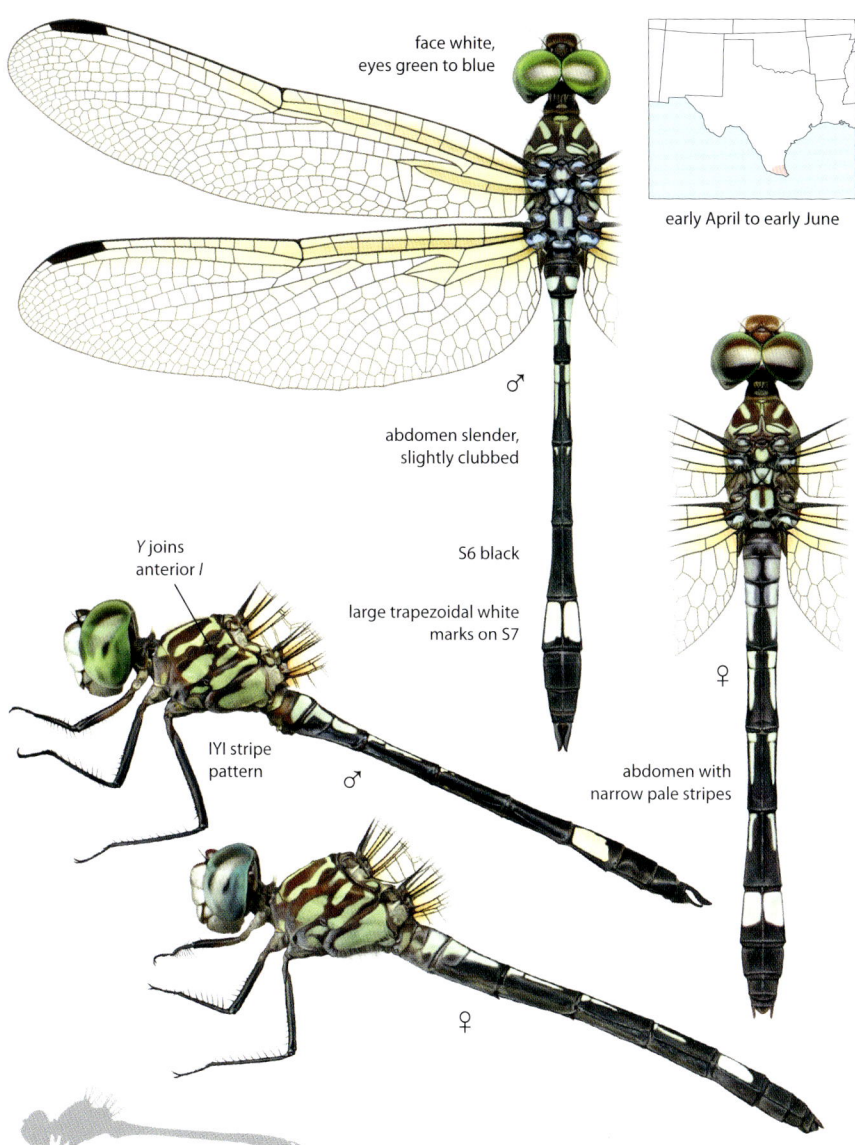

face white,
eyes green to blue

early April to early June

♂

abdomen slender,
slightly clubbed

Y joins
anterior *I*

S6 black

large trapezoidal white
marks on S7

IYI stripe
pattern

♂

♀

abdomen with
narrow pale stripes

♀

Similar species: Thornbush Dasher lateral thoracic pattern IYI, but *Y* joined to posterior *I* at top; abdomen shorter, S7 pale markings smaller and square-shaped, male with pale dashes on S3-6, female abdomen stout with wider pale markings. **Spot-tailed Dasher** with IIW lateral thoracic pattern, S7 spots triangular. **Three-striped** and **Blue Dashers** have III lateral thoracic pattern; Blue Dasher lacks large pale markings on S7.

Spot-tailed Dasher • *Micrathyria aequalis*

28–33 mm, 1.1–1.3 in. Very small tropical dasher found at vegetated ponds, marshes, ditches, sloughs, mainly in South Texas and Florida. Face white, eyes green. Thorax with complex WII lateral stripe pattern. Pale triangular markings on male S7, female S5-7 (S7 mark largest). Mature male thorax and abdomen pruinose gray-blue, abdomen tip black.

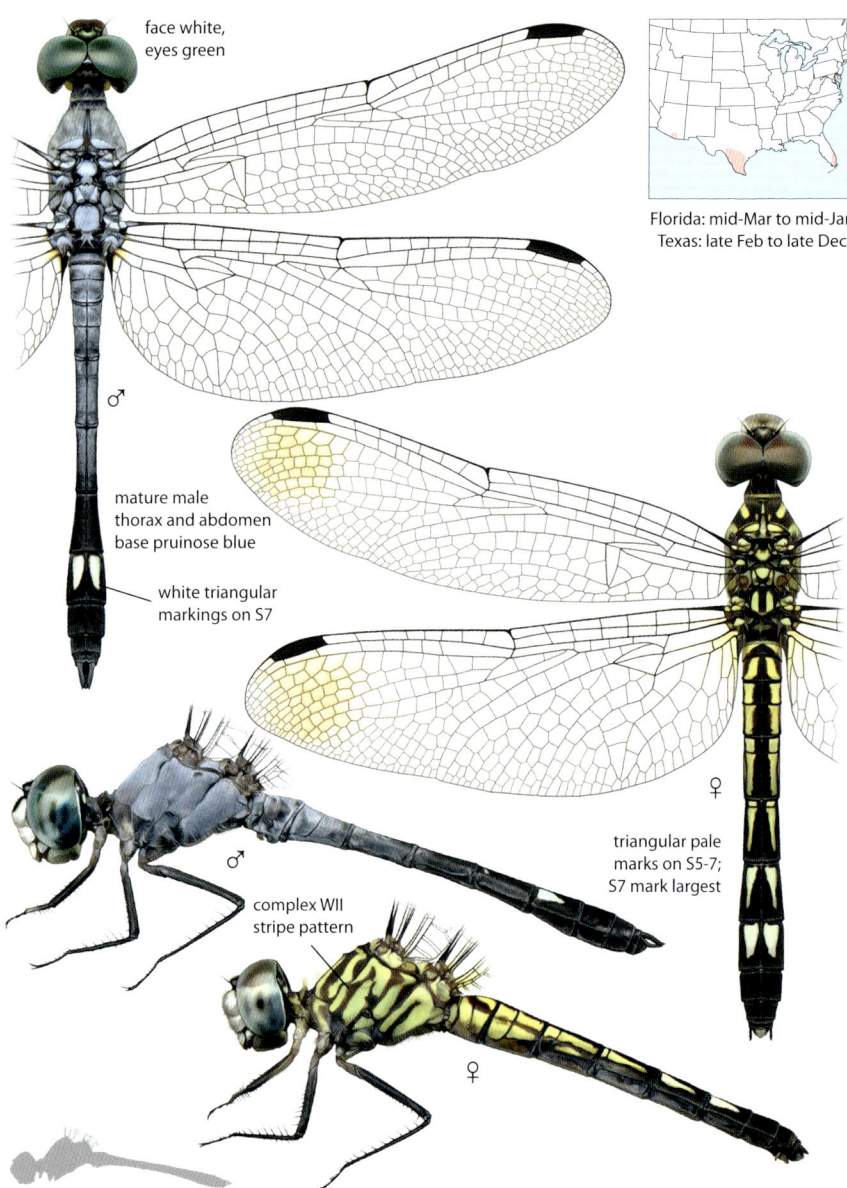

face white, eyes green

Florida: mid-Mar to mid-Jan
Texas: late Feb to late Dec

mature male thorax and abdomen base pruinose blue

white triangular markings on S7

triangular pale marks on S5-7; S7 mark largest

complex WII stripe pattern

Similar species: Thornbush and **Caribbean Dashers** have IYI lateral thoracic pattern, S7 pale markings larger and squarer in shape. **Three-striped Dasher** has III lateral thoracic pattern, S7 pale marking rectangular. **Blue Dasher** has III lateral thoracic pattern, narrow dashes on S7.

Three-striped Dasher • *Micrathyria didyma*

35–41 mm, 1.4–1.6 in. Small tropical dasher found at shaded ponds, canals, and swamps, primarily in South Texas and Florida. Face pale, eyes green. Thorax green with three dark brown lateral stripes. Abdomen black with narrow pale markings, top of S7 with large rectangular pale green markings.

face pale, eyes green

Florida: late Apr to mid-Dec
Texas: early May to mid-Nov

abdomen slender, slightly clubbed

♂

thorax green with three dark lateral stripes

rectangular pale markings on S7

abdomen with narrow pale green markings; S6 almost all black

♀

♂

♀

Similar species: Blue Dasher abdomen shorter and stouter, lacks pale rectangular markings on S7. **Spot-tailed Dasher** has more complex lateral thoracic pattern, S7 pale markings triangular.

Black Setwing • *Dythemis nigrescens*

40–45 mm, 1.6–1.8 in. Dark, slender southwestern setwing inhabiting streams and rivers, also ponds and lakes. Mature male all black, immatures heavily striped dark brown and pale. Mature female loses pattern contrast, may appear light overall or darken like male but usually retains some patterning. Wings mostly clear with dark at tips.

male face and eyes all dark

early March to early January

wings clear, often with tips darkened

mature male all black, abdomen slender

abdomen slender, pale markings on all segments

thorax heavily striped

Similar species: Swift Setwing has more contrasting pattern, lateral thoracic stripes in a YIY pattern; female lacks pale stripes on side of abdomen, has S8-10 black, darker at wing tips. **Pin-tailed Pondhawk** has dark patch at base of hindwing, base of abdomen more expanded, appendages pale. **Seaside Dragonlet** smaller with shorter, more rounded wings.

Swift Setwing • *Dythemis velox*

41–48 mm, 1.6–1.9 in. Boldly marked setwing found at southern streams and rivers, occasionally pond and lakes. Thorax heavily striped with YIY lateral pattern. Abdomen with large white markings on top of S7, lacks lateral pale markings on S4-10; S8-10 black. Wings with short brown streaks at base and large dark tips. Eyes red over gray.

eyes red over gray

mid-March to late November

wings with dark tips

♂

abdomen slender,
slightly clubbed

mature male
pattern darkens

♀

abdomen with large
pale marks on top of
S7, tip black

♂

♀

bold contrasting pattern,
thorax with YIY stripe pattern

Similar species: Black Setwing immature and female have HII stripe pattern on side of thorax, abdomen with pale lateral markings, smaller dark wing tips. **Blue Dasher** has III lateral stripe pattern, abdomen shorter and wider with yellow dorsal dashes. **Tropical dashers** smaller, thorax greener with different lateral stripe patterns; males with green or blue eyes.

Checkered Setwing • *Dythemis fugax*

44–50 mm, 1.7–2 in. Distinctly patterned southern setwing found at slow streams, rivers, open lakes, and ponds. Wing bases with large, lacy, brown and amber patches. Thorax heavily striped light gray and dark brown; mature male thorax browner, face red. Abdomen black, contrastingly patterned with white dorsal and lateral markings. Eyes red over gray.

male face and eyes bright red

♂

early Mar to late December

wings with large amber and brown patches at base

♀

abdomen with contrasting white markings; S7 marking large

male thorax browner, stripes obscured

♂

♀

bold, contrasting pattern

Similar species: Black and **Swift Setwings** lack large patches at wing bases. **Marl Pennant** wing patches more uniform in color; mature male with dark face and eyes, female with top of abdomen mostly yellow-brown.

Mayan Setwing • *Dythemis maya*

43–45 mm, 1.7–1.8 in. Brightly colored southwestern desert setwing inhabiting rocky streams with pools, typically in canyons. Mature male face, thorax, and abdomen brilliant red. Female plain brown. Both sexes with large amber hindwing patch, male's patch with red veins; female usually with dark wing tips. Stigmas black. Eyes red over gray.

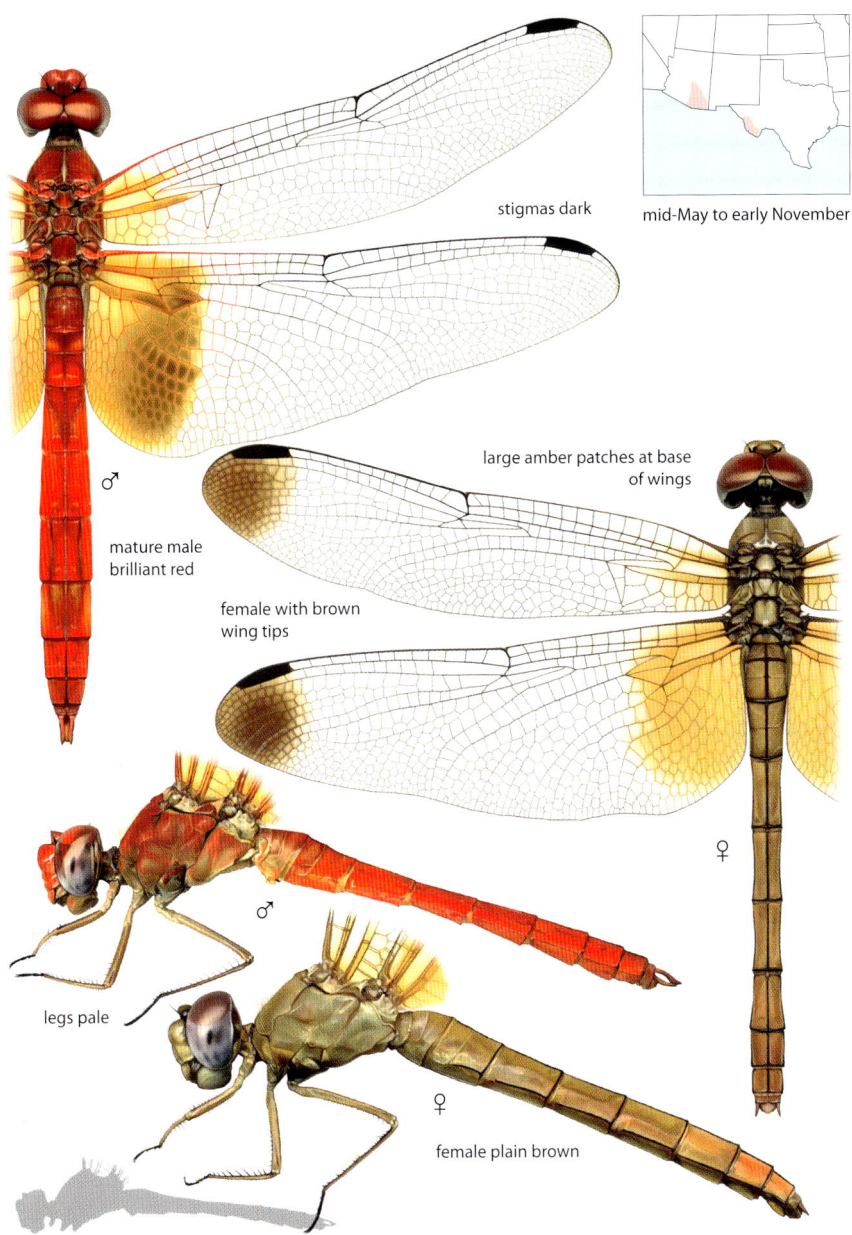

stigmas dark

mid-May to early November

large amber patches at base of wings

mature male brilliant red

female with brown wing tips

♂

legs pale

♀

♂

♀

female plain brown

Similar species: Flame and **Neon Skimmers** larger, heavy-bodied with much wider abdomen, stigmas pale, females with lateral flanges on S8; lack dark wing tips.

Red Rock Skimmer • *Paltothemis lineatipes*

47–54 mm, 1.9–2.1 in. Dark mottled skimmer inhabiting rocky southwestern upland streams, where it habitually perches on rocks. Male with red face and eyes, thorax and abdomen intricately patterned in red and black, base of wings amber. Female complexly patterned in gray, black, and brown, wings mostly clear.

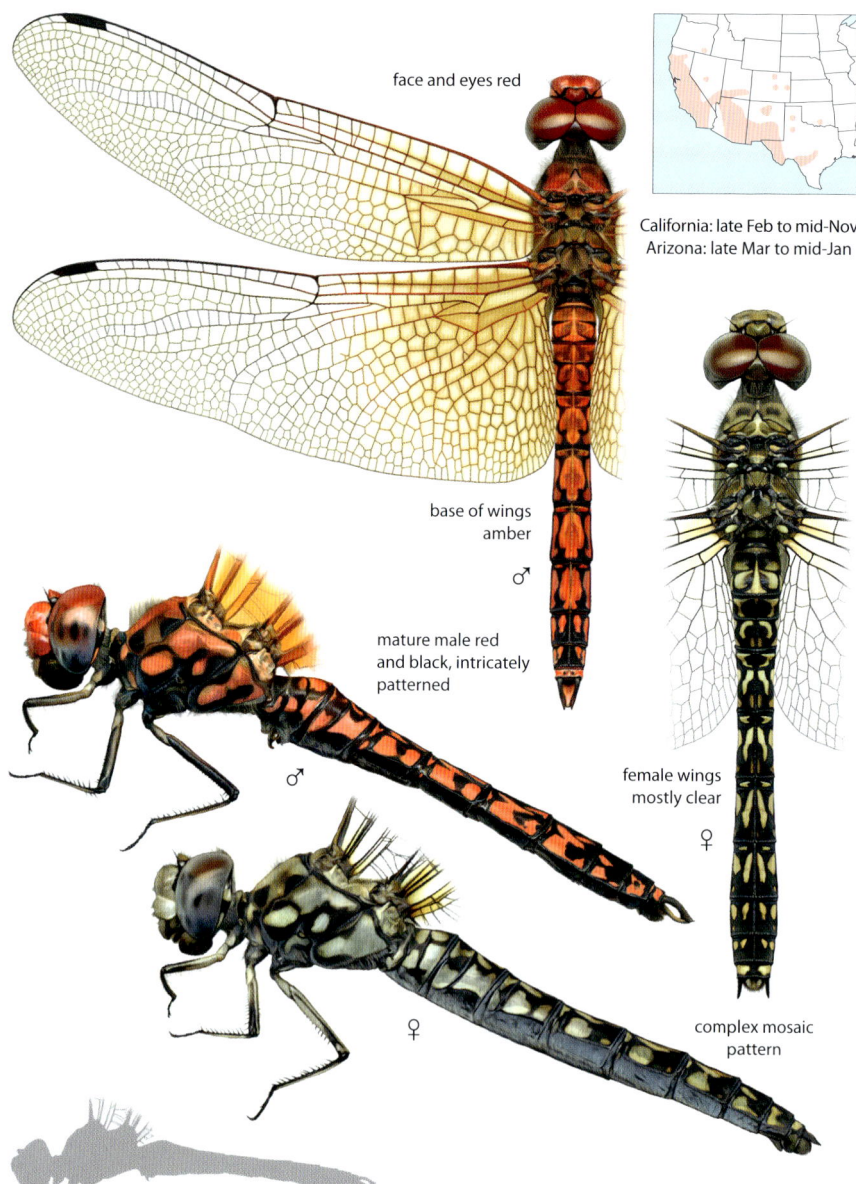

face and eyes red

California: late Feb to mid-Nov
Arizona: late Mar to mid-Jan

base of wings
amber

♂

mature male red
and black, intricately
patterned

♂

female wings
mostly clear

♀

complex mosaic
pattern

♀

Similar species: Pattern unique. **Filigree Skimmer** pattern features fine pale lines, has dark wing markings. Male **Flame** and **Neon Skimmers** have wider, brighter red abdomen, orange brown stigmas, do not perch on rocks.

Filigree Skimmer • *Pseudoleon superbus*

38–45 mm, 1.5–1.8 in. Dark southwestern skimmer found at clear, rocky streams. Often perches on rocks and flat on the ground. Both sexes dark brown, intricately patterned with fine pale stripes; eyes vertically striped. Wings with scattered dark markings; dark band between nodus to stigma on male, female band narrower. Stigmas partially pale.

stigma pink

scattered dark markings

dark band between nodus and stigma

flies all year

♂

abdomen with complex V-shaped pattern of fine white lines

female wing markings sparer than male

♀

eyes vertically striped

♂

wings and body often darker when mature

♀

Similar species: Intricate pattern unique. **Black-winged Dragonlet** male slenderer; dark wing bands do not reach stigmas.

Evening Skimmer • *Tholymis citrina*

48–53 mm, 1.9–2.1 in. Largely unmarked brown skimmer found at vegetated ponds and lakes. Southern. Crepuscular; hangs in forest vegetation during day. Sexes similar; eyes grayish, subtly banded above, thorax and abdomen plain grayish brown, hindwing with amber spot at nodus. Legs pale brown.

eyes grayish with subtle dark bands

early March to late January

both sexes plain gray-brown

amber spot below hindwing nodus

♂

♀

female cerci downturned

Similar species: Twilight and **Bar-sided Darners** larger, with longer abdomens; wings lack amber spot. **Shadowdragons** have amber or brown dots along leading edge of wings (or wings more heavily marked); most species with small pale spot on side of thorax.

Pale-faced Clubskimmer • *Brechmorhoga mendax*

53–62 mm, 2.1–2.4 in. Our most common clubskimmer; found at clear, rocky, southwestern streams and rivers. Thorax appears grayish with three pale gray lateral stripes, two wide, middle one narrow. Abdomen slightly clubbed with pale markings on S7 forming large, conspicuous rounded spot. Face pale. Eyes gray; male bluer, female browner.

face pale

Texas: late Feb to early Dec
Arizona: early Apr to late Dec

female has tinted wing tips

S7 pale marks form rounded spot

thorax with three pale lateral stripes, middle stripe short and narrow

side of S2-3 mostly pale

♂

♀

♂

♀

Similar species: Slender Clubskimmer smaller, more delicate, abdomen less clubbed; S7 dorsal pale markings spatulate. **Masked Clubskimmer** has top of face dark, thorax with two narrower pale lateral stripes, S7 dorsal pale markings smaller and separated. **Sylphs** are smaller with proportionally shorter abdomens; top of face dark.

Masked Clubskimmer • *Brechmorhoga pertinax*

50–52 mm, 2 in. Local southwestern clubskimmer inhabiting clear, rocky streams and rivers. Thorax dark brown with two narrow pale lateral stripes. Abdomen slightly clubbed, mostly black with pair of separated pale markings on top of S7. Top of face dark. Eyes pale blue. Outer half of wings milky in appearance, best seen against a dark background.

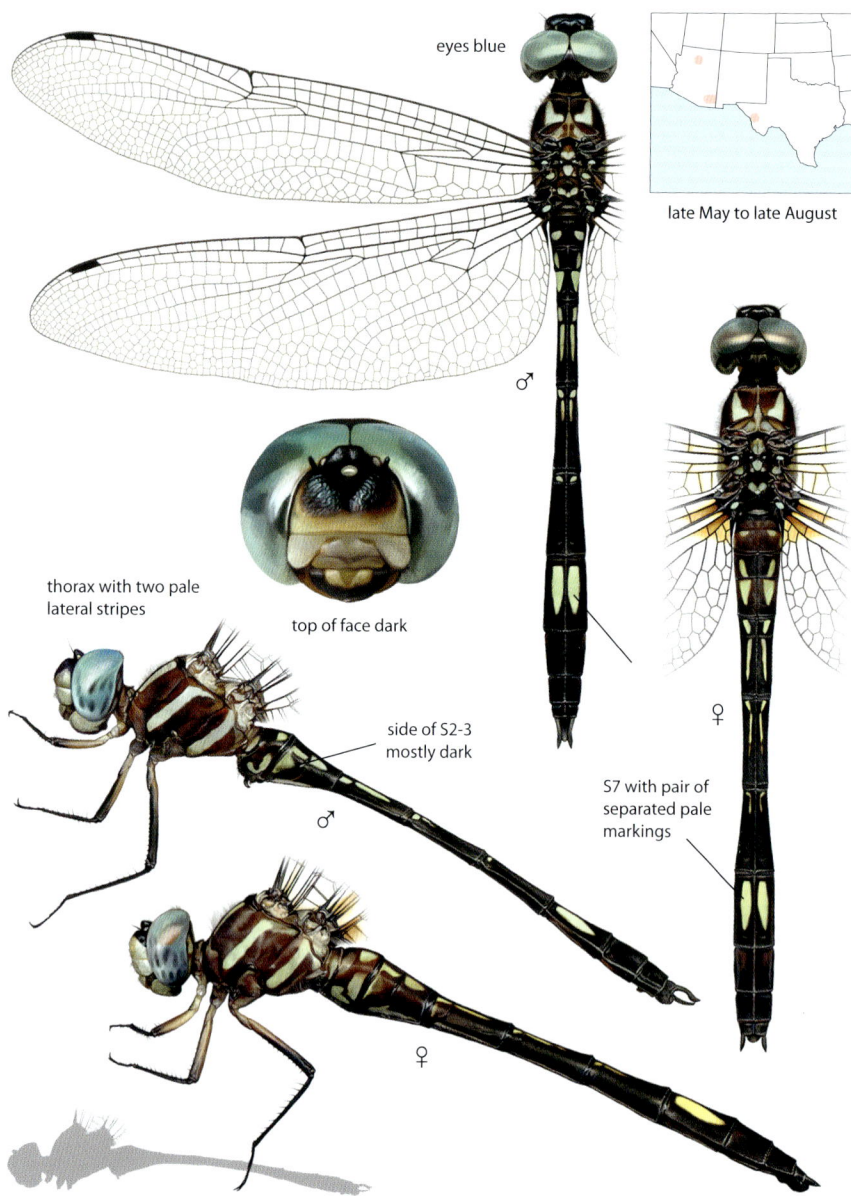

eyes blue

late May to late August

thorax with two pale lateral stripes

top of face dark

♂

side of S2-3 mostly dark

♂

♀

S7 with pair of separated pale markings

♀

Similar species: Pale-faced and **Slender Clubskimmers** have pale face, narrow pale lateral thoracic stripe between two wider ones; S7 dorsal pale markings larger and meet at middle. **Sylphs** are smaller with proportionally shorter abdomens.

Slender Clubskimmer • *Brechmorhoga praecox*

47–53 mm, 1.9–2.1 in. Rare southwestern clubskimmer found at clear streams and rivers. Thorax brown with three pale lateral stripes, middle one narrow and broken. Abdomen slightly clubbed in male; large pale markings on top of S7 spatulate-shaped, appear as single large mark covering most of segment. Face pale. Male eyes blue.

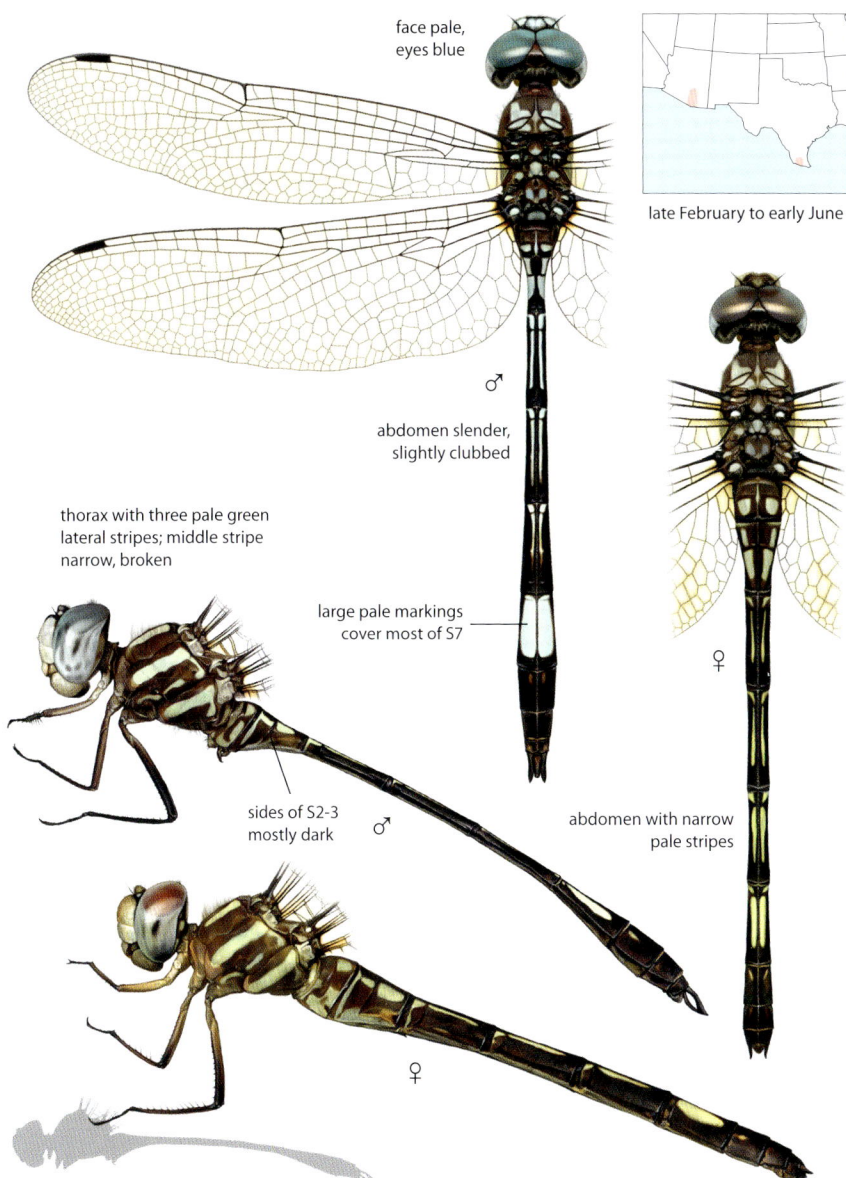

face pale,
eyes blue

late February to early June

abdomen slender,
slightly clubbed

thorax with three pale green
lateral stripes; middle stripe
narrow, broken

large pale markings
cover most of S7

♂

sides of S2-3
mostly dark   ♂

abdomen with narrow
pale stripes

♀

♀

Similar species: Pale-faced Clubskimmer larger; pale lateral thoracic stripes wider, sides of S2 and 3 more extensively pale, S7 pale dorsal marking oval-shaped. **Masked Clubskimmer** has top of face dark, two narrow pale lateral thoracic stripes; S7 pale markings small and separated. **Sylphs** are smaller with shorter abdomen, top of face dark.

Straw-colored Sylph • *Macrothemis inacuta*

42–48 mm, 1.7–1.9 in. Slender brown sylph found at southwestern rivers and streams. Feeds and patrols in constant flight. Thorax brown with pale lateral spots, pale L-shaped frontal stripes. Abdomen long and slender, pale yellow-brown. Mature male eyes bluish. Female often with amber patch at hindwing base, amber spot at forewing tip.

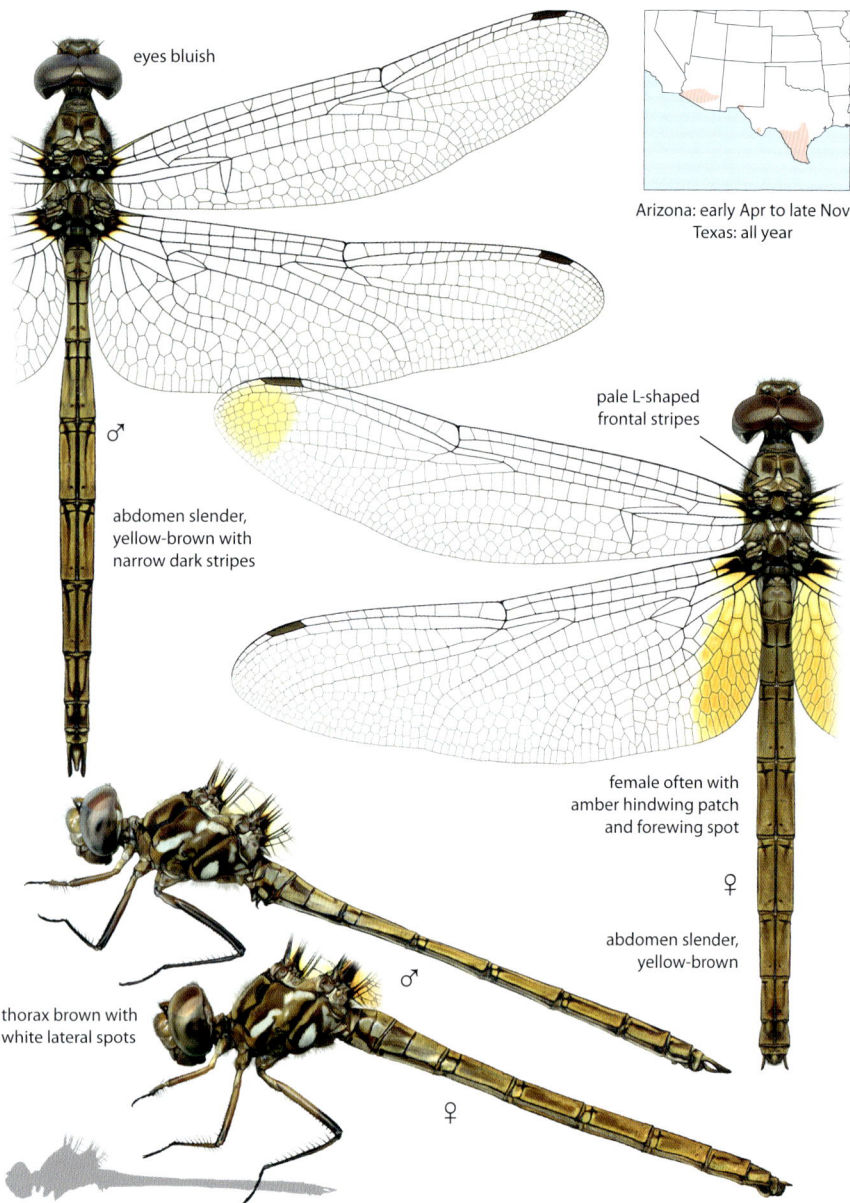

eyes bluish

Arizona: early Apr to late Nov
Texas: all year

pale L-shaped frontal stripes

♂

abdomen slender, yellow-brown with narrow dark stripes

female often with amber hindwing patch and forewing spot

♀

abdomen slender, yellow-brown

♂

thorax brown with white lateral spots

♀

Similar species: Slender, yellow-brown abdomen distinctive. **Other sylphs** have shorter clubbed abdomens. **Hyacinth Glider** has dark patch at base of hindwing. **Evening Skimmer** stouter, lacks pale markings on side of thorax.

Jade-striped Sylph • *Macrothemis inequiunguis*

33–36 mm, 1.3–1.4 in. Small, slender skimmer, local at some forested Texas streams. Feeds and patrols in flight, often hangs while perched. Thorax dark brown with pale green L-shaped frontal stripes and three greenish lateral stripes. Abdomen black with pale streaks, pair of large pale spots on S7. Male abdomen clubbed. Female with dark wing tips.

eyes blue

mid-March to mid-November

pale L-shaped frontal stripes

male abdomen short and clubbed

female with large dark wing tips

squarish white spots on S7

thorax with three green lateral stripes ♂

female abdomen slender, white marks on S7

Similar species: Ivory-striped Sylph thorax has frontal pale spots and only two pale lateral stripes with the posterior stripe broken. **White-tailed Sylph** thorax has long, triangular, pale frontal stripes and four pale lateral spots. **Clubskimmers** larger and elongated.

Ivory-striped Sylph • *Macrothemis imitans*

33–37 mm, 1.3–1.45 in. Small, slender, constant-flying skimmer scarce at some Texas streams. Side of thorax with pale anterior stripe and two posterior pale spots. Thorax front with conspicuous pale spots, smaller or absent in female. Abdomen dark with paired pale dorsal markings, S7 spots widest. Both sexes clubbed, male club wider.

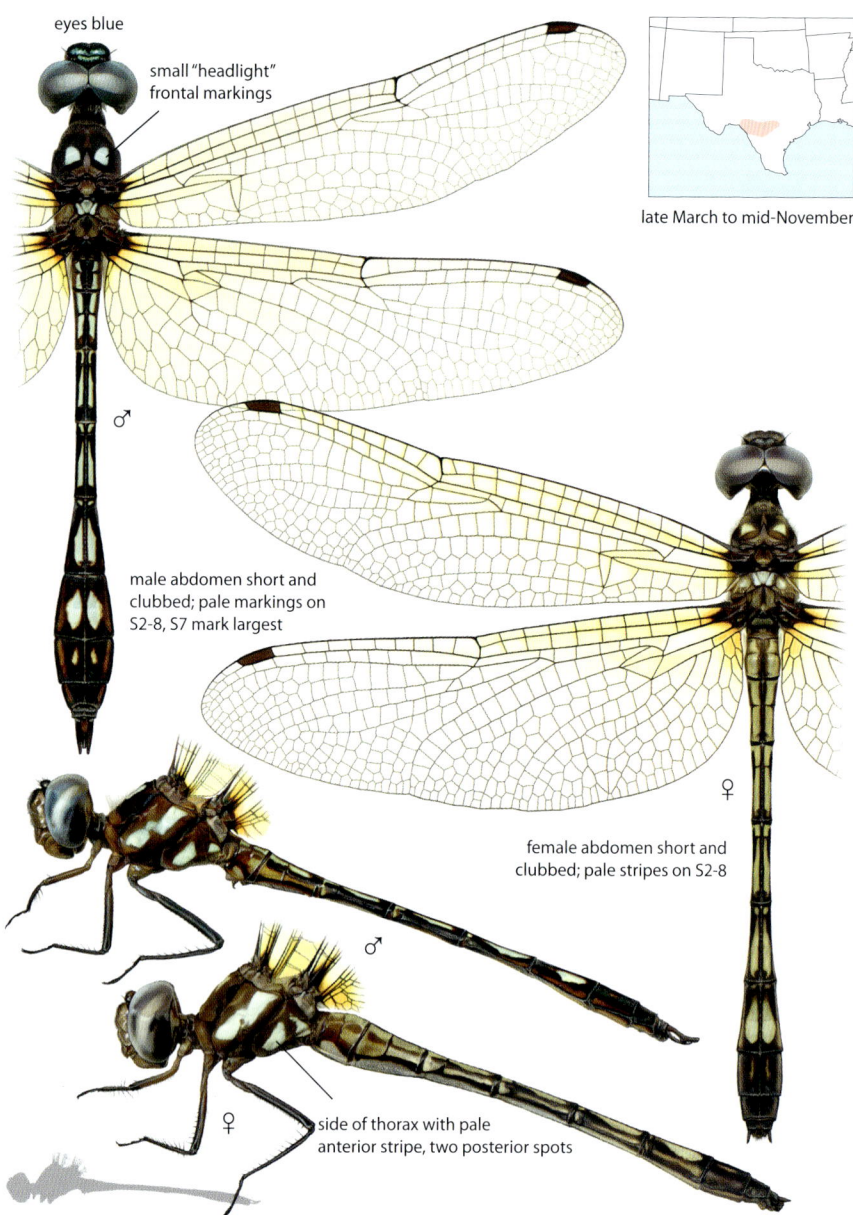

eyes blue

small "headlight" frontal markings

late March to mid-November

♂

male abdomen short and clubbed; pale markings on S2-8, S7 mark largest

♀

female abdomen short and clubbed; pale stripes on S2-8

♂

♀

side of thorax with pale anterior stripe, two posterior spots

Similar species: White-tailed Sylph thorax has four pale lateral spots; pale frontal stripes longer. **Jade-striped Sylph** thorax has L-shaped pale frontal stripes and three pale lateral stripes. **Clubskimmers** larger and elongated; thorax with at least two pale lateral stripes.

White-tailed Sylph • *Macrothemis pseudimitans*

40–43 mm, 1.6–1.7 in. Rare, small, slender southwestern skimmer inhabiting streams, often seen in flight. Thorax dark brown with large full-length pale green triangles at front and four pale spots laterally. Abdomen dark with short pale stripes. Male with conspicuous pale tear-shaped marks on top of S7; female markings smaller. Male abdomen clubbed.

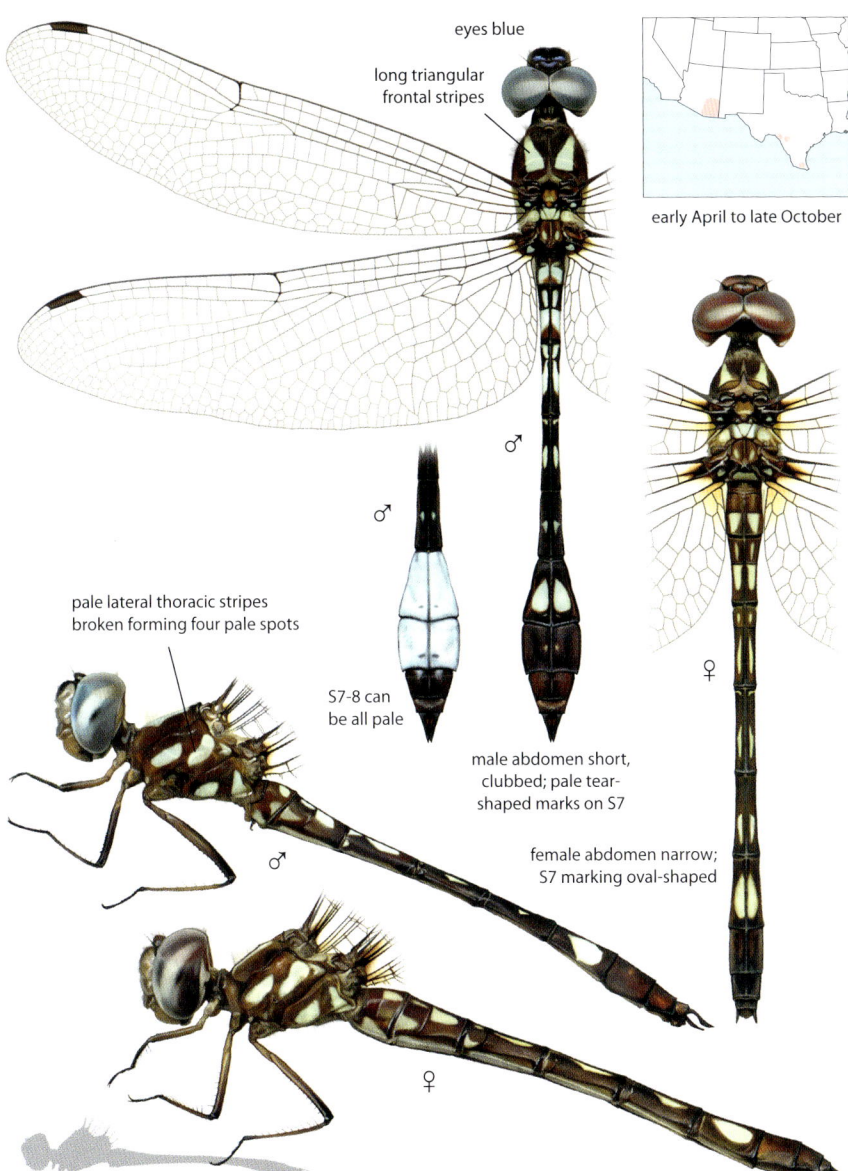

eyes blue

long triangular frontal stripes

early April to late October

♂

♂

pale lateral thoracic stripes broken forming four pale spots

S7-8 can be all pale

male abdomen short, clubbed; pale tear-shaped marks on S7

female abdomen narrow; S7 marking oval-shaped

♀

♂

♀

Similar species: Ivory-striped Sylph thorax has an unbroken pale anterior lateral stripe; pale frontal stripes short. **Jade-striped Sylph** thorax has L-shaped pale frontal stripes and three pale lateral stripes. **Clubskimmers** larger and elongated; thorax with complete pale lateral stripes.

Wandering Glider • *Pantala flavescens*

47–50 mm, 1.9–2 in. Widespread yellow glider seen in constant flight; breeds primarily in temporary ponds. Migratory. Sexes similar. Body tapers from thorax to abdomen tip; yellow to orange overall, with narrow dorsal dark stripe on abdomen. Face pale yellow to orange, eyes red over gray. Wings unmarked, base of hindwing very broad.

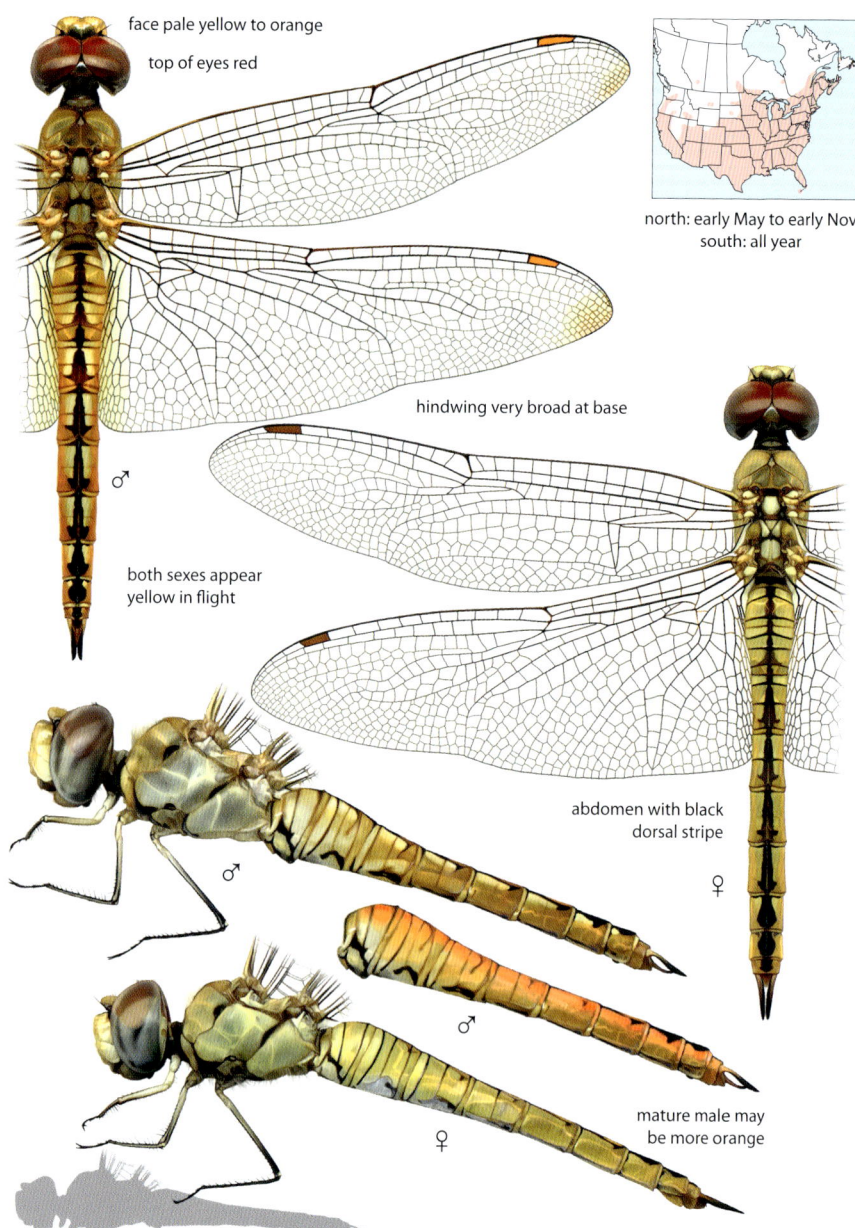

face pale yellow to orange

top of eyes red

north: early May to early Nov
south: all year

hindwing very broad at base

♂

both sexes appear
yellow in flight

abdomen with black
dorsal stripe

♀

♂

♀

mature male may
be more orange

Similar species: Spot-winged Glider darker, gray-brown in color; hindwing with dark spot at base. Face of mature male red.

Spot-winged Glider • *Pantala hymenaea*

45–50 mm, 1.8–2 in. Gray-brown glider seen in constant flight; breeds in temporary pools and fishless ponds. Sexes similar; thorax gray-brown with vague pale lateral stripe, abdomen gray-brown with dark middorsal stripe and thin cross-stripes. Hindwing with small dark spot at base, observable when overhead. Mature male face red.

top of eyes red

small dark hindwing spot

♂

north: late May to mid-Oct
south: early Mar to early Nov

both sexes appear gray-brown in flight

♀

male face red

♂

female face reddish

♀

abdomen with black dorsal stripe, narrow dark bands

Similar species: Wandering Glider yellower, lacks hindwing spot. **Saddlebags** have larger hindwing markings.

Arch-tipped Glider • *Tauriphila argo*

42–47 mm, 1.7–1.9 in. Pasture glider seen in constant flight at vegetated ponds and sloughs; recorded in South Texas. Abdomen with narrow black cross-bands. Both sexes with dark hindwing patch; female patch variable, large to almost absent. Male abdomen spindle-shaped, red when mature. Immature and female brown.

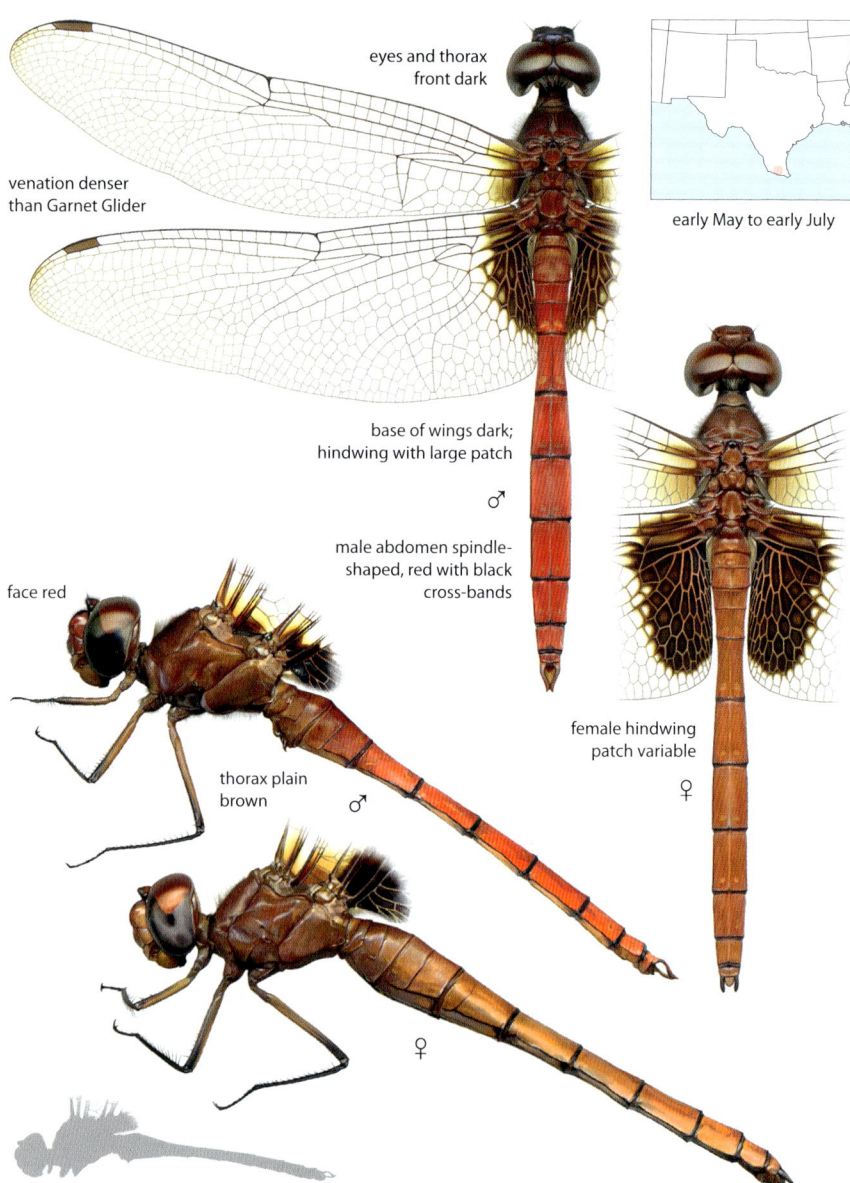

eyes and thorax front dark

venation denser than Garnet Glider

early May to early July

base of wings dark; hindwing with large patch

♂

male abdomen spindle-shaped, red with black cross-bands

face red

thorax plain brown

♂

female hindwing patch variable

♀

♀

Similar species: Garnet Glider has black dorsal stripe on S8-9; wing venation sparer, male cerci straighter (see page 436). **Aztec Glider** has yellow-orange abdomen with black bands. **Hyacinth Glider** has pale lateral thoracic stripes, black dorsal stripe on abdomen. **Saddlebags** have more black on end abdominal segments, longer appendages.

Garnet Glider • *Tauriphila australis*

42–47 mm, 1.7–1.9 in. Dark pasture glider, local in Florida at ponds and lakes with floating vegetation. Flier. Abdomen with narrow black cross-bands, black middorsal stripe on S8-9. Male abdomen constricted toward base, red when mature; hindwing with dark patch at base. Immature and female brown, female with reduced hindwing markings.

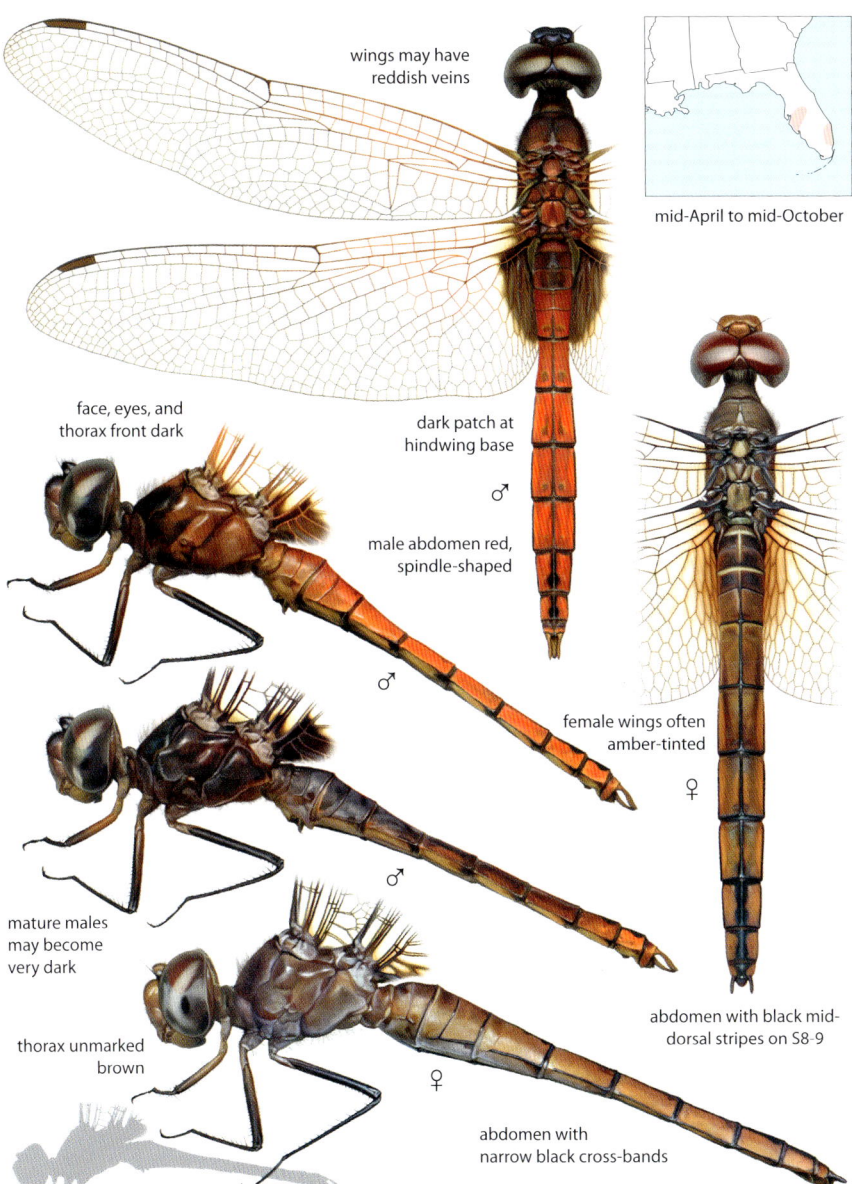

wings may have reddish veins

mid-April to mid-October

face, eyes, and thorax front dark

dark patch at hindwing base

♂

male abdomen red, spindle-shaped

♂

female wings often amber-tinted

♀

mature males may become very dark

♂

thorax unmarked brown

abdomen with black mid-dorsal stripes on S8-9

♀

abdomen with narrow black cross-bands

Similar species: Arch-tipped Glider nearly identical, lacks black dorsal stripe on abdomen; wing venation denser, male cerci distinctly arched (see page 436). **Aztec Glider** has wide black abdominal bands. **Hyacinth Glider** has pale lateral thoracic stripes. **Saddlebags** have more black at abdomen tip, longer appendages.

Aztec Glider • *Tauriphila azteca*

42–50 mm, 1.7–2 in. Dark pasture glider inhabiting ponds and lakes with floating vegetation; recorded in Texas and Florida. Constant flier. Sexes similar. Thorax and eyes dark. Abdomen constricted toward base, yellow-orange with black bands, black stripe on top of S8-10. Hindwing with dark patch at base.

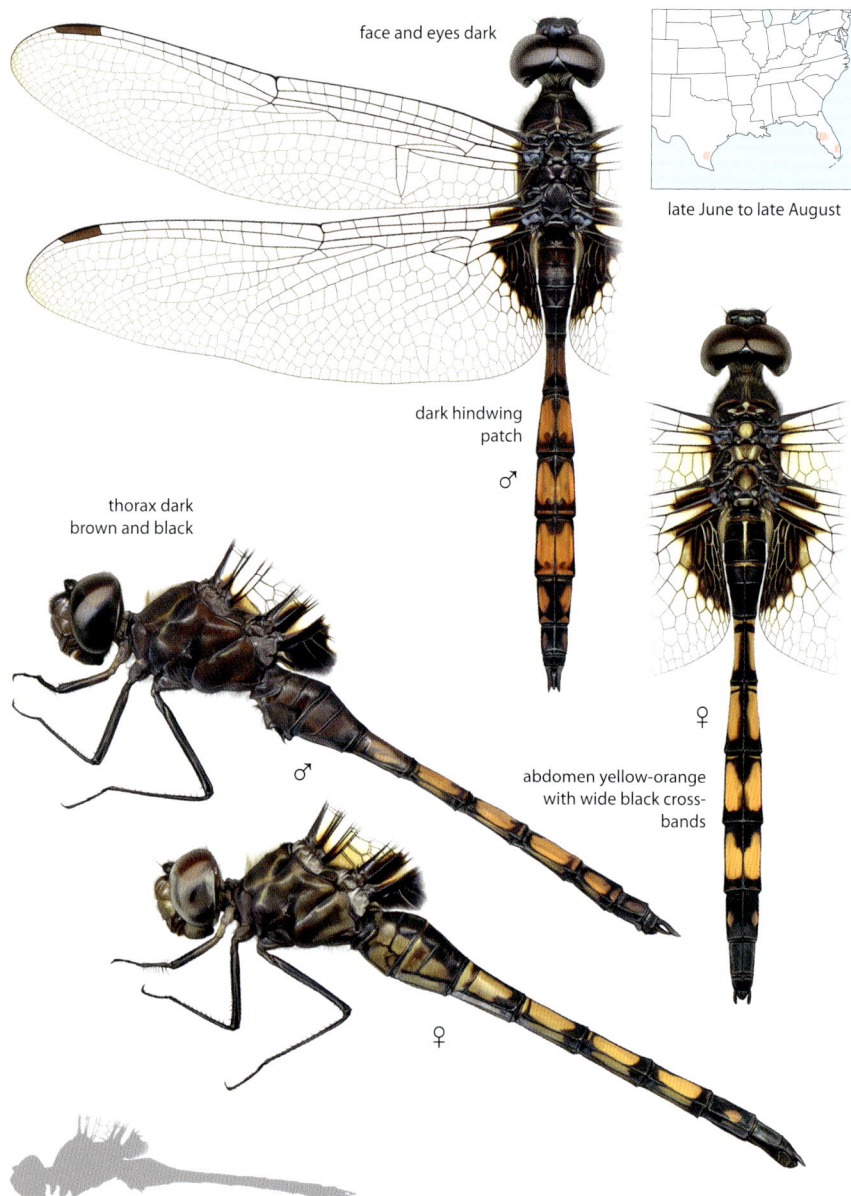

face and eyes dark

late June to late August

dark hindwing patch

♂

thorax dark brown and black

♂

abdomen yellow-orange with wide black cross-bands

♀

♀

Similar species: Hyacinth Glider has pale lateral thoracic stripes, abdomen with more conspicuous black middorsal stripe and narrower black cross-lines. **Saddlebags** lack wide dark cross-bands.

Hyacinth Glider • *Miathyria marcella*

37–40 mm, 1.5–1.6 in. Small, dark southeastern glider found at ponds and lakes with water hyacinth or water lettuce. Flier. Hindwing patch narrow. Thorax brown with two lateral diagonal pale stripes. Abdomen with black middorsal stripe. Adult male has top of face metallic purple, eyes magenta, thorax pruinose purple, abdomen yellow.

male eyes magenta

wing veins orange at base

Texas: mid-Mar to late Dec
Florida: all year

narrow dark hindwing patch

top of face metallic purple

♂

female hindwing patch shorter
♀

thorax pruinose purple

♂

mature male abdomen yellow with black tip

pale stripe extends forward of first thoracic suture

side of female and immature thorax has two pale diagonal stripes

♀

abdomen with black dorsal stripe

Similar species: Striped Saddlebags has black only on top of S8-10, lacking black stripes on top of S5-7, pale anterior lateral stripe does not extend across first thoracic suture. Other **saddlebags** also lack black stripes on top of S5-7 and lack pale lateral thoracic stripes. **Aztec Glider** has triangular dark hindwing patch, abdomen with black bands.

Black Saddlebags • *Tramea lacerata*

51–55 mm, 2–2.2 in. Our most widespread saddlebags; common at vegetated ponds and lakes, seen in constant flight. Hindwing with very large dark patch at base. Mature male face, eyes, and body dark. Immature and female dark brown with yellow abdominal markings that fade with age.

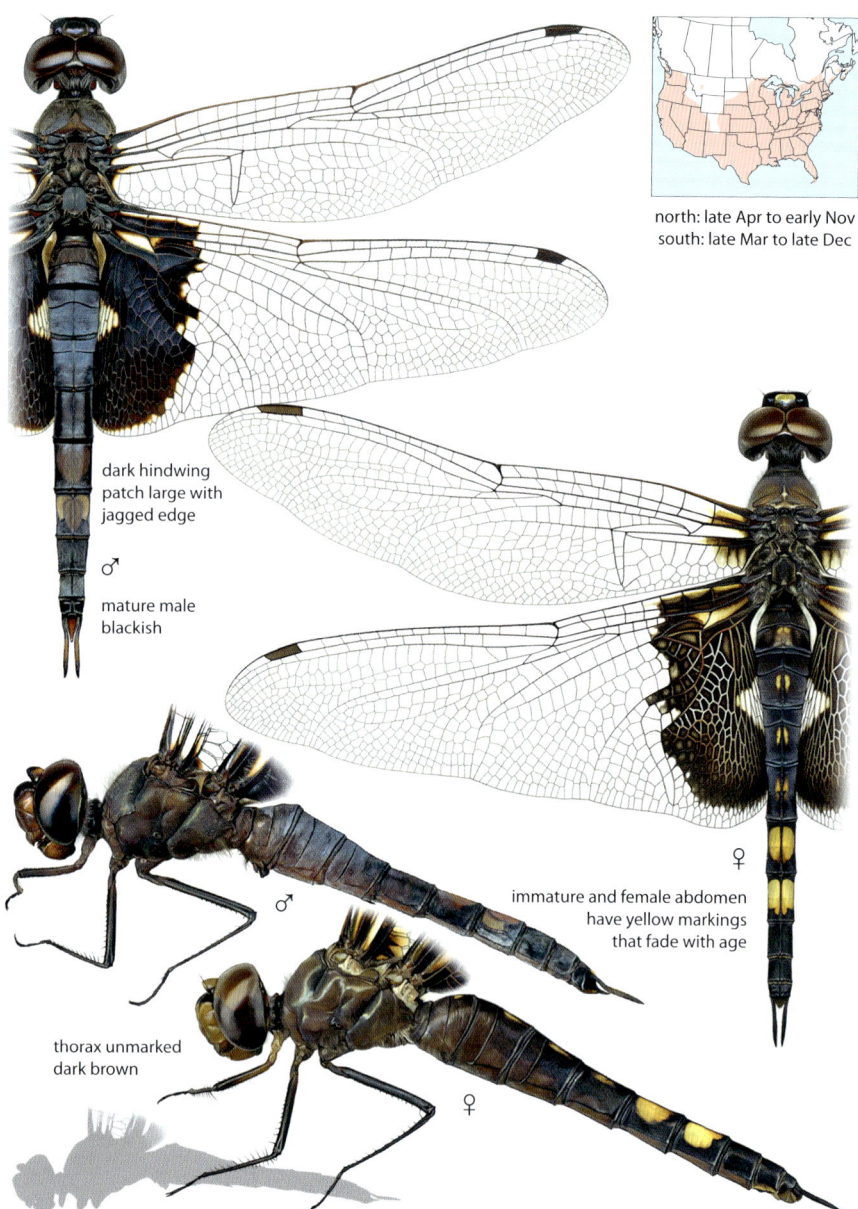

north: late Apr to early Nov
south: late Mar to late Dec

dark hindwing patch large with jagged edge

♂

mature male blackish

♂

thorax unmarked dark brown

♀

immature and female abdomen have yellow markings that fade with age

♀

Similar species: Carolina and **Red Saddlebags** lack pale spots on top of abdomen; immatures and females lighter brown, mature males red. **Sooty Saddlebags** has narrow dark hindwing band. **Marl Pennant** smaller, with rounded dark hindwing patch.

Sooty Saddlebags • *Tramea binotata*

45–51 mm, 1.8–2 in. Very rare, dark, southern vagrant saddlebags inhabiting ponds and lakes. Constant flier. Dark hindwing patch narrow. Mature male face, eyes, and body mostly black. Immature and female brown, abdomen with S8-10 mostly black. Maturing male with red abdomen.

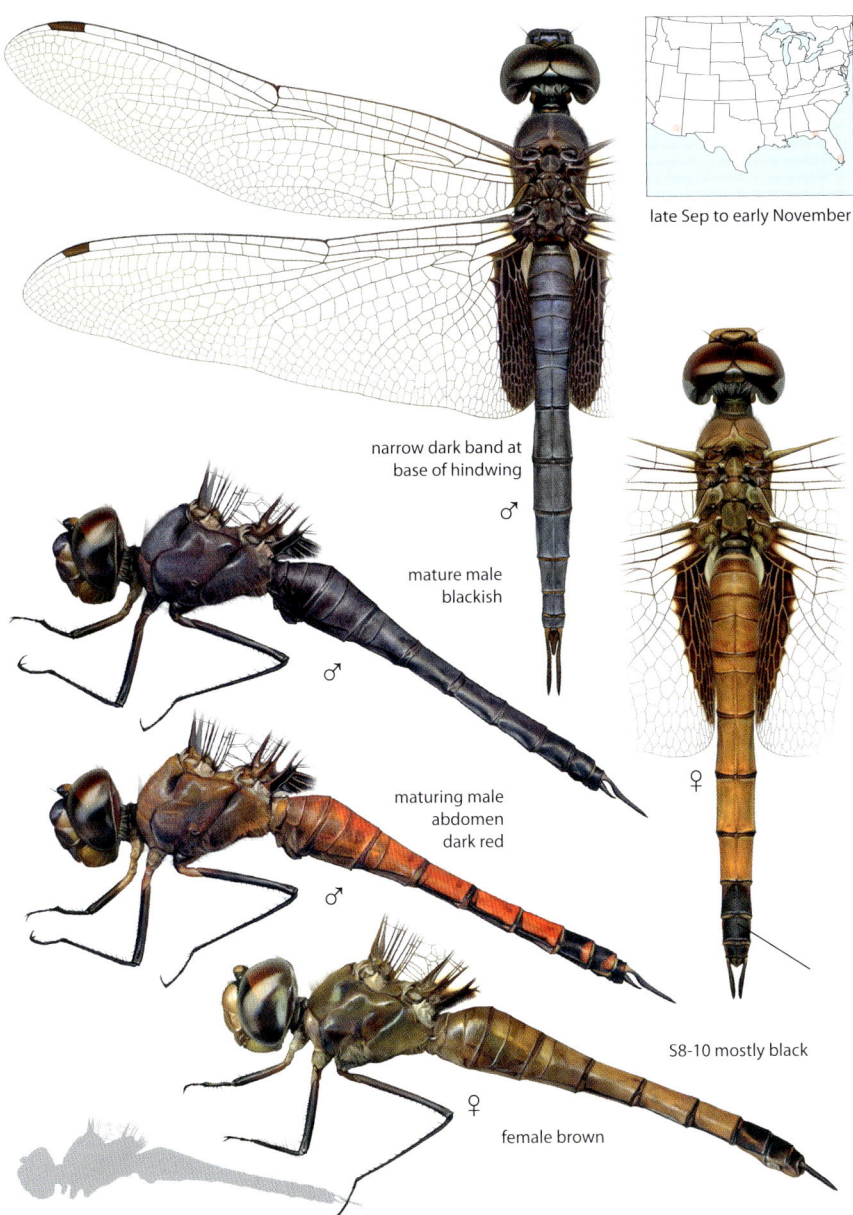

late Sep to early November

narrow dark band at
base of hindwing

♂

mature male
blackish

♂

maturing male
abdomen
dark red

♂

♀

S8-10 mostly black

♀

female brown

Similar species: Black Saddlebags has larger dark hindwing patch. **Antillean** and **Vermilion Saddlebags** lack black markings on sides of S8-10. **Striped Saddlebags** has white lateral thoracic stripes.

Carolina Saddlebags • *Tramea carolina*

48–53 mm, 1.9–2.1 in. Common eastern saddlebags inhabiting lakes and ponds, including temporary ones. Constant flier. Hindwing with very large dark patch at base. Abdomen with sides of S8-9 darkened. Mature male has face, abdomen, and basal wing veins red, top of face dark metallic purple. Immature and female brown.

north: late Apr to early Nov
Florida: all year

mature male abdomen and basal wing veins red

top of face dark purple

♂

♂ short hamules

♀

very large hindwing patch, often covers entire anal loop

small clear window

♀

sides of S8-9 darkened

Similar species: Red Saddlebags has smaller hindwing patch that does not cover anal loop; clear window at base of patch larger, nearly reaching anal loop; sides of S8-9 not darkened. Male lacks dark purple at top of face, has long, projecting hamules. **Antillean, Vermilion,** and **Striped Saddlebags** have narrow dark hindwing bands.

Red Saddlebags • *Tramea onusta*

41–49 mm, 1.6–1.9 in. Widespread common saddlebags found at ponds, lakes, swamps, and slow streams. Constant flier. Hindwing with large dark patch at base. Abdomen with black marks on top of S8-10. Mature male with face, abdomen, and wing veins red. Immature and female brown.

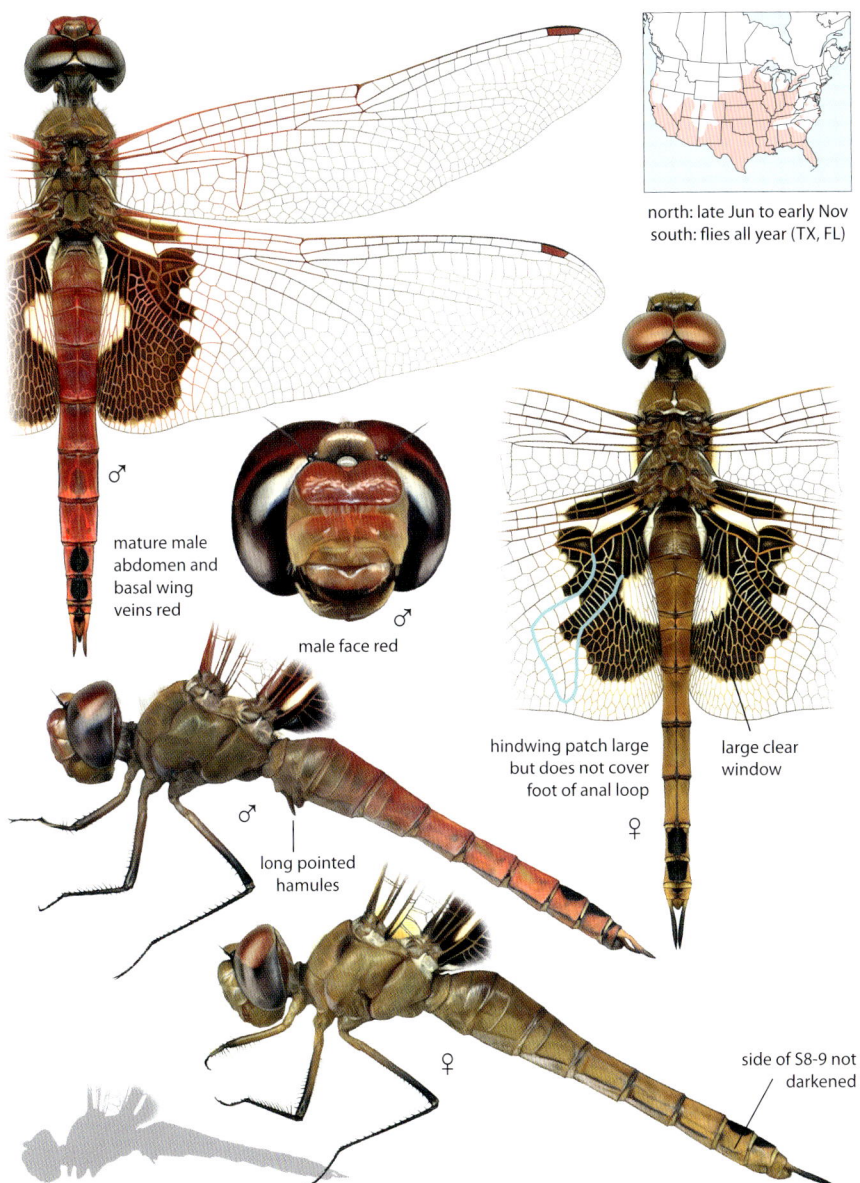

north: late Jun to early Nov
south: flies all year (TX, FL)

♂

mature male
abdomen and
basal wing
veins red

male face red ♂

hindwing patch large
but does not cover
foot of anal loop

large clear
window

♀

♂

long pointed
hamules

♀

side of S8-9 not
darkened

Similar species: Carolina Saddlebags has larger hindwing patch often covering entire anal loop, clear window at base of patch smaller, sides of S8-9 usually darkened; male with top of face dark purple, lacks long, projecting hamules. **Antillean, Vermilion,** and **Striped Saddlebags** have narrow dark hindwing bands.

Antillean Saddlebags • *Tramea insularis*

41–49 mm, 1.6–1.9 in. Far-southern saddlebags found at ponds, lakes, ditches, canals, and other human-made water bodies. Constant flier. Hindwing with narrow dark band. Abdomen with black marks on top of S8-9. Mature male abdomen red, top of face dark metallic purple. Immatures and female brown, female with small amount of purple on face.

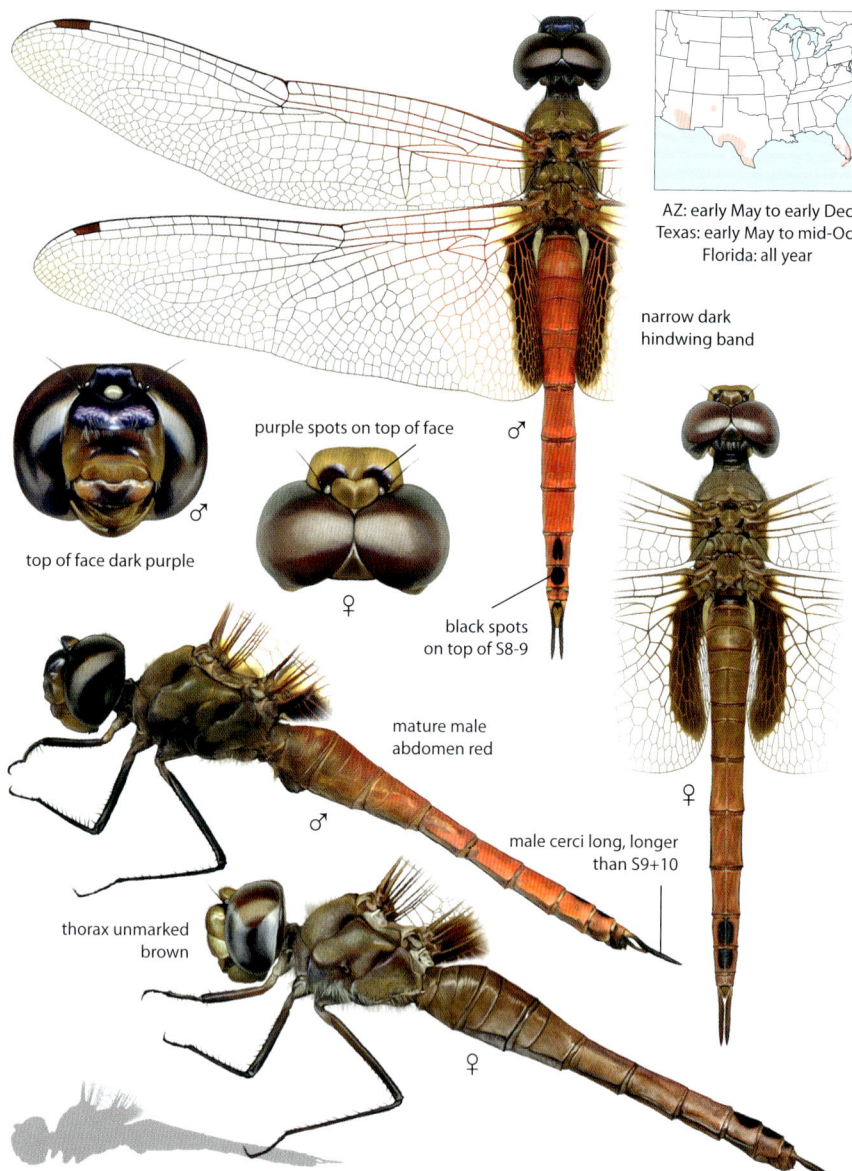

AZ: early May to early Dec
Texas: early May to mid-Oct
Florida: all year

narrow dark
hindwing band

purple spots on top of face

♂

top of face dark purple

♀

black spots
on top of S8-9

mature male
abdomen red

♂

male cerci long, longer
than S9+10

thorax unmarked
brown

♀

♀

Similar species: Vermilion Saddlebags male has red face, top of face lacks dark purple, cerci shorter; female largely identical, lacks dark markings on face, separable by longer subgenital plate, page 436. **Striped Saddlebags** has pale lateral stripes on thorax, black markings on sides of S8-9.

Vermilion Saddlebags • *Tramea abdominalis*

44–50 mm, 1.7–2 in. Far-southern saddlebags, rare stray elsewhere, inhabiting vegetated ponds, ditches, canals, and stream pools. Constant flier. Hindwing with narrow dark band. Abdomen with small dark dorsal markings on S8-9. Mature male with red abdomen, all-red face. Immature and female brown.

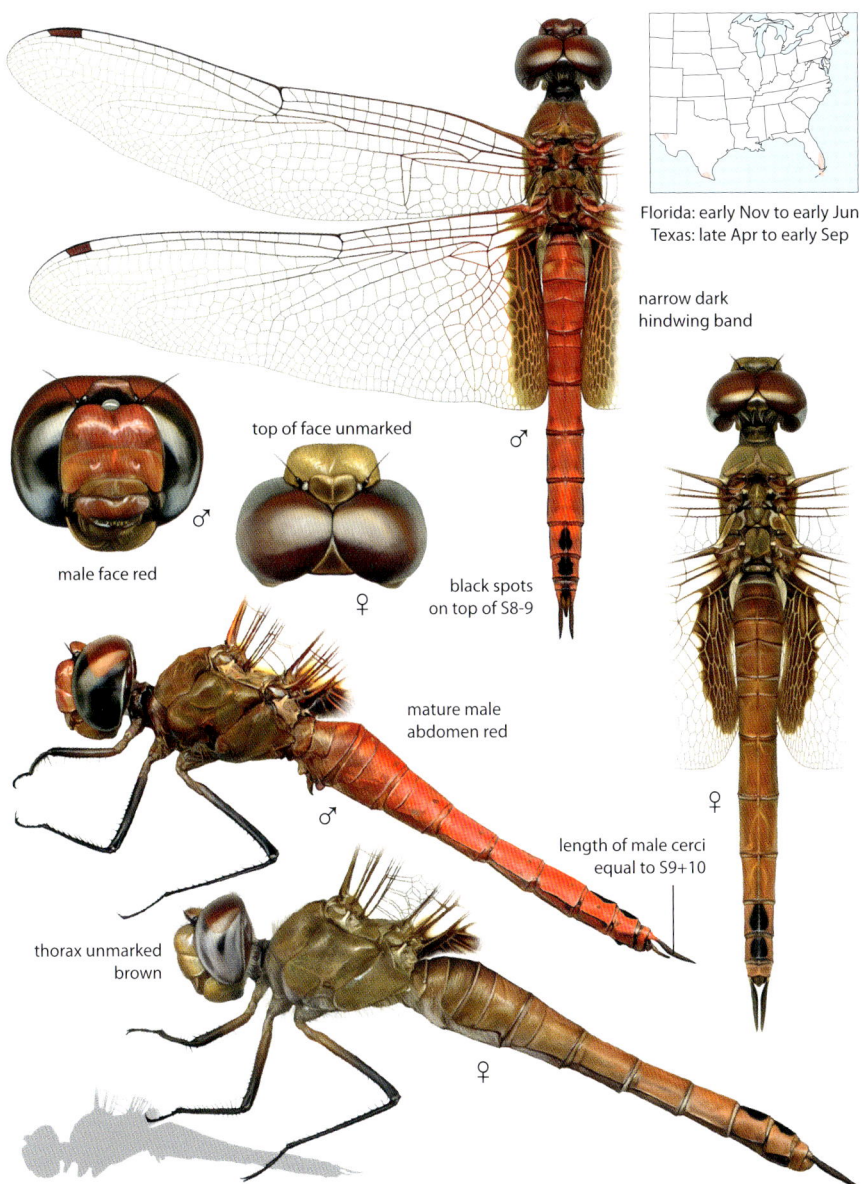

Florida: early Nov to early Jun
Texas: late Apr to early Sep

narrow dark hindwing band

♂

top of face unmarked

male face red

♂

♀

black spots on top of S8-9

mature male abdomen red

♂

♀

length of male cerci equal to S9+10

thorax unmarked brown

♀

Similar species: Antillean Saddlebags male has top of face dark purple, cerci longer; female nearly identical, has some dark purple on top of face, subgenital plate slightly shorter (see page 436). **Striped Saddlebags** often shows pale lateral thoracic stripes, abdomen with black markings on side of S8-9.

Striped Saddlebags • *Tramea darwini*

45–49 mm, 1.8–1.9 in. Our only saddlebags with a striped thorax; inhabits ponds including temporary ones, primarily in Texas, but strays widely, usually in fall. Constant flier. Hindwing with narrow dark band. Thorax brown with two pale lateral stripes. Abdomen with S8-10 mostly black. Mature male with red abdomen and face, top of face dark purple.

north: late Jun to mid-Oct
Texas: all year

narrow dark
hindwing band

♂

mature male
abdomen red

mature male
face red

♂

S8-10 mostly black

♀

thorax with two
white lateral stripes,
fade with age

♀

Similar species: Antillean and **Vermilion Saddlebags** lack pale lateral thoracic stripes and black markings on sides of S8-10. **Hyacinth Glider** has dorsal black stripes on S5-7, mature male with yellow abdomen. **Sooty Saddlebags** lacks pale lateral thoracic stripes.

In-Hand Characters

Despite recent advances in field identification, some species remain very difficult to separate without close examination of anatomical structures. These structures are typically very small and hard to see well in the field, even with modern optics. Some, like female subgenital plates, are located on the underside of the dragonfly, where they are almost always hidden from view. The most difficult species require capture and examination in the hand with the aid of magnification.

Canada and Green-striped Darners

These striped darner (*Aeshna*) species are usually separable by differences in thoracic and abdominal pattern, but with some individuals, the pattern may be inconclusive.

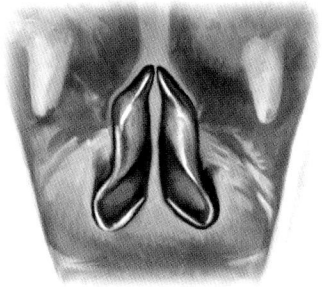 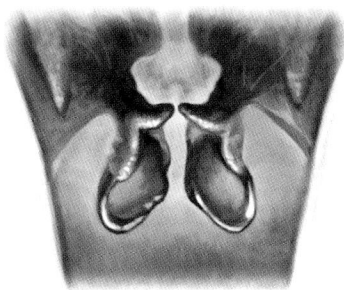

Canada Darner (*A. canadensis*) **male hamules**
(ventral view)

Green-striped Darner (*A. verticalis*) **male hamules**
(ventral view)

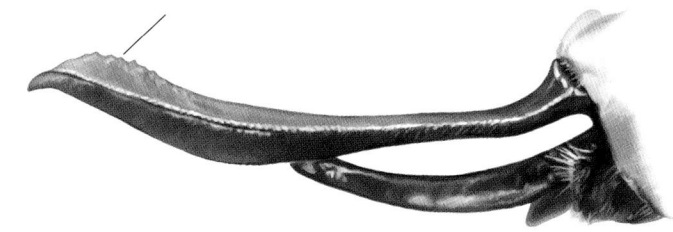

Canada Darner male appendages (lateral view)
cercus has minute teeth, which Green-striped Darner lacks

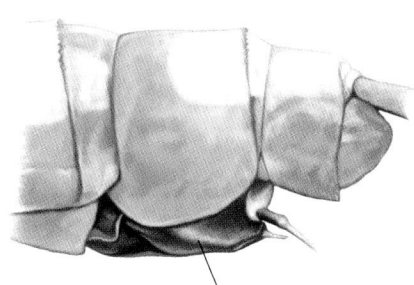 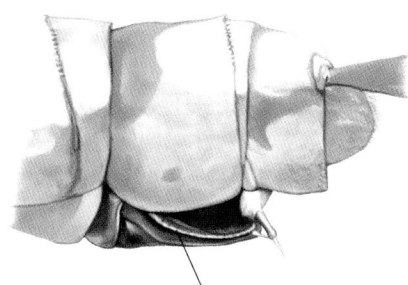

Canada Darner female abdomen tip
ovipositor with weak lateral ridge

Green-striped Darner female abdomen tip
ovipositor with sharp lateral ridge

Comet and Blue-spotted Comet Darner female cerci

Comet Darner *(Anax longipes)*
slightly wider toward base, tip blunt

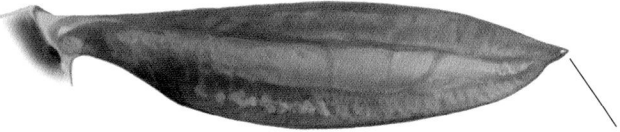

Blue-spotted Comet Darner *(Anax concolor)*
wider toward middle, tip with sharp tooth

Malachite and Secretive Darner male appendages

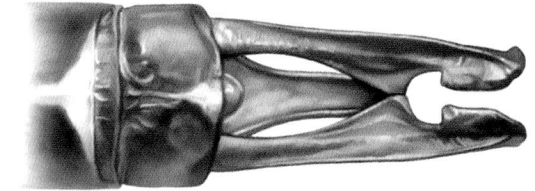

Malachite Darner *(Remartinia luteipennis)*

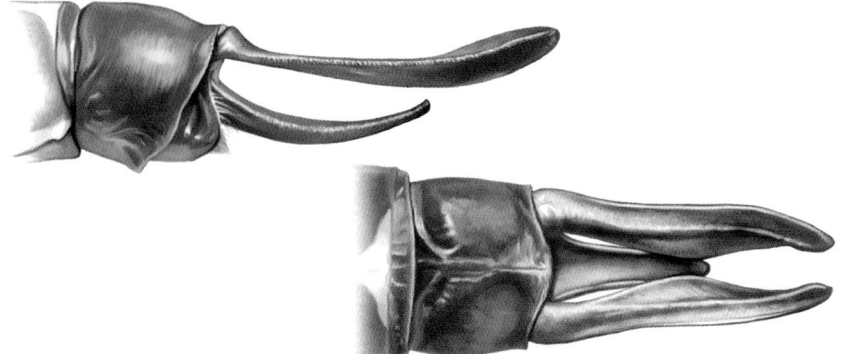

Secretive Darner *(Remartinia secreta)*

Some Neotropical darners *(Rhionaeschna)* male appendages

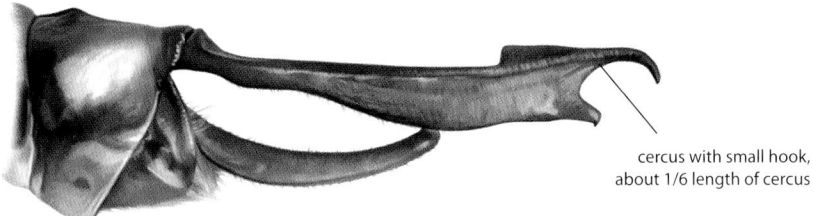

cercus with small hook,
about 1/6 length of cercus

Spatterdock Darner *(Rhionaeschna mutata)*

cercus with large hook,
about 1/4 length of cercus

Blue-eyed Darner *(Rhionaeschna multicolor)*

cercus not hooked

Arroyo Darner *(Rhionaeschna dugesi)*

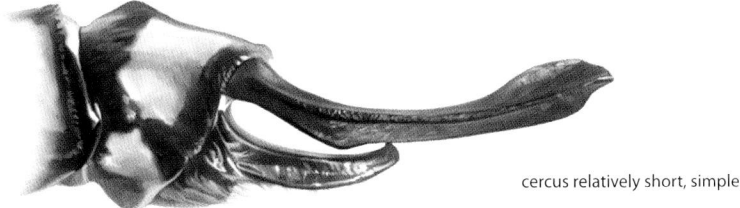

cercus relatively short, simple

California Darner *(Rhionaeschna californica)*

Some pond clubtails *(Arigomphus)* male appendages

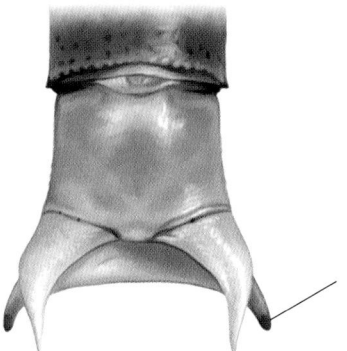

Jade Clubtail *(A. submedianus)*
epiproct spreads wider than cerci

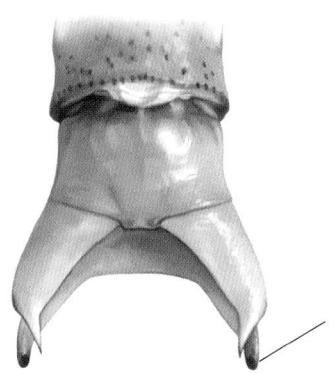

Stillwater Clubtail *(A. lentulus)*
epiproct longer than cerci, cerci wider

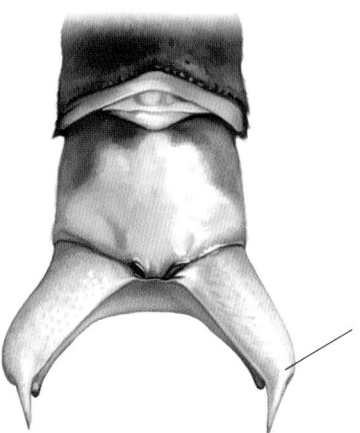

Unicorn Clubtail *(A. villosipes)*
cerci wide

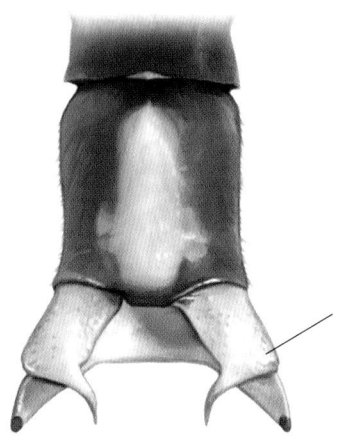

Lilypad Clubtail *(A. furcifer)*
cerci bent with tips convergent

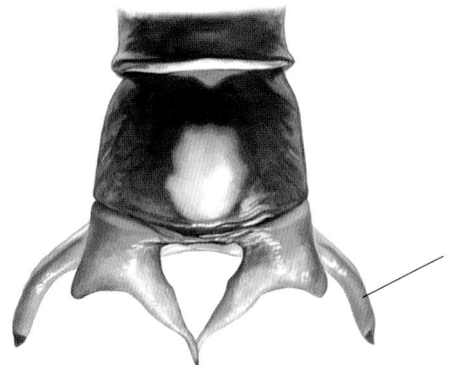

Horned Clubtail *(A. cornutus)*
cerci branched, epiproct very wide

Bantam clubtails (*Hylogomphus*) male appendages

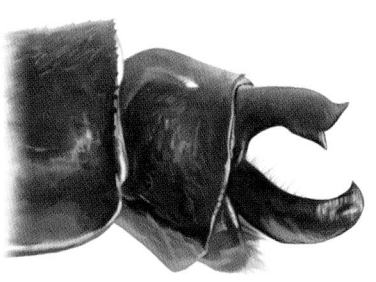 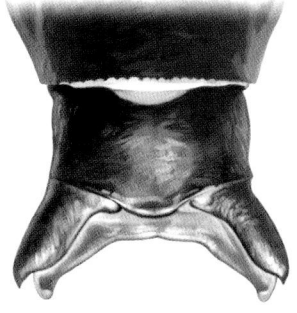

Mustached Clubtail *(H. adelphus)*

 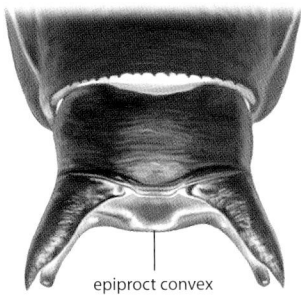

epiproct convex

Green-faced Clubtail *(H. viridifrons)*

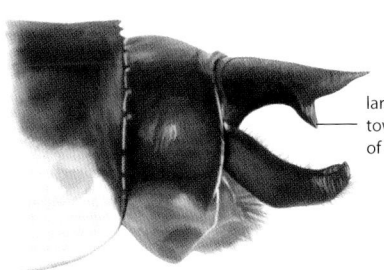 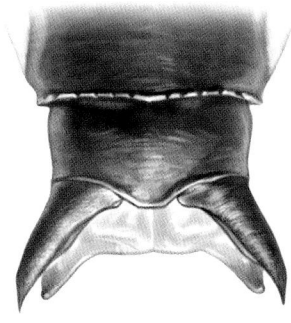

large ventral spine
toward middle
of cercus

Spine-crowned Clubtail *(H. abbreviatus)*

 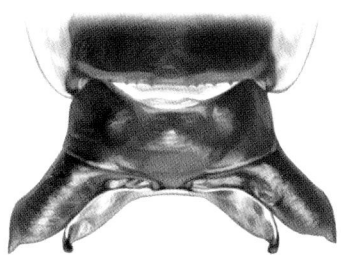

Banner Clubtail *(H. apomyius)*

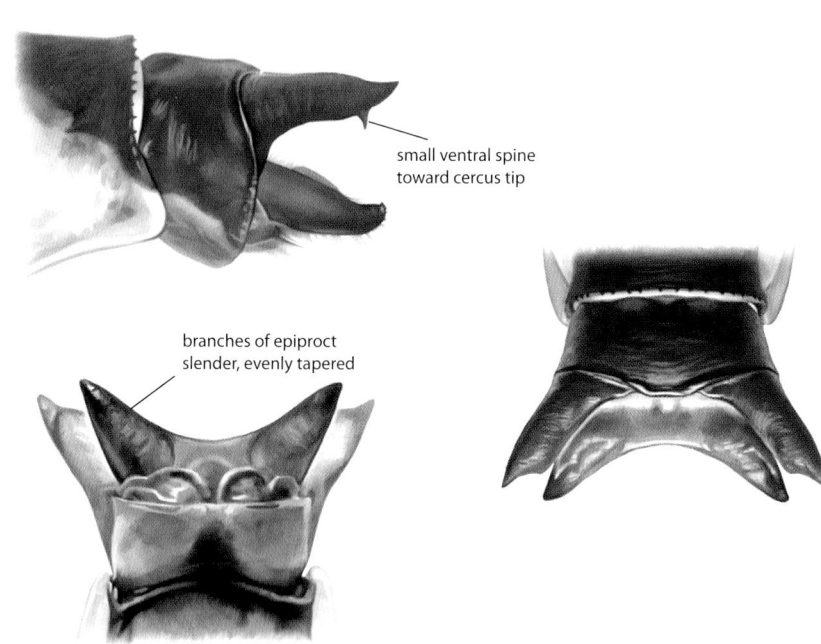

small ventral spine
toward cercus tip

branches of epiproct
slender, evenly tapered

Twin-striped Clubtail *(H. geminatus)*

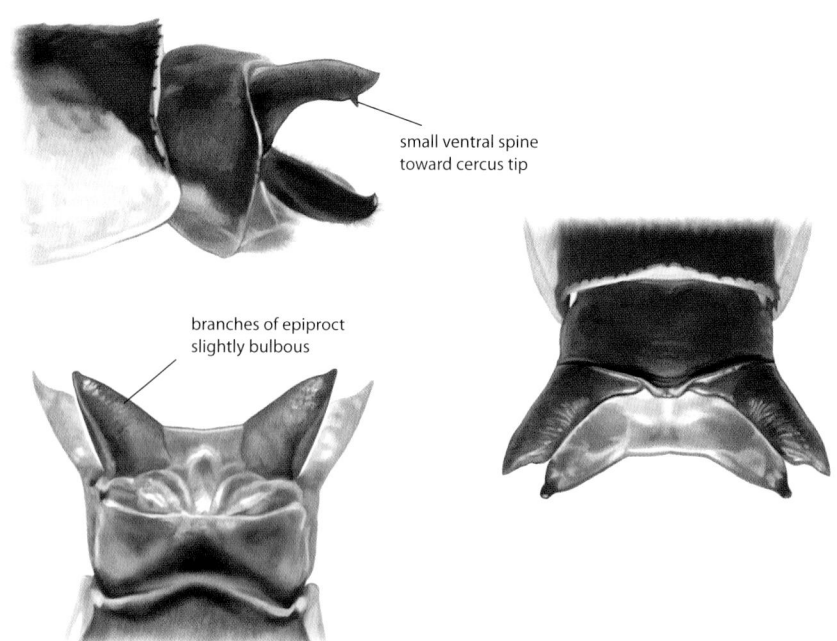

small ventral spine
toward cercus tip

branches of epiproct
slightly bulbous

Piedmont Clubtail *(H. parvidens)*

Bantam clubtails (*Hylogomphus*) females

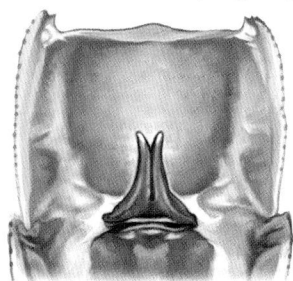

Mustached Clubtail *(H. adelphus)*
subgenital plate narrow trough

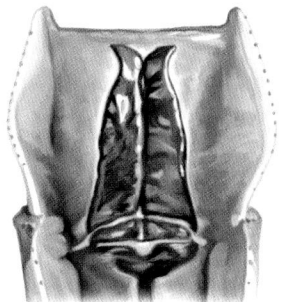

Green-faced Clubtail *(H. viridifrons)*
subgenital plate long, equal to S9

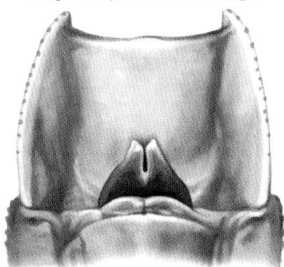

Spine-crowned Clubtail *(H. abbreviatus)*
subgenital plate less than half of S9

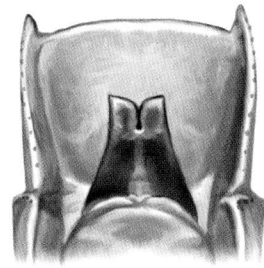

Banner Clubtail *(H. apomyius)*
subgenital plate about half of S9

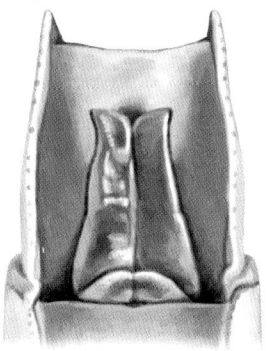

Twin-striped Clubtail *(H. geminatus)*
subgenital plate 2/3 length of S9

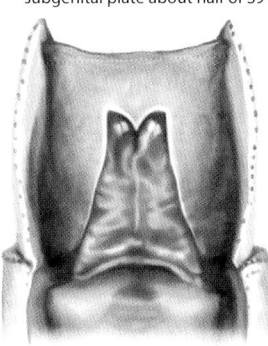

Piedmont Clubtail *(H. parvidens)*
subgenital plate 2/3–3/4 length of S9

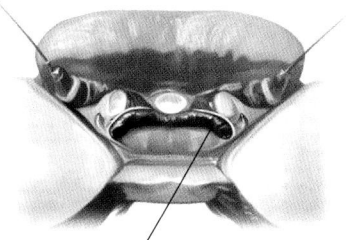

Twin-striped Clubtail *(H. geminatus)*
postocellar ridge long, extends to compound eyes

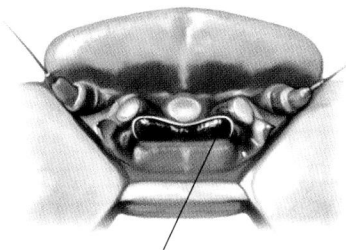

Piedmont Clubtail *(H. parvidens)*
postocellar ridge short, ends behind simple eyes

American clubtails (*Phanogomphus*) male appendages

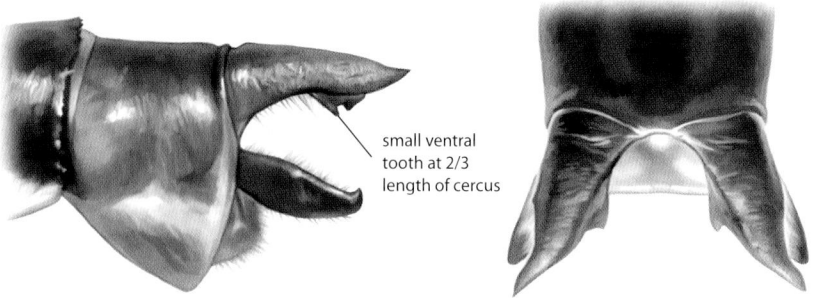

small ventral tooth at 2/3 length of cercus

Rapids Clubtail *(P. quadricolor)*

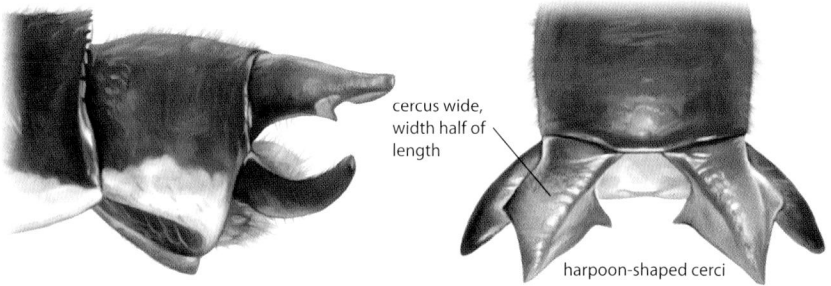

cercus wide, width half of length

harpoon-shaped cerci

Beaverpond Clubtail *(P. borealis)*

cercus width less than half of length

harpoon-shaped cerci

Harpoon Clubtail *(P. descriptus)*

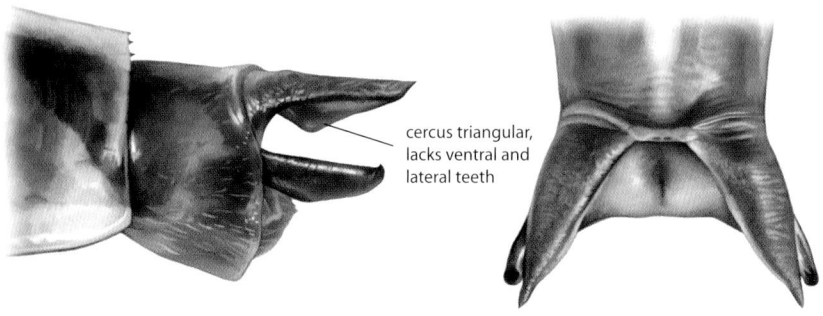

cercus triangular, lacks ventral and lateral teeth

Lancet Clubtail *(P. exilis)*

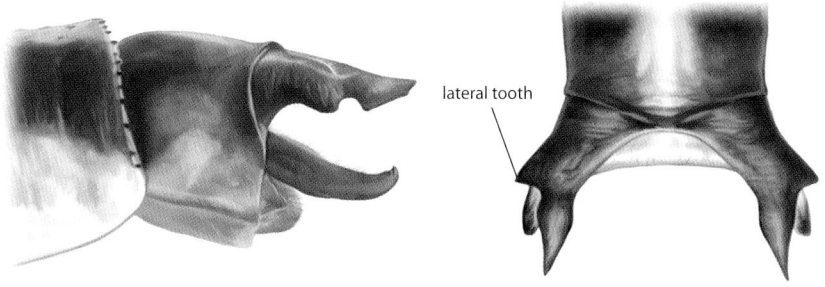

lateral tooth

Oklahoma Clubtail *(P. oklahomensis)*

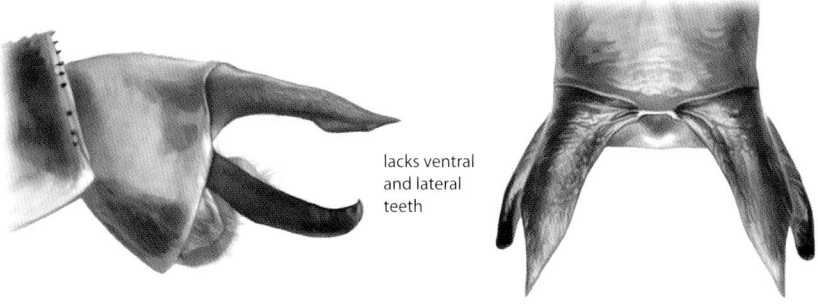

lacks ventral and lateral teeth

Ashy Clubtail *(P. lividus)*

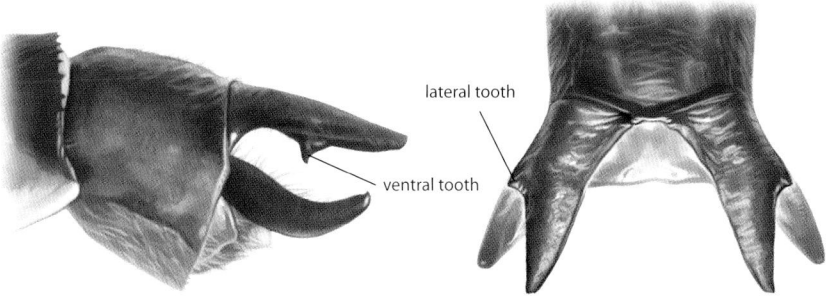

lateral tooth

ventral tooth

Dusky Clubtail *(P. spicatus)*

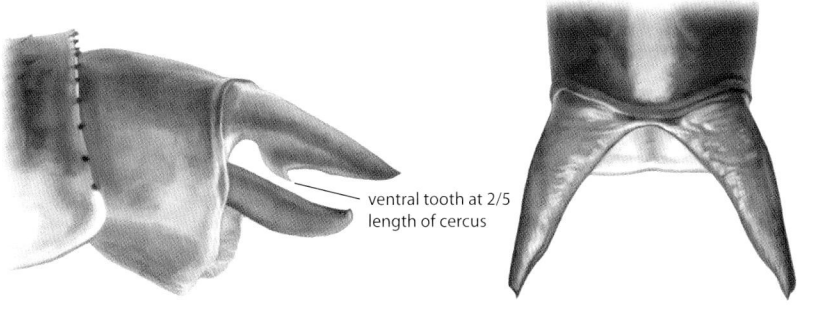

ventral tooth at 2/5 length of cercus

Cypress Clubtail *(P. minutus)*

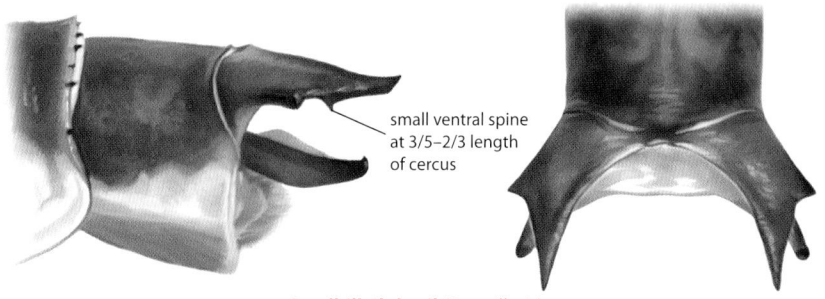

small ventral spine
at 3/5–2/3 length
of cercus

Sandhill Clubtail *(P. cavillaris)*

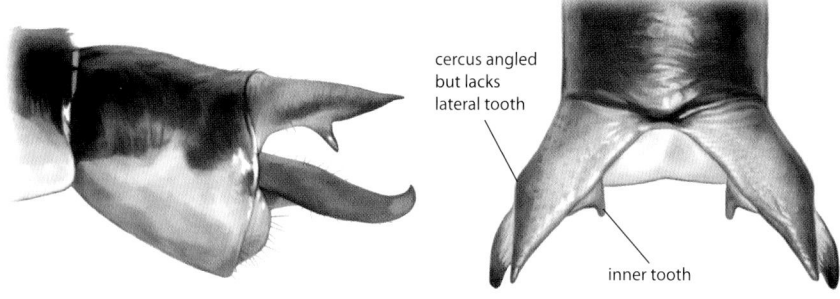

cercus angled
but lacks
lateral tooth

inner tooth

Westfall's Clubtail *(P. westfalli)*

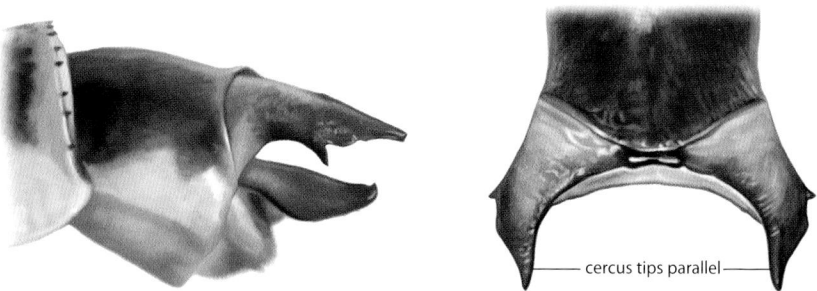

cercus tips parallel

Hodges' Clubtail *(P. hodgesi)*

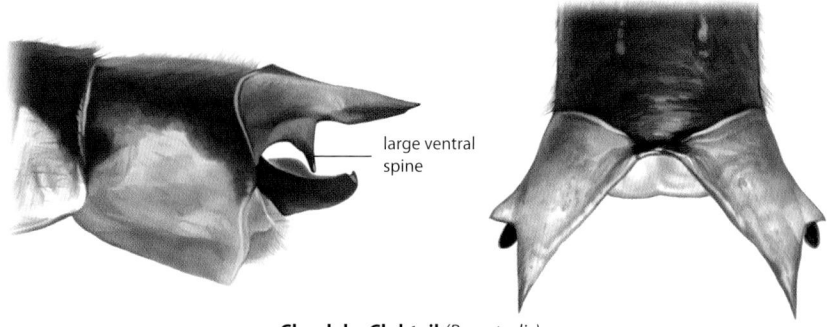

large ventral
spine

Clearlake Clubtail *(P. australis)*

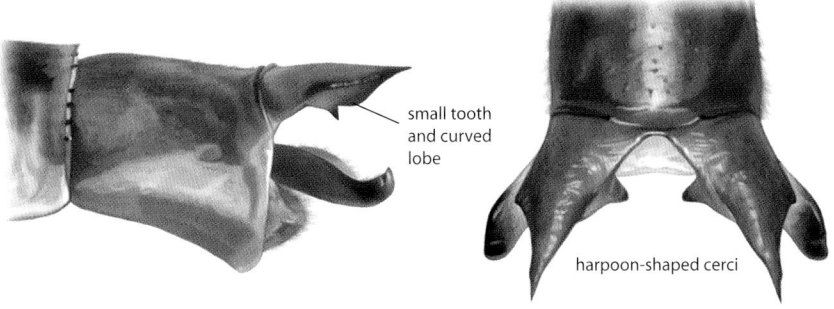

Diminutive Clubtail (*P. diminutus*)

small tooth and curved lobe

harpoon-shaped cerci

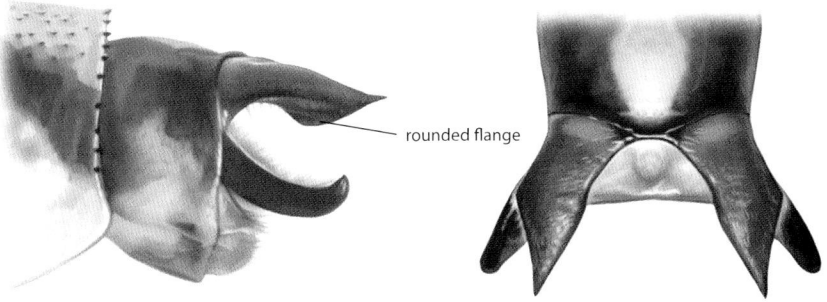

Tennessee Clubtail (*P. sandrius*)

rounded flange

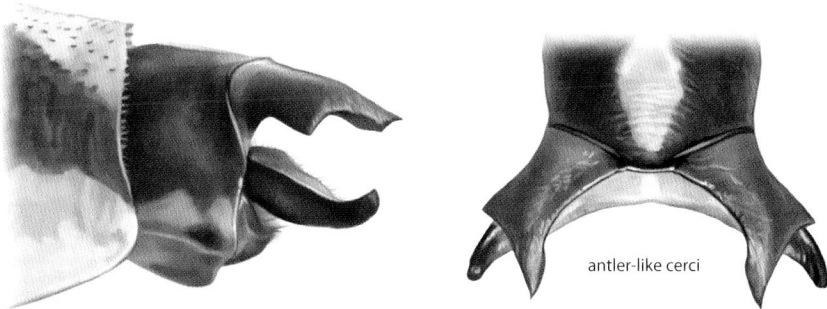

Pronghorn Clubtail (*P. graslinellus*)

antler-like cerci

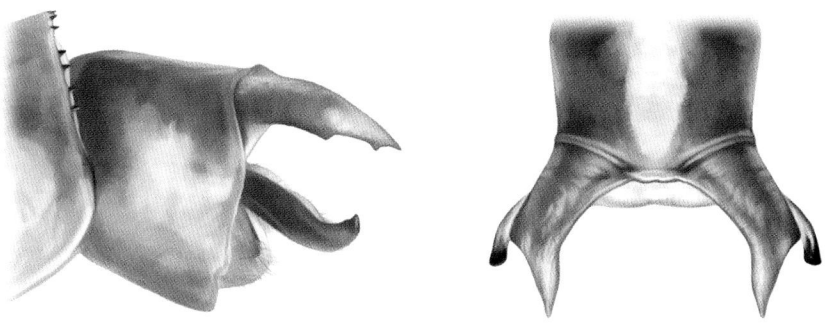

Sulphur-tipped Clubtail (*P. militaris*)

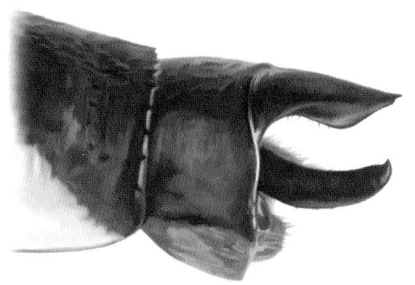 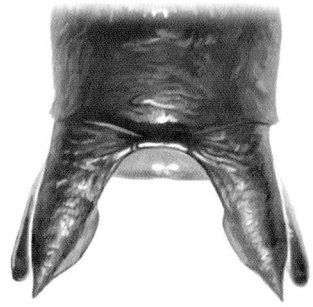

Pacific Clubtail *(P. kurilis)*

American clubtails (*Phanogomphus*) female subgenital plates

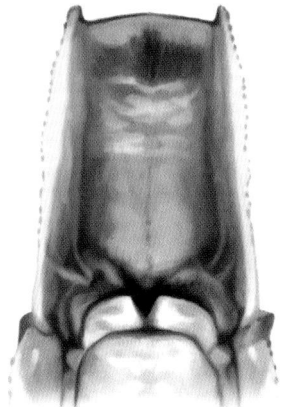

Rapids Clubtail *(P. quadricolor)*
plate short with tips pointed downward

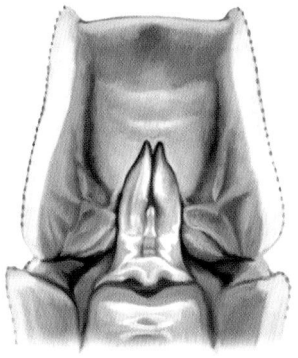

Beaverpond Clubtail *(P. borealis)*
plate half length of S9

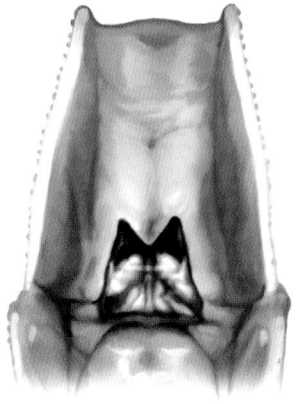

Harpoon Clubtail *(P. descriptus)*
plate 1/3 length of S9, V-notched

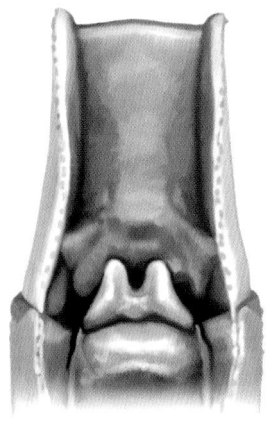

Lancet Clubtail *(P. exilis)*
plate 1/5–1/4 length of S9, V-notched

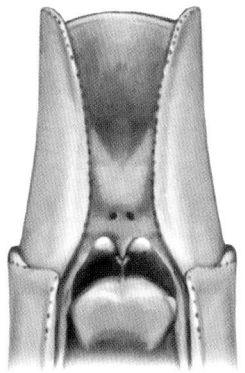

Oklahoma Clubtail *(P. oklahomensis)*
plate 1/6 length of S9, V-notched

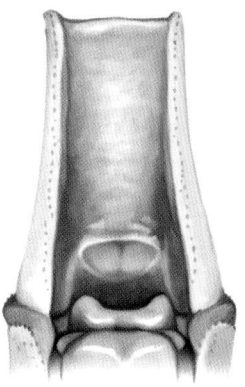

Ashy Clubtail *(P. lividus)*
plate 1/9 length of S9, lobes bulbous

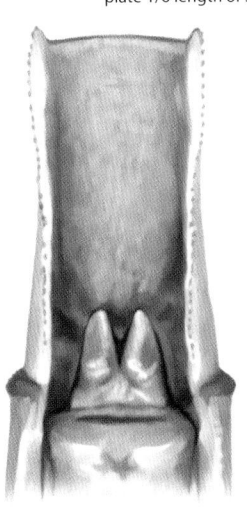

Dusky Clubtail *(P. spicatus)*
plate 3/8 length of S9

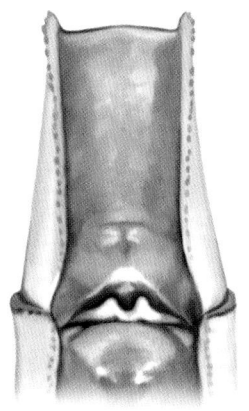

Cypress Clubtail *(P. minutus)*
plate 1/10 length of S9

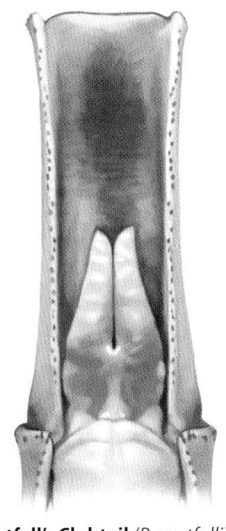

Westfall's Clubtail *(P. westfalli)*
plate almost half length of S9

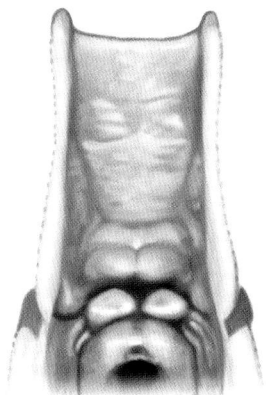

Sandhill Clubtail *(P. cavillaris)*
short rounded lobes

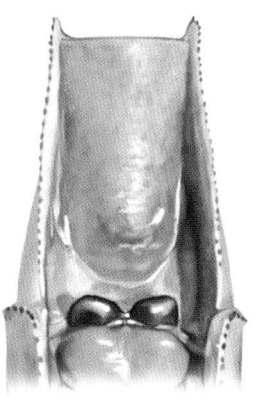

Hodges' Clubtail *(P. hodgesi)*

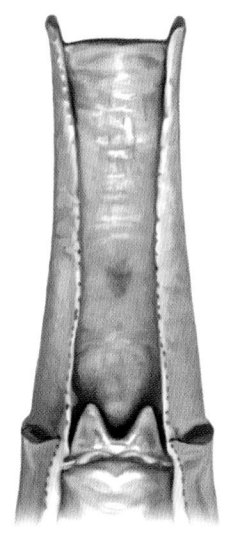

Clearlake Clubtail *(P. australis)*
plate 1/7 length of long S9

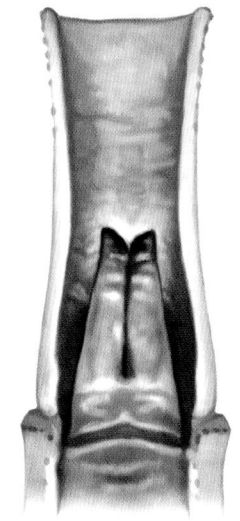

Diminutive Clubtail *(P. diminutus)*
plate half length of S9

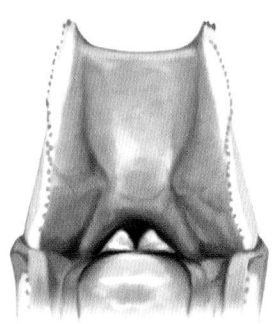

Tennessee Clubtail *(P. sandrius)*

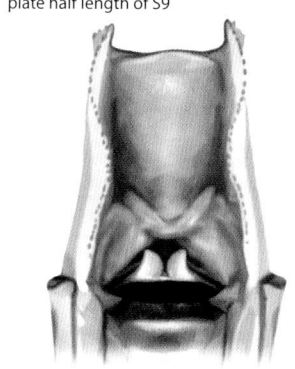

Pronghorn Clubtail *(P. graslinellus)*

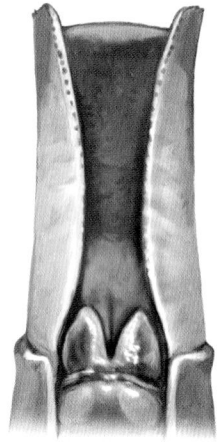

Pacific Clubtail *(P. kurilis)*

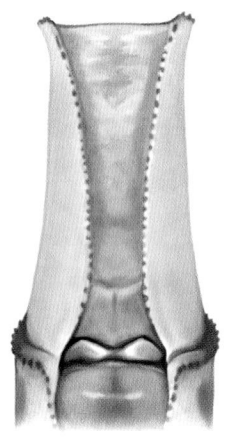

Sulphur-tipped Clubtail *(P. militaris)*
plate 1/10 length of S9

Majestic clubtails (*Gomphurus*) male appendages

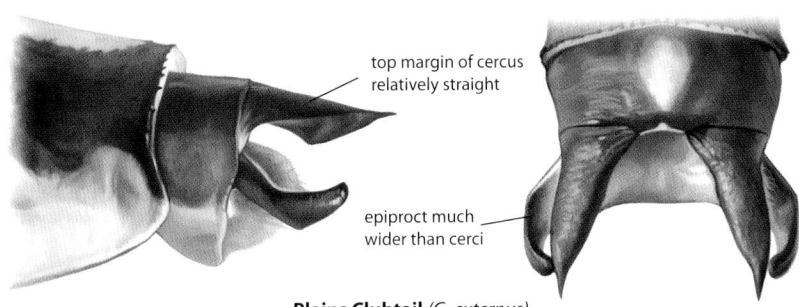

top margin of cercus relatively straight

epiproct much wider than cerci

Plains Clubtail (*G. externus*)

top margin of cercus relatively straight

epiproct slightly wider than cerci

Tamaulipan Clubtail (*G. gonzalezi*)

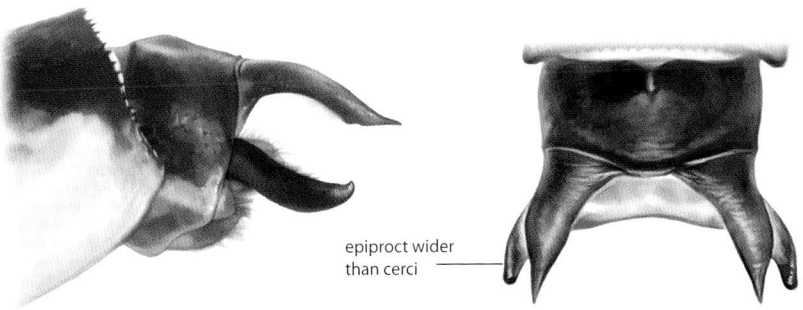

epiproct wider than cerci

Midland Clubtail (*G. fraternus*)

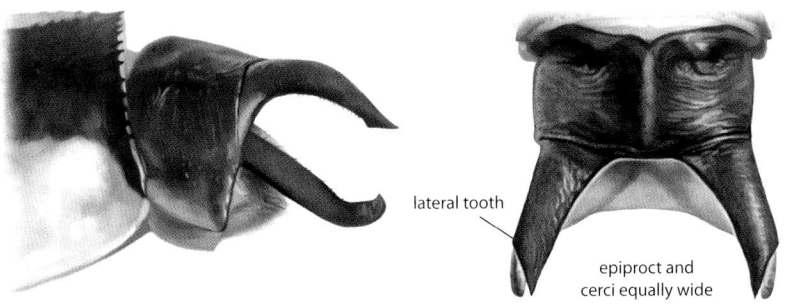

lateral tooth

epiproct and cerci equally wide

Handsome Clubtail (*G. crassus*)

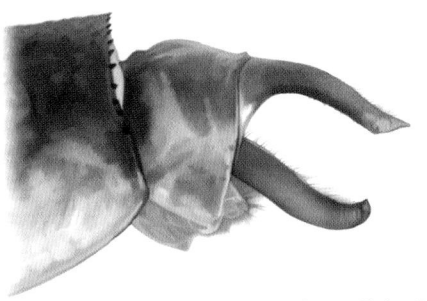

Cocoa Clubtail *(G. hybridus)*

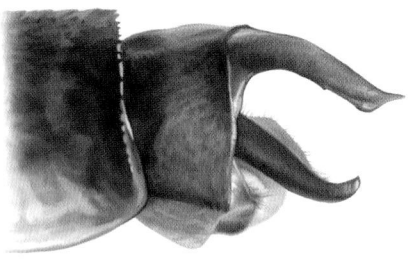

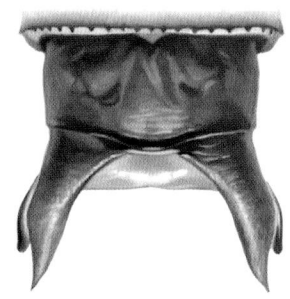

Septima's Clubtail *(G. septima)*

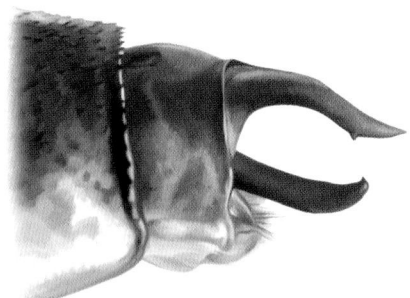

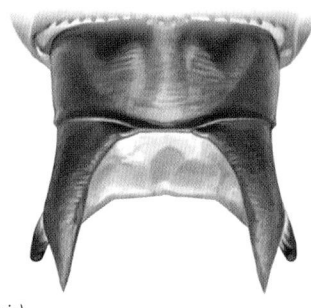

Ozark Clubtail *(G. ozarkensis)*

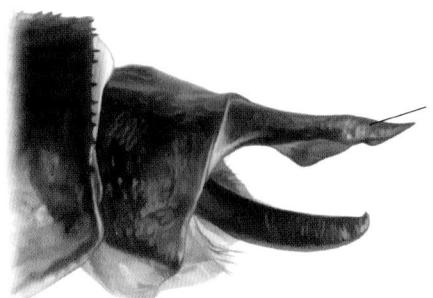

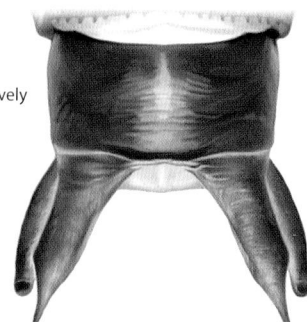

cercus relatively straight

Columbia Clubtail *(G. lynnae)*

 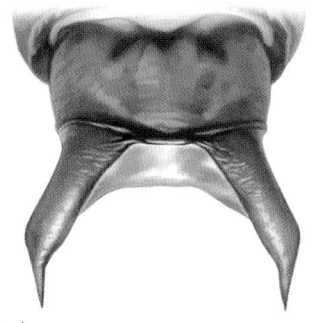

Cobra Clubtail *(G. vastus)*

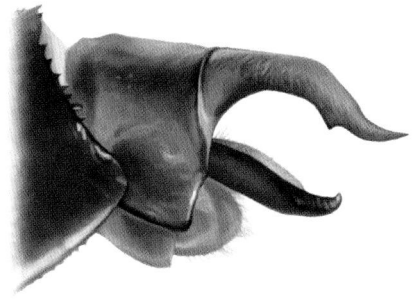 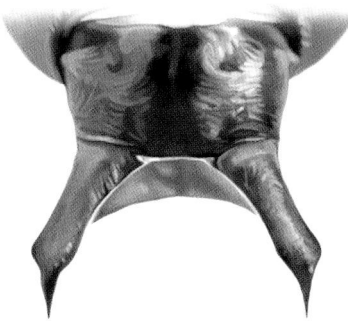

Blackwater Clubtail *(G. dilatatus)*

 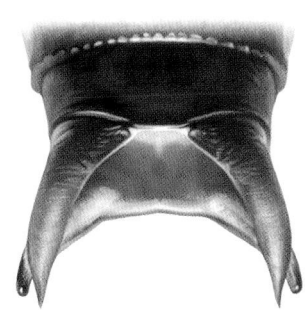

Splendid Clubtail *(G. lineatifrons)*

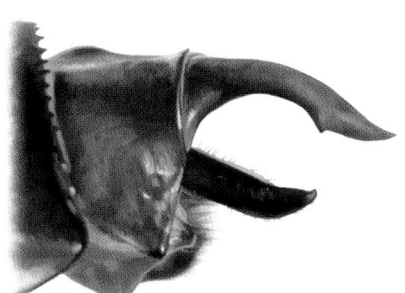 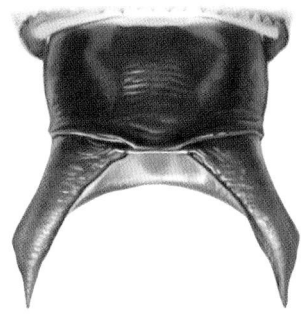

Gulf Coast Clubtail *(G. modestus)*

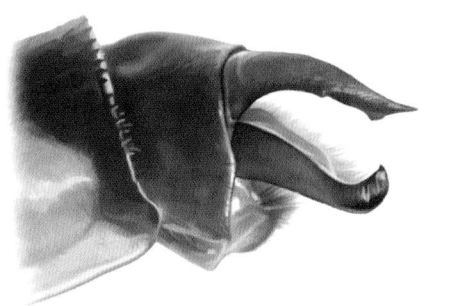
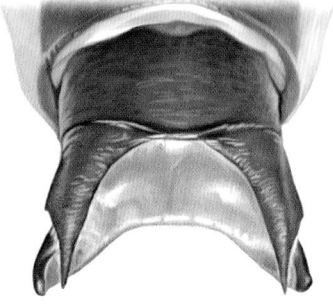

Skillet Clubtail *(G. ventricosus)*

Majestic clubtails (*Gomphurus*) female subgenital plates

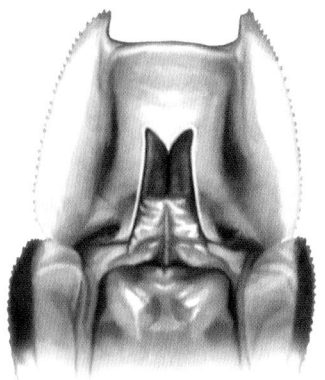

Plains Clubtail *(G. externus)*
plate length more than half of S9

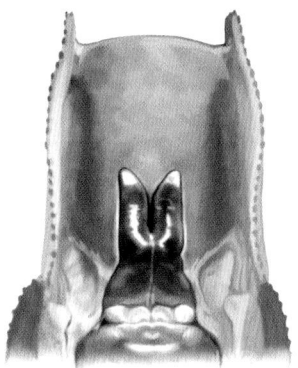

Tamaulipan Clubtail *(G. gonzalezi)*

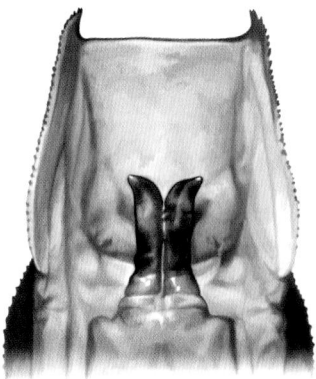

Midland Clubtail *(G. fraternus)*

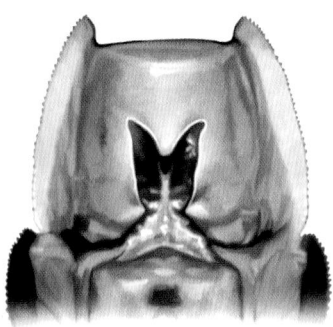

Handsome Clubtail *(G. crassus)*

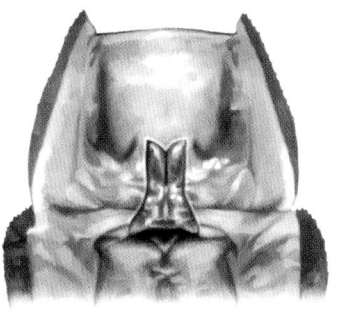

Cocoa Clubtail (*G. hybridus*)

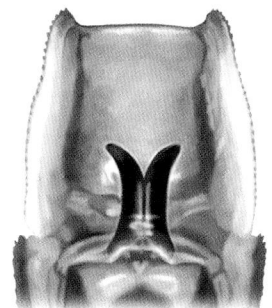

Septima's Clubtail (*G. septima*)
plate length half of S9, tips divergent

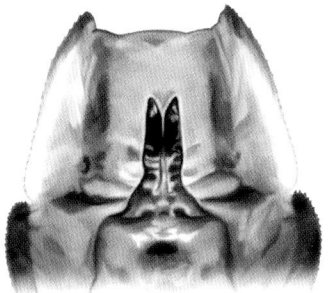

Ozark Clubtail (*G. ozarkensis*)

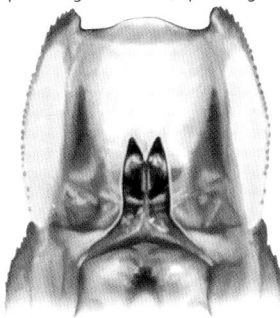

Columbia Clubtail (*G. lynnae*)

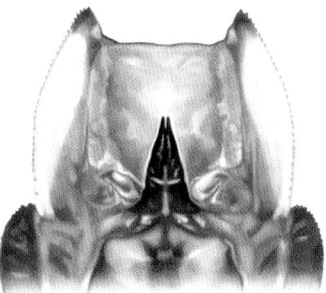

Cobra Clubtail (*G. vastus*)
narrow plate, length half of S9

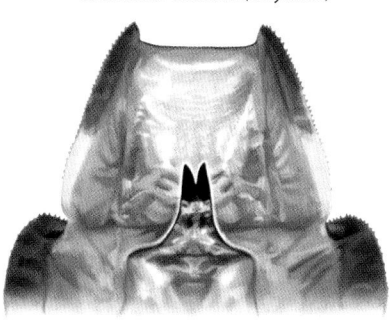

Blackwater Clubtail (*G. dilatatus*)
narrow plate, length less than half of S9

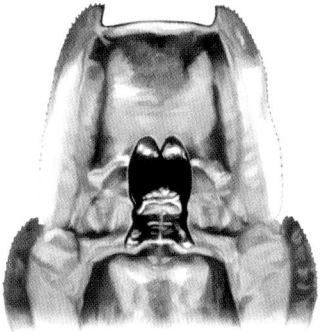

Splendid Clubtail (*G. lineatifrons*)
plate wide, length half of S9, tips rounded

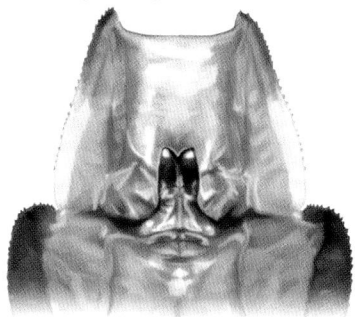

Gulf Coast Clubtail (*G. modestus*)
plate length less than half of S9, tips blunt

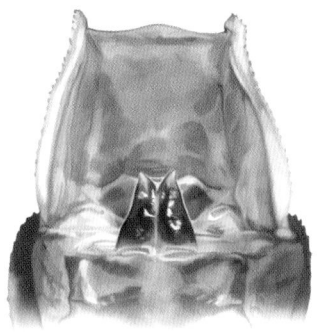

Skillet Clubtail *(G. ventricosus)*
plate length 1/3 of S9

Least clubtails (*Stylogomphus*) male appendages

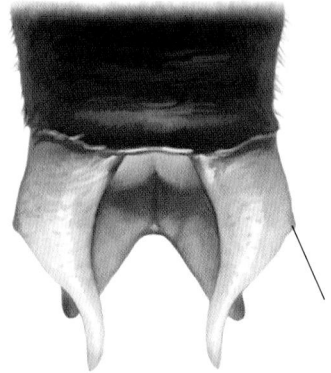

Eastern Least Clubtail *(S. albistylus)*
cerci angled at half length

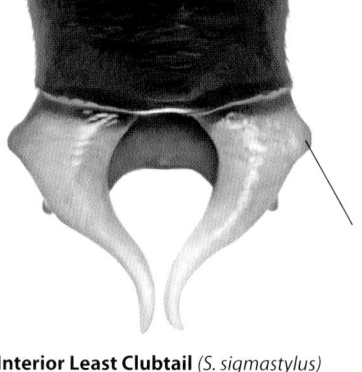

Interior Least Clubtail *(S. sigmastylus)*
cerci angled closer to base, 1/4–1/3 length

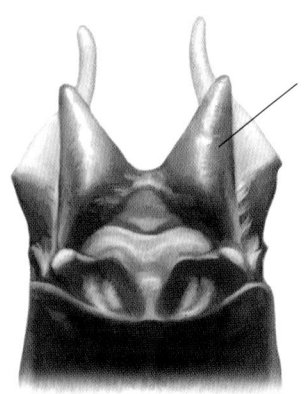

Eastern Least Clubtail *(S. albistylus)*
branches of epiproct long

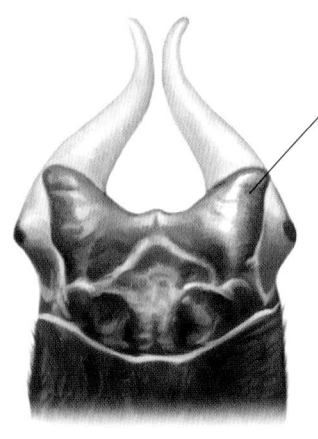

Interior Least Clubtail *(S. sigmastylus)*
branches of epiproct short, blunter, spread wider

Snaketails (*Ophiogomphus*) male appendages

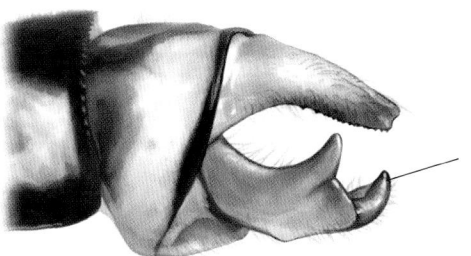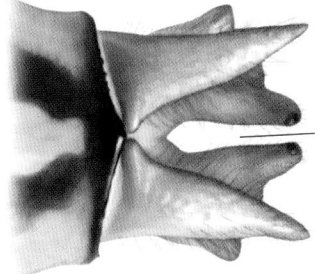

Maine Snaketail (*O. mainensis*)
epiproct deeply cleft, branches strongly upturned

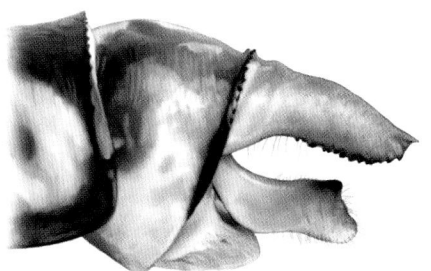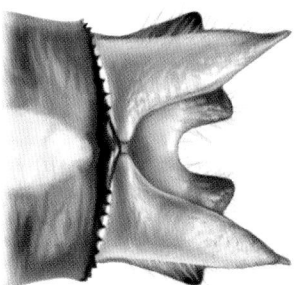

Riffle Snaketail (*O. carolus*)

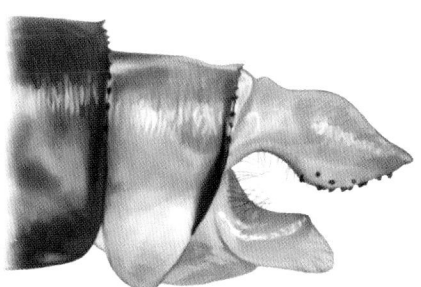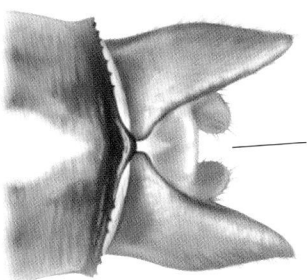

Brook Snaketail (*O. aspersus*)
epiproct with small squarish cleft

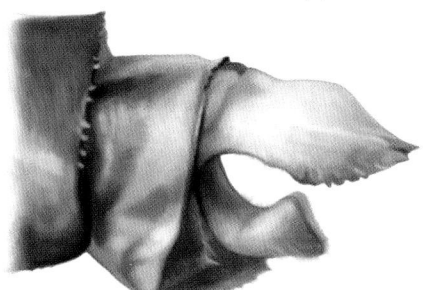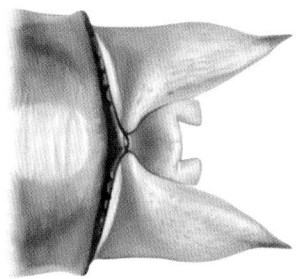

Sioux Snaketail (*O. smithi*)

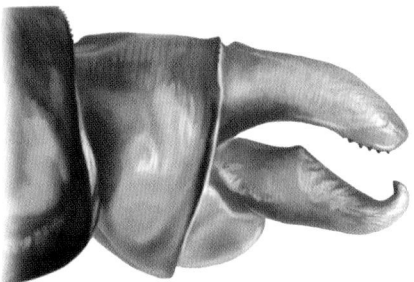
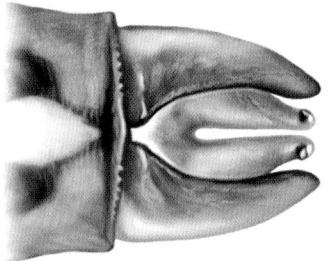

Boreal Snaketail *(O. colubrinus)*

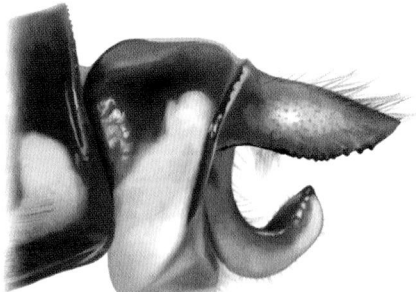
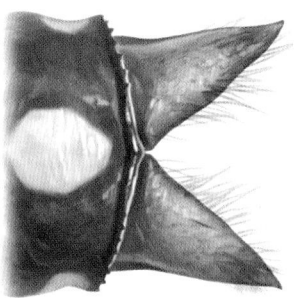

Extra-striped Snaketail *(O. anomalus)*

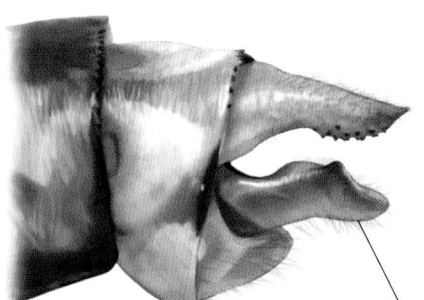
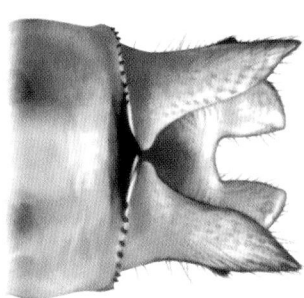

Edmund's Snaketail *(O. edmundo)*
epiproct bent downward

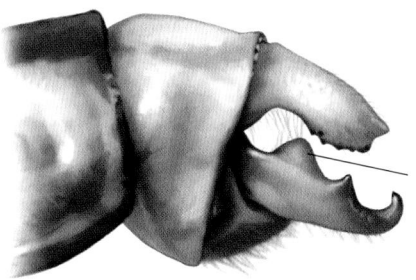
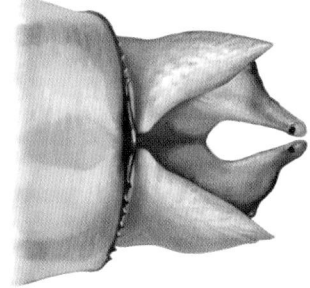

St. Croix Snaketail *(O. susbehcha)*
epiproct with large dorsal bump

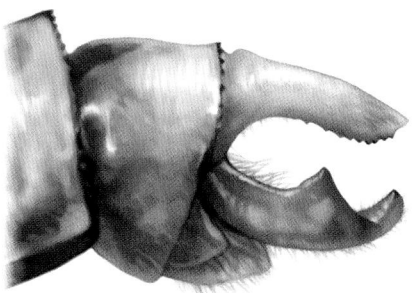
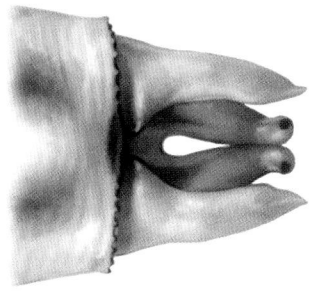

Appalachian Snaketail (*O. incurvatus incurvatus*)
epiproct upcurved

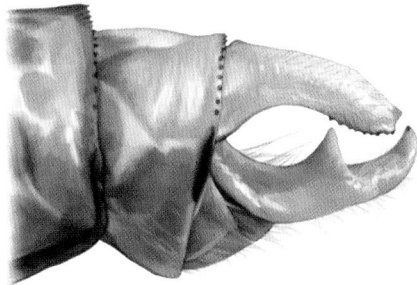
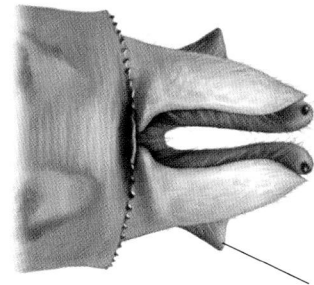

Appalachian Snaketail (*O. incurvatus alleghaniensis*)
lateral spine on epiproct more prominent

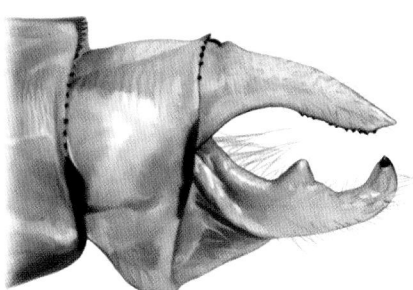
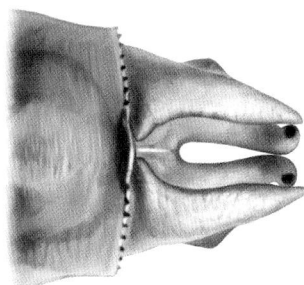

Southern Snaketail (*O. australis*)

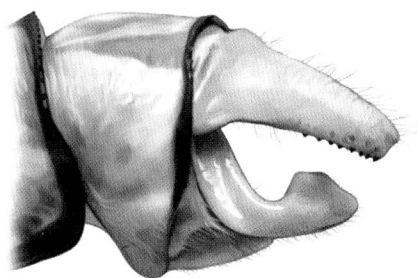
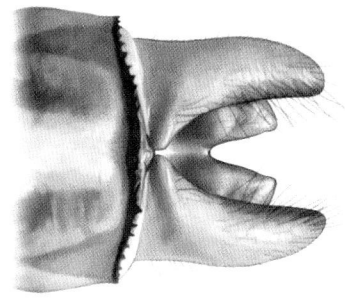

Rusty Snaketail (*O. rupinsulensis*)
cercus tip blunt, epiproct lacks tooth in middle

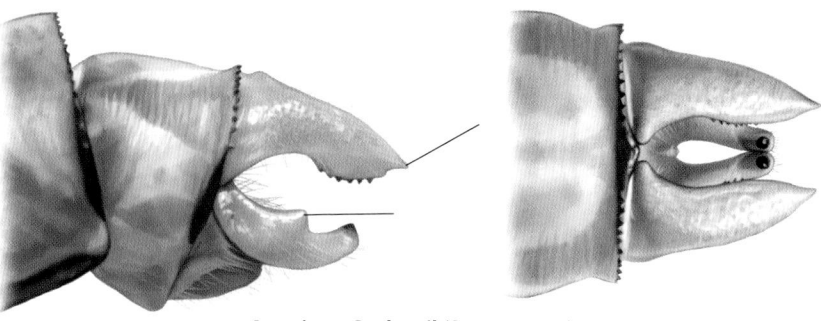

Acuminate Snaketail *(O. acuminatus)*
cerci tip pointed, epiproct with tooth in middle

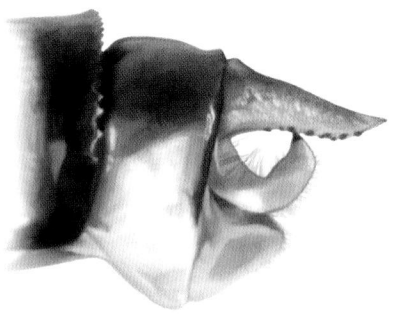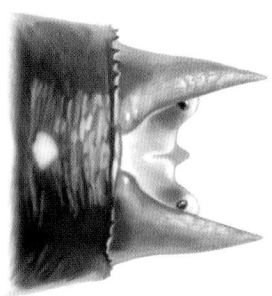

Pygmy Snaketail *(O. howei)*

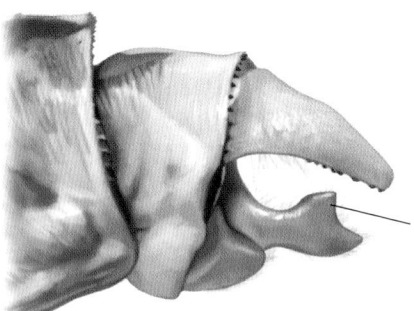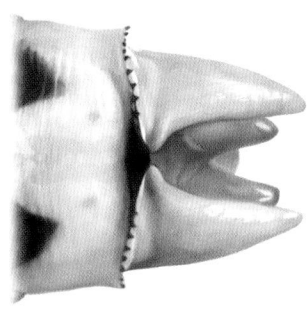

Westfall's Snaketail *(O. westfalli)*
epiproct with large dorsal tooth

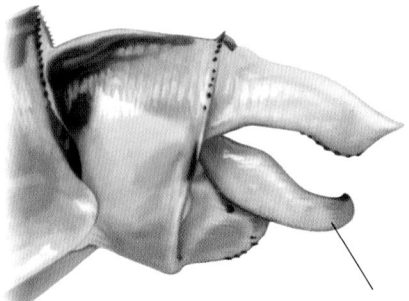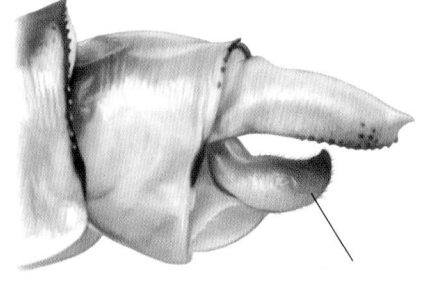

Pale Snaketail *(O. severus)*
epiproct long, 3/4–6/7 length of cerci

Arizona Snaketail *(O. arizonicus)*
epiproct short, 1/2–2/3 length of cerci

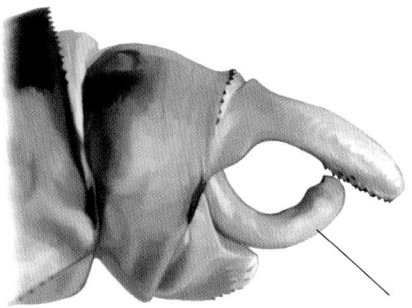
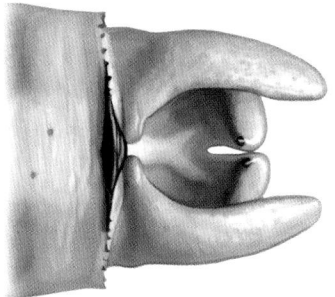

Sinuous Snaketail *(O. occidentis)*
epiproct smoothly upcurved

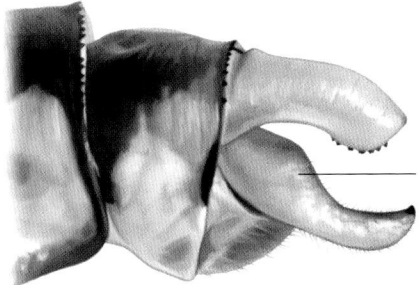
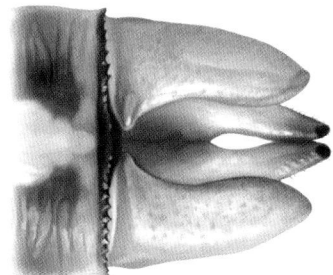

Great Basin Snaketail *(O. morrisoni)*
epiproct slightly curved, expanded at base, longer than cerci

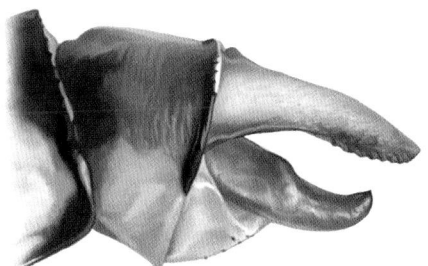
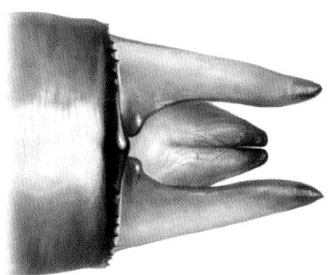

Bison Snaketail *(O. bison)*

Snaketails (*Ophiogomphus*) female subgenital plates and other features

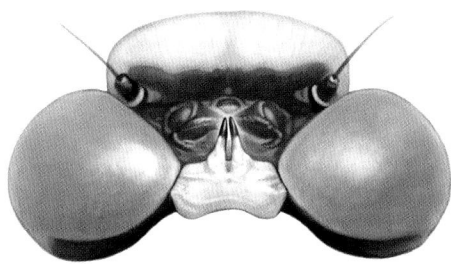

Maine Snaketail *(O. mainensis)* **head**
large forward-facing occipital horns

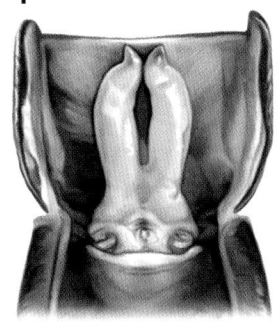

Maine Snaketail *(O. mainesis)*

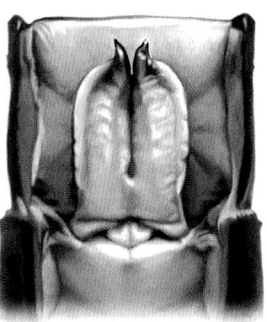

Riffle Snaketail *(O. carolus)*

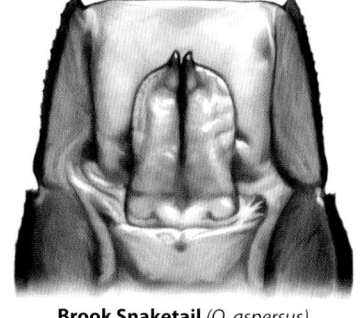

Brook Snaketail *(O. aspersus)*
plate 3/4 length of S9

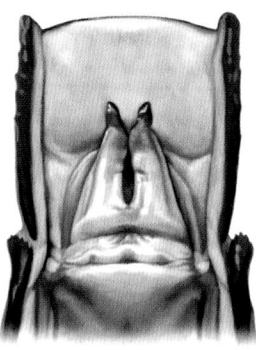

Sioux Snaketail *(O. smithi)*
plate less than 3/4 length of S9

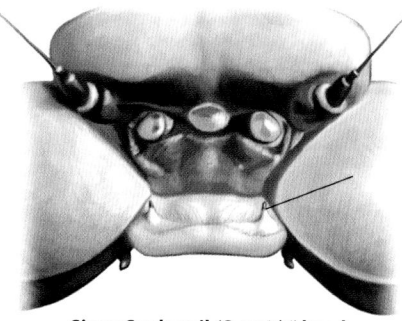

Sioux Snaketail *(O. smithi)* **head**
small occipital and postoccipital horns

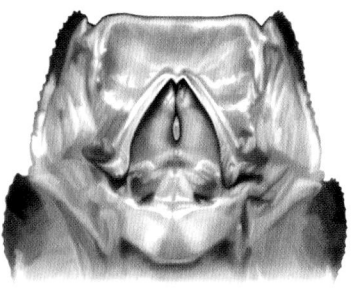

Boreal Snaketail *(O. colubrinus)*

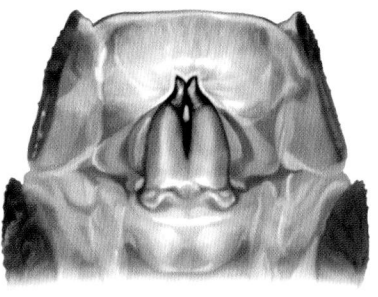

Extra-striped Snaketail *(O. anomalus)*

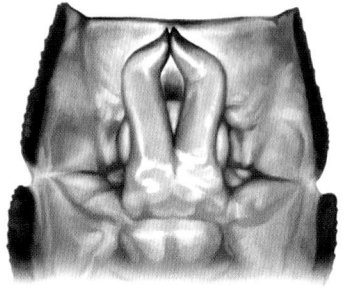

Edmund's Snaketail *(O. edmundo)*

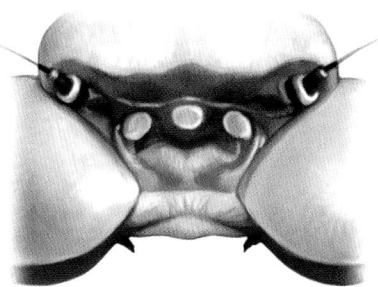

St. Croix Snaketail *(O. susbehcha)* **head**
small postoccipital horns

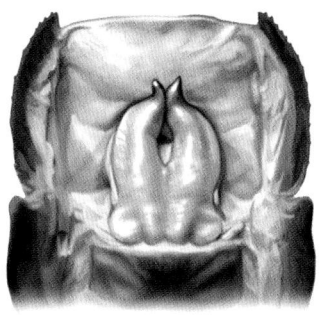

St. Croix Snaketail *(O. susbehcha)*
plate 3/5 length of S9, divided half its length

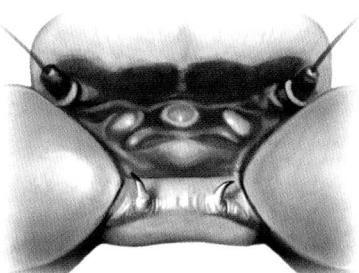

Appalachian Snaketail *(O. i. incurvatus)*
small, separated occipital horns

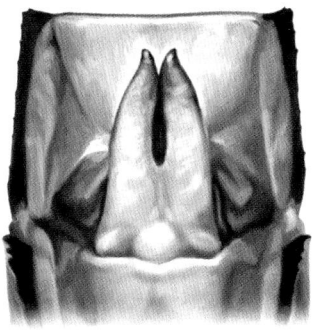

Appalachian Snaketail *(O. incurvatus)*
plate 3/4 length of S9

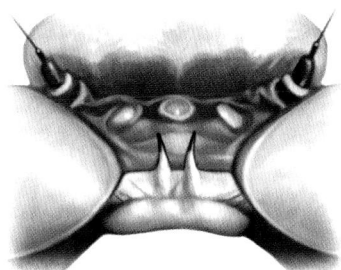

Appalachian Snaketail *(O. i. alleghaniensis)*
pair of horns near center of occiput

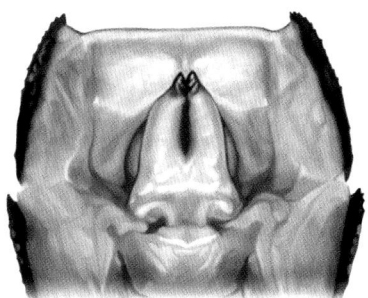

Southern Snaketail *(O. australis)*

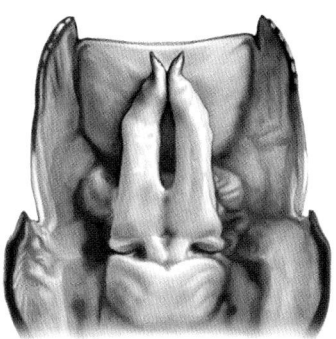

Rusty Snaketail *(O. rupinsulensis)*

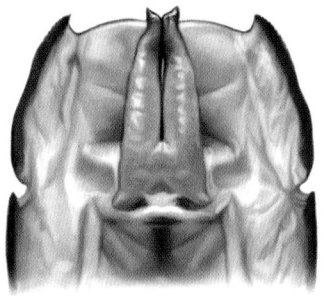

Acuminate Snaketail *(O. acuminatus)*
plate longer than S9

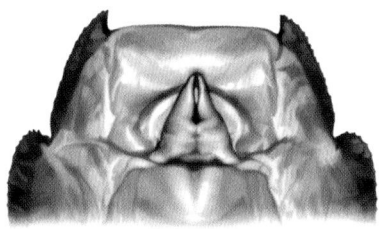

Pygmy Snaketail *(O. howei)*

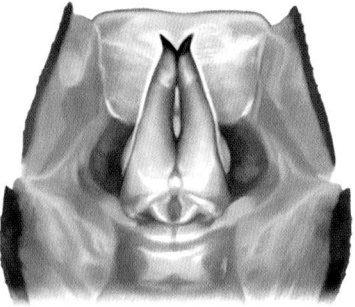

Westfall's Snaketail *(O. westfalli)*

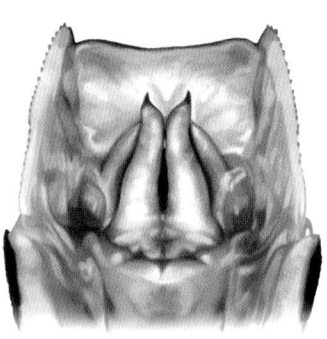

Pale Snaketail *(O. severus)*

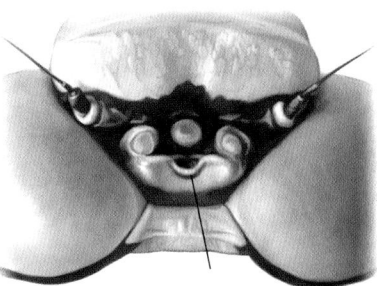

Pale Snaketail *(O. severus)*
V-shaped undulation in ridge behind ocellus

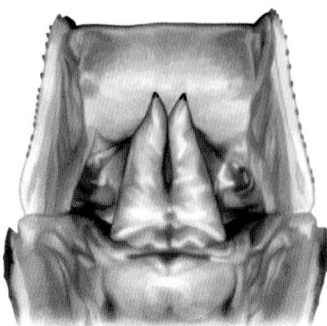

Arizona Snaketail *(O. arizonicus)*

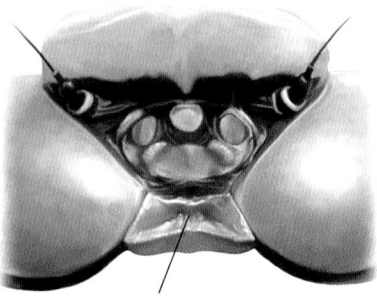

Arizona Snaketail *(O. arizonicus)*
occiput strongly concave

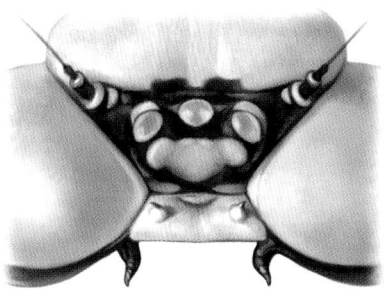

Sinuous Snaketail *(O. occidentis)*
small occipital horns, postoccipital horns larger, crooked

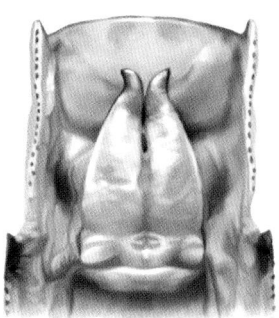

Sinuous Snaketail *(O. occidentis)*
subgenital plate 3/4 length of S9

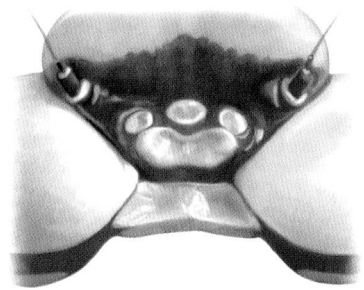

Great Basin Snaketail *(O. morrisoni)*
lacks postoccipital horns, may have small occipital horns

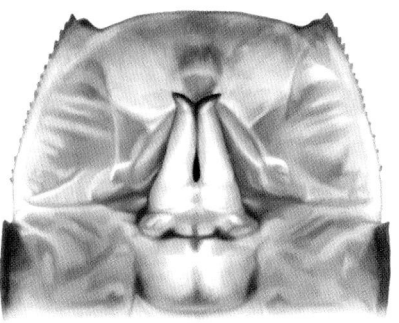

Great Basin Snaketail *(O. morrisoni)*
subgenital plate 6/10 length of S9

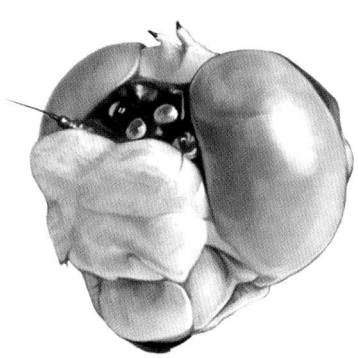

Bison Snaketail *(O. bison)*
prominent occipital horns

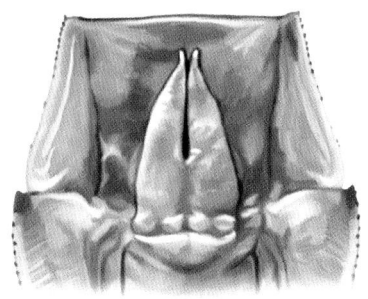

Bison Snaketail *(O. bison)*

Delta-spotted and Ouachita Spiketails male appendages

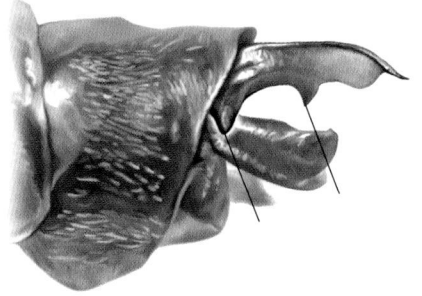 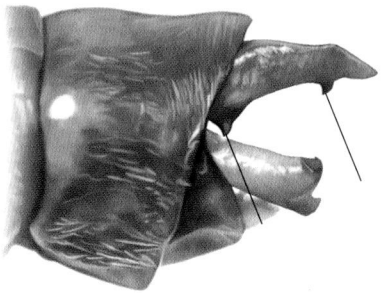

Delta-spotted Spiketail *(Cordulegaster diastatops)*
distance between teeth on cercus shorter

Ouachita Spiketail *(Cordulegaster talaria)*
distance between teeth on cercus longer

Striped emeralds (*Somatochlora*) male appendages

 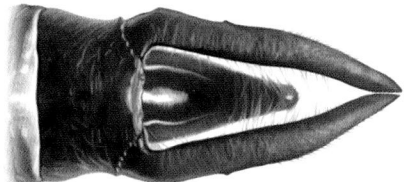

Ocellated Emerald *(S. minor)*
cerci ski-type, tips convergent

 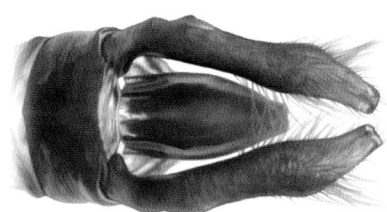

Brush-tipped Emerald *(S. walshii)*
posterior 1/3 of cerci covered with dense hairs

 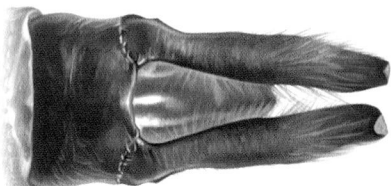

Williamson's Emerald *(S. williamsoni)*
cerci ski-type with basal lateral tooth, hairy

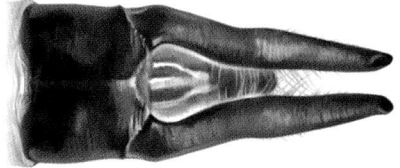

Ski-tipped Emerald *(S. elongata)*
cerci ski-type

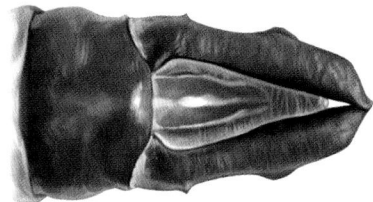

Mocha Emerald *(S. linearis)*
cerci with long ventral spine near tip

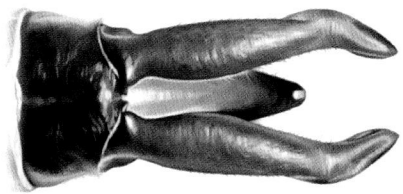

Delicate Emerald *(S. franklini)*
cerci pincer-type

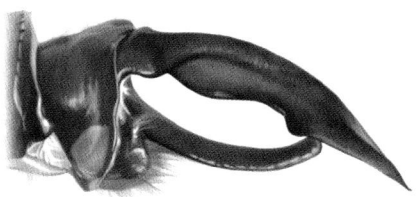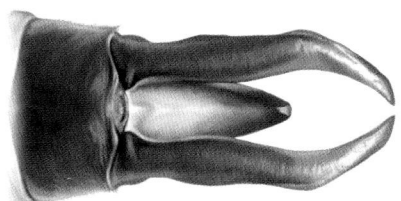

Forcipate Emerald *(S. forcipata)*
cerci pincer-type, tips sharply pointed

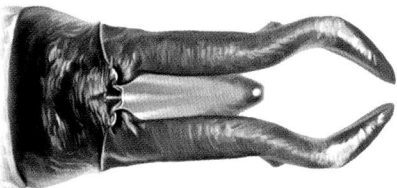

Incurvate Emerald *(S. incurvata)*
cerci pincer-type, straight in lateral view, tips strongly convergent

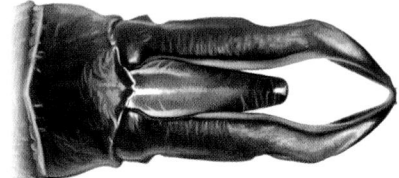

Kennedy's Emerald *(S. kennedyi)*
cerci pincer-type

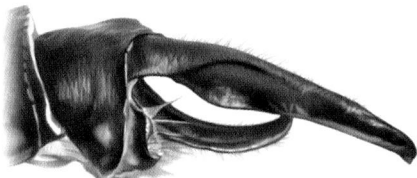
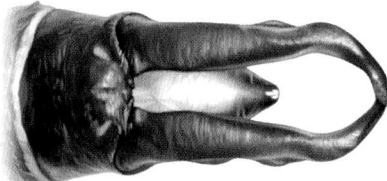

Mountain Emerald *(S. semicircularis)*
cerci pincer-type

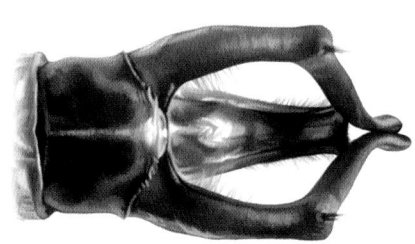

Clamp-tipped Emerald *(S. tenebrosa)*
cerci bent-type, spread wide in dorsal view, epiproct strongly upcurved

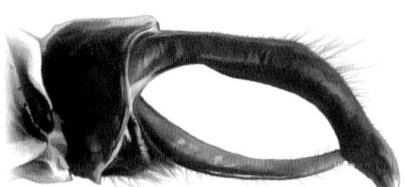
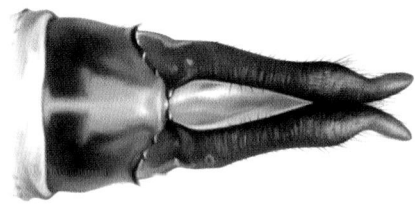

Calvert's Emerald *(S. calverti)*
cerci bent downward at mid-length

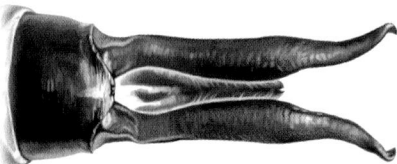

Fine-lined Emerald *(S. filosa)*
cerci relatively straight

 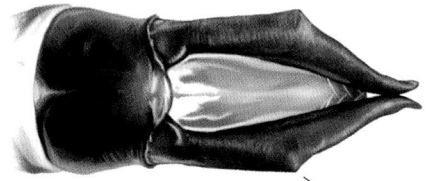

Treetop Emerald *(S. provocans)*
cerci bent downward 30° at mid-length, with pointed angle at mid-length in dorsal view

 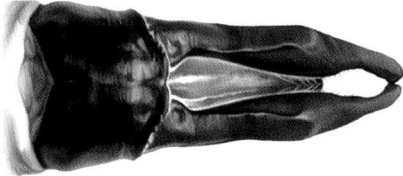

Texas Emerald *(S. margarita)*
cerci bent downward 35° at mid-length, with lateral ridge extending from base to tip

 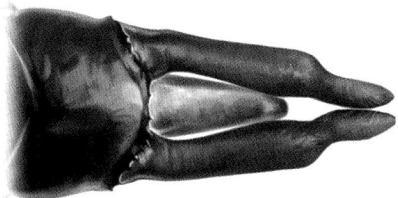

Ozark Emerald *(S. ozarkensis)*
cerci bend downward 40° at mid-length, with lateral ridge ending at mid-length angle

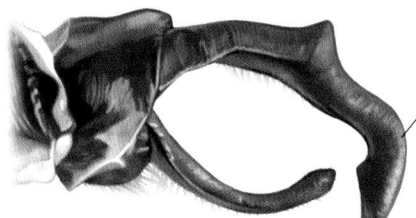 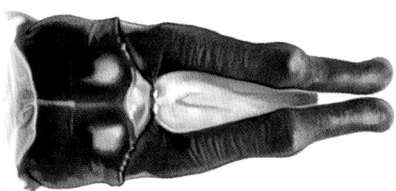

Hine's Emerald *(S. hineana)*
cerci with large dorsal spine at mid-length, strongly bent downward at tips

 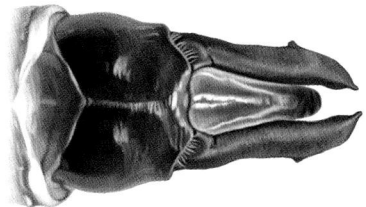

Plains Emerald *(S. ensigera)*
cerci relatively straight with ventral spine at tips

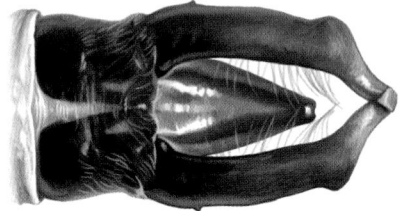

Quebec Emerald *(S. brevicincta)*
cerci bent-curled type, single small angle near base

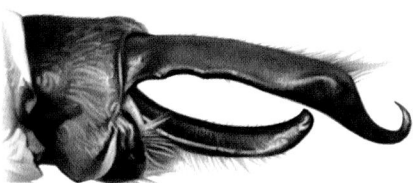
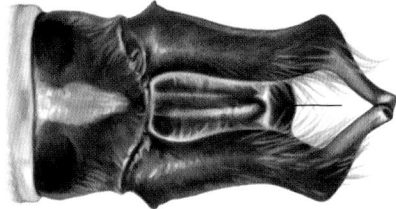

Lake Emerald *(S. cingulata)*
cerci bent-curled type, epiproct rectangular in ventral view (other striped emeralds triangular)

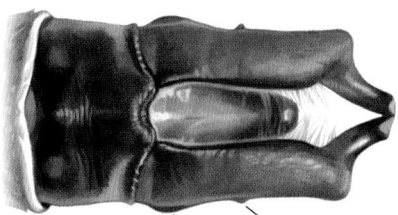

Ringed Emerald *(S. albicincta)*
cerci bent-curled type, with lateral angle at base and mid-length

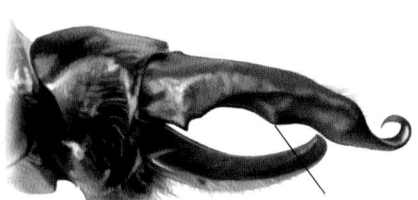
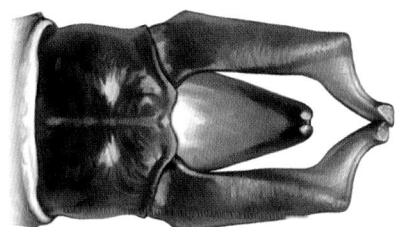

Hudsonian Emerald *(S. hudsonica)*
cerci bent-curled type, with large ventral angle at mid-length

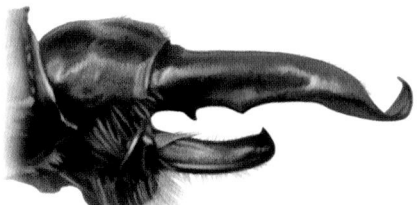
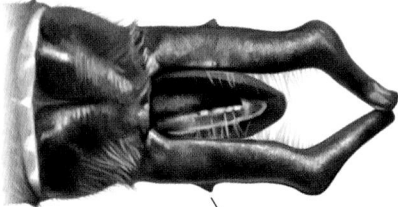

Muskeg Emerald *(S. septentrionalis)*
cerci bent-curled type, with lateral angle at 1/3 length

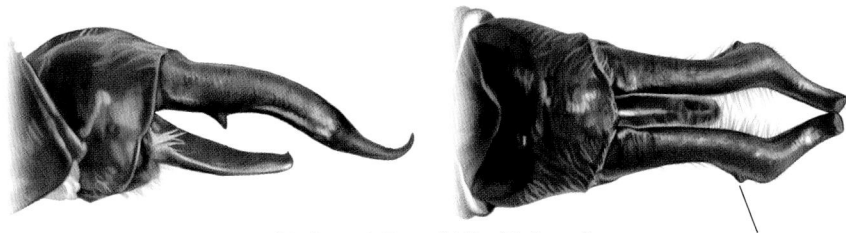

Whitehouse's Emerald *(S. whitehousei)*
cerci bent-curled type, with lateral angle at 3/5–2/3 length

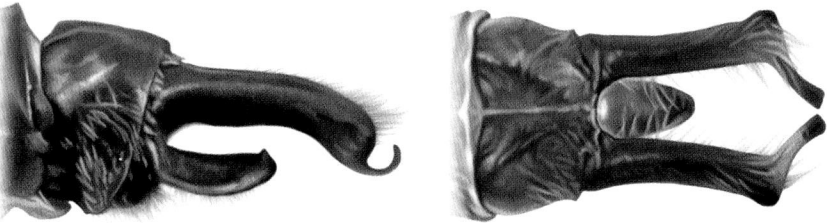

Treeline Emerald *(S. sahlbergi)*
cerci bent-curled type, lack spines or angles, epiproct relatively short

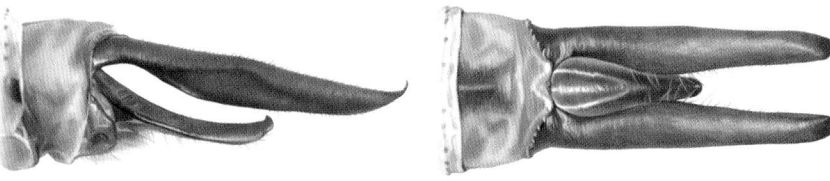

Coppery Emerald *(S. georgiana)*
cerci ski-type

Striped emeralds (*Somatochlora*) female subgenital plates

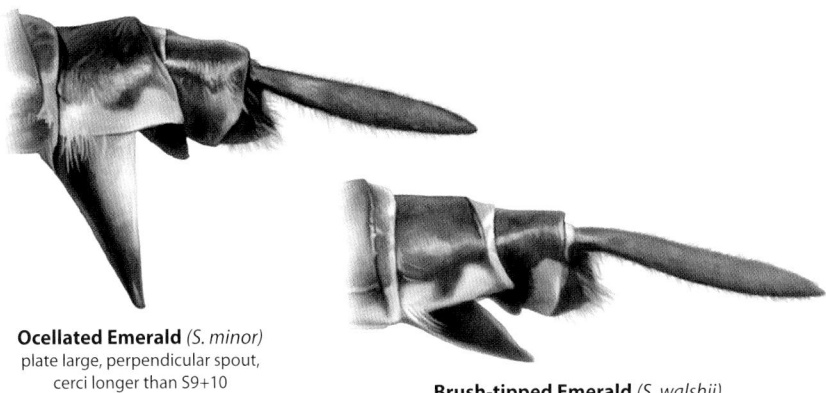

Ocellated Emerald *(S. minor)*
plate large, perpendicular spout,
cerci longer than S9+10

Brush-tipped Emerald *(S. walshii)*
plate oblique, triangular scoop, longer than S9,
cerci much longer than S9+10

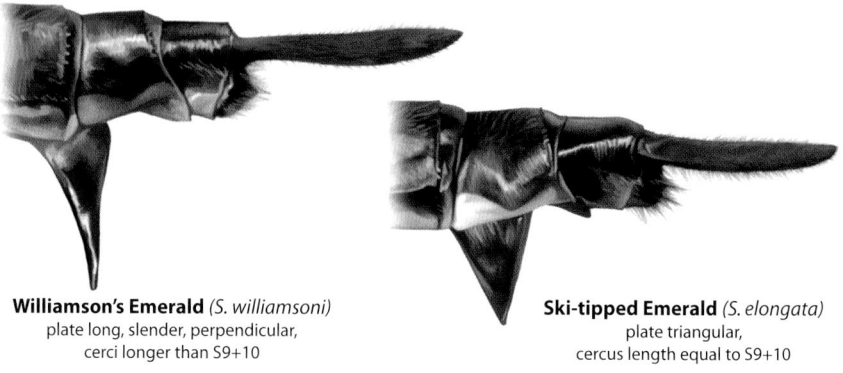

Williamson's Emerald *(S. williamsoni)*
plate long, slender, perpendicular,
cerci longer than S9+10

Ski-tipped Emerald *(S. elongata)*
plate triangular,
cercus length equal to S9+10

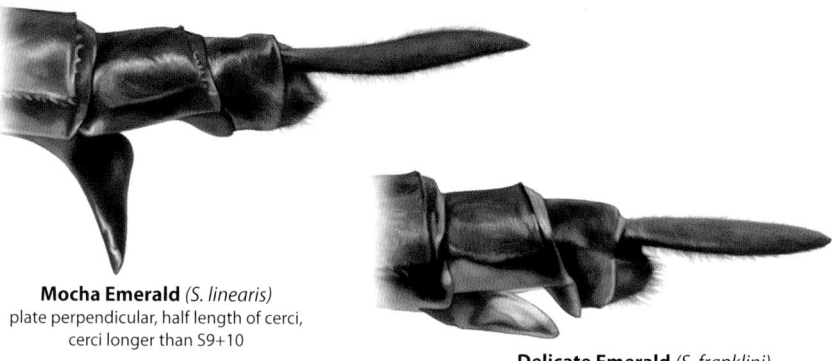

Mocha Emerald *(S. linearis)*
plate perpendicular, half length of cerci,
cerci longer than S9+10

Delicate Emerald *(S. franklini)*
plate horizontal scoop equal to S9 in length, cerci
longer than S9+10

Forcipate Emerald *(S. forcipata)*
plate horizontal scoop longer than S9,
cerci slightly longer than S9+10

Incurvate Emerald *(S. incurvata)*
plate horizontal scoop extending under S10,
cerci slightly longer than S9+10

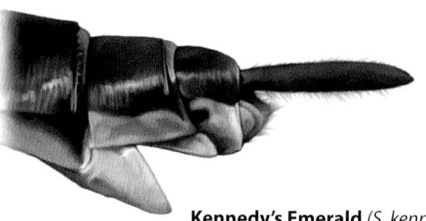

Kennedy's Emerald *(S. kennedyi)*
plate horizontal scoop equal to S9 in length,
cerci equal or slightly longer than S9+10

Mountain Emerald *(S. semicircularis)*
plate notched, half length of S9,
cerci longer than S9+10

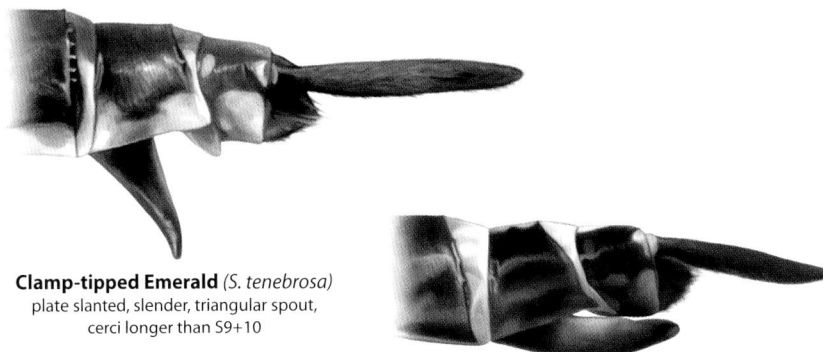

Clamp-tipped Emerald *(S. tenebrosa)*
plate slanted, slender, triangular spout,
cerci longer than S9+10

Calvert's Emerald *(S. calverti)*
plate horizontal, extends beyond S10,
cerci slightly longer than S9+10

Fine-lined Emerald *(S. filosa)*
plate long, skid-like, extends under S10, cerci
longer than S9+10

Treetop Emerald *(S. provocans)*
plate horizontal trough, extends to tip of S10,
cerci shorter than S9+10

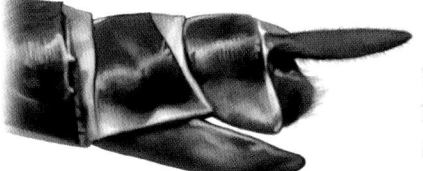

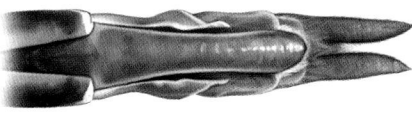

Texas Emerald *(S. margarita)*
plate horizontal trough, extends to tip of S10, narrow in ventral view,
cerci shorter than S9+10

Ozark Emerald *(S. ozarkensis)*
plate horizontal scoop, wider than Texas Emerald in ventral view,
cerci shorter than S9+10

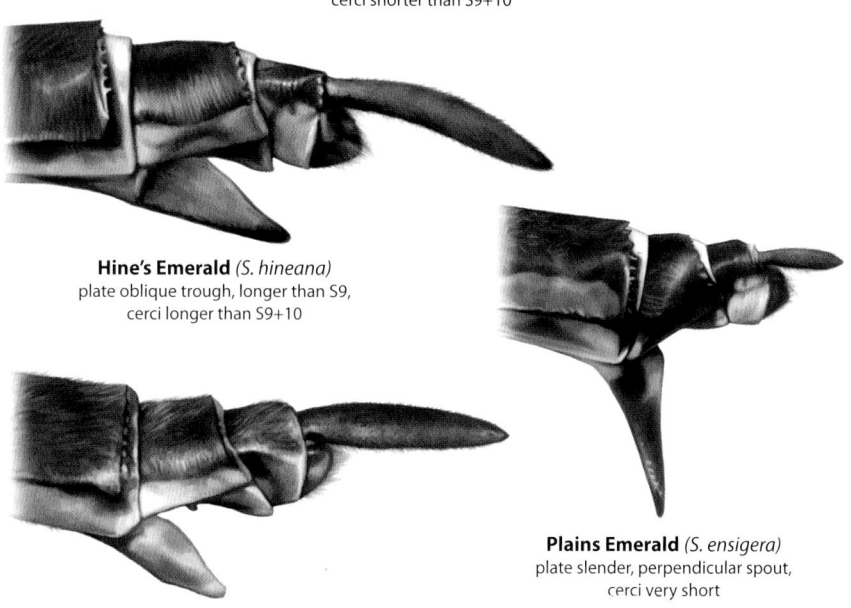

Hine's Emerald *(S. hineana)*
plate oblique trough, longer than S9,
cerci longer than S9+10

Plains Emerald *(S. ensigera)*
plate slender, perpendicular spout,
cerci very short

Quebec Emerald *(S. brevicincta)*
plate equal in length with S9,
cerci longer than S9+10

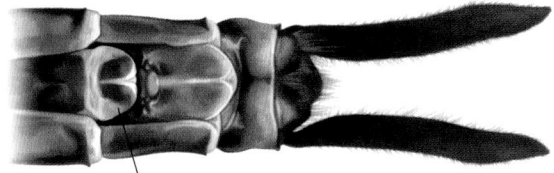

Lake Emerald *(S. cingulata)*
plate 1/3 length of S9,
cerci much longer than S9+10

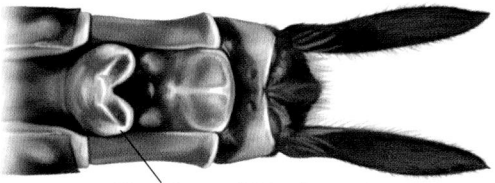

Ringed Emerald *(S. albicincta)*
plate length less than half of S9, notched or bilobed,
cerci slightly longer than S9+10

Hudsonian Emerald *(S. hudsonica)*
plate oblique, 2/3 length of S9,
cerci about equal in length to S9+10

 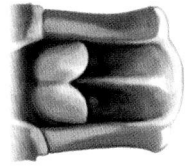

Muskeg Emerald *(S. septentrionalis)*
plate 1/4–1/3 length of S9, deeply notched,
cerci longer than S9+10

Whitehouse's Emerald *(S. whitehousei)*
plate oblique spout, 3/4 length of S9,
cerci longer than S9+10

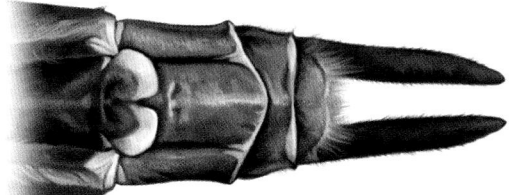

Treeline Emerald *(S. sahlbergi)*
plate less than half length of S9, notched,
cerci about equal in length to S9+10

Coppery Emerald *(S. georgiana)*
plate slender, triangular spout,
cerci much shorter than S9+10

Baskettails (*Epitheca*) male appendages

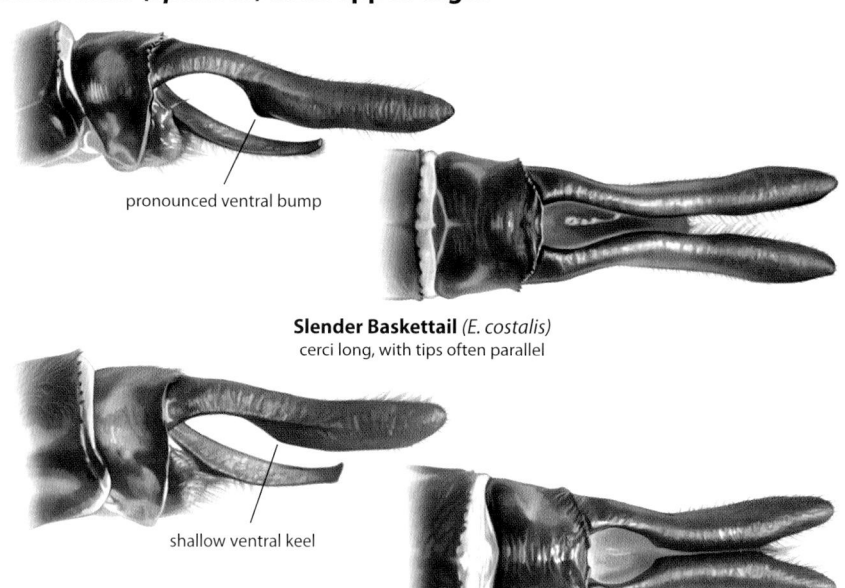

pronounced ventral bump

Slender Baskettail *(E. costalis)*
cerci long, with tips often parallel

shallow ventral keel

Dot-winged Baskettail *(E. petechialis)*
cerci long, with tips often divergent

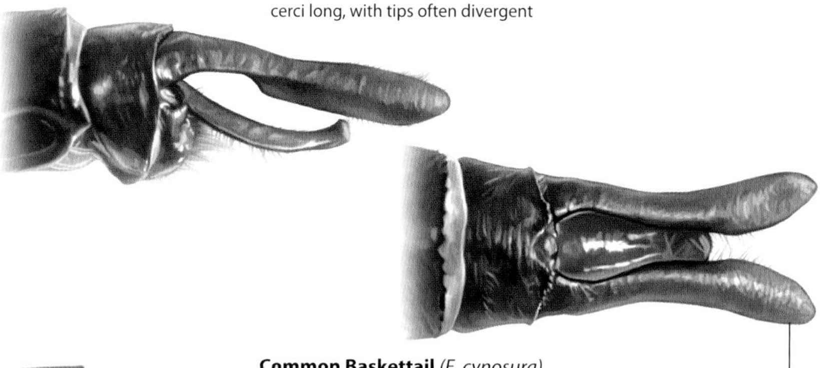

Common Baskettail *(E. cynosura)*
cerci tips usually divergent

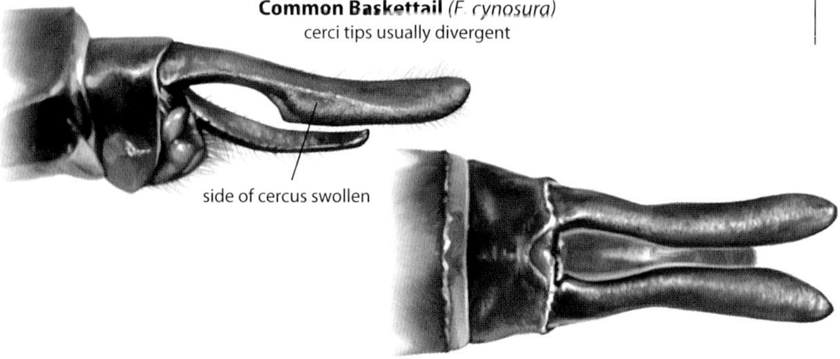

side of cercus swollen

Mantled Baskettail *(E. semiaquea)*

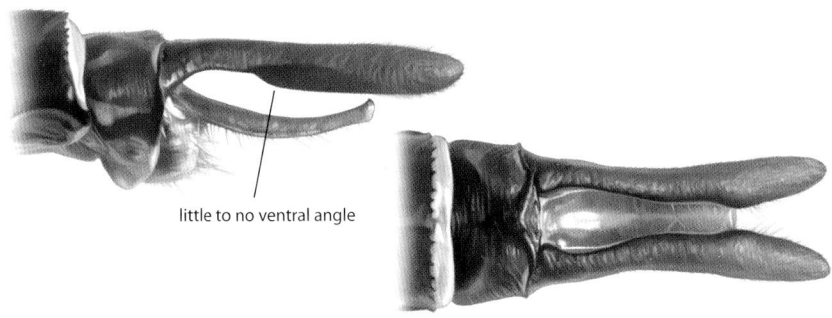

little to no ventral angle

Florida Baskettail *(E. stella)*
cerci long and straight

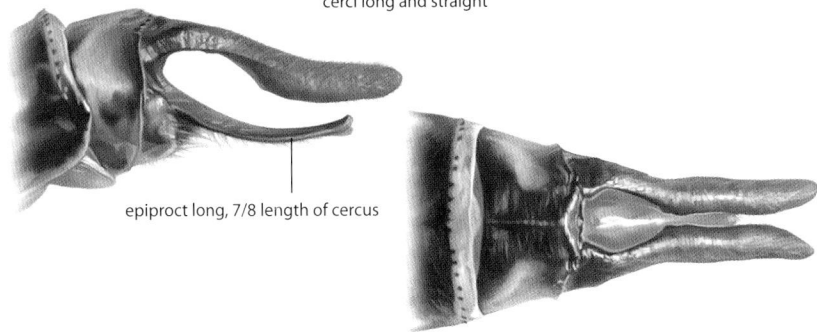

epiproct long, 7/8 length of cercus

Sepia Baskettail *(E. sepia)*

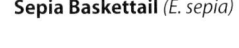

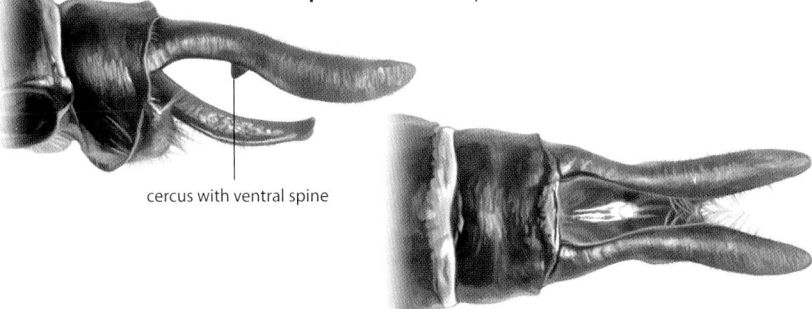

cercus with ventral spine

Spiny Baskettail *(E. spinigera)*
cerci long

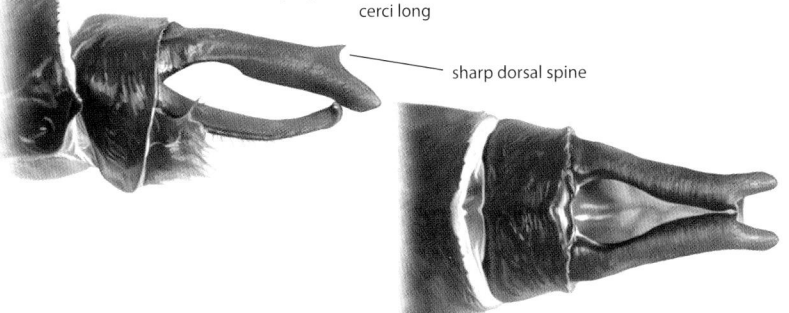

sharp dorsal spine

Robust Baskettail *(E. spinosa)*
cerci tips bent downward

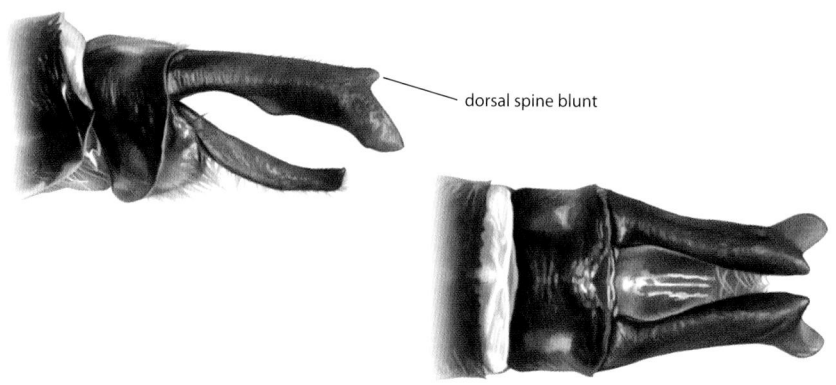

Beaverpond Baskettail *(E. canis)*
cerci tips bent downward

Some whiteface (*Leucorrhinia*) male hamules
(lateral view, inverted)

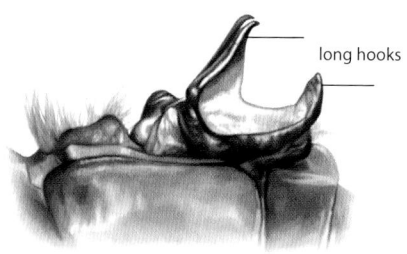

long hooks

Crimson-ringed Whiteface *(L. glacialis)*

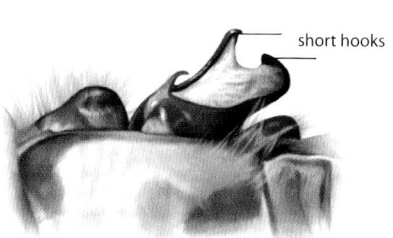

short hooks

Frosted Whiteface *(L. frigida)*

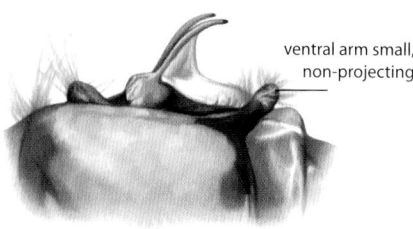

ventral arm small, non-projecting

Canada Whiteface *(L. patricia)*

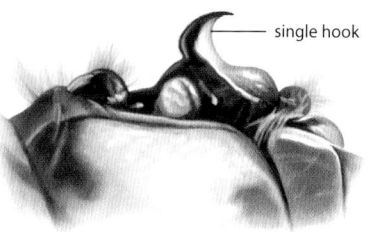

single hook

Belted Whiteface *(L. proxima)*

Cherry-faced, White-faced, and Ruby Meadowhawks

These meadowhawk (*Sympetrum*) species are typically not separable in the field.

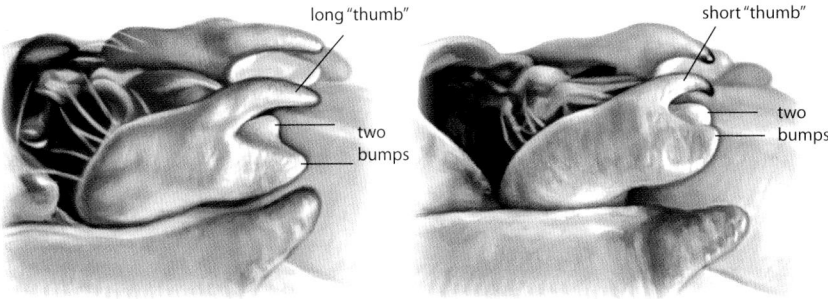

Cherry-faced Meadowhawk (*S. internum*)
male hamules (oblique view, inverted)

White-faced Meadowhawk (*S. obtrusum*)
male hamules (oblique view, inverted)

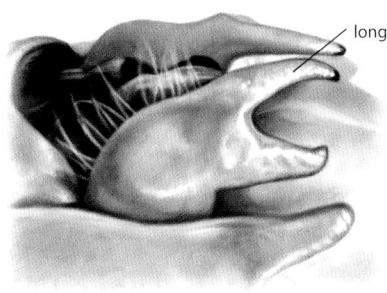

Ruby Meadowhawk (*S. rubicundulum*)
male hamules (oblique view, inverted)

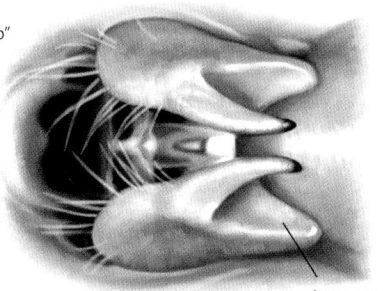

Ruby Meadowhawk (*S. rubicundulum*)
male hamules (ventral view)

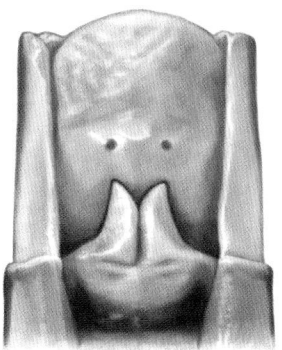

Cherry-faced Meadowhawk
(*S. internum*)
female subgenital plate:
lobe tips often divergent

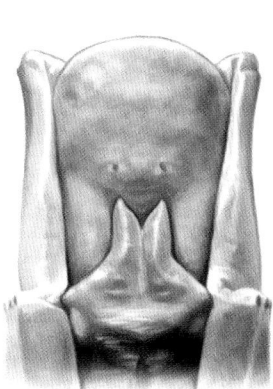

Ruby Meadowhawk
(*S. rubicundulum*)
lobes of subgenital plate bulbous

White-faced Meadowhawk (*S. obtrusum*)
female subgenital plate: lobe tips relatively straight

Arch-tipped and Garnet Gliders male appendages

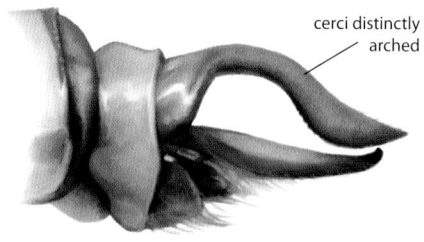

cerci distinctly
arched

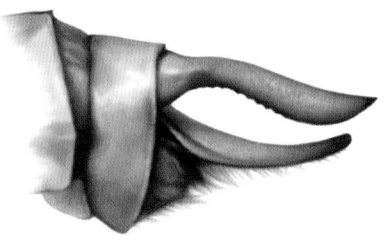

Arch-tipped Glider *(Tauriphila argo)*

Garnet Glider *(Tauriphila australis)*

Antillean and Vermilion Saddlebags female subgenital plates

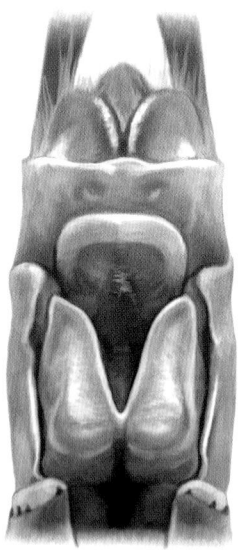

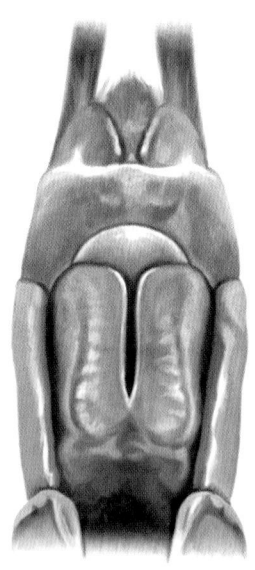

Antillean Saddlebags *(Tramea insularis)*
plate shorter than S9, lobes divergent

Vermilion Saddlebags *(Tramea abdominalis)*
plate longer than S9

Acknowledgments

The road to this guide's completion has been a very long one. The project was initially suggested to me by Russell Galen, my soon-to-be literary agent, in the autumn of 2005. My first identification guide, *Damselflies of the Northeast*, had been published the previous year, and Russell sensed there was an opportunity in the field guide universe for a more ambitious venture. A year later, we agreed with Houghton Mifflin to produce the book as part of the Peterson Field Guide series.

The first years of this project were devoted to gathering the resource materials needed to produce the illustrations. This included travel to all corners of the continent to collect and scan living specimens. I am grateful to many fine people for their company and guidance, including Jerrell Daigle, Giff Beaton, Greg Lasley, Steve and Mary Jane Krotzer, Marion Dobbs, John and Kendra Abbott, Bryan Pfeiffer, Mike Blust, Chris Hill, Dave Fitch, Pam Hunt, Ken Tennessen, Steve Valley, Randy Emmitt, Jeff Pippen, George and Phoebe Harp, Jim Bangma, Jim Johnson, Allen Barlow, Fred Sibley, Gayle and Jeanell Strickland, Michael Veit, Dave Halstead, Paul Bedell, Dan Bogar, Kathy and Dave Biggs, Dave Hanson, David and Jan Dauphin, Jeff Peter, Martin Reid, Josh Rose, Gil Quintanilla, Dave Small, Tim Vogt, Erik Pilgrim, Denis Doucet, Ed Keppner, Charlie and Ginger Brown, Hal White, Tom Young, Carl Cook, Steve and Marcia Hummel, Charlie Muise, Glenn Corbiere, Roy Morris, Troy Hibbitts, Jessica Ware, Eric Isley, Steve and Annie Collins, Ann Johnson, Doug Danforth, Rich and Elaine Bailowitz, Sandy Upson, Linda LaPan, Ted Mack, and Sheryl Chacon. Many graciously shared their hard-won specimens and images.

Many thanks to Blair Nikula, Paul Brunelle, Michael May, Nick Donnelly, Dennis Paulson, Colin Jones, Steve Roble, Bill Mauffrey, and Peter Burke for contributing their vast expertise. The inspiration I received from the works of Sidney Dunkle, Richard Lewington, Edmund Walker, Minter Westfall, and James Needham cannot be understated. Thanks to Mitch Sternberg for access to and permission to collect at Santa Anna LRGV NWR, Anne Morkill at Deer Key NWR, David Riskind at Texas State Parks, and Beth Herr at Ward Pound Ridge Reservation. The resources provided by the International Odonata Research Institute at the Florida State Collection of Arthropods, Gainesville, FL, and the Texas Natural Science Center, University of Texas at Austin, were invaluable.

I wish to particularly acknowledge my "hometown" circle for their long friendship, support, and shared love of the natural world: Nick Wagerik, Marc Weinberg, Michael Bonifanti, Marie Winn, Guy Tudor, Sheila and Lew Rosenberg, Brad Klein, Steve Walter, Fabrice de Lacour, Tom Fiore, and Kristine Wallstrom.

Thanks to Russell Galen for pitching this project so long ago. I agree that creating a painted field guide is for crazy people indeed. Much appreciation to Lisa White of Houghton Mifflin. Thanks to Robert Kirk and the staff at Princeton University Press for taking on this wayward project and guiding it to completion.

Lastly, I thank my wife, Anne O'Connor, and our boys, Nathaniel, Jonah, and Lucas, for their love and patience.

I hope I have not forgotten anyone. Reading through the names, I am saddened by the number of friends and colleagues who have passed on. This book is dedicated to them.

Resources

Dragonfly Society of the Americas
The DSA serves North, Central, and South America, but membership is open to everyone. It holds annual and regional meetings and publishes the quarterly newsletter *Argia*, and the peer-reviewed *Bulletin of American Odonatology*.
dragonflysocietyamericas.org

Odonata Central
This citizen science website provides access to the most comprehensive database of North American odonata distribution. Users can view distribution maps and species lists and contribute observations. Records are peer-reviewed.
odonatacentral.org

The Dragonfly

Today I saw the dragonfly
Come from the wells where he did lie.
An inner impulse rent the veil
Of his old husk: from head to tail
Came out clear plates of sapphire mail.
He dried his wings: like gauze they grew;
Thro' crofts and pastures wet with dew
A living flash of light he flew.

Alfred, Lord Tennyson, 1833

Species Index

Common Names

Scientific Names